The Earth's Dynamic Systems

The Earth's Dynamic Systems

A Textbook in Physical Geology

Second Edition

W. Kenneth Hamblin
Brigham Young University
Provo, Utah

Illustrated by
William L. Chesser
and
Dennis Tasa

Burgess Publishing Company
Minneapolis, Minnesota

Cover photos, National Space Data Center, NASA

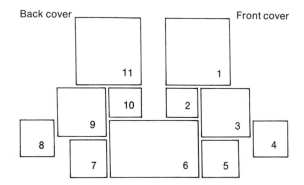

Back cover   Front cover

1. Folded mountains in northern Mexico
2. A cyclonic storm in the central Pacific
3. The earth as seen from space
4. The Sinai peninsula and the Red Sea
5. The Ganges River
6. The Labrador fold belt, Canada
7. The Mississippi delta
8. A view of the moon as seen by Apollo 8 astronauts
9. Volcanoes of the Salar de Coipasa region (Andes Mountains) of Bolivia and Chile
10. A view of Mars as photographed from a Viking satellite
11. The Grand Canyon of the Colorado River, Arizona

This book was set in Times Roman by Typographic Arts. The cover design and new illustrations were done by Dennis Tasa. Other illustrations were done by William L. Chesser. The book was printed by the Colwell Press and bound by Midwest Editions, Inc.

Copyright © 1978 by Burgess Publishing Company
Printed in the United States of America
Burgess Publishing Company
7108 Ohms Lane, Minneapolis, Minnesota 55435
Library of Congress Catalog Card Number 77-93824
ISBN 0-8087-0890-2

0 9 8 7 6 5 4

Preface

The second edition of *The Earth's Dynamic Systems* is a major revision with the addition of new chapters on energy and matter, minerals, rocks, earthquakes, and volcanism. Other chapters have been extensively rewritten, but the original approach and emphasis have been preserved. The objective is to use the unifying theory of plate tectonics and to consider the planet earth as a system of material constantly changing toward a state of equilibrium.

The most obvious improvement in the second edition is the addition of colored plates and diagrams. Thirty-two space photographs have been selected to provide the students with an accurate concept of the earth's surface and tectonic features. It is intended that students use these as a basic visual reference while studying the facts and concepts presented in each chapter. Full color has been used in many of the major diagrams to emphasize geologic features that cannot be shown in photographs.

Sincere thanks go to many colleagues and others who have helped with this book by discussing ideas and problems, critically reading the manuscript, and supplying unpublished information. For the first edition, these colleagues included James L. Baer, Willis H. Brimhall, Myron G. Best, H. M. Davis, Thomas Hendrix, Douglas Hill, Robert Lakemaker, Wade E. Miller, Joseph R. Murphy, John Reid, James A. Rhodes, and Morris S. Petersen. Many of their suggestions have been incorporated into the second edition. In connection with the second edition, further thanks are due to Kenneth Van Dellen, of Macomb County Community College, who read the entire manuscript and offered many helpful suggestions. While others reviewed portions of the manuscript at an early stage, the following reviewed the entire manuscript: E. Julius Dash, Oregon State University; Douglas Brookins, University of New Mexico; Roger Hoggen, Ricks College; and Chester Johnson, Gustavus Adolphus College. Early reviewers included H. Wayne Leimer, Tennessee Technical University; Arthur L. Reesman, Vanderbilt University; and Donald C. Haney, Eastern Kentucky University.

The assistance and persistence of Michele West and Susan Burns, who typed the manuscript and assisted in preliminary editing, are also appreciated. Special appreciation is extended to Marcia Bottoms, Ann Seivert, and Gary Brahms of Burgess Publishing Company, whose efforts extended well beyond their normal call of duty.

To the Student

Our Approach

In studying the processes in the earth's system, we are continually confronted with the problem of scale. How can we begin to understand the nature and inter-relation of things as large as rivers, mountains, continents, and ocean basins? One approach is to use models, a common method in all branches of science and engineering for studying and analyzing things too large to see or handle conveniently. A model is a formulation that stimulates a real-world phenomenon and has value in analyzing and predicting behavior of the real thing. Models may be physical or abstract representations of the structure and function of real systems. Throughout this text, we have made extensive use of graphic models (diagrams, photographs, and maps) in an effort to provide you with an accurate image of the features and processes described. These models, especially the block diagrams, are as important as the text in providing information, and as much time should be devoted to studying them as to reading the text.

Organization of the Chapters

One of the most difficult problems faced by a student beginning a new course in a subject with which he has little or no experience is to identify the fundamental facts and concepts and separate them from supportive material. This problem is often expressed by the question "What do I need to learn?" We have attempted to overcome this problem by presenting the material in each chapter in a manner that will help you recognize immediately the essential ideas and concepts. In each chapter the major subdivisions are introduced under the subtitle *Statement*. The statement presents the major

concept, principle, or theory. It is not intended as a summary or abstract of what the section contains but is an expression of the facts and concepts of the subject in one all-embracing point of view. The statement may be difficult for you to comprehend fully the first time you read it, but you will gain further insight from the paragraphs under the subtitle *Discussion*. Here, a more complete discussion of the ideas or principles is presented and illustrated. Evidence supporting the statement is discussed, and, where it is pertinent to the understanding of the concept, a brief history of how it developed is presented. This material is designed to help you understand the ideas present in the statement.

The idea is to learn the material in the statement and clearly understand it. In this way you can use your study time effectively, and in reviewing for a test you need only make sure you understand the statement.

The Physiographic Map of the Earth

The physiographic map of the surface of the earth (an optional accessory to the book) was compiled from the best sources available and represents, in perspective, the surface features of our planet. These features are the net result of processes operating in the earth's systems, and you should study them carefully as you read the text. For the most effective results, hang this "model" on the wall of your study and identify and analyze each feature as it is discussed. Much can be gained if you annotate the map as you study.

The shaded relief maps of the moon and Mars are on the reverse side of the physiographic map of the earth. They are accurate models showing regional features in perspective. Continued reference to these models is absolutely essential in mastering the chapter on the moon, Mars, and Mercury.

Contents

The Planet Earth

One of the most significant advances made in the twentieth century concerns man's view of the earth and the dynamics of how it changes. Until recently, most of man's knowledge of the earth had to be obtained from what he could see by looking horizontally from viewpoints at, or near, the surface. This perspective was nearsighted and limited, and, in an attempt to overcome it, studies of minute details were plotted on maps and charts that served as models of the real world. Rivers, mountains, shorelines, and weather patterns were surveyed and studied from hundreds of observation points, but observers were never able to see the earth in a regional, panoramic view or to observe how it functions as a planet in space. Now, "for the first time in all of time men have seen the earth from the depths of space—seen it whole and round and beautiful and small." (Archibald MacLeish, 1968.) The horizontal no longer dominates man's vision, and, with this new vertical dimension, he has extended his limits and now can look down on the earth in a way that seems almost godlike.

Space photography permits man to see the earth as it truly is and to observe the large-scale structures of the atmosphere, the oceans, and the continents and the regional relationships of river systems, mountain belts, and coastal features. Instead of seeing political divisions of the land, man sees the physical features of the planet and the earth functioning as a dynamic system on a scale that may at first be difficult to grasp. The surface fluids (air and water) can be seen in spectacular motion as they travel from the oceans to the atmosphere, to the continents, and back to the oceans through the great river systems of the world.

Man also can see motion in the crust of the earth. Even though the earth's surface appears to be firm and stationary, there is convincing evidence that the materials that form the solid rock on the surface of the planet are moving. Man can see motion in the folded mountain belts, which have been deformed by compression, and he can see the great rifts, which split and separate the continents. There also is evidence that the seafloors are moving and are continually being created and destroyed. Never before has man had such vivid evidence that the earth is a dynamic, continually changing planet. This chapter will utilize some of the newly acquired space photographs in an attempt to point out the major geologic characteristics of the earth. From this study, you should develop a new awareness of the features seen on the surface of the planet and an appreciation of the earth as a dynamic system.

Major Concepts

1. The surface fluids of the earth (the atmosphere and water) are in constant motion. The tremendous volume of water circulating at the surface greatly modifies the landscape.
2. The continents and ocean basins are the principal surface features of the earth.
3. The continents consist of three major components: shields, stable platforms, and belts of folded mountains. All of these components show the mobility of the crust.
4. The ocean floor contains several major topographic and structural divisions: (a) the oceanic ridge, (b) abyssal floors, (c) seamounts, (d) trenches, and (e) continental margins.
5. The earth is a differentiated planet with these major internal structural units: (a) a central core, (b) a thick mantle, (c) a soft upper mantle (asthenosphere), (d) a rigid lithosphere, and (e) surface fluids (water and atmosphere).

Interpreting the Surface Features of a Planet

Statement

The surface features of a planet preserve an imprint of the processes by which they were formed. In some cases, the record is clear and easily understood. Sand dunes of the great desert regions obviously are formed by wind action, the beaches of the Atlantic coast are the result of wave action and the islands of Hawaii are a record of volcanic eruptions. The origin of other landforms, however, can be more obscure and can be only partly understood, but the fascinating thing about the surface of a planet is that it can be explained by natural, physical, chemical, and geologic processes. As Einstein once said, "The eternal mystery of the world is its comprehensibility."

To help you understand and appreciate the earth, this chapter includes a section of space photographs that illustrate many of the landforms and structural features of the earth (*plates 1* through *32*). These photographs are, in essence, a pictorial essay on the earth's dynamic systems. They are intended to be a visual reference for material discussed in this and all the chapters that follow. Space photographs of the moon and Mars are included so that you can compare and contrast the features of the earth with those of other planets. This should help you recognize the things that make the earth unique. The serious student will study these illustrations carefully and learn to read the geologic history they record.

Discussion

The Moon.   Consider first the moon and what its cratered surface tells about its history and internal dynamics (*plate 1*). It is obvious from telescopic views, as well as from satellite photographs, that the moon's surface is practically saturated with craters and that the major geologic process on the moon has been the impact of meteorites. The moon does not have water or an atmosphere, so there is no evidence of windblown sand or stream valleys. The only way the surface of the moon can be modified is by the impact of meteorites coming from space and by the internal energy that deforms the crust and produces volcanic products. We now know from rock samples brought back from the Apollo missions, and from close-up satellite photographs, that the smooth, dark areas called **maria** (seas) are areas covered by basalt resulting from vast floods of lava. As can be seen in *plate 1,* the lava flows filled the large, circular basins and the adjacent lowlands and spread out over parts of the more rugged, densely cratered terrain. This clearly indicates that the densely cratered terrain was formed before the lava flows—an obvious, but fundamental, age relationship between two major events in lunar history.

Two additional facts are apparent from *plate 1:* (1) the maria, or lava plains, have relatively few craters, indicating that, since the lava was extruded, the rate of meteorite impact has decreased greatly; and (2) the outlines of all craters, both young and old, remain circular and undeformed, indicating that the crust of the moon has never been compressed to the extent that it would buckle and fold. The crust of the moon has remained fixed and undeformed from internal movement throughout its history.

From these observations of the surface features of the moon, the following sequences of events can be recognized.

1. A period of intense bombardment by meteorites to form the densely cratered terrain.
2. A major thermal event resulting in the volcanic activity that produced the great floods of lava.
3. The subsequent impact of relatively few meteorites.

Except for the impact of a few meteorites, the moon has not changed significantly since the formation of the lava plains. The crust has not been deformed by internal forces, nor have the surface features been modified by the activity of wind or water. Nothing moves on the moon, except during the rare occasion when a meteorite strikes the surface. Except for the motion they themselves created, no motion was observed by the Apollo astronauts. The footprints they made and the equipment they left will remain fresh and unaltered for literally millions of years, except as they are modified by particles from space.

Mars.   The surface of Mars is somewhat similar to that of the moon; yet, it is different. As can be seen in *plate 2,* much of the surface is pockmarked with craters; there are large, circular basins; and there are vast floods of lava, which form the smooth plains in the Northern Hemisphere. Mars, however, has an atmosphere, and the Martian winds have eroded significantly the crater rims and have piled sand into large dune fields (see *figure 24.41*). There is also water on Mars that now is frozen in the pore spaces of the soil and rock and locked up in the polar ice caps. However, at times in the past, conditions must have been different, since there is evidence that water flowed on the surface and eroded numerous stream channels (see *figure 24.38*). In addition, the crust of Mars has been fractured by tensional forces, and some of the fractures have been enlarged by erosion to form huge canyons visible in the upper left-hand part of *plate 2*. Mars also has a number of huge volcanoes that are younger than the lava plains and may still be active. These younger volcanoes testify to continued thermal activity in the planet.

It is clear from the cratered surface and the lava plains visible in *plate 2* that Mars experienced early events similar to those that occurred on the moon. A period of intense bombardment by meteorites produced a densely cratered surface. A period of volcanic activity produced floods of lava that formed the smooth plains in the north, and subsequent bombardment by meteorites has been relatively light. But Mars also has been modified by wind activity, which continually shifts the sand over the planet's surface. Running water has eroded networks of stream channels, and relatively recent volcanic activity has built up huge volcanic mountains. In addition, the crust of Mars has been deformed by upwarps and fracturing.

Mars, then, is a more dynamic planet than the moon, but neither Mars nor the moon has continents and ocean basins or folded mountain belts and great river systems like those on the earth. As will be shown, these features are the result of much greater dynamic activity.

The Earth from a Perspective in Space

Statement

A view of the earth from space, like the one in *plate 3*, shows many features of the earth and how they are related. The **atmosphere,** the thin, gaseous envelope that surrounds the earth, is the most conspicuous feature as seen from space and appears as bright, swirling clouds. The **lithosphere,** the outer solid part of the earth, is visible as continents and islands. The **hydrosphere,** the discontinuous layer of water covering the planet, is seen in the vast surface of the oceans. Even parts of the **biosphere,** the organic realm that includes all of the earth's living things, can be seen from space in the dark green tropical forests of equatorial Africa. Each of these major realms of the earth's surface is in constant motion and is continually changing. The motion of the atmosphere and the hydrosphere is dramatic and readily visible. Movement, growth, and change in the biosphere can be readily appreciated and easily understood. But the solid, rigid lithosphere—the earth's crust—is also in motion and has been throughout most of the earth's history. Evidence for this, as will be shown, is vivid in perspective from space.

Discussion

The Atmosphere and the Hydrosphere. Everyone is aware that air and water are the fundamental elements of the earth's environment, but few people give much thought to the fact that air and water are as much parts of the planet as the solid rocks and that the continually moving surface fluids blanket the globe and are responsible for so many of the landforms and surface features of the continents. Without these fluids, the surface of the land would be nothing like the familiar one. There would be no rivers, valleys, glaciers, or even continents as they are known today, because the movement of surface waters has played a major role in the development of all of these features, and many more.

The circulation patterns of the atmosphere are indicated in *plate 3* by the shape and orientation of the clouds. At first glance, the patterns may appear confusing, but, upon close examination, one finds they are well organized. When the details of local weather systems are smoothed out, the global atmospheric circulation becomes apparent. Solar heat is greatest in the equatorial regions and causes the water in the oceans to evaporate and the moist air to rise. The warm, humid air forms an equatorial cloud belt bordered by relatively cloud-free zones to the north and south, where this "dehydrated" air descends into the middle latitudes. At higher latitudes, low-pressure systems develop where the warm air from the low latitudes meets the polar air masses. The pattern of circulation around the resulting low-pressure cells produces winds moving in a counterclockwise direction in the Northern Hemisphere and a clockwise direction in the Southern Hemisphere. Swirls of clouds showing this circulation are readily apparent in *plate 3*.

The Continents and the Ocean Basins. If there were a special telescope that could see through clouds and water, from space the surface of the planet could be seen as two principal regions, continents and ocean basins. These major structural and topographic divisions of the earth differ markedly, not only in elevation, but in rock types, density, chemical composition, age, and history. The largest is the ocean basin, which occupies 60% of the earth's surface and is characterized by a variety of spectacular topographic forms, most of which are due to extensive volcanic activity and earth movements that continue today. The continents rise above the ocean basins as large platforms, but the waters of the oceans more than fill their basins and flood approximately 11% of the

Figure 1.1 *Graph comparing the elevations of continents and ocean basins. The average height of the continents is 800 m above sea level, and the average depth of the ocean floor is 3700 m below sea level. Only a small percentage of the earth's surface rises above the average elevation of the continents or below the average elevation of the ocean basins.*

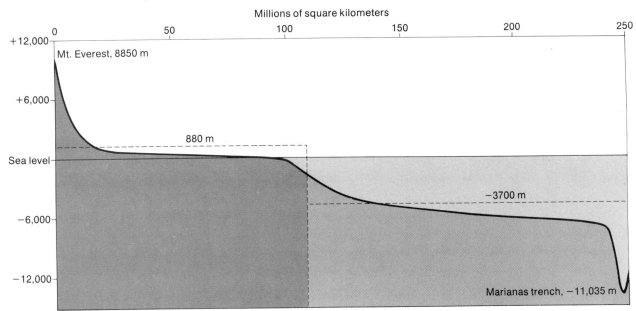

continental surface. The present shoreline, which is so important geographically and has been mapped in great detail, has fluctuated greatly throughout the earth's history and has no great structural significance. From a geologic point of view, the fact that the continents rise 6 km above the ocean basin is much more significant than the position of the shore. The difference in elevation between continents and ocean basins represents a fundamental difference in rock density. Rocks of the continents have a lower density than rocks of the ocean basins; that is, a given volume of continental rocks weighs less than the same volume of oceanic rocks. It is this difference in density that causes the continents to float on, or above, the denser oceanic crust.

The elevation and area of the continents and ocean basins have been mapped with precision, and the data can be summarized in various forms. The data presented graphically in *figure 1.1* shows that the continents have an average elevation of 880 m above sea level and the ocean floor has an average depth of about 3700 m below sea level. Only a relatively small percentage of the earth's surface rises above the average elevation of the continents or below the average depth of the ocean floor. This is significant because, if the continents did not rise so high above the ocean floor, the entire surface of the earth would be covered with water.

Why does the earth have ocean basins and continents instead of a cratered surface like that of the moon and Mars? By their very existence, the continents pose one of the most fundamental questions about the earth. Theoretically, a large, rotating planet like the earth with a strong gravitational field would mold itself into a smooth spheroid and would be covered with a layer of water approximately 2.4 km deep. Continents, as they are known today, have not been found on other planets in this solar system. The origin and evolution of continents and ocean basins will be studied in chapters 21 and 22, but it should be emphasized at this point that the continents and ocean basins constitute the two fundamental topographic and structural features of the planet, features not found on the moon, Mercury, or Mars.

Major Features of the Continents

Statement

The broad, flat continental platforms that rise above the ocean basins contain a great diversity of surface features with an almost endless variety of hills and valleys, plains and plateaus, and high mountain peaks, some of which rise nearly 9000 m above sea level. From a regional perspective, however, the continents are remarkably flat, with most of the surface being within a few hundred meters of sea level. Extensive geologic studies over the past hundred years have shown that several striking facts stand out about the continents.
1. Their shapes are roughly triangular.
2. They are concentrated in the Northern Hemisphere.
3. Although each continent may appear to be unique, they all have three basic components:
   (a) A large area of very ancient rocks known as a **shield.**

(b) Broad, **stable platforms** where the shield is covered with a veneer of sedimentary rock.
(c) Young, folded mountain belts located along the continental margins.
(Geologic differences between continents are mostly in size, shape, and proportions of these components.)
4. Continents consist of rock that is less dense than the rock in the ocean basins.
5. The rocks that compose the continents are old, some as old as 3.8 billion years.
6. The climatic zone occupied by the continent usually determines the style and variety of landforms developed on it.

Discussion

Shields. Without some firsthand knowledge of the shield, it is difficult to visualize the nature and significance of this very important part of the continental crust. *Plate 4* shows part of the Canadian Shield of the North American continent and will help you comprehend the extent, the complexity, and some of the typical features of shields, but you also should study *figures 6.1, 22.1,* and *22.3.* The most important characteristic of a shield is that it forms the vast expanse of the lowlands. Throughout an area of hundreds of thousands of square kilometers, the surface of the shield is within a few hundred meters of sea level. The only features that stand out in relief are the resistant rock formations that rise 50 to 100 m above the adjacent surface. The structure and rock types of the shield are complex. Many rock bodies were once liquid; others have been compressed and extensively deformed. The shields probably provide the best indication of the true nature of the continental crust.

Stable Platforms. Large parts of the continental shields are covered with a veneer of sedimentary rocks. Throughout the last 600 to 700 million years, these areas have been relatively stable; that is, they have never been uplifted a great distance above sea level or submerged far below it. Hence, the term *stable platform.* In North America, the stable platform is located between the Appalachian Mountains and the Rocky Mountains and extends northward to the Lake Superior region and into western Canada. Throughout most of this area, the sedimentary rocks that cover the shield are nearly horizontal, but locally they have been warped into large domes and basins (*plates 5* and *6*). Rocks of the underlying shields rarely are exposed in the stable platforms, but their presence at depth is known from thousands of wells that penetrate the sedimentary cover.

Folded Mountains. One of the most significant aspects of the continents is the belts of young, folded mountains that typically occur along their margins. Generally, most people think of a mountain as simply a high, more or less rugged landform, in contrast to flat plains and lowlands. Mountains, however, are much more than high country. To a geologist, the term *mountain belt* refers to a long, linear zone in the earth's crust where the rocks have been intensely deformed by great horizontal stresses and generally intruded by molten rock material. The topography can be high and rugged, or it can be worn down to a surface of low relief. The important features about mountain belts, regardless of the topography, are the extent and style of deformation. The great folds and fractures in mountain belts, therefore, provide evidence that the earth's crust is, and has been, in motion.

Plate 7 is a space photograph that illustrates some of the characteristics of folded mountains and the extent to which the margins of continents have been deformed. The rocks in this photograph are mostly sedimentary rocks originally deposited in a shallow sea much like the one that now covers the edge of the continental platform off the coast of the eastern United States. These rocks have been deformed by compression like wrinkles in a rug. Erosion has removed the upper part of the folds so that the resistant layers form zigzag patterns similar to those that would be produced if the crests of wrinkles in a rug were cut off.

The mobility of the earth's crust appears to have started early in the planet's history and has continued to the present. We now know that the moon, Mars, and Mercury lack this type of deformation, because all impact craters (regardless of their age) are circular and have not been deformed by compressive forces. The crusts of these planets appear to have been fixed and immovable throughout their histories, whereas the earth's crust has been, and continues to be today, in motion.

Major Features of the Ocean Floor

Statement

The ocean floor, not the continents, is the typical surface of the solid earth, and it is the ocean floor that holds the key to the evolution of the earth's crust. Yet, it was not until the 1960s that enough data about the ocean floor was obtained to present a clear picture of its regional characteristics. This new knowledge caused a revolution in our ideas about the nature and evolution of the crust. Before 1947, most geologists believed that the ocean floor was simply a submerged version of the continents, with huge areas of flat abyssal plains covered with sediment derived from the landmass. Although as early as 1922 echo sounders (sonar) were used to determine depths, they were limited to shallow water and required an observer to listen with headphones and time the interval between transmission and return to the source. A major breakthrough occurred in 1953 with the development of a precision depth recorder that could plot automatically a continuous profile of the ocean floor in any depth of water. Since then, millions of kilometers of profiles across the ocean floor have been made, and the information has been used to make maps of the topography of the ocean basin.

The profiles of the ocean floor show that the submarine topography is as varied as that on the continents and, in some respects, more spectacular. Yet, the topographic forms of the ocean floor are in many ways unique. There are no highly folded mountain belts, and, in contrast to the continents, erosion is not the major process that forms the topography of the ocean floor. Among the most significant facts that we have learned about the oceanic crust are these.

1. The oceanic crust is composed mostly of basalt, a dense volcanic rock, and the major topographic fea-

tures are related in some way to volcanic activity. The oceanic crust, therefore, is entirely different from that of the continents.
2. The rocks of the ocean floor are young in a geologic time frame. Most are less than 150 million years old, compared to the ancient rocks of the shield that are more than 700 million years old.
3. The rocks of the ocean floor have not been deformed by compression, so their undeformed structure stands out in marked contrast to that in the folded mountains and the shields of the continents.

The major provinces of the ocean floor are:
1. The oceanic ridge.
2. The abyssal floor.
3. Seamounts.
4. Trenches.
5. Continental margins.

Discussion

Although most of the topography of the ocean floor can be seen only indirectly through profile records, some features can be detected from satellite photographs. In *plate 8,* a deep channel on the ocean floor north of Cuba can be seen cutting across the shallow Bahamas Platform. In the Pacific, submarine volcanoes commonly are expressed at the surface by circular coral reefs called **atolls** (*plate 9*). The physiographic maps shown in *figures 1.2* and *1.3,* however, provide the best visual reference for regional features of the ocean basins.

The Oceanic Ridge. The oceanic ridge is probably the most striking feature on the ocean floor and is certainly one of the most important. It extends as a continuous feature from the Arctic basin down the center of the Atlantic Ocean, into the Indian Ocean, and across the eastern Pacific Ocean. It is essentially a broad, fractured swell, generally more than 1400 km wide, with higher peaks rising as much as 3 km above the ocean floor. A huge, cracklike valley, called the **rift valley,** is located along the axis of the ridge and extends throughout most of its length. In addition, great fracture systems—some as long as 4000 km—trend across the ridge.

The Abyssal Floor. The oceanic ridge divides the Atlantic and Indian roughly in half but is located in the southern and eastern part of the Pacific. On both sides of the ridge, vast areas of the deep-ocean basin consist of broad, relatively smooth surfaces known as the **abyssal floor.** This surface extends from the flanks of the oceanic ridge to the continental margins and generally lies at depths of about 3000 m. The abyssal floor can be subdivided into two sections: (1) the **abyssal hills** and (2) the **abyssal plains.**

The abyssal hills are relatively small hills rising up to 900 m above the surrounding ocean floor. In the Pacific, they cover 80% to 85% of the seafloor and thus are the most widespread landform on the earth. Near the continental margins, the abyssal hills are completely buried with sediment derived from the land and form the flat, smooth abyssal plains.

Seamounts. Some of the most fascinating features on the seafloor are the isolated peaks of submarine volcanoes called **seamounts.** Some rise above sea level and form islands, but most are completely submerged and

Figure 1.2 Physiographic map showing landforms on a portion of the floor of the Atlantic Ocean.

Figure 1.3 Physiographic map showing landforms on a portion of the floor of the Pacific Ocean.

are known only from oceanographic soundings. Although many appear to occur at random, others form groups, or chains, along well-defined lines. Islands and seamounts testify to the extensive, continuing volcanic activity throughout the ocean basins and provide an important insight into the type of dynamics operating within the earth.

Trenches. The deep-sea trenches are the lowest areas on the earth's surface and have attracted the attention of geologists for years, because they represent a fundamental structural feature of the earth's crust. As illustrated in *figure 1.4*, the trenches are invariably located adjacent to island arcs or coastal mountain ranges of the continents. As will be shown in subsequent chapters, the trenches are involved in the most intense volcanic and earthquake activity on the planet.

Continental Margins. The submerged part of the continent is referred to as the **continental shelf** and is geologically part of the continent, not the ocean basins. Presently, the continental shelves are equal to 18% of the earth's total surface, but, at times in the geologic past, the ocean covered much more of the continental platform.

The seafloor descends from the outer edge of the continental shelf as a long, continuous slope to the deep-ocean basin. This slope is appropriately called the **continental slope,** because it marks the edge of the continental mass of granitic rock. Continental slopes are found around the margin of every continent and around smaller fragments of continental crust, such as Madagascar and New Zealand. If you study closely the continental slopes shown on a physiographic map (see *figures 1.2* and *1.3*)—especially those surrounding North America, South America, and Africa—you should note that the slopes form one of the major topographic features on the earth's surface. On a regional scale, they are by far the steepest, longest, and highest slopes on the earth. Within this zone, 20 to 40 km wide, the average relief above the seafloor is 4000 m; adjacent to the marginal trenches, relief is as great as 10,000 m. In contrast to the shorelines of continents, the continental slopes are remarkably straight over distances measured in thousands of kilometers.

In many areas, the continental slopes are cut by deep **submarine canyons** remarkably similar to canyons cut by rivers into continental mountains and plateaus. As shown on the physiographic map, the submarine canyons cut across the edge of the continental shelf and terminate on the deep abyssal floor some 5000 to 6000 m below sea level.

Figure 1.4   *Outline map showing the major features of the ocean floor.*

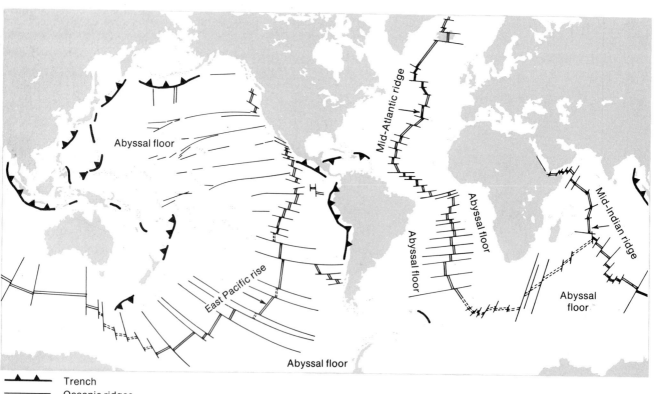

Trench
Oceanic ridges
Fracture zones

Plate 1 *A view of the moon as seen by Apollo 8 astronauts. Two types of surface features stand out in sharp contrast: the densely cratered highlands called terra and the dark, smooth areas called maria. We know from rock samples brought back from the Apollo missions that the maria resulted from great floods of lava that filled many large craters and spread out over the surrounding lowlands. Thus, the volcanic activity is younger than the densely cratered terrain. From these relationships, it is apparent that the moon had a history involving a period of intense bombardment by meteorites, a period of volcanic activity, and subsequent impact by relatively few meteorites (resulting in young, bright-rayed craters). The lunar surface has a very low level of erosion and has not been modified by wind, water, or glaciers.*

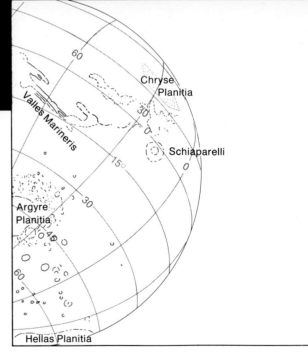

Plate 2 *A view of Mars as photographed from a Viking satellite. The surface of Mars is somewhat similar to that of the moon. There are two major provinces: densely cratered highlands in the Southern Hemisphere and smooth plains (presumably vast floods of lava) covering the floors of large basins and the lowlands in the Northern Hemisphere. Unlike the moon, Mars has many other features that indicate that its surface has been modified by atmospheric processes, recent volcanic activity, and crustal deformation. Mars, then, has experienced early events similar to those on the moon (a period of intense bombardment and extrusion of lava floods), but it has been modified by subsequent surface processes, crustal deformation, and volcanism.*

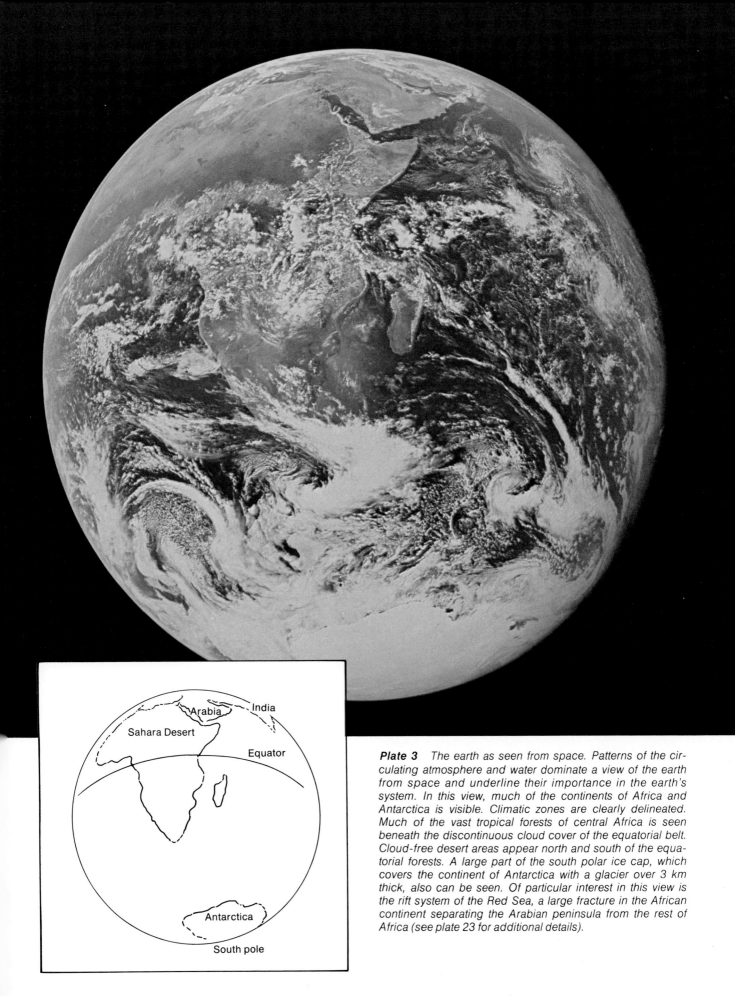

Arabia

India

Sahara Desert

Equator

Antarctica

South pole

Plate 3 *The earth as seen from space. Patterns of the cir-culating atmosphere and water dominate a view of the earth from space and underline their importance in the earth's system. In this view, much of the continents of Africa and Antarctica is visible. Climatic zones are clearly delineated. Much of the vast tropical forests of central Africa is seen beneath the discontinuous cloud cover of the equatorial belt. Cloud-free desert areas appear north and south of the equa-torial forests. A large part of the south polar ice cap, which covers the continent of Antarctica with a glacier over 3 km thick, also can be seen. Of particular interest in this view is the rift system of the Red Sea, a large fracture in the African continent separating the Arabian peninsula from the rest of Africa (see plate 23 for additional details).*

Plate 4 The Canadian Shield as seen from a satellite about 640 km above the earth's surface. The shields are fundamental structural components of the continents. They are composed of complexly deformed crystalline rock bodies eroded down to a nearly flat surface near sea level. Throughout much of the Canadian Shield, the topsoil has been removed by glaciers, and differences between rock bodies are etched out in relief by erosion. The resulting depressions commonly are filled with water and form lakes and bogs that emphasize the structure in the rock bodies. Areas of dark tones are metamorphic rock. Areas of light pink tones are granitic rock.

Granite rocks

Metamorphic rocks

Granite

Granite

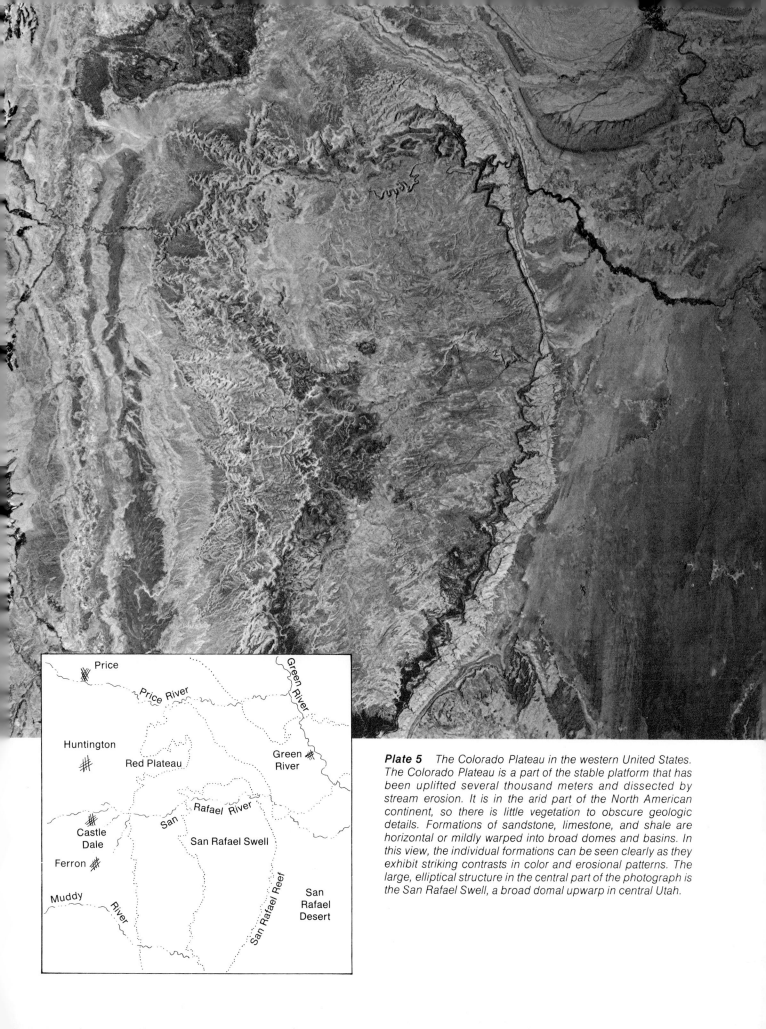

Price

Price River

Green River

Huntington

Red Plateau

Green River

Rafael River

San

Castle Dale

San Rafael Swell

Ferron

San Rafael Reef

Muddy River

San Rafael Desert

Plate 5 *The Colorado Plateau in the western United States. The Colorado Plateau is a part of the stable platform that has been uplifted several thousand meters and dissected by stream erosion. It is in the arid part of the North American continent, so there is little vegetation to obscure geologic details. Formations of sandstone, limestone, and shale are horizontal or mildly warped into broad domes and basins. In this view, the individual formations can be seen clearly as they exhibit striking contrasts in color and erosional patterns. The large, elliptical structure in the central part of the photograph is the San Rafael Swell, a broad domal upwarp in central Utah.*

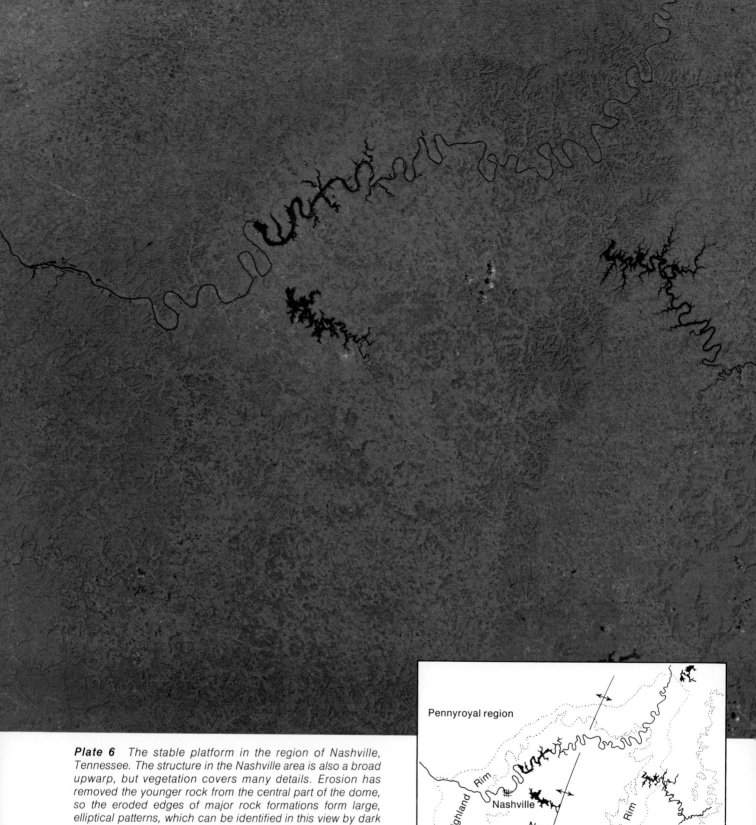

Plate 6 *The stable platform in the region of Nashville, Tennessee. The structure in the Nashville area is also a broad upwarp, but vegetation covers many details. Erosion has removed the younger rock from the central part of the dome, so the eroded edges of major rock formations form large, elliptical patterns, which can be identified in this view by dark tones and a delicate erosional texture. This ridge is known as the Highland Rim and rises 30 to 60 m above the surrounding area. The older rocks exposed in the central part of the structural dome are nonresistant limestone and have been eroded into lowland, so that the structural dome is expressed topographically as a lowland or basin. In the southeastern part of the area, younger sedimentary rocks form the western margins of the Cumberland Plateau, a dissected flatland rising 150 m above the surrounding area.*

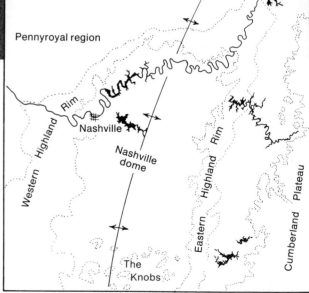

Pennyroyal region

Western Highland Rim

Nashville

Nashville dome

Eastern Highland Rim

Cumberland Plateau

The Knobs

Plate 7 The folded Appalachian Mountains in central Pennsylvania. This view of the Appalachian Mountains shows the style of deformation typical of a folded mountain belt. The folded strata are expressed by ridges of resistant sandstone that rise about 300 m above the surrounding area. Erosion has removed the upper part of many of the folds, so that the limbs that form the resistant ridges form long, narrow, elliptical patterns or zigzag across the area. This is the famous Valley and Ridge Province of North America, which extends from central Alabama to New England. Similar styles of deformation are found in the Ouachita Mountains of Oklahoma, the Marathon Mountains of Texas, and in most mountain belts of the world. Deformation such as this is an important structural feature of the earth's crust and clearly shows that the crust has been mobile and has moved throughout much of geologic time.

Plate 8 *The Bahamas Platform. The ocean floor is not flat and featureless but contains many fascinating and important landforms. These have been mapped by special instruments on oceanographic ships that plot a profile of the seafloor. Some submarine landforms can be seen in satellite photo- grpahs; this view of the Bahamas Platform, which rises abruptly from depths below 3700 m, is an example. The light tones of blue and purple are the shallow Bahama Banks, covered by water only 20 to 100 m in depth. The dark tones reflect deep water. The shallow platform is the site of exten- sive deposition of carbonate sediment, much of which is produced by marine organisms. Shallow seas such as this existed over much of the continental platforms in past geo- logic ages and resulted in the deposition of extensive lime- stone formations.*

Great Bahama Banks
(less than 10 meters deep)

N.

Old Bahama Channel

Cayo Coco

Cayo

Romano

Lagana de Leche

Moron

Cuba

0   40 km

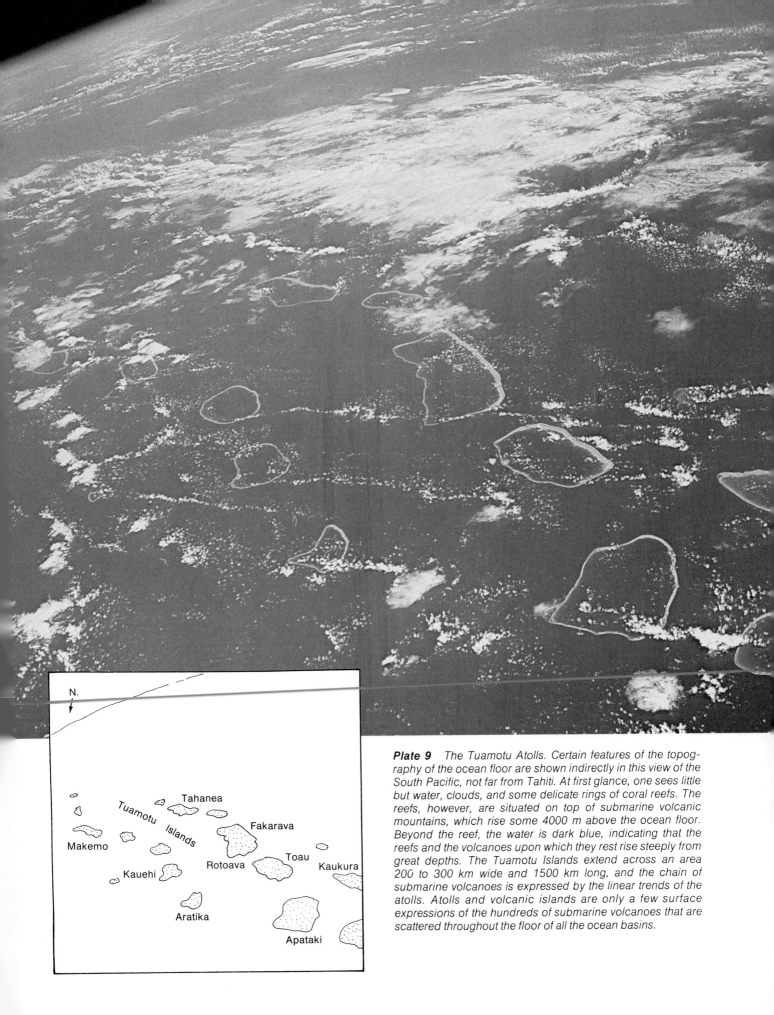

Plate 9 The Tuamotu Atolls. Certain features of the topography of the ocean floor are shown indirectly in this view of the South Pacific, not far from Tahiti. At first glance, one sees little but water, clouds, and some delicate rings of coral reefs. The reefs, however, are situated on top of submarine volcanic mountains, which rise some 4000 m above the ocean floor. Beyond the reef, the water is dark blue, indicating that the reefs and the volcanoes upon which they rest rise steeply from great depths. The Tuamotu Islands extend across an area 200 to 300 km wide and 1500 km long, and the chain of submarine volcanoes is expressed by the linear trends of the atolls. Atolls and volcanic islands are only a few surface expressions of the hundreds of submarine volcanoes that are scattered throughout the floor of all the ocean basins.

N.

Tahanea

Tuamotu Islands

Fakarava

Makemo

Rotoava
Toau

Kaukura

Kauehi

Aratika

Apataki

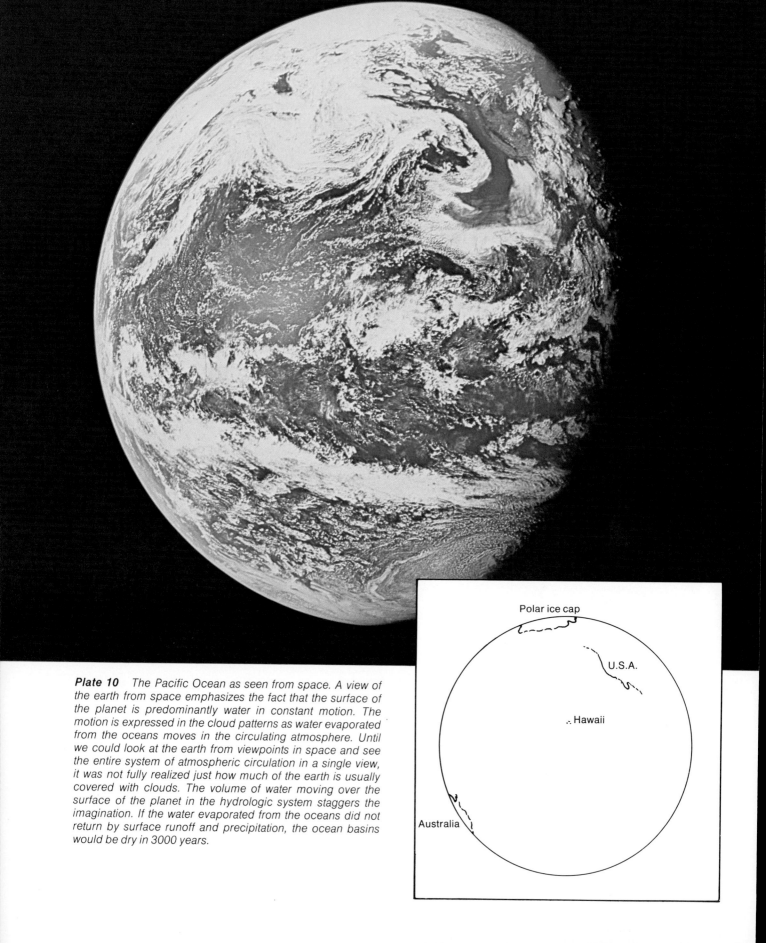

Plate 10 The Pacific Ocean as seen from space. A view of the earth from space emphasizes the fact that the surface of the planet is predominantly water in constant motion. The motion is expressed in the cloud patterns as water evaporated from the oceans moves in the circulating atmosphere. Until we could look at the earth from viewpoints in space and see the entire system of atmospheric circulation in a single view, it was not fully realized just how much of the earth is usually covered with clouds. The volume of water moving over the surface of the planet in the hydrologic system staggers the imagination. If the water evaporated from the oceans did not return by surface runoff and precipitation, the ocean basins would be dry in 3000 years.

Polar ice cap

U.S.A.

Hawaii

Australia

Spiral cloud pattern

Plate 11 *A cyclonic storm in the central Pacific. When viewed from space, the operation of the hydrologic system is especially vivid, since one can see a complete storm system spiraling over hundreds of square kilometers and pumping huge quantities of water from the ocean into the atmosphere. The centers of the cyclonic storms are marked by vigorous updrafts with winds of over 200 km per hour. When large cyclonic storms move over the land, they precipitate abnormally large quantities of water and generally cause widespread flooding.*

Plate 12 *Drainage systems in southern Arabia. One of the important effects of the hydrologic system is the development of intricate drainage systems upon the continents. On the moon, Mercury, and Mars, craters dominate the landscape. But, on the earth, stream valleys are the most abundant landform. Indeed, river systems constitute a clear record of how surface runoff has sculptured the land and testify to the magnitude at which the hydrologic system operates. Very few areas of the land are untouched by stream erosion, and, in this photograph of a desert region, details of the delicate network of tributaries are clearly shown.*

Hadramaut Plateau

Ras Baghashwa

Sharma Baj

N.

Gulf of Aden

Plate 13 *The Ganges River. The great Ganges River begins in the snow-covered Himalaya Mountains (to the left in this view) and flows parallel to the range across the Ganges plain to the ocean. As it flows, it carries vast quantities of sediment eroded from the world's highest mountains. The large sediment load is indicated by the brown color of the river, its braided pattern, and the numerous islands and sandbanks deposited along the river's course. Much of the sediment is deposited in the Ganges plain, where a total thickness of 7800 m has accumulated; but most debris carried by the river reaches the ocean, and this debris has built up the huge Ganges delta and the world's largest submarine fan off the eastern coast of India.*

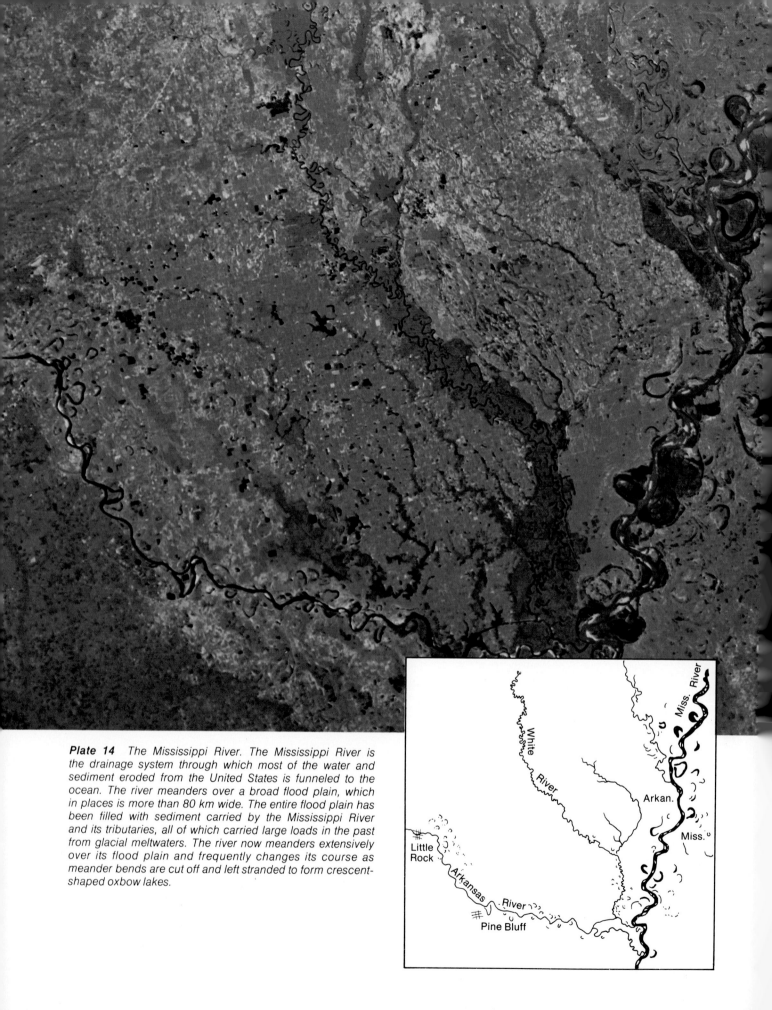

Plate 14 *The Mississippi River. The Mississippi River is the drainage system through which most of the water and sediment eroded from the United States is funneled to the ocean. The river meanders over a broad flood plain, which in places is more than 80 km wide. The entire flood plain has been filled with sediment carried by the Mississippi River and its tributaries, all of which carried large loads in the past from glacial meltwaters. The river now meanders extensively over its flood plain and frequently changes its course as meander bends are cut off and left stranded to form crescent-shaped oxbow lakes.*

White River

Miss. River

Arkan.

Miss.

Little Rock

Arkansas River

Pine Bluff

Plate 15 *The Grand Canyon of the Colorado River, Arizona. The capacity of a stream to erode through solid rock is illustrated more vividly in the Grand Canyon than perhaps in any other place in the world. The Colorado River drains a large part of the Rocky Mountains and flows across the Colorado Plateau to the Gulf of California. The plateau has been uplifted several thousand meters, so the river has been able to cut a system of deep canyons across the region. Where uplift is greatest, the canyons are deepest. The Grand Canyon is incised into essentially horizontal sedimentary rock and is roughly 1.6 km deep and 19 km wide. The innermost gorge is cut into the ancient crystalline rocks of the shield. Two volcanic fields are seen in this view of the San Francisco field to the southeast and the Uinkaret field to the west, where frozen lava falls cascade 1000 m into the inner gorge of the canyon.*

Toroweap fault

W. Kaibab fault

Echo Cliffs

Kaibab Plateau

Hurricane fault

Grand Canyon

Coconino Plateau

San Francisco volcanics

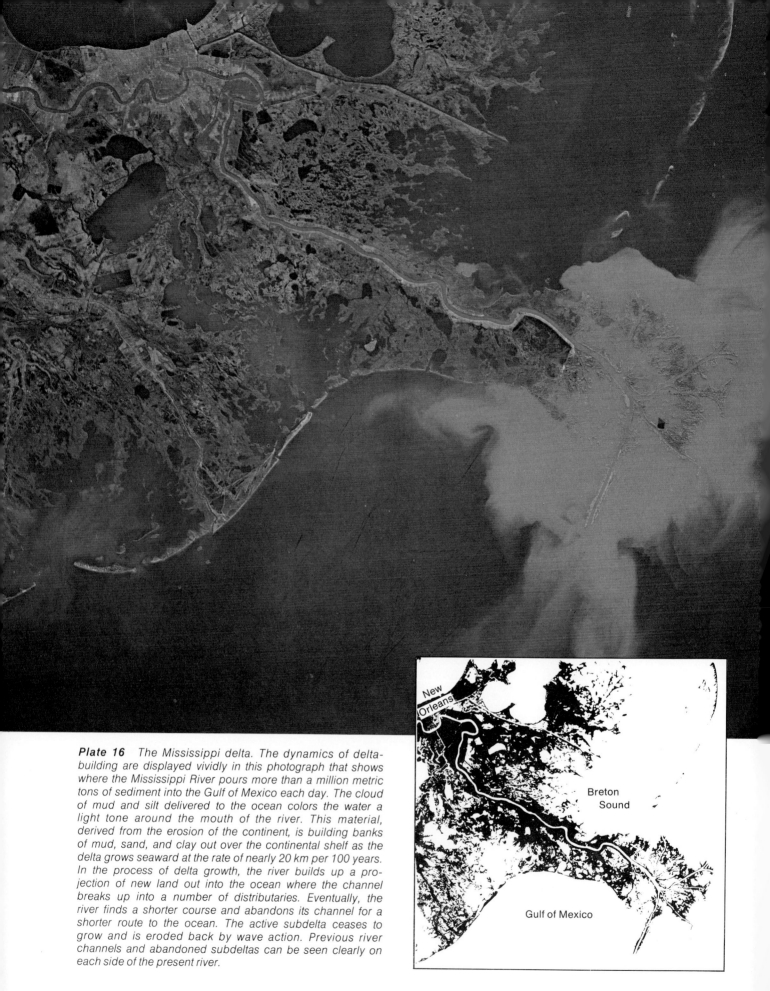

Plate 16 *The Mississippi delta. The dynamics of delta-building are displayed vividly in this photograph that shows where the Mississippi River pours more than a million metric tons of sediment into the Gulf of Mexico each day. The cloud of mud and silt delivered to the ocean colors the water a light tone around the mouth of the river. This material, derived from the erosion of the continent, is building banks of mud, sand, and clay out over the continental shelf as the delta grows seaward at the rate of nearly 20 km per 100 years. In the process of delta growth, the river builds up a projection of new land out into the ocean where the channel breaks up into a number of distributaries. Eventually, the river finds a shorter course and abandons its channel for a shorter route to the ocean. The active subdelta ceases to grow and is eroded back by wave action. Previous river channels and abandoned subdeltas can be seen clearly on each side of the present river.*

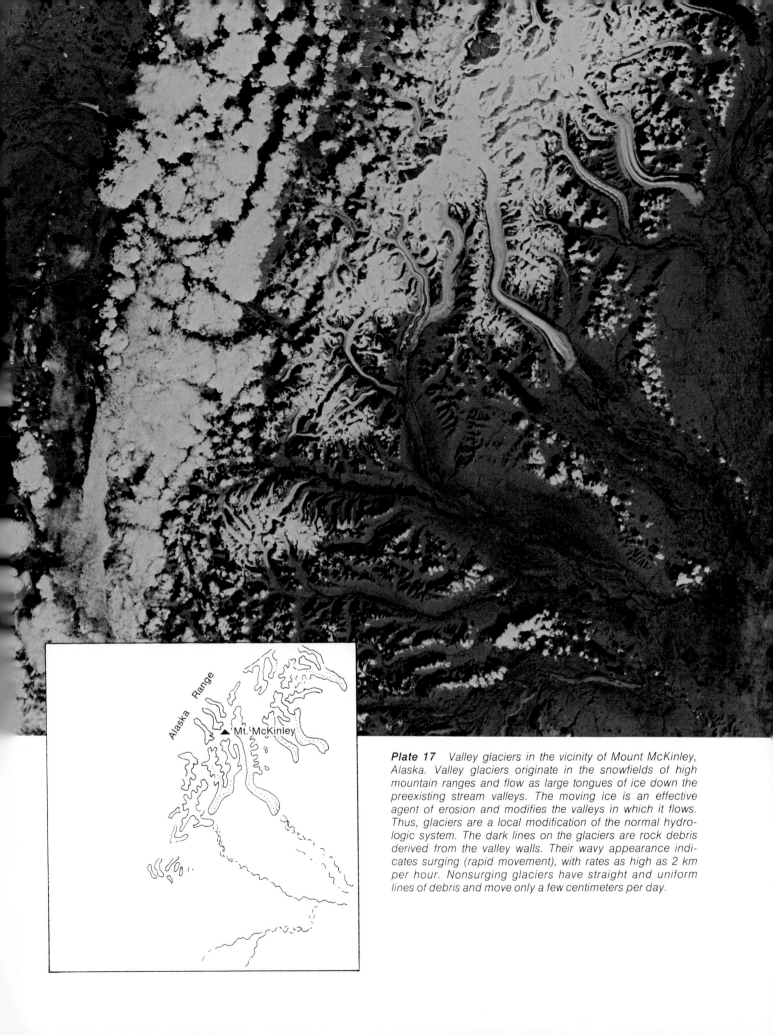

Plate 17 *Valley glaciers in the vicinity of Mount McKinley, Alaska. Valley glaciers originate in the snowfields of high mountain ranges and flow as large tongues of ice down the preexisting stream valleys. The moving ice is an effective agent of erosion and modifies the valleys in which it flows. Thus, glaciers are a local modification of the normal hydrologic system. The dark lines on the glaciers are rock debris derived from the valley walls. Their wavy appearance indicates surging (rapid movement), with rates as high as 2 km per hour. Nonsurging glaciers have straight and uniform lines of debris and move only a few centimeters per day.*

Plate 18 The Finger Lakes, New York. At least three advances of continental glaciers occurred in this area during the recent ice age and modified the topography in a variety of ways. In this area, many preexisting stream valleys were further enlarged by erosion from ice and meltwaters. Later, more thick ice masses occupied some of these valleys and deepened them. As the ice melted, the deep troughs filled with water to form the Finger Lakes. The plains area north of these lakes is covered with glacial sediment and has distinctive glacial landforms, which are not defined clearly on the image.

Plate 19 Cape Canaveral, Florida. Ground water is largely an invisible part of the hydrologic system, since it occupies the pore spaces in the soil and rocks beneath the surface. It can, however, dissolve soluble rocks such as limestone and form complex networks of caves and subterranean passageways. As the caverns enlarge, their roofs may collapse to form circular depressions called sinkholes. This creates a pockmarked surface called karst topography. The hundreds of lakes shown in this photograph occupy sinkholes and testify to the effectiveness of ground water as a geologic agent. The linear features along the coast are offshore bars and islands formed by waves and currents along the southern Atlantic coast.

Atlantic Ocean

Cape Canaveral

Old shorelines

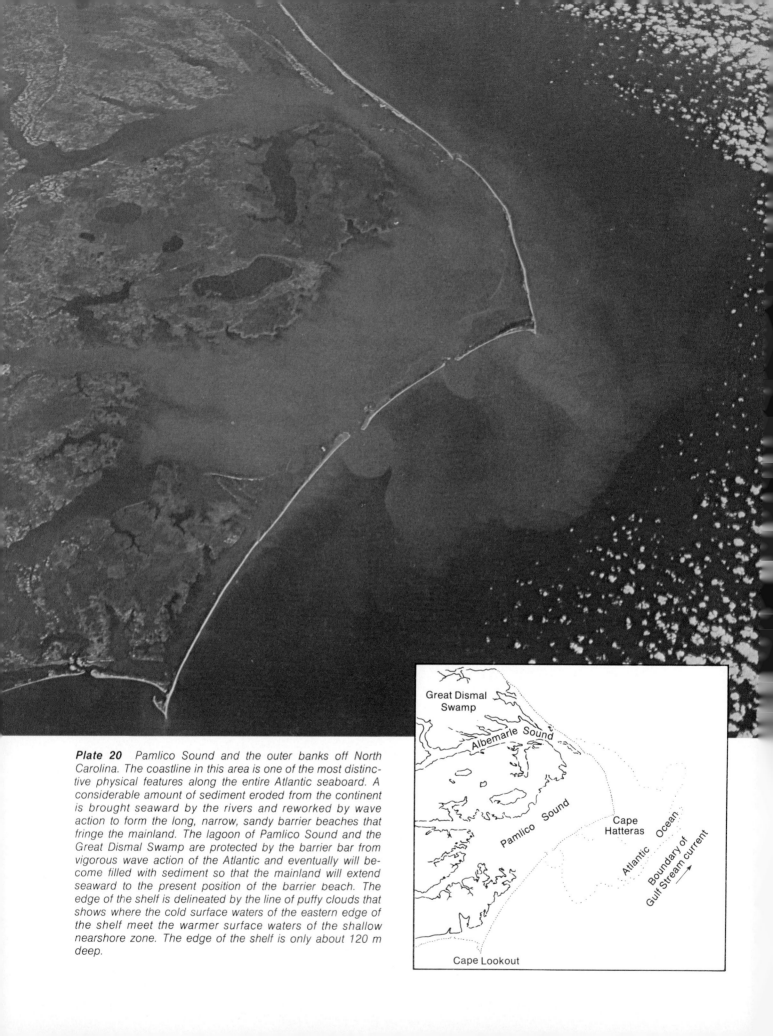

Plate 20 *Pamlico Sound and the outer banks off North Carolina. The coastline in this area is one of the most distinctive physical features along the entire Atlantic seaboard. A considerable amount of sediment eroded from the continent is brought seaward by the rivers and reworked by wave action to form the long, narrow, sandy barrier beaches that fringe the mainland. The lagoon of Pamlico Sound and the Great Dismal Swamp are protected by the barrier bar from vigorous wave action of the Atlantic and eventually will become filled with sediment so that the mainland will extend seaward to the present position of the barrier beach. The edge of the shelf is delineated by the line of puffy clouds that shows where the cold surface waters of the eastern edge of the shelf meet the warmer surface waters of the shallow nearshore zone. The edge of the shelf is only about 120 m deep.*

Great Dismal Swamp

Albemarle Sound

Pamlico Sound

Cape Hatteras

Atlantic Ocean

Boundary of Gulf Stream current

Cape Lookout

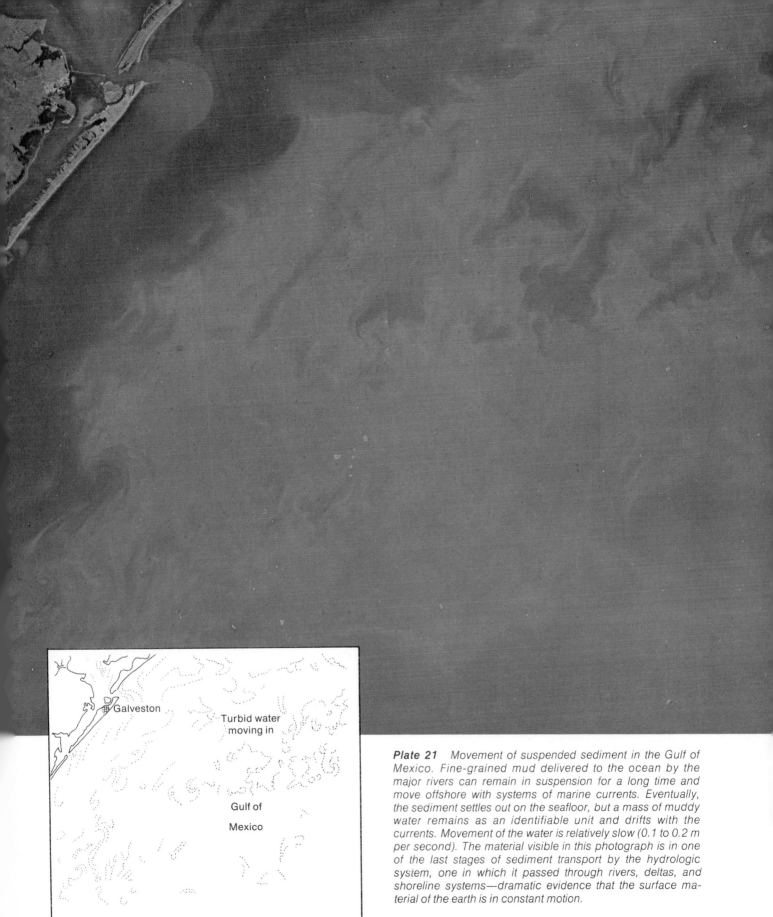

Plate 21 Movement of suspended sediment in the Gulf of Mexico. Fine-grained mud delivered to the ocean by the major rivers can remain in suspension for a long time and move offshore with systems of marine currents. Eventually, the sediment settles out on the seafloor, but a mass of muddy water remains as an identifiable unit and drifts with the currents. Movement of the water is relatively slow (0.1 to 0.2 m per second). The material visible in this photograph is in one of the last stages of sediment transport by the hydrologic system, one in which it passed through rivers, deltas, and shoreline systems—dramatic evidence that the surface material of the earth is in constant motion.

Galveston

Turbid water moving in

Gulf of

Mexico

Plate 22 *Longitudinal sand dunes in the Arabian peninsula. This view is to the southwest, toward the Hadramaut Plateau. The long parallel stripes of dunes are part of one of the largest "sand seas" of the world. The dunes are typical of much of the Empty Quarter of Arabia and are remarkable in that some extend more than 200 km without a break. When we consider the vast areas of migrating sand dunes in the deserts of the world, we recognize that the circulating atmosphere alone continually transports enormous quantities of sediment over the surface of the earth.*

Plate 23 *The Sinai peninsula and the Red Sea. The Red Sea is a rift in the earth's crust that separates Africa from Asia. This is an important part of the world rift system in that it represents the incipient stages of crustal fragmentation and the movement of continental plates. The rift extends up the Red Sea and splits at its northern end, with one branch forming the Gulf of Suez and the other extending up the Gulf of Aqaba, into the Dead Sea, and up the Jordan Valley. The Arabian plate has moved in a northwesterly direction away from Africa, creating the beginning stages of a new ocean basin. As the rift widens, the edges of the continental block break off and form a series of steps leading down toward the depression. These step blocks can be seen along the Gulf of Suez as distinct parallel lines in the bedrock, as well as linear trends in the offshore islands.*

Map labels: Beirut, Euphrates, Jordan River, Tel Aviv, Dead Sea, Mediterranean Sea, Nafud Desert, Suez Canal, Faults, Wadi Araba, Gulf of Suez, Red Sea, Faults, Wadi Kena, N

Plate 24 *The Wasatch fault zone in central Utah. A large segment of the earth's crust in Nevada and eastern Utah is being uparched and pulled apart. This causes large crustal blocks to subside or collapse to form structural valleys. In this photograph, the valleys occupied by Great Salt Lake and Utah Lake are the downdropped blocks and the adjacent Wasatch Mountains are the flanks of the uparched areas. The major fracture is the Wasatch fault, which zigzags along the base of the mountain front. It has moved as much as 67 m in the last 20,000 years.*

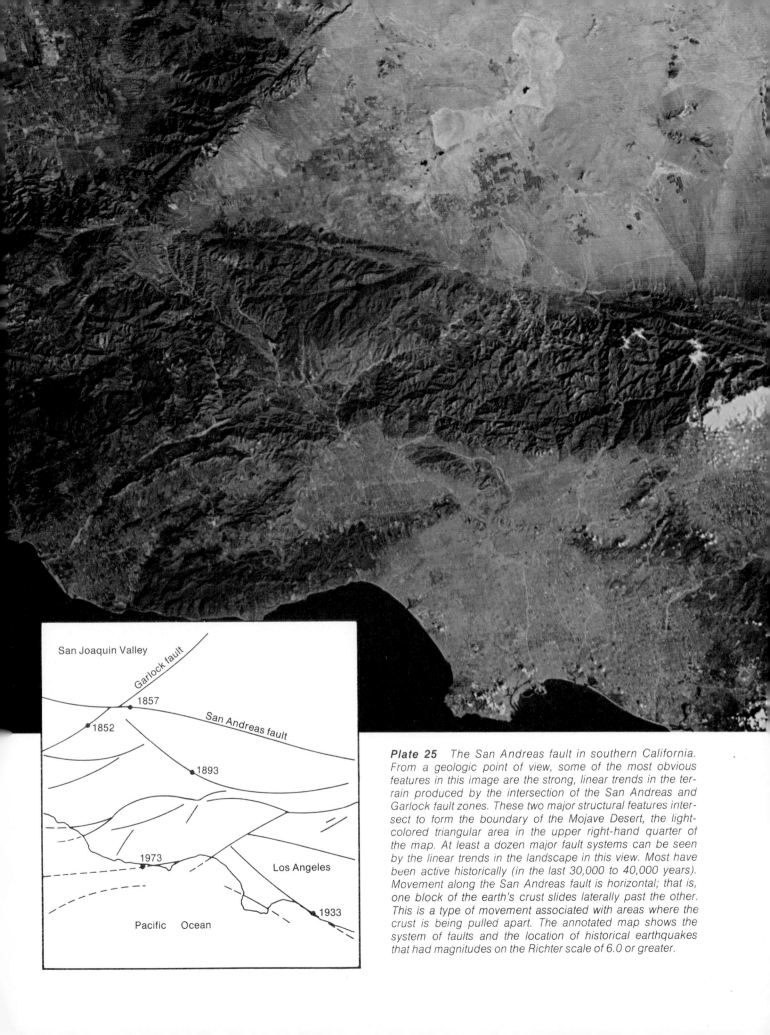

Plate 25 *The San Andreas fault in southern California. From a geologic point of view, some of the most obvious features in this image are the strong, linear trends in the terrain produced by the intersection of the San Andreas and Garlock fault zones. These two major structural features intersect to form the boundary of the Mojave Desert, the light-colored triangular area in the upper right-hand quarter of the map. At least a dozen major fault systems can be seen by the linear trends in the landscape in this view. Most have been active historically (in the last 30,000 to 40,000 years). Movement along the San Andreas fault is horizontal; that is, one block of the earth's crust slides laterally past the other. This is a type of movement associated with areas where the crust is being pulled apart. The annotated map shows the system of faults and the location of historical earthquakes that had magnitudes on the Richter scale of 6.0 or greater.*

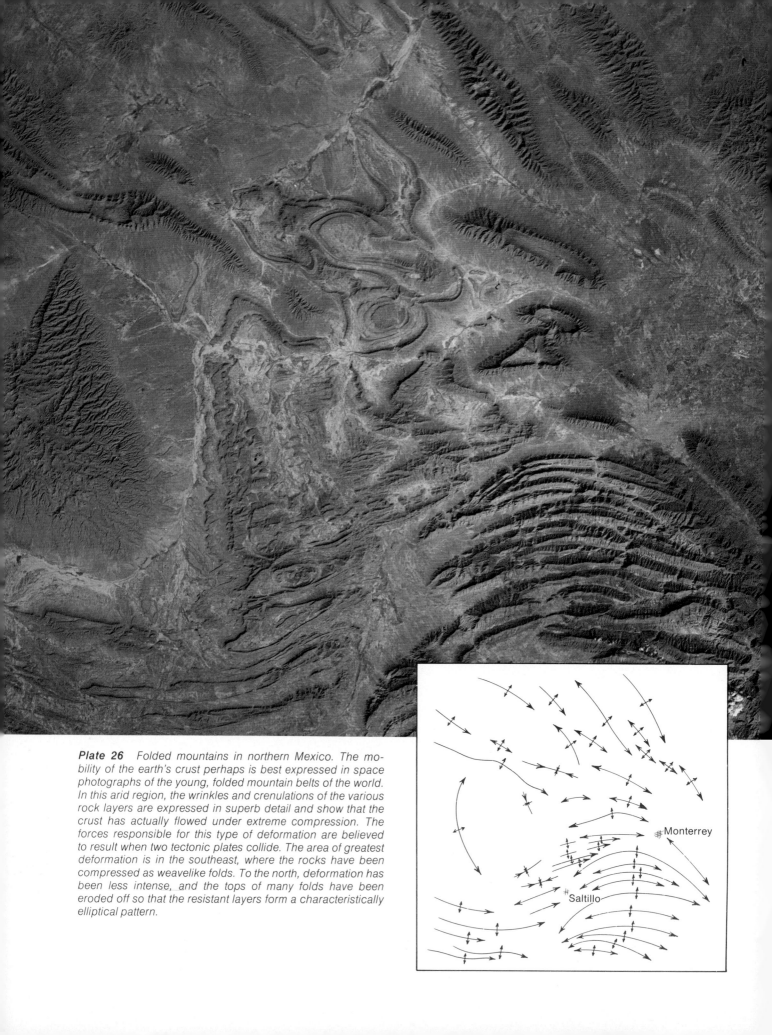

Plate 26 *Folded mountains in northern Mexico. The mobility of the earth's crust perhaps is best expressed in space photographs of the young, folded mountain belts of the world. In this arid region, the wrinkles and crenulations of the various rock layers are expressed in superb detail and show that the crust has actually flowed under extreme compression. The forces responsible for this type of deformation are believed to result when two tectonic plates collide. The area of greatest deformation is in the southeast, where the rocks have been compressed as weavelike folds. To the north, deformation has been less intense, and the tops of many folds have been eroded off so that the resistant layers form a characteristically elliptical pattern.*

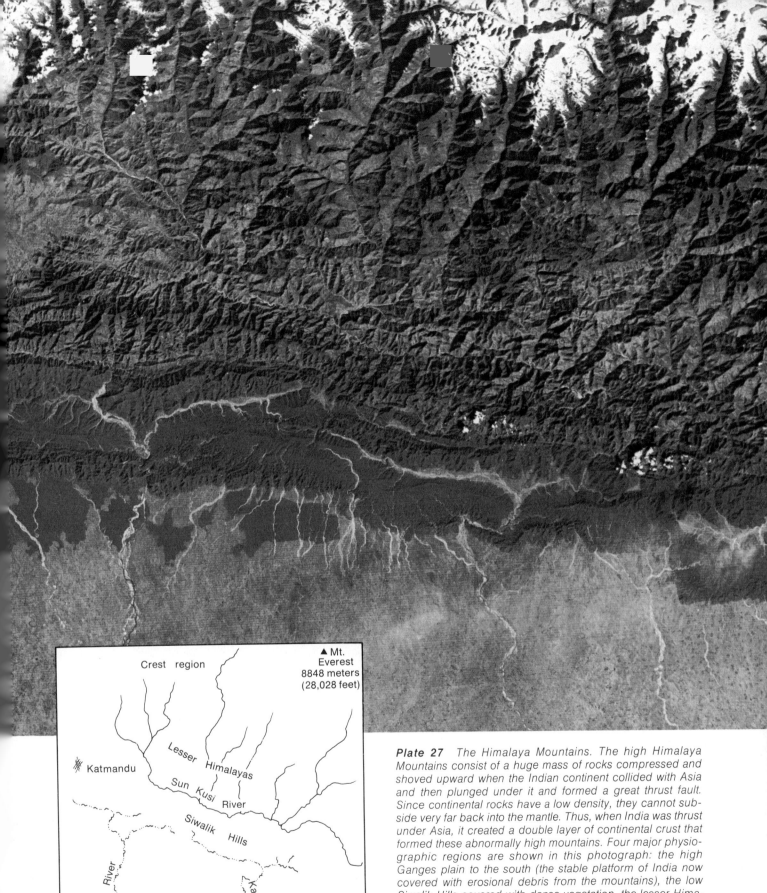

Plate 27 The Himalaya Mountains. The high Himalaya Mountains consist of a huge mass of rocks compressed and shoved upward when the Indian continent collided with Asia and then plunged under it and formed a great thrust fault. Since continental rocks have a low density, they cannot subside very far back into the mantle. Thus, when India was thrust under Asia, it created a double layer of continental crust that formed these abnormally high mountains. Four major physiographic regions are shown in this photograph: the high Ganges plain to the south (the stable platform of India now covered with erosional debris from the mountains), the low Siwalik Hills covered with dense vegetation, the lesser Himalayas, and the crest region. Mount Everest, the highest point on the earth, is in the upper right-hand corner of the photograph.

Plate 28 The Labrador fold belt, Canada. The deformed rocks shown in this photograph are examples of the style and degree of deformation that occurs in the deeper parts of a folded mountain belt. The sedimentary strata that were deposited originally in horizontal layers have been compressed and squeezed into tight folds and have been displaced along numerous faults. A great variety of folds of different sizes and styles are obvious, and intrusions of granitic rocks appear as small patches with a greenish gray tone. The Labrador fold belt represents the roots of a former mountain system similar to the Alps or the Appalachians, although the high mountainous topography has been eroded down to a surface of low relief near sea level.

Deformed rock strata

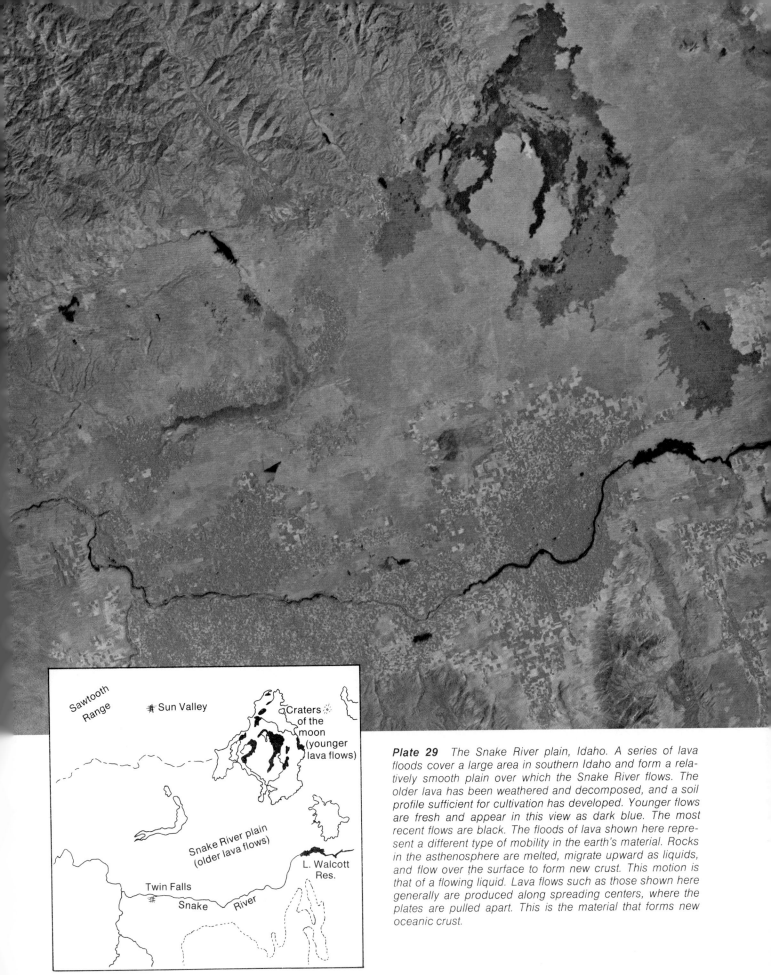

Plate 29 The Snake River plain, Idaho. A series of lava floods cover a large area in southern Idaho and form a relatively smooth plain over which the Snake River flows. The older lava has been weathered and decomposed, and a soil profile sufficient for cultivation has developed. Younger flows are fresh and appear in this view as dark blue. The most recent flows are black. The floods of lava shown here represent a different type of mobility in the earth's material. Rocks in the asthenosphere are melted, migrate upward as liquids, and flow over the surface to form new crust. This motion is that of a flowing liquid. Lava flows such as those shown here generally are produced along spreading centers, where the plates are pulled apart. This is the material that forms new oceanic crust.

Sawtooth Range

⚏ Sun Valley

Craters of the moon (younger lava flows)

Snake River plain (older lava flows)

L. Walcott Res.

Twin Falls

Snake River

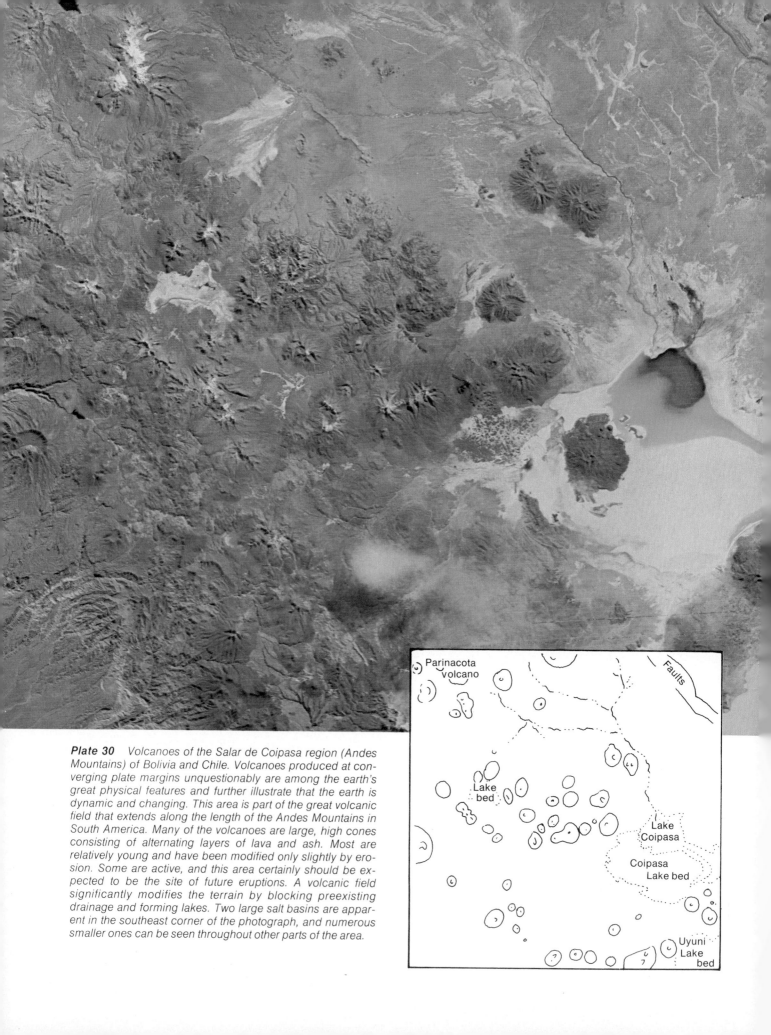

Plate 30 *Volcanoes of the Salar de Coipasa region (Andes Mountains) of Bolivia and Chile. Volcanoes produced at converging plate margins unquestionably are among the earth's great physical features and further illustrate that the earth is dynamic and changing. This area is part of the great volcanic field that extends along the length of the Andes Mountains in South America. Many of the volcanoes are large, high cones consisting of alternating layers of lava and ash. Most are relatively young and have been modified only slightly by erosion. Some are active, and this area certainly should be expected to be the site of future eruptions. A volcanic field significantly modifies the terrain by blocking preexisting drainage and forming lakes. Two large salt basins are apparent in the southeast corner of the photograph, and numerous smaller ones can be seen throughout other parts of the area.*

Parinacota volcano

Faults

Lake bed

Lake Coipasa

Coipasa Lake bed

Uyuni Lake bed

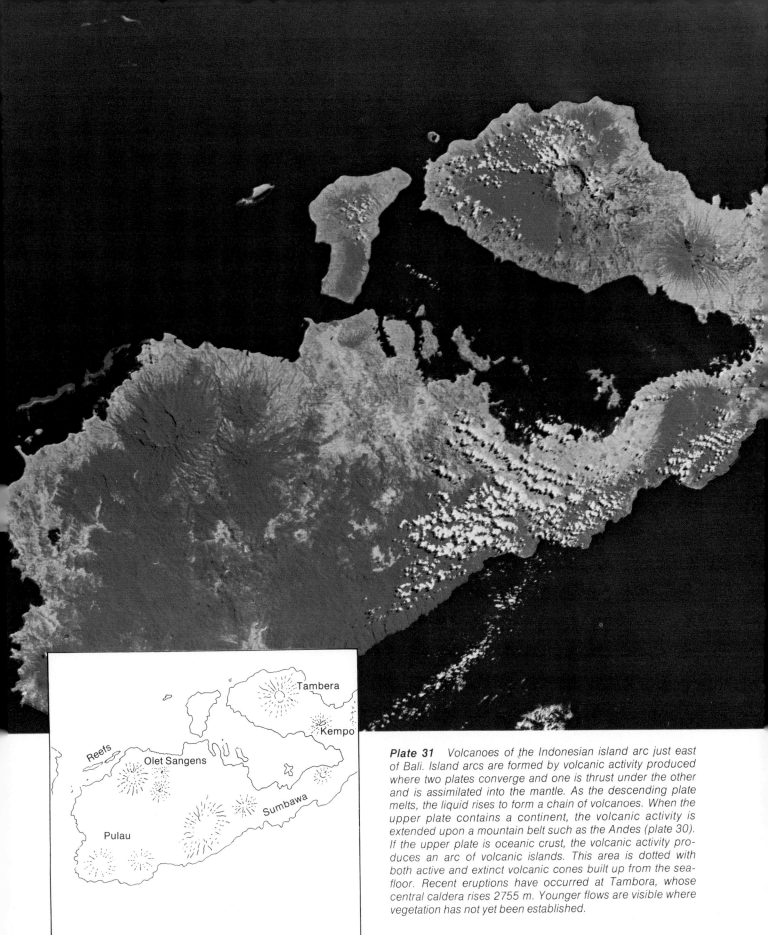

Plate 31 *Volcanoes of the Indonesian island arc just east of Bali. Island arcs are formed by volcanic activity produced where two plates converge and one is thrust under the other and is assimilated into the mantle. As the descending plate melts, the liquid rises to form a chain of volcanoes. When the upper plate contains a continent, the volcanic activity is extended upon a mountain belt such as the Andes (plate 30). If the upper plate is oceanic crust, the volcanic activity produces an arc of volcanic islands. This area is dotted with both active and extinct volcanic cones built up from the sea-floor. Recent eruptions have occurred at Tambora, whose central caldera rises 2755 m. Younger flows are visible where vegetation has not yet been established.*

Plate 32 *The Atlas Mountains of southern Morocco. This view shows the major dynamic systems of the earth. The swirling white clouds over the Atlantic illustrate the operation of the hydrologic system, which is powered by heat from the sun. The hydrologic system continually pumps water from the ocean and carries it over the land, where the water precipitates and returns to the ocean as runoff in rivers. The result is erosion of the land. The operation of the tectonic system is recorded in the structure of the deformed rocks of the Atlas Mountains, where the folded rocks testify that the earth's crust is mobile and is being compressed in some areas and pulled apart in others. The dynamic systems make the earth unique among the inner planets of the solar system. If they did not operate, the surface of the earth would have remained, essentially, unchanged since its early period of accretion and it would be saturated with craters, like the surface of the moon.*

The Interior of the Earth

Statement

The atmosphere, the oceans, and the surface of the land are known in considerable detail, because they can be studied by direct observation, but the internal structure of the earth presents some of the most difficult problems faced by geologists and geophysicists. The deepest bore holes penetrate no more than 8 km, and erosion exposes rocks that were created no more than 20 to 25 km below the surface. Volcanic eruptions provide samples of material that comes from greater depths, possibly as much as 200 km. Aside from this limited data, we have no direct knowledge about the nature of the earth's interior. How, then, are we able to determine the structure and composition of the earth's interior? The evidence comes largely from studies of the physical characteristics of the earth—its density, the nature of its magnetic field, and the way in which it transmits seismic (earthquake) waves. These evidences will be considered in later chapters; the objective here is to summarize the present-day understanding of the structure, composition, and physical properties of the earth's interior.

The earth is a planet in which the materials are differentiated; that is, the heavy material is concentrated near the center and the lighter material near the surface. The major structural units are the following.

1. A central core, composed predominantly of iron and nickel.
2. A thick, surrounding mantle, composed of silicate minerals rich in iron and magnesium (the upper mantle, called the **asthenosphere,** is near the melting point and yields to plastic flow).
3. A rigid lithosphere, composed of relatively light silicate minerals, that includes the crusts of the continents and the ocean basins (*figure 1.5*).

Discussion

The Core. The core of the earth is a central mass about 7000 km in diameter. Its density increases with depth but averages about 10.78 g/cm³. It is nearly twice as dense as the mantle; thus, although it has only 16.2% of the earth's volume, it has 31.5% of the earth's mass. Most scientists believe that the core is composed mostly of iron and nickel and consists of two distinct parts—a solid inner core and a liquid outer core. The rotation of the earth probably causes the liquid core to circulate; this would generate the earth's magnetic field.

The Mantle. The next major structural unit of the earth surrounds or covers the core and therefore is called the **mantle.** This zone constitutes the great bulk of the earth (82.3% of the volume and 67.8% of the mass).

The Asthenosphere. The asthenosphere is the upper part of the mantle. It is approximately 200 km thick, and its upper boundary is 65 to 100 km below the earth's surface. The asthenosphere is distinctive in that the temperature and pressure in it are in delicate balance so that most of its material is near the melting point. The asthenosphere is, therefore, believed to be partly molten and structurally weak, and so it is capable of flow. It is movement within this layer that is responsible for volcanic activity and crustal deformation observed at the surface.

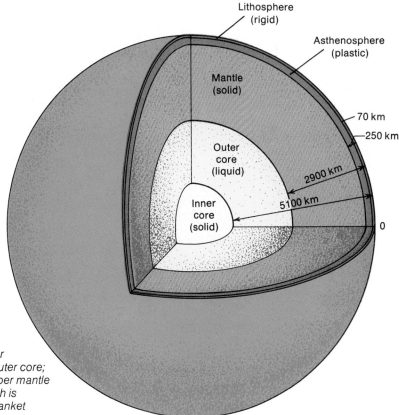

Figure 1.5 *Diagram summarizing the present-day understanding of the structure of the earth. The major components are a dense, solid inner core; a liquid outer core; a thick, solid mantle; and a rigid lithosphere. The upper mantle contains a significant layer: the asthenosphere, which is soft and plastic. Surrounding the entire planet is a blanket of surface fluids—water and air.*

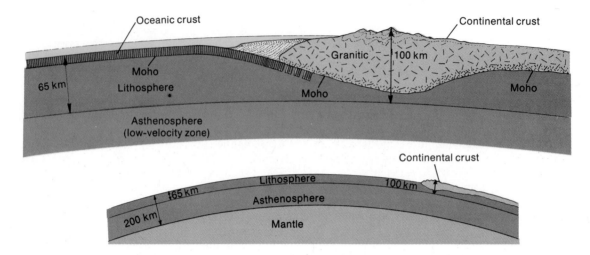

Figure 1.6 *Structure of the outer layers of the earth. The asthenosphere is near the melting point and is capable of flow. The lithosphere above it is rigid. It includes two types of crust: a thin, dense oceanic crust and a thick, lighter continental crust.*

The Lithosphere. The lithosphere is the outermost shell of the earth. It is a rigid, solid layer about 65 to 100 km thick that rests upon the weak, partly molten asthenosphere. The continental crust can be thought of as raftlike slabs of granitic rock embedded in the lithosphere (*figure 1.6*). The rocks of the continental crust have been highly deformed by compression and include the oldest rocks on the surface of the earth. The oceanic crust, in contrast, is composed mostly of relatively young volcanic rocks that are essentially undeformed.

The Surface Fluids. The gases and liquids of the atmosphere and oceans are referred to as the earth's surface fluids and are as much a part of the earth's systems as the rocks of the crust. Not only do they constitute part of the material of the planet, but, as was shown in previous sections, they are of prime importance in the origin and development of the landscape. The atmosphere and water of the earth's system are concentrated above the solid part of the earth because of their lighter density.

The earth's atmosphere consists of a mixture of highly mobile gases held to the earth by gravitational attraction. The gases are densest close to sea level, and they thin rapidly at higher altitudes, so that 97% of the mass of the atmosphere lies within 29 km of the surface. It is difficult to draw a sharp boundary for the outer limit of the atmosphere, because the density of gas molecules decreases almost imperceptibly into interplanetary space. Many scientists consider the outer boundary to be about 9500 km above the earth's surface, a distance nearly as great as the diameter of the earth. Only the lower part of the atmosphere is visible as clouds formed by water vapor. Water on the earth is present as oceans, lakes, rivers, glaciers, ground water, and vapor in the atmosphere.

Summary

Perhaps the most overwhelming impression one gains from seeing the earth from space and reviewing the major features seen on its surface is that the earth is a dynamic planet and has been changing throughout all of geologic time. Not only are the surface fluids extremely mobile, but the crust itself is in motion and has been deformed by mountain-building forces, volcanic activity, and earthquakes. There has been no such activity on the moon or Mercury, and the geodynamics of Mars has been much less than that of the earth.

The facts about the earth that have been discussed and shown by means of space photographs can be summarized as follows.

1. The atmosphere and the hydrosphere constitute a system of moving surface fluids in which tremendous volumes of water are in constant motion.
2. The two major structural and topographic features of the earth are (a) the continents and (b) the ocean basins.
3. The continents consist of three major components: (a) the shields, which are areas composed of complexly deformed rocks eroded down to near sea level; (b) stable platforms, or areas of the shield covered with a veneer of horizontal sedimentary rocks; and (c) mountain belts, in which sedimentary rocks have been compressed into folds.
4. The major features on the ocean floors are (a) the oceanic ridge, (b) the abyssal floor, (c) seamounts, (d) trenches, and (e) continental margins.
5. The major structural units of the earth are (a) a solid inner core, (b) a liquid outer core, (c) a thick mantle, (d) a soft asthenosphere, and (e) a rigid lithosphere.

Additional Readings

Bates, D. R. 1964. The Planet Earth. New York: Pergamon Press.

Bolt, D. A. 1973. "The Fine Structure of the Earth's Interior." Sci. Amer. 228 (3): 24-33.

Bott, M. H. P. 1971. The Interior of the Earth. New York: St. Martin's Press.

The Earth's Dynamic Systems

2

In the previous chapter, you saw that the earth is a differentiated planet; that is, the materials that make up the their earth are segregated and separated according to density into a series of concentric layers, or shells. The material within the outer layers is moving in two major systems.

The hydrologic system involves movement of the surface fluids (air and water). Drawing energy from the sun, water circulates from the oceans, to the atmosphere, and back to the earth and the oceans. The precipitation of water on the earth furnishes a fluid medium for many processes that shape the earth's surface features. The rivers systematically carve away the land as they flow to the seas. Ground water percolating through the pore spaces of rocks carries with it dissolved minerals from the earth. Where the temperature drops low enough, glaciers form and move over large parts of the continents, modifying the surface by erosion and deposition as they move outward from centers of accumulation.

The tectonic system results from the earth's internal energy and involves movement in the asthenosphere, presumably in convection cells. As the material rises and moves laterally, the overlying lithosphere is uparched and split apart. Thus, the lithosphere is broken into a series of plates that are in constant motion. Where plates converge, one is thrust down into the mantle and is consumed. Where the plates pull apart, new oceanic crust is developed by volcanic activity. Movement of the lithospheric plates results in seafloor spreading, continental drift, mountain-building, volcanic activity, and earthquakes.

This chapter summarizes the basic elements of these two fundamental geologic systems that continually modify the earth. Throughout the remainder of this book, each chapter is concerned largely with the physical processes of these systems.

Major Concepts

1. The earth is a dynamic planet, as evidenced by the fact that the materials are differentiated and segregated into distinct layers or zones (the core, the mantle, the lithosphere, and the surface fluids). The most dynamic layers are (a) the surface fluids (water and atmosphere) and (b) the soft asthenosphere in the upper mantle.
2. The system of moving water on the earth's surface (the hydrologic system) involves movement of water in rivers, as ground water, in glaciers, and in oceans. The volume of water in motion is almost incomprehensibly large, and, as it moves, it erodes, transports, and deposits the sediment. The source of energy for the earth's hydrologic system is heat from the sun.
3. The tectonic system involves movement of the material in the earth's interior, which results in seafloor spreading, creation of new crust, continental drift, volcanism, earthquakes, and mountain-building. Radiogenic heat in the upper mantle is probably the source of energy for the tectonic system.
4. The earth's lithosphere is buoyed up and floats on the denser, plastic mantle beneath and rises and sinks in attempts to establish isostatic equilibrium.

The Hydrologic System

Statement

Surface processes (such as erosion and deposition by running water, **ground water, glaciers,** and waves) are the result of the global system of moving fluids at, or near, the earth's surface. The basic elements of this system are very simple and are shown diagrammatically in *figure 2.1.* In the broadest sense, the **hydrologic system** includes all possible paths of water movement. Of primary concern are the major patterns shown by the blue arrows. The system operates as heat from the sun evaporates water from the oceans, the principal reservoir for the earth's water. Most of the water returns directly to the oceans as rain; the rest drifts over the continents and is precipitated as rain or snow. The water that falls on the land can take a variety of paths back to the ocean. The greatest quantity returns to the atmosphere by evaporation, but the most obvious return is by surface runoff in river systems as they funnel water back to the ocean. Some water also seeps into the ground and moves slowly through the pore spaces of the rocks. Plants use part of the ground water and then expel it back into the atmosphere, but much ground water slowly seeps into streams and lakes or migrates through the subsurface back to the ocean. Water can be trapped temporarily upon a continent as glacial ice, but glaciers move from cold centers of accumulation into warmer areas and ultimately melt, returning their water to the system of surface runoff.

Water in the hydrologic system—moving as surface runoff, ground water, glaciers, and waves and currents—erodes and transports surface rock material and ultimately deposits it as deltas, beaches, and other types of sedimentary deposits. In this way, the surface material is in motion—motion that results in a continually changing landscape.

Discussion

The idea of a complete cycle in the movement of water was recorded in Biblical times (Eccles. 1:7), but it was not demonstrated as a fact until the mid-seventeenth century, when two French scientists, Pierre Perrault and Edme Mariotte, independently measured precipitation in the drainage basin of the Seine River and then measured the discharge into the ocean during a given interval of time. Their measurements proved that precipitation alone could produce not only enough water for river flow but also enough for springs and seeps as well. Precipitation then was recognized as the basic source of all surface water.

The importance of the hydrologic system is so great that it might well be considered the most fundamental and significant geologic system operating on the surface of the earth. Your problem as a student new to geology will be not so much in understanding the process but in conceiving the worldwide scope of this system, which influences in both subtle and dramatic fashion the development of the earth's surface features.

The Hydrologic System Seen from Space. One of the best ways to gain an accurate concept of the magnitude of the hydrologic system is to study the space photographs in *plates 10* and *11,* because viewing the earth from perspectives in space permits you to see the system in operation on a global scale. A traveler arriving from space would observe that the surface of the earth is predominantly water (a conclusion obvious from the view of the Pacific Ocean in *plate 10*). The movement of water from the ocean to the atmosphere is expressed in the flow patterns of the clouds. This motion is one of the most distinctive features of the earth when it is viewed from space, and it stands out in marked contrast to views of the moon, Mars, and Mercury.

In a close view, the swirling motion of the atmosphere pumping water from the ocean during a cyclonic storm is

Figure 2.1 *Schematic diagram showing the circulation of water in the hydrologic system. Water evaporates from the oceans, circulates around the globe with the atmosphere, and ultimately returns to the surface by precipitation in the form of rain or snow. The water that falls upon the land returns to the ocean by surface runoff and ground-water seepage. Variations in the major flow patterns of the system include the temporary storage of water in lakes and glaciers.*

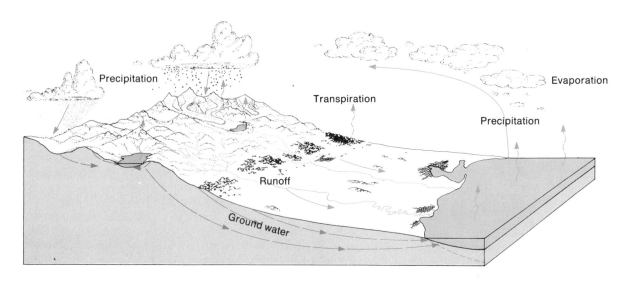

Precipitation

Transpiration

Evaporation

Precipitation

Runoff

Ground water

most impressive (*plate 11*). The kinetic energy produced by a hurricane amounts to roughly 100,000 million kilowatt hours per day, which is considerably more than the energy used by man in one day.

Another way of grasping the magnitude of the hydrologic system is to consider the volume of water involved. From measurements of rainfall and stream discharge, together with measurements of heat and energy transfer into bodies of water, scientists have calculated that 400,000 km^3 of water evaporate each year (or 1100 km^3 per day). Of this, about 336,000 km^3 are taken from the oceans, and 63,000 km^3 evaporate from water bodies on the land. Of the total 400,000 km^3 of water evaporated each year, about 101,000 km^3 falls as rain or snow on the continents. Most of the precipitation (about 60% to 80%) returns directly to the atmosphere by evaporation, with about 38,000 km^3 of water flowing back to the ocean over the surface or beneath the ground. These figures may or may not be impressive in themselves, but, when related to more familiar volumes and rates, they give some idea of the magnitude of this system.

From these rates, we know that, if the hydrologic system were interrupted and water did not return to the oceans (by precipitation back to the ocean and by surface runoff from the continents), sea level would drop 1 m per year and all the ocean basins would be completely dry within 4000 years. The recent glacial epoch demonstrates this point very clearly, since it partly interrupted the hydrologic system by freezing much of the water that fell upon the Northern Hemisphere and preventing it from returning immediately to the ocean. Consequently, during the ice age, sea level dropped over 100 m.

As a final observation, consider the volume of water running off the earth's surface. At any given instant, there is an average of 1260 km^3 of water flowing in the world's rivers. Gauging stations on the major rivers of the world indicate that, if the ocean basins were empty, enough runoff from the continents would occur to fill them completely in 40,000 years. (This would not take into consideration precipitation that went directly back into the ocean.) On the other hand, if all the evaporated water returned by runoff with no return to the atmosphere by immediate evaporation and evapotranspiration, the ocean basins would be filled in slightly more than 3000 years.

River Systems. As would be expected, such vast volumes of water moving over the surface produce a landscape dominated by features formed by running water. From viewpoints on the ground, the dominance of stream channels on the earth's surface cannot be appreciated; but, from perspectives in space, we see that stream valleys are the most abundant landform on the planet. In arid regions, where vegetation and soil cover do not obscure details, the intricate network of stream valleys is most impressive (*plate 12*). Stream valleys of this type dominate the surface of the earth. On every continent, the entire surface of the land is related in some way to a slope of a stream valley that collects and funnels surface runoff toward the ocean.

The oblique view in *plate 13* provides another clear example of the dominance of stream valleys on the

earth's surface. This view shows the Ganges plain beginning at the base of the snow-covered Himalayas and extending across the entire photograph. The plains are crossed by numerous tributary rivers that contribute to the waters of the main Ganges. Most of the channels, even the small tributaries, can be traced in detail, and the meanders and braided patterns show that the river system is moving large quantities of both sediment and water. The entire landscape, including the great Himalaya Mountains, is related to the Ganges River and its system of tributaries.

Plate 14 shows similar relationships between landforms and river systems in North America, where the Mississippi River and its tributaries carry most of the surface runoff to the Gulf of Mexico. In the view of the river in *plate 14,* the surface of the land slopes gently toward the Gulf of Mexico and the river meanders across a wide flood plain.

In contrast, the Colorado River in the western United States flows through high plateau country and has eroded great systems of canyons. The Grand Canyon (*plate 15*) is the most spectacular of many canyons in this area. Here again, all the landforms are intimately related to the drainage system.

The important point these examples illustrate is that each continent has one or more major drainage systems and the valley slopes formed by the network of tributary streams are the dominant landforms. We see from these space photographs that the surface drainage of a continent is an important segment of the hydrologic system and is a vivid record of how the system has continually modified and shaped the surface of the land.

Another very important aspect of the hydrologic system is that it provides a fluid medium by which huge amounts of sand, silt, and mud are transported to the ocean. This material forms the great deltas of the world and is a record of how the hydrologic system has operated in the past. The deltas are a record of the amount of material washed off from the continents as the rivers flow to the ocean. They illustrate the amount of erosion accomplished by running water.

The Mississippi delta (*plate 16*) is a classic example. The river is confined to a single channel far downstream from New Orleans. It then breaks up into a series of distributaries that build extensions of new land out into the ocean. A large cloud of sediment suspended in the ocean forms where the river empties into the Gulf of Mexico and is actively building new land seaward. Ultimately, the main channel shifts its course to seek a more direct route to the ocean, and the extension of land (subdelta) is eroded back by waves and currents. Previous courses of the Mississippi River can be seen on both sides of the present river.

Glacial Systems. In colder climates, precipitation is in the form of snow, which remains frozen and does not return immediately to the ocean as surface runoff. When this occurs, huge bodies of ice build up glaciers that modify the operation of the hydrologic system. In *plate 17,* the glacial field in the vicinity of Mount McKinley in Alaska is pictured. Large valley glaciers originate from snowfall in the high country and slowly flow down previous stream valleys as rivers of ice. They melt at their lower end and return their water to the hydrologic system as surface runoff.

At the present time, the continent of Antarctica is covered with a **continental glacier,** a sheet of ice 2.0 to 2.5 km thick, covering an area of 13 million km²—an area larger than the United States and Mexico. A large part of the Antarctica glacier can be seen in *plate 3.* When a continental glacier forms, it completely modifies the normal hydrologic system, because the water does not return immediately to the ocean as surface runoff but moves slowly as flowing ice. An ice sheet similar to that now on Antarctica recently covered a large part of North America. As the ice moved, it modified the landscape, creating numerous lakes and other landforms (*plate 18*).

Ground-Water Systems. Another segment of the hydrologic system is the water that seeps into the ground and moves slowly through the pore spaces in the soil and rocks. As it moves, ground water dissolves soluble rocks (such as limestone) and creates caverns and caves that enlarge and collapse to form depressions called **sinkholes.** This type of solution-generated landform is common in Kentucky and Florida and is easily recognized on space photographs (*plate 19*). The sinkholes are filled with water and create a pockmarked surface somewhat reminiscent of the cratered surface of the moon.

Shoreline Systems. The hydrologic system also operates along the coasts, where water and sediment are moved by waves and currents to form beaches, bars, lagoons, and other coastal features (*plate 20*). The sediment is derived from erosion of the continents and is reworked and deposited in the shallow offshore areas.

Patterns of suspended sediment in the ocean are most striking when seen from space (*plate 21*), and their motion can be monitored by Landsat space photographs, which take pictures of the same area every 18 days.

Eolian (Wind) systems. The hydrologic system also operates in the desert regions of the world. In many deserts, river valleys are still the dominant landforms, because there is no completely dry place on earth. Even in the most arid regions, some rain falls and climatic patterns change over the years. River valleys can be obliterated by the great seas of sand that cover the landscape (*plate 22*). But, in the broadest sense, the wind itself is part of the hydrologic system, a moving "fluid" on the planet's surface. As the wind blows, it moves the loose sand and dust, leaving a distinctive record of its operation.

Before this subject is left, it should be emphasized again that the source of energy for the hydrologic system is heat from the sun that evaporates the water from the ocean and causes the atmosphere to circulate. The water then moves with the circulating atmosphere and ultimately condenses, under proper conditions of temperature and pressure, and falls as rain or snow. Under the force of gravity, it then flows back to the ocean in several subsystems (rivers, ground water, and glaciers), all of which involve gravity flow from higher to lower levels. Without solar heat, there would be no hydrologic system, because solar heat provides the mechanism for transporting water from the ocean to the highlands of the continents.

Thus, the motion involved in the hydrologic system is surely one of the most distinguishing features of the earth. Water and air move in a great global system, and,

as they move, they transport an enormous amount of loose sand, silt, and clay-size particles. This is evident from the volume of sediment in rivers, deltas, coastal waters, and desert regions.

The surface materials of the earth are in constant motion that continually changes the landscape on a scale so great that it is difficult to comprehend. Although the surface of the earth may appear to be stable and unchanging from year to year, or even throughout a human lifetime, it is continually being changed by the hydrologic system. Indeed, change is a fundamental law of the universe.

The Tectonic System

Statement

The motion involved in the hydrologic system can be easily observed, especially from viewpoints in space, but there is convincing evidence that the earth's crust is also in motion and that it has moved throughout all of geologic time. The evidence comes from many phenomena, including earthquakes, volcanoes, huge fractures in the crust, folded mountain belts, and characteristics of the ocean floor. But it was not until the late 1960s that a unifying theory of the earth's dynamics was developed. This theory, known as **plate tectonics,** provides for the first time a master plan into which everything we know about the earth seems to fit. It explains the origin of the major features of the planet (such as continents, ocean basins, mountain belts, and volcanoes) as the results of a simple system of slow-moving material in the asthenosphere. The theory was developed during the 1960s largely as a result of a newly acquired ability to study the characteristics of the ocean floor, map its surface features, and measure its magnetic and seismic properties. Chapters 17 through 20 will consider and evaluate the evidence for the plate tectonic theory. The purpose here is to summarize the principal concepts involved in the theory.

The basic elements of the plate tectonic theory are quite simple and can be understood by studying the diagram in *figure 2.2.* The lithosphere, which includes the earth's crust, is rigid, but the underlying asthenosphere yields to plastic flow. The principle behind plate tectonics is that the rigid lithosphere moves in response to flow in the asthenosphere below it. Although there are many unanswered questions, it is believed that **radiogenic heat** in the upper mantle is the basic source of energy for tectonic movement. The heat causes the material in the asthenosphere to move slowly in a convection cell, with the hot material rising to the base of the lithosphere, where it then moves laterally, cools, and descends to become reheated, thus beginning the cycle again. (A familiar example of convection can be seen in the heating of a pan of soup. Heat applied to the base of the pan warms the soup, causing it to expand and become less dense [*figure 2.3B*]. The warm fluid rises to the top and is forced to move laterally. It then cools, becomes more dense, and sinks. The regular flow circuit of rising warm fluid and sinking cold fluid is called a **convection cell.**)

As illustrated in *figure 2.3A,* where the convecting

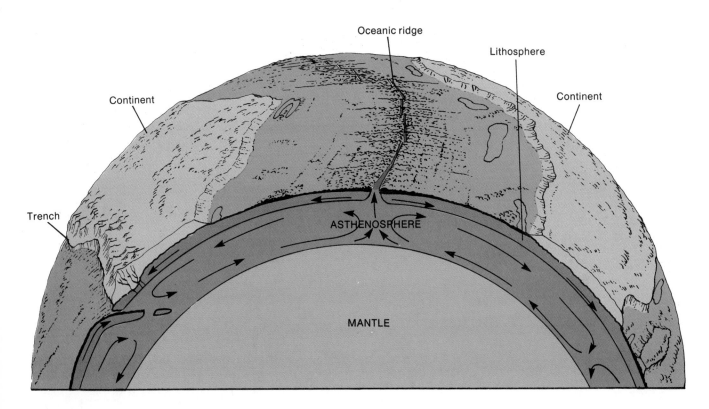

Figure 2.2 *Schematic diagram showing the major elements of the tectonic system. The material in the asthenosphere is thought to be moving in a series of convection cells resulting from heat generated by radioactivity. Where the convecting mantle rises and moves laterally, it pulls the overlying rigid lithosphere apart and creates new crust in the rift zone. The lithosphere, which may contain a block of continental crust, is carried by the convection cell, moves as a single mechanical unit, and descends back into the mantle at the deep-sea trenches. The continents carried by the plates can split and drift apart, but, being less dense than the mantle, they cannot sink back into the asthenosphere. As a result, where two plates collide, the continental margins are deformed into mountain ranges. The plate margins are the most active areas and are the sites of the most intense volcanism, earthquake activity, and crustal deformation.*

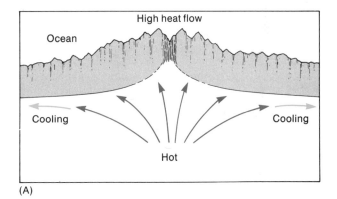

(A)

mantle rises, it causes the lithosphere to arch upward and form the oceanic ridge. As the asthenosphere flows laterally, it pulls the rigid lithosphere apart. Thus, the lithosphere is broken into a series of fragments (or **plates**) that are several thousand kilometers across (*figure 2.4*). As the plates move apart, molten rock from the hot asthenosphere rises into the rift zone and cools to form new oceanic crust. The continental blocks, composed of relatively light granitic rock, float passively on the denser, lower part of the lithosphere, sometimes splitting (when a convection cell rises beneath a continent) and sometimes colliding. The continents do not drift through the lower lithosphere; they are carried by it. Since the earth is a sphere, the shifting plates are in collision with each other. Plates containing dense oceanic crust move down into the asthenosphere at the deep-sea trenches and are consumed. By contrast, plates containing light con-

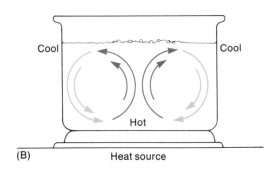

(B)

Figure 2.3 *Convection. (A) Convection in the upper mantle of the earth. (B) Convection cells in a pan of soup.*

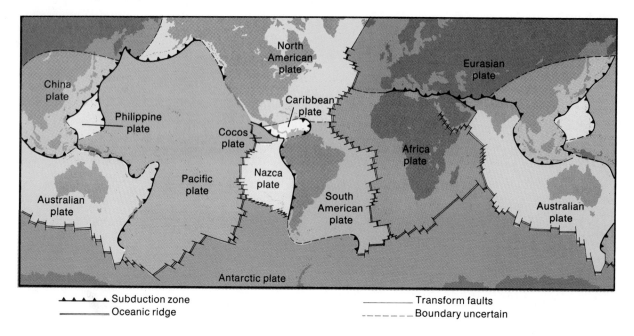

Subduction zone
Oceanic ridge

Transform faults
Boundary uncertain

Figure 2.4   *Map showing the major tectonic plates.*

tinental crust cannot sink back into the mantle. Instead, continental margins adjacent to the descending plates are deformed into linear, folded mountain belts. It should be emphasized again that most plate boundaries are not associated with boundaries between oceans and continents and that plate boundaries, instead, coincide with oceanic ridges, trenches, and young mountain belts and are characterized by zones of earthquakes and volcanic activity.

Discussion

The evidence for such a revolutionary theory of crustal movement comes from many sources and includes data on the structure, topography, and magnetic patterns of the ocean floor, the locations of earthquakes, the patterns of heat flow in the crust, the locations of volcanic activity, the structure and geographic fit of the continents, and the nature and history of mountain belts. (See chapters 17 through 22 for details.)

Consider now the present structural features of the earth and how they fit into the plate tectonic theory. The boundaries of the plates are delineated with dramatic clarity by the belts of active earthquakes and volcanoes (*figure 2.5*). Seven major lithospheric plates are recognized, together with several smaller ones. The spreading center, where the lithosphere is pulled apart, is marked by the oceanic ridge that extends from the Arctic down through the central Atlantic into the Indian and the Pacific. Movement of the plates is away from the crest of the ridge.

The North and South American plates are moving in a westward direction and are interacting with the eastern Pacific, Cocos, and Nazca plates along the west coast of the Americas. The Pacific plate consists only of oceanic crust, and it is moving from the oceanic ridge northwestward toward the system of deep trenches in the western Pacific basin. The Australian plate includes Australia,

India, and the northeastern Indian Ocean. It is moving northward, causing India to collide with Asia and producing the Himalaya Mountains. The African plate includes the continent of Africa, plus the southeastern Atlantic and western Indian Oceans. It is moving eastward and northward. The Eurasian plate is the largest and is moving eastward.

Many major tectonic features of the earth can be explained nicely by this theory. The mountain chains of the Rockies and Andes result from the encounter of the American and eastern Pacific plates. Earthquakes that frequently rock Chile, Peru, and Central America result from the encounter of the eastern Pacific and American plates. The mountain systems of the Alps and Himalayas result from the collision of the African and Australian plates with the Eurasian plate, which also produces the volcanism and earthquakes that torment the Mediterranean and Near East. The great system of deep-sea trenches in the Pacific marks the zone where the Pacific plate descends into the mantle. Earthquakes and volcanic activity also mark this plate boundary where the lithosphere is being destroyed.

Our new perspective from space permits us to see many of the regional structural features of the earth in one synoptic view and to see for the first time on a global scale the results of a moving crust. Several examples will be examined.

The great rift of the Red Sea, shown in *plate 23,* is an extension of the spreading center of the Indian Ocean, which splits the Sinai and Arabian peninsulas from Africa. The structure of the area shown in *plate 23* is dominated by two long, linear trenches. One is occupied by the Gulf of Aqaba, the Dead Sea, and the Jordan River; and the other, by the Gulf of Suez. This is a dramatic expression of tensional stresses in the earth's crust and their effects on the shape of the earth's surface. The long, straight lines extending through the Jordan River Valley and the Gulf of Aqaba are major fractures produced by the rift. Volcanic rocks have been extruded along the fracture zone and are especially evident near

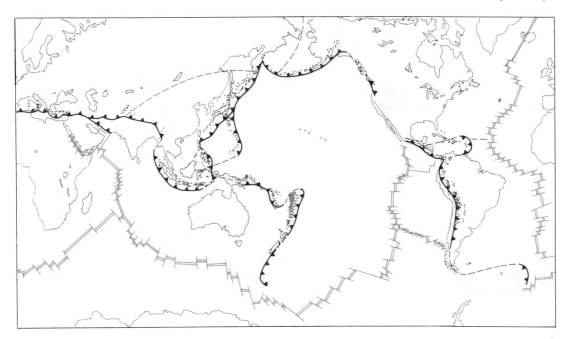

Figure 2.5 *Map showing the geologic phenomena along plate boundaries. The plate margins are outlined with remarkable fidelity by zones of earthquake activity and recent volcanic activity. Shallow earthquakes and submarine volcanic eruptions and tensional fractures occur along the oceanic ridge, where the plates are being pulled apart. Deep earthquakes, volcanoes, deep-sea trenches, and active mountain belts occur along margins where plates collide.*

the Sea of Galilee, in the upper part of the photograph. Oceanographic studies have shown that the Red Sea is an extension of the oceanic ridge that has split the upper part of Africa, causing the Arabian peninsula to drift to the north. It represents the initial stages of seafloor spreading and the development of a new ocean basin. As the Arabian peninsula moved northward, a large fracture developed along its left flank to form the long, narrow trench of the Jordan Valley and the Dead Sea.

Another example of tensional stresses that fractured the crust is in the western United States. There, the crust has been uparched and pulled apart and long, linear blocks have subsided to form basins (*plate 24*). The mountain ranges are segments of the uparched crust that have not collapsed.

Another type of fracture in the crust results when the plates slide past each other horizontally. The great San Andreas fault in California, parts of which are clearly shown in *plate 25*, results from this type of movement.

Where the moving plates collide, the crust is compressed and folded. The space photograph of the Appalachian Mountains (*plate 7*) shows the type of deformation that probably is the most vivid expression of motion in the crust. The sequences of sedimentary rocks that were deposited originally in horizontal layers now are tightly folded as a result of plate collision. The huge folds are like wrinkles in a rug. To understand what you are seeing in this view, you must visualize the tops of folds in a rug being eroded away so that limbs of the folds form zigzag ridges. The folded belt of the Appalachian Moun-

tains extends from central Alabama to Newfoundland, a distance of more than 4000 km, and is similar to many of the other folded mountain belts of the world (*plate 26*).

Where two plates containing continental crust collide, one may override the other to produce a double layer of continental crust. This results in an exceptionally high mountain range. The Himalaya Mountains, it is believed, are an example of this. According to the plate tectonic theory, India has moved northward and collided with Asia. The Indian plate was thrust under the Asian plate to produce the great east-west-trending Himalaya Mountains, the highest mountain range in the world (*plate 27*).

The mountain belts just discussed are all relatively young in a geologic time frame. Much older and more highly eroded mountain belts are found in the shield areas of the continents (*plate 28*). These areas are, in essence, the roots of ancient mountains. Their compressed and folded rock layers illustrate the mobility of the crust in vivid detail.

Volcanism is another expression of the motion of material in the earth. During volcanic activity, molten rock rises to the surface to form new crust. Evidence of recent floods of lava is widespread in southern Idaho (*plate 29*) and in Iceland, areas where the plates are splitting and moving apart. The greatest volcanic activity on the planet, however, occurs on the oceanic ridge beneath the ocean and so generally is not seen.

Where plates collide, the descending plate is partly melted and the lighter liquids rise to the surface to produce a distinctive type of volcanism. This type of volcanic activity occurs throughout the Andes Mountains (*plate 30*) and in the island arcs of the Pacific (*plate 31*). Most of the great volcanic mountains of the world are the result of the collision of tectonic plates.

In this summary of the tectonic system, the point that should be emphasized is that one can see quite vividly in space photographs the constant changes in the crustal structure of the earth—changes on a scale so great that it is difficult to comprehend from limited viewpoints on the ground. The great central theme that these and other examples illustrate is that the earth is a dynamic planet. Although its surface may appear to be stable and unchanging, it is continually being modified by the hydrologic and tectonic systems. Both of these systems can be seen in operation in *plate 32,* which is a satellite view of part of the northern coast of Africa. The hydrologic system is seen in the swirling mass of clouds, as heat from the sun continually pumps water from the ocean and carries it over the land, where it precipitates and returns as surface runoff in river systems. The result is erosion of the land and deposition of sediment in the ocean. The tectonic system is expressed by the folded rocks of the Atlas Mountains that result from the collision of tectonic plates. These two systems make the earth unique among the inner planets of the solar system. The objective in subsequent chapters is to detail the various processes operating in these two systems.

Gravity and Isostasy

Statement

In addition to changes brought about by the hydrologic and tectonic systems, the earth's crust is continually responding to the force of gravity in an effort to reach a gravitational balance, or **isostasy** (Greek: *isos,* equal; *stasis,* standing). Isostasy occurs because the lithospheric plates are buoyed up on the softer, plastic asthenosphere beneath them, with each portion of the crust displacing the mantle according to its bulk and density (*figure 2.6*). Denser crustal material sinks deeper into the mantle than crustal material of lower density.

Isostatic adjustment in the earth's crust can be thought of as though the surface layers were floating like ice on a lake. If you were to skate on the ice, your weight would cause the ice layer to bend down beneath you, displacing water equal to your weight. When you leave that point, the ice rebounds and the displaced water flows back.

As a result of isostatic adjustments, high mountains and plateaus having a large vertical thickness sink deeper into the mantle than areas of low elevation. Any change in an area of the crust—such as removal of material by erosion or addition of material by sedimentation, volcanic extrusion, or accumulations of large continental glaciers—causes an isostatic adjustment.

The concept of isostasy, therefore, is fundamental to studies of the major features of the crust, such as the continents, ocean basins, and mountain ranges and the response of the crust to erosion, sedimentation, and glaciation.

Discussion

Man-made Reservoirs and Isostasy.   The construction of Hoover Dam on the Colorado River provides an excellent, well-documented illustration of isostatic adjustment, because the added weight of water and sediment in the reservoir was sufficient to cause the crust to subside. From the time of its construction in 1935, 24 billion metric tons of water, plus an unknown amount of sediment, accumulated in Lake Mead. In a matter of years, this weight added to the earth's surface caused the crust to subside in a roughly circular area centered around the lake. Maximum subsidence was 1.7 m.

Figure 2.6   *The concept of isostasy. Isostasy is neither a force nor a process but the universal tendency to establish a condition of gravitational balance between segments of the earth's crust. (A) J. H. Pratt proposed that mountains were high because they were composed of lighter materials than the surrounding lowland. (B) G. B. Airy concluded that mountains are of the same density as adjacent lowland but are high because they are thicker.*

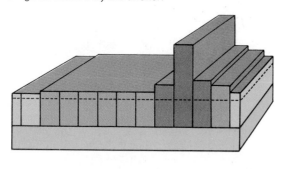

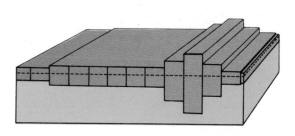

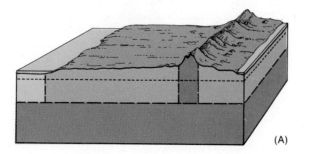

(A)

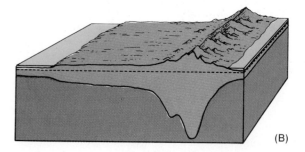

(B)

Ice Sheets and Isostasy. Continental glaciers provide another clear example of isostatic adjustment of the crust, since the weight of an ice sheet several thousand meters thick disrupts the crustal balance and causes the crust beneath to be depressed. The weight of the ice in both Antarctica and Greenland has depressed the central part of the landmasses below sea level. A similar isostatic adjustment occurred during the ice age, when continental glaciers existed in Europe and North America. Parts of both areas are still below sea level; but, now that the ice is gone, the crust is rebounding at a rate of 5 to 10 m per 1000 years. Geologists can measure the extent and rates of rebound by mapping the tilted shorelines of ancient lakes (see *figure 14.24*).

Ancient Lakes and Isostasy. The tilted shorelines of ancient lakes provide a means of documenting isostatic rebound (*figure 2.7*). An example is Lake Bonneville, a large lake that existed in Utah and Nevada during the ice age but has since dried up to small remnants, such as Utah Lake and Great Salt Lake. The shorelines of Lake Bonneville were level when they were formed, but they have been tilted in response to unloading as the water has been removed. The lake was 305 m deep and covered a much smaller area than a continental glacier, but still this relatively small weight was sufficient to depress the crust. Since it has been removed, the shorelines near the deepest part of the lake have rebounded nearly 60 m.

These examples illustrate some very important facts about isostatic adjustments.

1. Gravity is the driving force for all isostatic adjustments. Therefore, all types of loading and unloading cause vertical movements. Isostasy is a universal concept involving all the processes that shift material on the earth's surface. Some of the most obvious isostatic adjustments to be expected are:
   (a) In mountains and highlands as erosion removes material (these areas should rebound).
   (b) In deltaic areas where sediment is deposited (the added weight should cause these areas to subside).
   (c) In areas of volcanic activity (the added weight should cause the crust to subside).
   (d) In regions of glaciation (the advance of the ice should cause the crust to subside; as the ice is removed, the crust should rebound).
2. Very small loads, such as water a few hundred meters deep, are sufficient to cause isostatic adjustments.
3. Isostatic adjustments can occur very rapidly (60 m in less than 20,000 years as indicated from the case of Lake Bonneville).
4. Gravity plays a fundamental role in earth's dynamics, since it is intimately involved with:
   (a) Isostatic adjustments.
   (b) The differentiation of the planet's interior.
   (c) The continental crust's floating on the denser lithospheric plates.
   (d) Gravity-flow systems such as rivers, glaciers, ground water, atmospheric circulation.
It is thus an underlying force in all geologic systems.

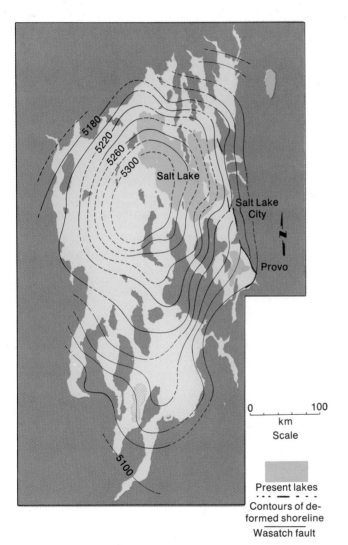

0   100

km

Scale

Present lakes

Contours of deformed shoreline

Wasatch fault

Figure 2.7 *Isostatic rebound of the crust after the load from the water of ancient Lake Bonneville was removed. The map shows the maximum size of Lake Bonneville and the present remnants, Great Salt Lake and Utah Lake. The contour lines show the present elevations of the shoreline features of ancient Lake Bonneville. Note that the shorelines in the central part of the lake are now nearly 60 m higher than those along the margins. This is interpreted as rebound in the crust that occurred after the lake dried up.*

Summary

In this chapter, the mechanics of the two major dynamic systems that continually modify the earth's surface have been outlined. The hydrologic system involves all possible paths of water movement through the atmosphere and oceans, over the lands, and below the ground. The system operates as a gravity-flow system, with solar heat being the source of energy.

The tectonic system is an internal energy system; the upper mantle is in motion as a result of radiogenic heat inherited from the time of the earth's formation. Heat within the upper mantle produces convection. As it moves, it splits the rigid lithosphere and carries it laterally. New crust is created as molten mantle material moves into the fracture zone and cools. As the convection current descends, it carries with it the lithosphere, which ultimately is consumed in the hotter mantle.

This is an irreversible (one-way) system in which energy inherited from the beginning of the earth is continually being dissipated into space. Ultimately, this energy will run out, just as the sun's energy must be consumed. When this happens, the earth will cease to change by its own internal system of heat, but, until then, the tectonic system will continue to create ocean basins, cause continental drift, build mountains, and produce new crust by volcanic activity.

As these systems operate, isostatic adjustment in the earth's crust continues to occur in an effort to establish gravitational equilibrium.

In order for the dynamic systems of the earth to operate, matter is changed back and forth from solid to liquid or vapor states. Rocks are created and destroyed, a process that is basically one involving the growth and destruction of minerals. A knowledge of the nature and origin of minerals and rocks is, therefore, fundamental, not only to understanding the composition of the earth, but also to understanding the processes that operate upon, and within, the earth.

Additional Readings

Bates, D. R. 1964. The Planet Earth. New York: Pergamon Press.

Cailleux, A. 1968. Anatomy of the Earth. New York: McGraw-Hill.

Gaskell, T. F. 1970. Physics of the Earth. New York: Funk and Wagnalls.

Phillips, O. M. 1968. The Heart of the Earth. San Francisco: Freeman, Cooper and Company.

Sumner, J. S. 1969. Geophysics, Geologic Structures and Tectonics. Dubuque, Iowa: William C. Brown and Company.

Takeuchi, H., S. Uyeda, and H. Kanamori. 1970. Debate about the Earth. San Francisco: Freeman, Cooper and Company.

Matter and Energy

All of the complex geologic processes operating on, or within, the earth that cause it to change involve transformations of energy and matter. Without energy, there could be no change, no motion. If you are to understand the planet earth, you must understand something about the various forms of energy and how energy transformations occur. You also must understand some of the fundamentals of the structure of matter and how it changes from one state to another.

Although the scope of the study of energy and matter is immense and involves every branch of physical and biological science, an elementary survey of some of the fundamental concepts is a necessary first step in a study of geology.

Major Concepts

1. The forms of energy most important in geologic processes are (a) kinetic energy, (b) gravitational potential energy, (c) thermal energy, (d) chemical energy, and (e) nuclear energy.
2. Energy can change from one form to another, but energy can neither be created nor destroyed.
3. Energy is necessary for the operation of all geologic processes.
4. The atom is the smallest particle of an element.
5. An atom consists of a nucleus of protons and neutrons that is surrounded by a cloud of electrons.
6. The distinguishing feature of an atom of a given element is the number of protons in its nucleus.
7. Isotopes are varieties of an element produced by variations in the number of neutrons in the nucleus.
8. Ions are electrically charged atoms produced by the gain or loss of electrons.
9. In many minerals, atoms combine by ionic bonding.
10. Matter exists in three states: (a) solid, (b) liquid, and (c) gas. The difference in the three states is related to the degree of ordering of the atoms.

Energy

Statement

All of the processes in the hydrologic and tectonic systems result from the flow of energy on, or within, the earth. Indeed, without energy, the materials of the earth would be totally inert and motionless. There would be no change and no rain, rivers, earthquakes, or volcanoes. There would be no changes in the states of matter. But what is energy? In simplest terms, it is a measure of motion or the capacity for producing motion. The motion can be visible (like a moving train) or invisible (like the motion in an atom), but all motion is associated with energy.

The forms of energy most important in geologic processes are:

1. Kinetic energy—the energy of motion.
2. Gravitational potential energy—the energy resulting from position in a gravitational field.
3. Thermal energy—the energy resulting from the random movement of molecules and atoms.
4. Chemical energy—the energy resulting from the transfer of outer electrons in atoms and molecules.
5. Nuclear energy—the energy resulting from changes in the arrangement of protons and neutrons in nuclei.

One of the most fundamental laws of natural science is the law of conservation of energy, also known as the first law of thermodynamics, which states that *energy can be neither created nor destroyed. Energy can be changed from one form to another, but the total amount of energy remains constant.* This law has been tested and observed over a wide range of conditions and always has held true.

Discussion

Although the study of energy is fairly abstract, the relationships between energy and geologic processes can be better understood through intuitive reasoning and through citing familiar examples of events that are observed on the earth's surface.

Kinetic Energy.   Kinetic energy, or the energy of motion, is probably the most familiar form of energy. The word comes from the Greek *kinetikos* (to move). Every moving object has kinetic energy. The earth possesses kinetic energy as it moves around the sun and spins on its axis. All of the processes in the hydrologic system involve obvious kinetic energy, such as running water, moving glaciers, rushing ocean waves, and even the tiny rock particles swirling in a duststorm. All involve matter in motion and all have the capacity to change and modify the earth.

The amount of kinetic energy possessed by an object is proportional to the quantity of mass multiplied by the square of its velocity. For example, a large boulder rolling down a hill has more energy than a small pebble moving at the same velocity. However, a small meteorite moving at a high velocity has much more energy than a large boulder moving only a few meters per second.

Gravitational Potential Energy.   Gravitational potential energy is potential, or stored, energy in an object associated with its position in a gravitational field. It is equal to the kinetic energy the object would attain if it were allowed to fall under the influence of gravity. The moisture in the atmosphere has potential energy because of its position. As rain falls, it loses gravitational potential energy and gains kinetic energy. When it finally hits the earth, part of the kinetic energy is converted to heat and some is transferred into the soil particles with which it collides. If the water falls on highland or mountain, it still retains some potential energy that will be transformed into kinetic energy as it flows downslope in the drainage system.

Thermal Energy.   Thermal energy can be thought of as a special type of internal kinetic energy rather than the external form seen in a moving mass. It is the energy involved in the motion of atoms. All atoms are in constant motion, and, the faster they move, the greater their thermal energy. Atoms in a crystalline solid are locked into a fixed geometric position and their motion is one of vibration, without one atom moving past another. If the atoms move faster, they eventually break out of their fixed position to move freely and the crystal melts. Greater motion, which means more heat, causes the liquid to vaporize.

Heat is transferred in three principal ways: (1) conduction, (2) convection, and (3) radiation. *Conduction* is the process by which the vibration of an atom is transmitted to adjacent atoms. It is the method by which heat is transmitted through solids. *Convection* is a process in which the heated material expands and moves upward. Convection occurs in liquids and gases and transfers thermal energy much faster than conduction. As was seen in chapter 2, convection is the principal way in which thermal energy is transferred in both the hydrologic and tectonic systems. *Radiation* involves the emission of electromagnetic waves from the surface of a hot body to its cooler surroundings. The waves travel in a straight line through space at 186,000 miles per second and occur in a wide range of lengths that together constitute the electromagnetic spectrum. The transfer of energy from the sun by radiation obviously is one of the most important energy transformations in the earth's system.

Chemical Energy.   Chemical energy is the energy involved when chemical reactions take place. It is the energy that binds atoms together into molecules and becomes available when atoms have lost or gained electrons. The formation of rocks involving changes in state from liquid to solid releases large quantities of chemical energy. Energy from the sun is absorbed by molecules within plant cells and is used to combine carbon dioxide and water to form carbohydrates. The energy can be stored in carbohydrate compounds of plant tissue and released through burning (chemical oxidation) and radiated as heat to the surroundings.

Nuclear Energy.   Nuclear energy is the energy resulting from the arrangement of protons and neutrons in the atomic nuclei, which was acquired by the elements at the time of their formation. The radioactive elements (such as uranium, thorium, and potassium) release this type of energy spontaneously in the process of radioactive decay. The quantity of energy released in nuclear reactions is almost incomprehensibly larger than that released by ordinary chemical reactions.

Transformation and Conservation of Energy. In all natural processes, energy is transformed from one form to another, but it is never created or destroyed. Several examples of natural energy systems will illustrate this concept.

Example 1. Radiation from the sun is absorbed by land and water at the earth's surface. Air in contact with the surface is warmed by conduction and it rises. This results in winds, which may blow sand into migrating dunes. In this case, radiant thermal energy from the sun passes into thermal energy of the land to kinetic energy of moving air, which moves the sand and changes the earth's surface.

Example 2. Nuclear energy in the asthenosphere is transformed into thermal energy by radioactive disintegration of uranium or potassium. The heat melts some of the minerals in the rocks of the mantle, and the liquid rises in a convection system. As molten material reaches the surface, it is ejected in the form of a volcanic eruption. The explosion accelerates some of the lava fragments to a high speed (thermal energy to kinetic energy). As the fragment rises, it gains potential energy at the expense of kinetic energy. At its maximum height, all kinetic energy is transformed into potential energy and it stops rising. As it falls back to earth, the potential energy is converted back to kinetic energy. As the fragment strikes the ground, the kinetic energy is converted to heat energy and vibrations that pass through the ground and the air. As the lava cools, heat is lost to the surroundings through convection, conduction, radiation, and chemical energy as the minerals crystallize. In each transfer, some energy is removed from the system and is dispersed into the surroundings.

Example 3. Radiation from the sun strikes the surface of the ocean and heats the water. The thermal energy causes greater motion in the water molecules, and evaporation occurs. The heat from the water also warms the surrounding air, which rises by convection and carries the water in a vapor state. When precipitation occurs as rain or snow, the potential energy of the water in the atmosphere passes into kinetic energy as the water falls. Upon reaching the surface, some of the kinetic energy is transformed into thermal energy through impact. The water still retains some potential energy and flows to lower levels, washing down sand and mud with it. Gravitational potential energy of both the water and the solid particles is transformed into kinetic energy. Friction and resistance to flow transform some of the kinetic energy into heat energy, which is dispersed by conduction or radiation.

Matter

Statement

In order for you to understand the dynamics of the earth and how rocks and minerals are formed and changed through time, it is necessary to have some knowledge of the fundamental structure of matter and how it behaves under various conditions. The solid materials that make up the earth's outer shells are called **rocks.** Most rock bodies, however, are mixtures or aggregates of minerals. A mineral is a naturally occurring inorganic chemical compound with a definite chemical formula and a specific internal structure. Minerals, in turn, are composed of atoms, so it is necessary to understand something about atoms and the ways in which they combine to form minerals.

The atomic theory that is the basis for the present-day understanding of matter was developed by John Dalton in 1805. He proposed the idea that all matter is composed of tiny individual particles called atoms. Since then, many momentous discoveries have added greatly to the understanding of matter. A simplified description of matter today would include the following fundamental concepts.

1. An atom is the smallest fraction of an element that can exist and still show the characteristics of that element.
2. Although many subatomic particles have been identified in recent years, the most fundamental are protons, neutrons, and electrons.
3. An atom consists of a nucleus of protons and neutrons, surrounded by a cloud of electrons.
4. The distinguishing feature of an atom of an element is the number of protons in the nucleus. The number of electrons and neutrons in an element can vary, but the number of protons is constant.
5. Atoms are electrically neutral, because they have one negatively charged electron for every positively charged proton.
6. Electrically charged atoms are called **ions.** They are produced by the gain or loss of electrons.
7. Varieties of a given atom (element) are called **isotopes** and are produced by variations in the number of neutrons in the nucleus.
8. Atoms combine, mostly through **ionic bonding,** to form minerals.

Discussion

Atoms. An atom is far too small to be seen with any microscope, and it is best described by utilizing models contructed from abstract mathematical probabilities. In its simplest approximation, the atom is characterized by a relatively small nucleus of tightly packed protons and neutrons, with a surrounding cloud of electrons. The proton carries a positive electrical charge, and the mass of a proton is taken as the unit of mass. The neutron, as its name indicates, is electrically neutral and has approximately the same mass as the proton. The electron is a much smaller particle with a mass approximately $1/1850$ times that of a proton. It carries a negative electrical charge equal in intensity to the positive charge of the proton. Since the electron is so small, the entire mass of the atom, for practical purposes, is concentrated in the protons and neutrons that form the nucleus.

Hydrogen is the simplest of all atoms and consists of one proton in the nucleus and one orbiting electron (*figure 3.1*). The next heavier atom is helium, with two protons, two neutrons, and two electrons. Each subsequently heavier atom contains more protons, neutrons, and electrons. The distinguishing feature of an atom of an element is the number of protons in the nucleus. The

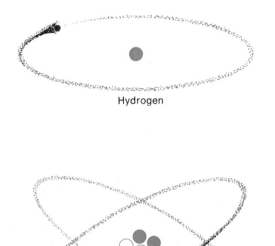

Hydrogen

Helium

Figure 3.1 *The structure of atoms of hydrogen and helium. Hydrogen has one proton and one orbiting electron. Helium has two protons, two neutrons, and two electrons.*

number of electrons and neutrons in an element can vary, but the number of protons remains constant.

Atoms have the same number of electrons as protons and do not carry an electrical charge. As the number of protons increases in the succeedingly heavier atoms, the number of electrons also increases and fills a series of energy-level shells, each of which has a maximum capacity.

Isotopes. Although the number of protons in the atoms of a given element is constant, the number of neutrons can vary. This means that atoms of any one element are not necessarily all alike. Iron, for example, has 26 protons but exists in forms with 28, 30, 31, and 32 neutrons. All varieties of iron have 26 protons and all have the properties of iron, but the mass number of each variety is different. Atoms related in this way are called *isotopes.* Most common elements exist in nature as a mixture of isotopes.

Ions. Atoms that have as many electrons as protons are electrically neutral, but many atoms can gain or lose electrons in their outermost shells. When this occurs, the atom loses its electrical neutrality and becomes charged. These electrically charged atoms are called *ions.* The loss of an electron results in a positively charged ion, since the number of protons exceeds the number of negatively charged electrons. If an electron is gained, the ion has a negative charge. Ions are important, because the attraction between positive and negative ions is the bonding force that holds matter together.

Bonding. Atoms are most stable when their outermost shells contain the maximum number of electrons. Neon, for example, has 10 protons in the nucleus and 10 electrons, 2 of which are in the first shell and 8 of which are in the second shell. The atom does not have an electrical charge, and its two electron shells are complete. As a result, neon is stable and does not interact chemically

with other atoms. Helium, argon, and the other noble gases also have 8 electrons in their outer shells and normally do not combine with other elements. Elements that have incomplete outer shells readily lose or gain electrons to achieve a structure like that of neon and the other inert gases, with 8 electrons in their outermost shells.

For example, an atom of sodium has only one electron in its outer shell but eight in the shell beneath (*figure 3.2*). If it could lose the one electron, it would have a stable pattern similar to that of the inert gas neon. Chlorine, in contrast, has seven electrons in its outer shell, and, if it could gain an electron, it too would attain a stable configuration. Thus, whenever possible, sodium will give up an electron and chlorine will gain an electron. When this is done, the sodium atom becomes a positively charged sodium ion and the chlorine atom becomes a negatively charged chlorine ion. With opposite electrical charges, the sodium and chlorine ions are attracted to each other and bond together to form the compound sodium chloride (common salt of the mineral halite). This type of bond between ions of opposite electrical charge is called an *ionic bond.*

Atoms also can attain stability with the electronic arrangement of a noble gas by sharing electrons. No electrons are lost or gained and no ions are formed. The electron simply orbits both nuclei in a figure-eight pattern (*figure 3.3*). This type of bond is called a **covalent bond.**

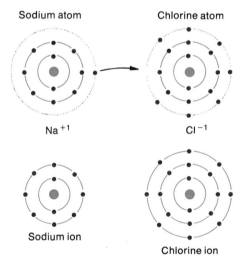

Sodium atom   Chlorine atom

Na^{+1} Cl^{-1}

Sodium ion   Chlorine ion

Figure 3.2 *The formation of sodium and chlorine ions by transfer of an electron from the outermost shell of sodium to the outermost shell of chlorine. This results in each atom becoming an ion with a stable outer shell.*

Figure 3.3 *Covalent bonding. In covalent bonding, one or more electrons orbit two nuclei, instead of only one nucleus.*

In a metal, each atom contributes one or more electrons to a sea of electrons that moves relatively freely throughout the entire aggregate of atoms. A given electron is not attached to a specific pair of atoms but moves about, orbiting one nucleus and then another. This sea of negatively charged electrons holds the positive metallic ions together in a crystalline structure and is responsible for the special characteristics of metals.

States of Matter

Statement

Groups of atoms can exist in one of three states of aggregation, that is, either as a gas, a liquid, or a solid.

Gases. The principal properties of gases are these.

1. Gases have no shape or definite volume; that is, they assume the shape and volume of their containers.
2. In gases, particles are in constant random motion and are widely spaced.
3. Gases can be compressed and expanded.
4. Gases exert pressure that increases as temperature increases.
5. Gases can be converted into liquids by subjecting them to increased pressure or decreased temperature or both.

Liquids. The principal properties of liquids are these.

1. Liquids have a definite volume but no definite shape. A liquid, therefore, tends to flow.
2. In liquids, the atoms or the molecules are packed more closely together than in gases, but they are not set in a rigid framework.
3. Liquids are more difficult to compress than gases.
4. The density of liquids is greater than that of gases.
5. Liquids can be changed into gases by an increase in temperature or a decrease in pressure.

Solids. The principal properties of solids are these.

1. Solids have a definite volume and a definite shape.

2. In solids, the atoms are bound tightly together into a fixed, rigid framework with a definite arrangement, or pattern, of atoms.
3. Most solids become liquids when heated, a process called melting, although solids can pass directly into a gaseous state. When the temperature of a solid is close to the melting point, a decrease in pressure will cause melting to occur.
4. Liquids can be converted to solids by a decrease in temperature. The process is called freezing, and the temperature at which solidification occurs is called the freezing point.

From this summary, it should be apparent that a given sample of matter can change from one state to another so that it has completely different properties and still retains the same chemical composition. To change matter from one state to another, there must be a gain or a loss of energy; that is, heat and/or pressure must be increased or reduced.

Discussion

The properties of the three states of matter can be understood by considering differences at the atomic level. The principal differences are related to the degree of ordering of the constituent atoms.

In a typical solid (*figure 3.4*), atoms are arranged in a rigid framework. The arrangement is a regular and repeating, three-dimensional pattern in crystalline solids, but it is random in amorphous solids. Each particle occupies a more or less fixed position but is in a vibrating motion. As the temperature rises, the vibrating motion increases, and eventually the particles are free and are able to glide and pass by one another. Melting ensues, and the matter passes into a liquid state.

Figure 3.4 *Schematic diagrams showing degrees of atomic ordering in the three states of matter.*

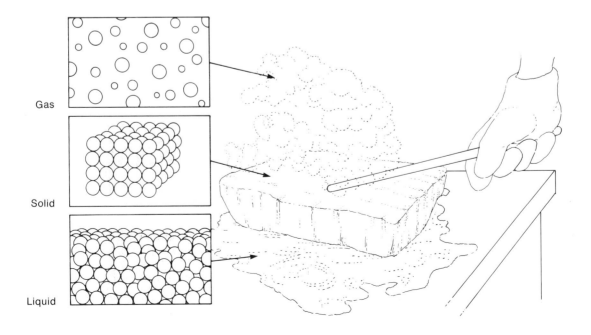

Gas

Solid

Liquid

In a liquid, the basic particles are in random motion, but they are packed closely together. They slip and glide past one another and then collide and rebound, but they are held together by forces of attraction greater than those in gases. This explains why density generally increases and compressibility decreases as matter changes from a gas to a liquid and then to a solid. If the liquid is heated, the motion of the particles increases, and ultimately the individual atoms become separated and move about at high speeds.

In a gas, the individual atoms or molecules are separated by empty spaces and are comparatively far apart. The atoms are in rapid motion, moving in straight lines until they collide and their direction is changed. This explains why a gas exerts pressure and can be compressed markedly. Gases have the ability to expand indefinitely, and the continuous rapid motion of the atoms results in rapid diffusion.

Water undoubtedly is the most familiar example of matter changing through the three basic states, because, at temperatures and pressures on the earth's surface, water changes from a solid to a liquid and a gas in a temperature range of only 100 °C. Other forms of matter in the solid earth are capable of similar changes, but their transition from solid to liquid to gas occurs at high temperatures. At normal room temperature and pressure, 93 of the 106 elements are solids, 2 are liquids, and 11 are gases.

Most people are familiar with the effects of temperature changes on the state of matter because of their experience with water as it freezes, melts, and boils. Less familiar are the effects of pressure. At high pressures, water remains liquid at temperatures as high as 371 °C. The combined effects of both temperature and pressure on water are shown in the phase diagram in *figure 3.5*. Similar diagrams could be constructed from laboratory

work on other minerals and would provide important insight into the processes operating at the higher temperatures and pressures below the earth's surface.

Summary

All change requires the interaction of energy and matter. The forms of energy most significant in geologic and planetary systems are:
1. Kinetic energy, or the energy of motion.
2. Gravitational potential energy, or the energy resulting from position in the gravitational field.
3. Thermal energy, or the energy resulting from random molecular or atomic motion.
4. Chemical energy, or the energy resulting from the transfer of outer electrons.
5. Nuclear energy, or the energy resulting from changes in the arrangement of protons and neutrons in nuclei.

In all natural processes, energy is transformed from one type to another, but it is never created or destroyed.

The atomic theory is the basis for the present-day understanding of matter. A simplified description of the fundamental ideas in this theory follows.
1. The atom is the smallest fraction of an element that can exist and still show the characteristics of that element.
2. Although many subatomic particles have been identified, the most fundamental are (a) protons, (b) neutrons, and (c) electrons.
3. An atom consists of a nucleus of protons and neutrons that is surrounded by a cloud of electrons.
4. The atom of each element has a unique number of protons in the nucleus. The number of electrons and neutrons in an element can vary, but the number of protons is constant.
5. "Normal" atoms are electrically neutral, because they have one negatively charged electron for every positively charged proton.
6. Electrically charged atoms are called ions. They are produced by the gain or loss of electrons.
7. Isotopes are varieties of an atom (element) produced by variations in the number of neutrons in the nucleus.
8. Atoms combine to form minerals by ionic bonding, covalent bonding, and metallic bonding.

There are three states of matter: gas, liquid, and solid. Each state of matter has completely different physical properties, but each state retains the same chemical composition. Processes in the earth's dynamics are involved mostly in changes among the three states of matter.

Additional Readings

Asimov, I. 1961. Building Blocks of the Universe. New York: Abelard-Schuman.

Azaroff, L. V. 1960. Introduction to Solids. New York: McGraw-Hill.

Fyfe, W. S. 1964. Geochemistry of Solids: An Introduction. New York: McGraw-Hill.

Krauskopf, K. B., and A. Beiser. 1971. Fundamentals of Physical Science. 6th ed. New York: McGraw-Hill.

Lapp, R. E. 1966. Matter. New York: Time-Life Books.

Wilson, M. 1966. Energy. New York: Time-Life Books.

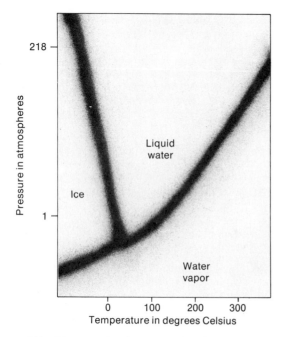

Figure 3.5   *Diagram showing the relationship of the gas, liquid, and solid phases of water to temperature and pressure.*

Minerals

In all processes involved in the earth's dynamic systems, matter changes from one state to another and minerals grow, melt, and dissolve or are broken and modified by physical forces. As the earth's surface is weathered and eroded, some minerals are destroyed and others grow in their place. As sediments are deposited in the oceans, new minerals grow from solution, while other mineral fragments are united into a coherent mass to form solid rock. During volcanic activity, minerals grow from a hot liquid melt. Even deep below the earth's surface, new minerals form under high pressure and temperature as atoms are removed from the crystal structure of one mineral and recombined in the structure of another, which is more stable in that environment. As tectonic plates move and continents drift, minerals are created and destroyed in a variety of chemical and physical processes. Some knowledge of the major minerals of the earth, therefore, is essential to an understanding of the earth's dynamics.

This chapter will explain what minerals are and how they can be identified. Then, the silicate mineral group (the major rock-forming minerals) will be explored in preparation for the study of the major rock types in chapters 5, 6, and 7.

Major Concepts

1. Minerals are natural, inorganic solids possessing a specific internal structure and a definite chemical composition that varies only within specific limits.
2. Minerals grow by the addition of atoms to the crystal structure as matter changes from a gaseous or liquid state to a solid state. Minerals dissolve or melt as atoms are removed from the crystal structure and are changed to a liquid or gaseous state.
3. All specimens of a given mineral type have a limited range of physical and chemical properties (such as crystal structure, cleavage or fracture, hardness, specific gravity, et cetera).
4. More than 95% of the earth's crust is composed of silicate minerals.
5. The most important rock-forming minerals are feldspars, olivines, pyroxenes, amphiboles, quartz, clay minerals, and calcite.

The Nature of Minerals

Statement

Commonly, **minerals** are thought of only as exotic crystals in a museum or as valuable gems and metals. But minerals are the major solid constituents of the earth. A grain of sand, a snowflake, and a particle of salt are also minerals and have much in common with gold and diamonds. To define the term *mineral* precisely, therefore, is somewhat complicated. But, for a substance to be considered a mineral, the following conditions must be met.

1. It occurs naturally as an inorganic solid.
2. It has a specific internal structure, that is, a definite arrangement of its constituent atoms into a crystalline solid.
3. It has a chemical composition that varies within specific limits and can be expressed by a chemical formula.
4. It has certain physical properties that result from its composition and crystalline structure.

Discussion

Inorganic Solids. By definition, only naturally occurring substances are minerals, that is, natural elements or compounds in a solid state. Thus, synthetic products such as artificial diamonds are not minerals, strictly speaking. Neither are organic compounds such as coal and petroleum considered to be minerals in the strict sense, because they are not crystalline solids.

Figure 4.1 *Cleavage of calcite. The large block and all the small fragments show three cleavage planes that do not intersect at right angles. A cleavage fragment thus will be in the form of a distorted cube.*

Minerals can consist of a single element (such as gold, silver, copper, and sulfur), but most minerals are compounds of two or more elements, because the abundant elements in the earth have a strong tendency to combine.

The Structure of Minerals. Perhaps the most important key words in the definition of a mineral are *internal structure*. This means that a mineral has a unique and orderly internal arrangement of its component atoms and that every specimen of an individual mineral has the same internal structure, regardless of when, where, and how it was formed. The orderly arrangement of the atoms that constitute the internal structure of a mineral was suspected long ago by mineralogists who observed the many expressions of order in crystals. Nicolaus Steno (1631-1687), a Danish monk, observed that, when a mineral is allowed to grow in an unrestricted environment, its crystals have a perfect geometric shape. He found from numerous measurements that each mineral has a characteristic crystal form and that, although the size and shape of the crystal may vary, similar pairs of crystal faces always meet at the same angle. This fact is known as the *law of constancy of interfacial angles*.

Later, René Hauy, a French mineralogist, accidently dropped a large crystal of calcite and observed that it broke only along three sets of planes so that all the fragments had a similar shape (see *figure 4.1*). He then proceeded to break all the calcite crystals in his own collection, plus many in the collections of his friends, and found that all the specimens of calcite broke in exactly the same manner and that all the fragments, however small, had the shape of a rhombohedron. To explain his observations, he concluded that calcite must be built of innumerable, infinitely small rhombohedra packed together in an orderly manner. Therefore, the cleavage of calcite would be related to the ease of parting of such units from adja-

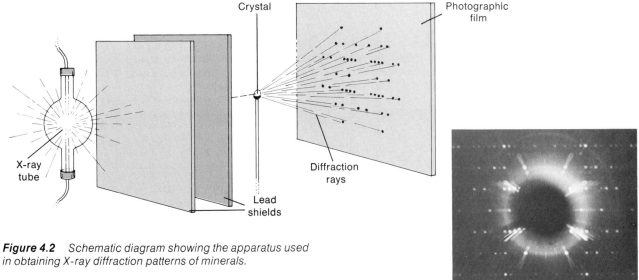

Figure 4.2 *Schematic diagram showing the apparatus used in obtaining X-ray diffraction patterns of minerals.*

X-ray diffraction pattern

cent layers. This was a remarkable advance in the understanding of crystals.

Today, we know that cleavage planes are planes of weakness in the crystal structure and are not necessarily parallel with the crystal face. However, they do constitute a striking expression of the orderly internal structure.

With modern methods of X-ray diffraction, it is possible to determine precisely the internal structure of a mineral. When a very thin beam of X rays is passed through a mineral, it is diffracted (or dispersed) by the framework of atoms. The dispersed rays produce an orderly arrangement of dots on photographic film placed behind the crystal (*figure 4.2*). By measuring the relationships among the dots, the systematic orientation of planes of atoms within the crystal can be reduced to a mathematical formula. From these patterns, detailed models of crystal structures can be constructed and analyzed. The X-ray instrument is now the most basic device for determining the internal structure of minerals, and it is used extensively for precise mineral identification and analysis.

The Composition of Minerals. A mineral has a definite chemical composition in which specific elements occur in definite proportions. The chemical composition of some minerals, however, can vary within specific limits, because in some minerals different kinds of ions can be substituted for others in the mineral structure. Ionic substitution results in a chemical change in the mineral without a change in the crystal structure, but there are definite limits within which the substitution can occur, and the composition of a mineral can be expressed by a chemical formula that expresses ionic substitution and how the composition can change. The suitability of an ion to substitute for another is determined by the size and the electrical charge of the ions in question (*figure 4.3*).

Figure 4.3 *The relative sizes of some common ions. The suitability of an ion to substitute for another in a crystal structure is determined by the size and the electrical charge of the ion in question. Thus, iron could be replaced by magnesium, sodium by calcium, silicon by aluminum, et cetera.*

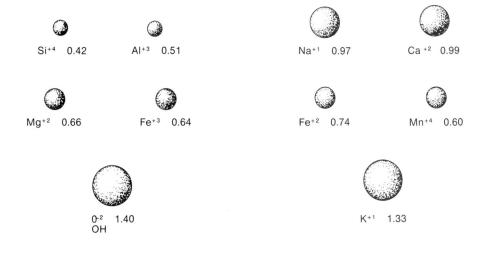

Si^{+4} 0.42 Al^{+3} 0.51 Na^{+1} 0.97 Ca^{+2} 0.99

Mg^{+2} 0.66 Fe^{+3} 0.64 Fe^{+2} 0.74 Mn^{+4} 0.60

O^{-2} 1.40
OH

K^{+1} 1.33

In a general way, the substitution of ions is analogous to the substitution of a plastic brick for a clay brick of equal size in a wall. Because the size of the substitute brick is the same as the original, the structure of the wall is not affected, but there is a change in composition.

Ionic substitution occurs widely in the common rock-forming minerals and is responsible for **mineral groups,** the members of which have the same structure but varying composition. For example, in the olivine group, with the formula $(Mg, Fe)_2SiO_4$, ions of Fe and Mg can substitute freely for one another. The total number of Fe and Mg atoms is constant relative to the number of Si and O atoms in the olivine, but the ratio of Fe to Mg can be different in different samples. The common minerals feldspar, pyroxene, amphibole, and mica each constitute a family (group of related minerals) in which atomic substitution produces a range of chemical composition.

Physical Properties of Minerals.

Because a mineral has a definite chemical composition and an internal crystalline structure, all specimens of a given mineral, regardless of when or where they were created, have the same set of physical and chemical properties that vary only within certain limits. This means that one piece of quartz, for example, will be as hard as any other piece of the same quartz mineral, that it will have the same specific gravity, and that it will break in the same manner.

The most significant and readily observable physical properties of minerals are **crystal form, cleavage, hardness, specific gravity, color,** and **streak.**

Crystal Form. In the previous section, it was illustrated that, when a crystal is allowed to grow in an unrestricted environment, it develops natural crystal faces and produces a crystal with a perfect geometric pattern. The shape or form of the crystal is a reflection of the internal structure and can be used as an identifying characteristic in many mineral specimens. Quartz, for example, typically forms elongate, hexagonal crystals, regardless of its size or when and where it was formed (*figure 4.4*). It should be remembered, however, that when the space for growth is restricted, perfect crystal faces cannot develop.

Cleavage. Cleavage is the tendency of a crystalline substance to split or break along smooth planes parallel to zones of weak bonding in the crystal lattice (*figure 4.5*). When the bonds are especially weak in a given direction (as in mica and halite), perfect cleavage occurs with ease and it is difficult to break the mineral in any direction other than along a cleavage plane. In other minerals, the difference in bond strength is not great, so cleavage may be poor or imperfect. Cleavage can occur in more than one direction, but the number and direction of cleavage planes in a mineral species is always the same.

Some minerals that have no weak planes in their crystalline structures lack cleavage and break along various types of fracture surfaces. Thus, the type of fracture can be useful in identifying some minerals. Quartz, for example, lacks cleavage and characteristically breaks along curved surfaces (conchoidal fracture) similar to the curved surface in chipped glass.

Hardness. Hardness is a property of a mineral that is easily recognized and is used widely in field identification of minerals. It is a measure of a mineral's resistance to abrasion. Over a century ago, Friedrich Mohs, a German mineralogist, assigned an arbitrary

Figure 4.4 *Quartz crystals. The angles between the sides of the prisms of all crystals of quartz always meet at 120°, regardless of the gross shape and size of the crystal or when and where the crystal was formed. This is an expression of the atomic structure of the mineral. Every mineral has a characteristic crystal form as a result of its crystalline structure.*

Figure 4.5 *Example of cleavage in mica in one direction.*

relative number to 10 common minerals according to their hardness. Diamond, the hardest mineral known, was assigned the number 10. Softer minerals were ranked in descending order, with talc—the softest mineral—assigned the number 1. Mohs' hardness scale (*table 4.1*) provides a standard for testing minerals for preliminary identification in the field.

Table 4.1. Mohs' Hardness Scale

Hardness	Mineral	Test
1	Talc	Fingernail
2	Gypsum	
3	Calcite	Copper coin
4	Fluorite	Knife blade or
5	Apatite	glass plate
6	K-feldspar	
7	Quartz	Steel file
8	Topaz	
9	Corundum	
10	Diamond	

Specific Gravity. Specific gravity is the ratio between the weight of a given substance and the weight of an equal volume of water. For example, if a liter of molten lead were poured into a mold and allowed to cool, the solid lead when cooled would weigh a little over 11 times more than a liter of water. Thus, the specific gravity of lead is 11.

Specific gravity is one of the more precisely defined properties. It depends upon the kinds of atoms making up the mineral and the closeness of atoms packed into the crystal structure. Clearly, the more numerous and compact the atoms, the higher the specific gravity. Most common rock-forming minerals have specific gravities from 2.65 (for quartz) to about 3.37 (for olivine).

Color. Color is one of the most obvious properties of a mineral, but it is *not* diagnostic because most minerals are found in various hues, depending on such things as subtle variations in composition and the presence of inclusions and impurities. Quartz, for example, ranges through a spectrum of colorless, clear crystals to purple, red, white, and jet black.

Streak. When a mineral is powdered, it usually exhibits a much more diagnostic color than when it is in large pieces. The color of the powdered mineral is referred to as streak. In the laboratory, streak is obtained by vigorously rubbing the mineral on an unglazed porcelain plate. The streak of a mineral can be different from the color seen in a hand specimen, so one should never anticipate the color of streak by visual examination of a mineral fragment. Furthermore, the hardness of porcelain is about that of glass, and minerals of greater hardness scratch the plate. Most minerals with a nonmetallic luster have a white or a pastel streak. For this reason, streak is not very useful in distinguishing nonmetallic minerals.

The Growth and Destruction of Minerals

Statement

One of the most important characteristics of a mineral is its susceptibility to chemical change. No mineral is "forever." Minerals grow and are destroyed as matter changes from a gaseous or liquid state to a solid state. All minerals came into being because of specific chemical and physical conditions in some part of the earth, and all are subject to change as the physical and chemical conditions change. In contrast, atoms remain unchanged by "normal" chemical reactions and are transferred from one mineral to another. Minerals, therefore, provide an important means of interpreting the changes that have occurred in the earth throughout its history.

Discussion

Crystal Growth. Growth occurs by the addition of atoms to the crystal face, which is possible because the outer layers of atoms on a crystal are never completed and can be extended indefinitely. Therefore, a single crystal containing all physical and chemical properties of the mineral may be so small that it cannot be identified with a high-power microscope.

An environment suitable for crystal growth includes (1) proper concentration of the kinds of atoms required for a particular mineral, such as elements in a liquid or gaseous state, and (2) the proper temperature and pressure. The atoms in solution are attracted to, and interlock with, the atoms on the surface of the crystal.

The time-lapse photographs in *figure 4.6* show the manner in which crystals grow in an unrestricted environment. Although the size of the crystal increases, the form and the initial structure remain the same. New atoms are added to the outer edge of a crystal, parallel to the place of atoms in the basic structure, thus producing a perfectly symmetrical crystal form.

However, some crystal faces grow faster than others

Figure 4.6 *Time-lapse photographs showing the growth of crystals.*

and, in an environment where spaces are restricted, a crystal face may not grow symmetrically. Where a crystal face encounters a barrier, it ceases to grow. This process is illustrated in *figure 4.7*. A crystal growing in a restricted area assumes the shape of the confining space, and well-developed crystal faces do not form. In such cases the external form of the crystal can be practically any shape, but the internal structure of the crystal is in no way modified because the growth of the crystal face is restricted. The internal structure of the mineral remains the same, the composition of the mineral is unaffected, and there is no change in the physical and chemical properties of the mineral. The only modification is in the relative size of the crystal faces.

The process of crystal growth in restricted space is especially important in rock-forming minerals; for, in a melt or in a solution, many crystals grow at the same time and must compete for space. As a result, in the latter stages of growth, crystals in rocks commonly lack unique, well-defined crystal faces and typically interlock with adjacent crystals to form a strong, coherent mass (*figure 4.8*).

Most crystals are rather small, measuring from a few tenths of a millimeter to several centimeters in diameter. Some are so small that they can be seen only with an electron microscope capable of enlargements of a million times actual size. On the other hand, in an unrestricted environment, crystals can grow to enormous sizes (*figure 4.9*).

Destruction of Crystals. Minerals can dissolve or melt by removal of the atoms from the crystal structure, permitting a return to a fluid state (liquid or gas). The heat that causes a crystal to melt increases atomic vibrations enough to break the bonds holding the atom to the crystal structure. Similarly, when solution occurs, atoms are "pried" loose by the solvent, usually water. The breakdown or dissolution of a crystal begins at the surface and moves inward.

The growth and destruction of crystals are of paramount importance in geologic processes, because rocks

Figure 4.7 *Diagrams showing the growth of crystals in unrestricted and restricted spaces. (A) Where growth is unrestricted, each crystal face grows with equal facility and the perfect crystal is enlarged by growth on each crystal face. (B) Restricted space can eliminate growth on certain crystal faces, whereas growth on faces w, x, y, and z is eliminated as available space is used. (C) The final shape of the mineral is determined by the geometry of the space available for growth. In the example, an octagonal crystal developed the form of a square. The internal structure of the crystal, however, remains the same, regardless of the growth space.*

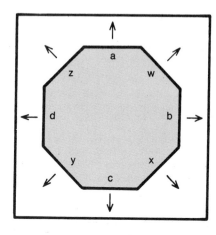
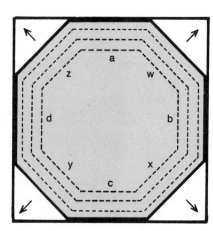
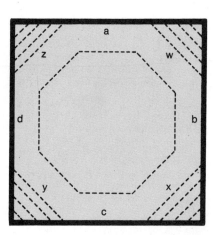

Figure 4.8 *Sketch showing how crystals can grow to form an interlocking texture.*

Figure 4.9 *Large crystals of gypsum.*

originate and change as minerals are created and destroyed. Crystal growth occurs in volcanic processes when lava cools and solidifies. It occurs in the ocean, where minerals grow from solution to form limestone, salts, and other mineral deposits. The destruction and growth of crystals also occur deep within the earth's crust, where heat and pressure cause crystal structures to break down and new minerals to form in their place. It also occurs at the surface, where minerals react chemically with elements in the atmosphere.

The Silicate Minerals

Statement

Although more than 2000 minerals have been identified, 95% of the volume of the earth's crust is composed of a group of minerals called the **silicates.** This should not be surprising since silicon and oxygen constitute nearly three-fourths of the mass of the earth's crust (*table 4.2*) and, therefore, would have to predominate in most rock-forming minerals. The silicate minerals are complex in both their chemistry and crystal structure,

Table 4.2. Concentrations of the Most Abundant Elements in the Earth's Crust (by weight)

Element	Percent
O	46.60
Si	27.72
Al	8.13
Fe	5.00
Ca	3.63
Na	2.83
K	2.59
Mg	2.09
Ti	0.44
H	0.14
P	0.12
Mn	0.10
S	0.05
C	0.03

After Mason, p. 48

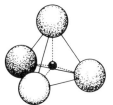

Figure 4.10 *Sketch of the silicon-oxygen tetrahedron. Four large oxygen atoms are arranged in the form of a pyramid (tetrahedron) with a small silicon atom fit into the central space between them. This is the most important building block in geology, because it is the basic unit for 90% of the minerals in the earth's crust.*

but all are composed of a basic building block called the **silicon-oxygen tetrahedron.** This is a complex ion [$(SiO_4)^{-4}$] in which four large oxygen ions [(O^{-2})] are arranged to form a four-sided pyramid with a small silicon ion [Si^{+4})] fit into the small cavity between them (*figure 4.10*). Differences between the major groups of silicate minerals result from the manner in which the silica tetrahedra are arranged to form a crystal structure.

Discussion

Perhaps the best way to understand the unifying characteristics of the silicate minerals, as well as the reasons for the differences, is to study the models shown in *figure 4.11,* which were built on the basis of X-ray studies of silicate crystals.

The sharing of oxygen ions by the silicon ions results in several fundamental configurations of tetrahedral groupings, such as single chains, double chains, hexagonal sheets, and three-dimensional frameworks. These structures define the major silicate mineral groups.

1. Single chains—pyroxenes.
2. Double chains—amphiboles.
3. Hexagonal sheets—micas, chlorites, and clays.
4. Three-dimensional frameworks—feldspars.
5. Single tetrahedra joined by other atoms such as iron and magnesium, olivines and garnets.

The unsatisfied electrons are balanced by various metallic ions such as calcium, sodium, potassium, magnesium, and iron. Thus, the silicate minerals contain silica tetrahedra linked together in various patterns with a variety of metallic ions.

It should be noted at this point that considerable ionic substitution can occur in the basic crystal structure, such as sodium substituting for calcium or iron substituting for magnesium. Thus, a major group of silicate minerals can differ chemically among themselves but have a common silicate structure.

Rock-forming Minerals

Statement

Fewer than 20 kinds of minerals account for the great bulk of the earth's crust and most of the upper mantle, and a knowledge of only 10 mineral types is adequate for a general understanding of most rocks. The most important minerals in each of the major rock types are as follows.

Figure 4.11 *Some examples of the arrangements of silica tetrahedra in silicate minerals.*

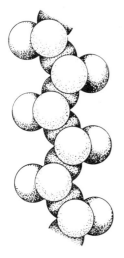

Single chain

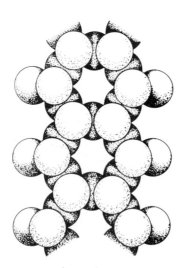

Double chain

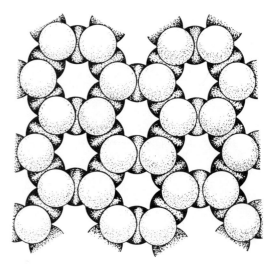

Sheet

1. Igneous rocks—feldspars, micas, amphiboles, pyroxenes, olivines, and quartz.
2. Sedimentary rocks—quartz, calcite, dolomite, clays, and feldspars.
3. Metamorphic rocks—quartz, feldspars, amphiboles, pyroxenes, micas, and chlorite.

The identification of minerals in rocks presents some special problems, because the mineral grains are usually small and rarely have well-developed crystal faces. You will find, however, that the distinguishing properties of the rock-forming minerals are easy to learn if you examine a specimen while studying the written description.

Discussion

A careful examination of the minerals that make up granite is a good beginning. *Figure 4.12A* illustrates a polished surface that shows that the rock is composed of a myriad of mineral grains having different sizes, shapes, and colors. Although the minerals interlock to form a tight, coherent mass, each one has definite distinguishing properties.

The Feldspars. A large part of granite is composed of a pink, porcelainlike mineral that has a rectangular shape and a milky white, porcelainlike mineral that is somewhat smaller but similarly shaped. These are feldspars (German: field crystals), the most abundant mineral group in the earth's crust. The feldspars have good cleavage in two directions, a porcelainlike luster, and a hardness of about 6 on Mohs' hardness scale.

The crystal structure of the feldspars permits considerable ionic substitution, giving rise to two major types: potassium feldspar and plagioclase feldspar.

Potassium feldspar ($KAlSi_3O_8$); most commonly pink colored in rocks, is shaded light gray in *figure 4.12*. Plagioclase (stippled gray) permits complete substitution of Na for Ca in the crystal structure, giving rise to a compositional range from $NaAlSi_3O_8$ to $CaAl_2Si_2O_8$. White plagioclase in granite is rich in Na.

Feldspars are common in most igneous rocks, in many metamorphic rocks, and in some sandstones.

The Micas. The tiny, black, shiny grains in *figure 4.12* are mica. This is the name of a group of minerals readily recognized by their perfectly one-directional cleavage, which permits the mineral to be broken easily into thin, elastic flakes. The mineral is a complex silicate with a sheet structure that is responsible for its perfect cleavage. Two common varieties occur in rocks: biotite, which is black mica, and muscovite, which is white or colorless. Mica is abundant in granites and in many metamorphic rocks and is also a significant constituent in many sandstones.

Quartz. The glassy grains with irregular shapes in *figure 4.12* are quartz. Because quartz is usually the last mineral to form in a granite, it lacks well-developed crystal faces. The mineral simply fills the spaces between early-formed feldspars and micas. Quartz is abundant in all three major rock types. It has the simple composition SiO_2 and is distinguished by its hardness (7), its conchoidal fracture, and its glassy luster. Pure quartz crystals are colorless, but slight impurities produce a variety of colors. Where crystals are able to grow freely, they

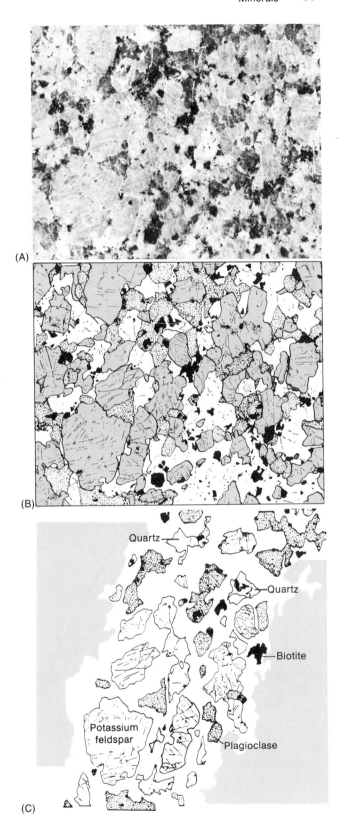

Figure 4.12 *Mineral grains in granite. (A) Polished surface of granite. The minerals have grown into a tight, interlocking texture. (B) Sketch emphasizing grains of individual minerals. K-feldspar is shaded gray, plagioclase is stippled gray, quartz is white, and biotite is black. (C) "Exploded" diagram of B showing size and shape of each mineral grain.*

form an elongate, six-sided crystal that terminates at a point (*figure 4.4*), but well-formed crystals rarely are found in rocks. In sandstone, the mineral is abraded into rounded sand grains. Quartz is stable both mechanically (it is very hard and lacks cleavage) and chemically (it will not react readily with elements at, or near, the earth's surface). Therefore, it is a difficult mineral to alter or destroy once it has formed.

Ferromagnesian Minerals.   In addition to the minerals found in a granite, there are a number of silicate minerals referred to as **ferromagnesian minerals,** because they contain appreciable amounts of iron and magnesium. These minerals are generally dark green to black in color and have a high specific gravity. Biotite is classified in this general group, together with the olivines, pyroxenes, and amphiboles. Biotite is common in granite, but the others are rare or absent in this particular rock type. The ferromagnesian minerals, however, are common in basalt (*figure 4.13*).

Olivine.   The only mineral clearly visible in the hand specimen in *figure 4.13* is the green, glassy mineral called olivine. The olivine family is a group of silicates in which iron and magnesium substitute in the crystal structure. The composition is expressed as $(Mg, Fe)_2SiO_4$. This hard mineral is characterized by an olive green color (if Mg is abundant) and a glassy luster and, in rocks, rarely forms crystals larger than a millimeter in diameter. Olivine is a high-temperature mineral and is common in basalt (volcanic rock) and is probably a major constituent of the material beneath the earth's crust.

Pyroxene.   Pyroxene and olivine commonly occur in the same types of rocks. In *figure 4.13*, pyroxene

(A)

Figure 4.13   *Mineral grains in basalt. (A) Hand specimen. (B) Sketch showing microscopic view of texture and mineral grains. (C) "Exploded" diagram showing sizes and shapes of individual grains.*

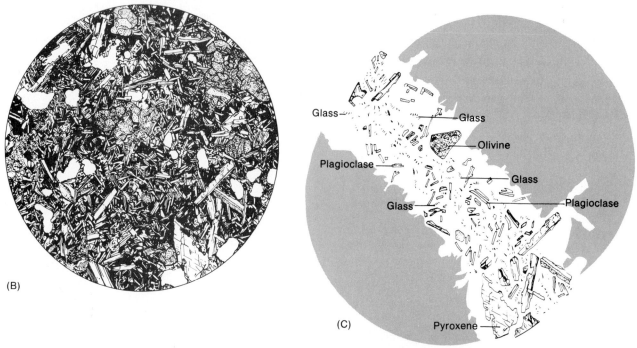

(B)

(C)

occurs as microscopic crystals, but some samples contain larger grains of this mineral, which is typically dark green to black. Pyroxenes are high-temperature minerals found in igneous and metamorphic rocks.

Amphiboles. *Figure 4.14* shows a rock that is similar to a granite but has less quartz and contains appreciable amounts of the mineral amphibole. The amphibole minerals have much in common with the pyroxenes. Their compositions are similar except that amphibole contains hydroxyl ions (OH^-) and pyroxene does not. The structure of the amphiboles, however, differs from the pyroxenes and produces elongate crystals that cleave perfectly in two directions, not at right angles. The color of amphibole ranges from green to black. This mineral is common in rocks closely related to granite and may be more abundant in that rock than biotite. It is especially common in some metamorphic rocks known as amphibolite. Hornblende is the most common variety of amphibole.

Clay Minerals. The clay minerals may be encountered more frequently than other minerals in everyday experience, because they form at the earth's surface through the interaction of air and water with the various silicate minerals. They constitute a major part of the soil and common dirt. Like the micas, the clay minerals are sheet silicates, but the crystals are microscopic and usually can be detected only with an electron microscope. More than a dozen clay minerals can be distinguished on the basis of their crystal structures and variations in composition.

Calcite. Calcite is calcium carbonate, the principal mineral in limestone. It can be precipitated directly from seawater, removed from seawater by organisms and used to make their shells, or dissolved by ground water and reprecipitated as new crystals in caves and fractures in rock. The color is usually white or transparent, but aggregates of calcite crystals that form limestone are mixed with various impurities, giving them gray or brown hues. Calcite, a common mineral at the earth's surface, is easy to identify. It is soft enough to scratch with a knife (hardness of 3) and effervesces in dilute hydrochloric acid. It has perfect cleavage in three directions but not at right angles, so cleaved fragments are in the form of rhombohedra (*figure 4.1*). In addition to being the major constituent of limestone, calcite is the major mineral in the metamorphic rock marble.

Dolomite. Dolomite is a carbonate of calcium and magnesium. Large crystals form rhombohedra, but most dolomite forms granular masses of small crystals. Dolomite is widespread in sedimentary rocks, having been altered from the original calcite through the activity of solutions of magnesium carbonate in seawater or ground water. Dolomite can be distinguished from calcite in that dolomite will effervesce in dilute hydrochloric acid only if it is in a powdered form.

Halite and Gypsum. Halite and gypsum are two of the most common minerals formed by evaporation of seawater or saline lakewater. Halite, common salt (NaCl), is easily identified by its taste, and it has one of the simplest of all crystal structures—the sodium and chlorine ions are united in a cubic structure. All the physical properties of halite are related to this structure. Halite crystals cleave in three directions at right angles

Figure 4.14 *Photograph showing crystals of dark-colored amphibole in a granitic rock.*

to form cubic or rectangular fragments. Salt, of course, is very soluble and readily dissolves in water.

Gypsum is composed of calcium sulfate and water and forms crystals that are generally clear white, with a glassy or silky luster. It is a very soft mineral and can be scratched easily with a fingernail. It cleaves perfectly in one direction to form thin, nonelastic plates. The mineral occurs as single crystals, as aggregates of crystals in compact masses (alabaster), and as a fibrous form (satinspar).

Summary

A mineral is a natural inorganic solid with a specific internal structure and a definite chemical composition that varies only within specific limits. As a result, all specimens of a given mineral, regardless of where, when, or how it was formed, have a specific set of physical properties. Some of the more readily observable physical properties are cleavage, crystal form, hardness, specific gravity, color, and streak, all of which aid in mineral identification.

Minerals are the building blocks of rocks and grow or are destroyed by chemical reactions as matter changes to and from a solid state. Minerals, therefore, provide an important means of interpreting the changes that have occurred in the earth throughout its history.

More than 95% of the earth's crust is composed of the silicate minerals, a group of minerals containing silicon and oxygen linked together in tetrahedral units of four oxygen atoms to one silicon atom. Several fundamental configurations of tetrahedral groupings—such as single chains, double chains, hexagonal sheets, and three-dimensional frameworks—result from the silicon ions sharing oxygen ions.

Fewer than 20 minerals form the great bulk of the earth's crust. These are the rock-forming minerals, the most important of which among the igneous rocks are feldspars, micas, amphiboles, pyroxenes, olivines, and quartz. The most important among the sedimentary rocks are quartz, calcite, dolomite, clay minerals, and feldspars. The most important among the metamorphic rocks are quartz, feldspars, amphiboles, pyroxenes, micas, chlorite, and garnet.

Additional Readings

Ahrens, L. H. 1965. Distribution of the Elements in Our Planet. New York: McGraw-Hill.

Deer, W. A., R. A. Howie, and J. Zussman. 1966. An Introduction to the Rock-forming Minerals. New York: John Wiley and Sons.

Ernst, W. F. 1969. Earth Materials. Englewood Cliffs, N.J.: Prentice-Hall.

Mason, B. 1958. Principles of Geochemistry. New York: John Wiley and Sons.

Simpson, B. 1966. Rocks and Minerals. New York: Pergamon Press.

U.S. Geological Survey. Atlas of Volcanic Phenomena. Washington, D.C.: U.S. Geological Survey.

Zim, H. S., and P. R. Shaffer. 1957. Rocks and Minerals. New York: Golden Press.

Igneous Rocks

Igneous rocks are formed by the cooling and crystallization of liquid rock material and, as a result, have a distinctive texture and composition. When mineral grains grow from a melt, they develop a tight, interlocking network of crystals that shows no signs of abrasion (such as is common in sedimentary rocks) nor evidence of stress (which characterizes metamorphic textures). The crystals that form igneous rocks are largely the silicate minerals—olivine, pyroxene, amphibole, feldspar, mica, and quartz. These, of course, are minerals that crystallize at relatively high temperatures, which range from 700 °C to 1200 °C.

The best-known examples of igneous activity are volcanic extrusions, where the liquid rock material works its way to the surface and erupts from volcanic vents and fissures and flows over the surface as lava. Less spectacular, but just as important, are the tremendous volumes of liquid rock that never make it to the surface but remain trapped in the crust, where they slowly cool and solidify. Granite* is the most common variety of this type of igneous rock. It is exposed in the cores of many eroded mountain belts and in the roots of ancient mountain systems in the shields.

Igneous rocks are important, because they are records of the thermal history of the earth. Their origin is closely associated with the movement of the tectonic plates, and igneous activity plays an important role in seafloor spreading, the origin of mountains, and the evolution of continents. In this chapter, the major types of igneous rocks and what they reveal about the thermal activity of the earth will be studied.

Major Concepts

1. Magma is a silica-rich melt capable of penetrating into, or through, the rocks of the crust.
2. Two major types of magma are recognized: (a) basaltic magma and (b) granitic magma.
3. The texture of a rock is important because it provides insight into the cooling history of the magma.
4. The major textures in igneous rocks are (a) glassy, (b) aphanitic, (c) phaneritic, (d) porphyritic, and (e) pyroclastic.
5. High-silica magmas produce rocks of the granite-rhyolite family, which are characterized by quartz, and K-feldspar, with minor amounts of Na-plagioclase, biotite, and amphibole.
6. Low-silica magmas produce rocks of the basalt-gabbro family, which are characterized by Ca-plagioclase and pyroxene, with minor amounts of olivine or amphibole.
7. Magmas with intermediate composition produce rocks of the diorite-andesite family.
8. The fluid basaltic magma and the viscous granitic magma produce contrasting types of eruptions.
9. Magma originates from the partial melting of the lower crust and the upper mantle at depths usually between 50 and 200 km.

*The term granite is used in the broadest sense to designate all granitic rocks, including quartz monzonites, monzonites, and granodiorites—rock types that introductory students usually do not need to distinguish.

The Nature of Magma

Statement

The nature of **magma** (molten rock material) where it originates in the earth's lower crust and upper mantle, of course, is not known from direct observation. Therefore, our understanding of the characteristics of magma is based on observations of volcanic activity and on studies of synthetic magmas made in the laboratory, together with observations of igneous rocks formed beneath the surface but now exposed by erosion. The data from such studies indicate that the essential characteristics of a magma are as follows.

1. Magma is molten silicate material including early-formed crystals and dissolved gases. It is a complex mixture of liquid, solid, and gas.
2. The principal elements in magmas are O, Si, Al, Ca, Na, K, Fe, and Mg. The two most important constituents controlling the properties of a magma are SiO_2 and H_2O.
3. Silica (SiO_2) is the principal constituent of a magma, ranging from 35% to 75%.
4. Dissolved gases can constitute a small percentage of a magma and are important in determining certain properties, such as viscosity and the explosive characteristics of volcanic eruptions. Water and carbon dioxide constitute more than 90% of the gases emitted from volcanoes.

Discussion

The term *magma* comes from the Greek word that means kneaded mixture, like a dough or paste. In its geologic application, magma refers to *any hot, mobile material* within the earth that is capable of penetrating into, or through, the rocks of the crust. This may be quite different from the concept most people have of the nature of magma, which is derived largely from spectacular illustrations of volcanic eruptions. In popular articles about volcanism, the sensational usually is emphasized and the idea that all magma is a very fluid, red-hot liquid closely resembling molten steel is impressed firmly on a reader's mind.

Most magmas, however, are not entirely liquid but are, in reality, a combination of liquid, solid, and gas. Early-formed crystals can make up a large portion of the mass, and a magma could be thought of more accurately as a slush, with liquid melt mixed with a mass of mineral crystals. The consistency of such a mixture would be similar to that of freshly mixed concrete, slushy snow, or thick oatmeal. Movement of such magmas is slow and sluggish.

Water vapor and carbon dioxide are the principal gases in a magma and can constitute as much as 14% of the volume. The gas phase is important in that it influences the mobility and melting point of a magma and the types of volcanic activity that can be produced.

Chemical analyses of many igneous rocks indicate that there are two principal kinds of magma: (1) basaltic magmas, which contain about 50% SiO_2 and have temperatures ranging from 900 °C to 1200 °C, and (2) granitic magmas, which contain between 60% and 70% SiO_2 and

generally have temperatures lower than 800 °C. Basaltic magmas characteristically are fluid, whereas granitic magmas are thick and viscous. This is because granitic magmas have lower temperatures and greater amounts of SiO_2. Even before crystallization occurs, silica tetrahedra are linked together and the linkages offer resistance to flow. Therefore, the higher the silica content, the higher the viscosity.

The Importance of Rock Textures

Statement

The texture of a rock refers to the size, shape, and arrangement of its constituent mineral grains and is an aspect of the rock apart from its composition. The importance of texture lies in the fact that the mineral grains bear record of the energy involved in the rock-forming process and the conditions that existed at the time the rock originated. The genetic imprint left upon the texture of a rock commonly is quite clear and easy to read. For example, rocks from the moon that were formed by the impact of meteorites have a texture characterized by various sizes of angular rock fragments with bits and pieces of glass fused by the impact. A sandstone formed on a beach, in contrast, consists of well-rounded, smooth grains, all approximately the same size, which result from the abrasion and washing action of the waves. Igneous rocks have distinctive textures that consist mostly of interlocking crystals that grew from a cooling magma.

The major textures of igneous rocks are related to the size of crystals and are summarized as follows.

1. **Glassy**—no crystals.
2. **Aphanitic**—crystals too small to be seen without the aid of a microscope.
3. **Phaneritic**—crystals large enough to be seen without the aid of a microscope.
4. **Porphyritic**—two different crystal sizes.
5. **Pyroclastic**—volcanic fragments.

Discussion

Five examples of rocks that have essentially the same chemical and mineralogical composition but different textures will illustrate the importance of texture. In each rock, a chemical analysis would disclose about 48% O, 30% Si, 7% Al, and 3% to 4% each of Na, K, Ca, and Fe. In glassy rocks, crystallization has not taken place, so the constituent atoms are random and are not arranged in a crystalline structure. In all other specimens, the elements occur in crystals of feldspar, quartz, mica, and amphibole. On the basis of chemical composition alone, these rocks would be considered the same—the difference is *only* in texture. It is the texture that provides the most information about how the rock was formed.

Glassy Texture.   The nature of volcanic glass is illustrated in *figure 5.1.* The hand specimen displays a conchoidal fracture with sharp edges typical of the manner in which glass breaks. No distinct crystals are visible, but, under the microscope, distinct flow layers are apparent from the uneven concentration of innumerable, minute, embryonic crystals. In the laboratory, it can be shown that melted rock or synthetic lava hardens to glass when

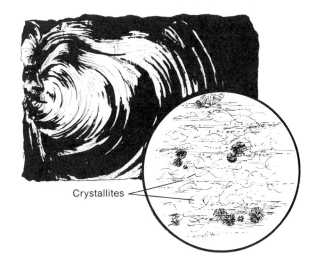

Figure 5.1 *Glassy texture (obsidian).*

it is quenched (or quickly cooled) when it is still above the temperature at which its crystals form.

It can be concluded that a glassy texture is produced by very rapid cooling and that the randomness of ions in a high-temperature melt is "frozen in" as a result of insufficient time for the ions to migrate and organize themselves into an orderly arrangement in a crystal structure. Field observations of glassy rocks in volcanic regions support this interpretation that rapid cooling produces glass. Small pieces of lava blown from a volcanic vent into the much cooler atmosphere harden to form glassy ash. A glassy crust forms on the surface of many lava flows, and glassy fragments form when a flow enters a body of water.

Aphanitic Texture. If the growth of crystals from a melt requires time for the ions to collect together and organize themselves into an orderly arrangement in the crystal structure, then a crystalline rock must indicate a slower rate of cooling than a glassy rock. The texture illustrated in *figure 5.2* is crystalline but extremely fine grained—a texture referred to as aphanitic (Greek: *a,*

not; *phaneros,* visible). In the hand specimen, few, if any, crystals can be detected in aphanitic textures, but, under a microscope, many crystals of feldspar and quartz can be recognized.

An aphanitic texture would indicate relatively rapid cooling but not nearly as rapid as that which would produce glass. Aphanitic textures are typical of the interior of lava flows, in contrast to the glassy texture that forms on the surface or crust.

A feature in many aphanitic and glassy rocks is the presence of numerous, small, spherical or ellipsoidal cavities called **vesicles**. These are produced by small gas bubbles trapped in the solidifying rock. When hot magma rises toward the surface of the earth, the confining pressure diminishes and dissolved gas (mainly steam) separates and collects into bubbles. The process is not unlike the effervescence of champagne and soda pop when their bottles are opened. Vesicular textures typically develop in the upper part of a lava flow, just below the solid crust, where the upward-migrating gas bubbles are trapped. The vesicles change the outward appearance of the rock and indicate the presence of gas in a rapidly cooling lava, but they do not change the basic aphanitic texture.

Phaneritic Texture. The specimen in *figure 5.3* is composed of crystals sufficiently large to be recognized without the aid of a microscope, a texture referred to as phaneritic (Greek: *phaneros,* visible). The grains are approximately equal in size and constitute an interlocking mosaic. The equigranular texture suggests a uniform rate of cooling, and the large size of the crystals shows that cooling was very slow.

In order for cooling to take place at such a slow rate, the magma must have been insulated under a cover of rock so that the entire body crystallized without reaching the surface. Field evidence supports this conclusion, since volcanic eruptions produce only aphanitic and glassy textures. Rocks with phaneritic textures are exposed only after deep erosion has removed the covering rock.

Porphyritic Textures. In some igneous rocks, two distinct sizes of crystals are apparent. The larger, well-formed crystals are referred to as **phenocrysts,** and the

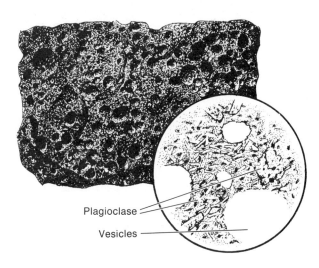

Figure 5.2 *Aphanitic texture (basalt).*

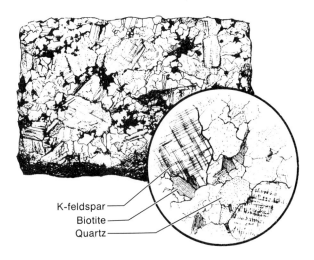

Figure 5.3 *Phaneritic texture (granite).*

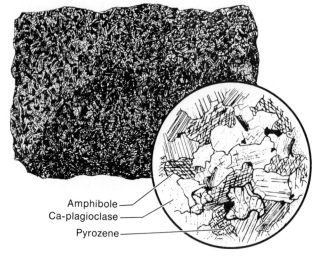

Figure 5.4 *Porphyritic-aphanitic texture (porphyritic basalt).*

crystals. Other fragments are bent and squeezed, and shards of glass fragments are twisted and deformed.

This is a pyroclastic texture (with fire fragments) and is produced by explosive eruptions in which ash and early-formed crystals are blown into the air as a mixture of hot fragments and glass. If the fragments are still hot when they are deposited, they may be welded (fused) together; otherwise, they may be cemented together later on.

In summary, the texture of an igneous rock records a great deal of information about the cooling history of a rock and how the rock was formed. All of the rocks described above had the same composition and could have been derived from the same magma, and yet each texture type was formed in a different way. Therefore, geologists pay particular attention to the texture of a rock and use it as one basis for classification in all major rock types.

Kinds of Igneous Rocks

Statement

The classification of igneous rocks is based upon texture and composition. Texture provides important information on the cooling history of the magma, and the composition provides insight into the nature and origin of the magma. A simple chart showing the major types of igneous rocks is shown in *figure 5.6.* Variations in composition are arranged horizontally, and variations in texture are arranged vertically. The names of the rocks are printed in bold type roughly proportionate to their abundance. Rocks in a vertical column have the same composition but a different texture. Rocks in a horizontal column have the same texture but a different composition. From this chart, it can be seen that granite, for example, has a phaneritic texture and is composed predominantly of quartz and K-feldspar. The bold type indicates that it is the most abundant intrusive igneous rock. Rhyolite has the same composition as granite but is aphanitic. Basalt has an aphanitic texture and is composed predominantly of Ca-plagioclase and pyroxene. It has the same composition as gabbro but is much more abundant.

smaller crystals constitute the **matrix** or the **groundmass.** This texture is referred to as porphyritic and can occur in either aphanitic or phaneritic rocks.

A porphyritic texture usually indicates two stages of cooling: an initial stage of slow cooling, in which the large crystals developed, and a subsequent period of more rapid cooling, in which the smaller crystals formed. The aphanitic matrix in *figure 5.4* indicates that the cooling melt had sufficient time for all the material to crystallize. The initial stage of relatively slow cooling produced the larger crystals, whereas the later stage of rapid cooling, when the magma was extruded, produced the smaller crystals. Similarly, a phaneritic matrix indicates two stages of cooling: an initial stage of very slow cooling followed by a second stage of more rapid cooling that was not rapid enough to form an aphanitic matrix.

Pyroclastic Texture. The texture shown in *figure 5.5* may appear at first to be a porphyritic rock with phenocrysts of quartz, but, under a microscope, the grains can be seen to be broken fragments rather than interlocking

Discussion

Rocks with Phaneritic Textures. The following is a brief description of the most common igneous rocks with phaneritic textures.

Granite. Granite is a coarse-grained igneous rock composed predominantly of feldspar and quartz (*figures 5.3* and *4.4*). K-feldspar is the most abundant mineral and usually is easily recognized by its pink color. Na/Ca-plagioclase is present in moderate amounts and usually is distinguished by its white color and its porcelainlike appearance. Mica is very conspicuous as black or bronze-like flakes, usually distributed evenly throughout the rock. The texture of the rock, together with laboratory experiments, indicates that plagioclase and biotite are the first minerals to crystallize from the magma. A seemingly insignificant, but very important, property of granite is its relatively low specific gravity of about 2.7. In contrast to basalt and related rocks, which have a specific

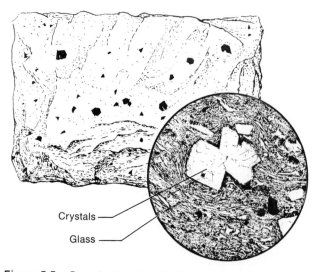

Figure 5.5 *Pyroclastic texture (tuff).*

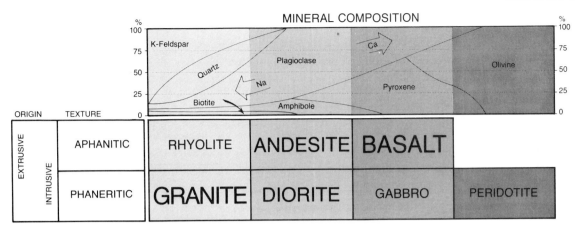

Figure 5.6 *Classification of igneous rocks.*

gravity of 3.2, granite is a light rock. This fact is very important in considering the nature of continents and the contrast between continental crust and oceanic crust. Granite and related rocks make up the great bulk of continental crust, whereas the oceanic crust is composed of basalt.

Diorite. Diorite is similar to granite in texture, but it differs in composition. Na-plagioclase feldspar is the dominant mineral, and quartz and K-feldspar are minor constituents. Amphibole is an important constituent, and some pyroxene may be present. The composition of diorite is intermediate between granite and basalt.

Gabbro. Gabbro is not an abundant rock, but it is important in some older intrusive bodies. It has a coarse-grained texture similar to granite, but it is composed almost entirely of pyroxene and calcium-rich plagioclase, with minor amounts of olivine. Gabbro is dark green, dark gray, or almost black because of the dominance of dark-colored minerals.

Peridotite. This phaneritic rock is composed almost entirely of two minerals, olivine and pyroxene. This type of rock is not common at the earth's surface nor, as far as we can tell, within the continental crust, but it is a very important rock type in the subcrustal part of the earth (the mantle). The high specific gravity, together with other physical properties, of the rock there indicates that the great bulk of the earth—the mantle—most likely is composed of peridotite and closely related rock types. The Alps and St. Paul's Rocks (islands in the Atlantic Ocean) are two areas where peridotite from the mantle appears to have been pushed through the crust to the earth's surface.

Rocks with Aphanitic Textures. The following is a brief description of the most common igneous rocks with aphanitic textures.

Basalt. Basalt is the most common aphanitic rock. It is a very fine grained, usually dark-colored rock that originates from lava flows. The mineral grains are so small that they rarely can be seen without the aid of a microscope. When a thin section (a thin, transparent slice of rock) is viewed through a microscope, the individual minerals can be seen and studied (*figures 5.2 and 4.13*).

Basalt is composed predominantly of calcium-rich plagioclase and pyroxene, with smaller amounts of olivine or amphibole. The plagioclase occurs as a mesh of elongate, lathlike crystals surrounding the more equidimensional pyroxene and olivine crystals. In some cases, large crystals of olivine or pyroxene form large phenocrysts, resulting in a porphyritic texture. Many basalts have some glass, especially near the tops of flows.

Andesite. Andesite is an aphanitic rock composed of Na-plagioclase, pyroxene, and amphibole and usually little or no quartz. Thus, it has the same composition as diorite. It is generally porphyritic, with phenocrysts of feldspar and ferromagnesian minerals. The term comes from the Andes Mountains, where volcanic eruptions have produced the rocks in great abundance. Probably the next most abundant lava type after basalt, it occurs most frequently along the continental margins but is not found in the ocean basins, nor is it abundant in the continental interiors. The origin of andesite along the continental margins probably is related to partial melting in the subduction zone.

Rhyolite. Rhyolite is an aphanitic rock with the same composition as a granite. It usually contains a few phenocrysts of feldspar, quartz, or mica but not always enough to be considered porphyritic. Rhyolite flows are viscous, and, instead of spreading out in a linear flow, rhyolite typically piles up in large, bulbous domes. Rhyolite and andesite are difficult to distinguish without the aid of a microscope and commonly are grouped together and referred to as feldsite.

Rocks with Pyroclastic Textures. The following is a brief description of the most common igneous rocks with pyroclastic textures.

Tuff. Volcanic eruptions of rhyolitic and andesitic lavas commonly produce large volumes of fragmental material. The fragments range from dust-size pieces to large blocks more than a meter in diameter. The rock resulting from the accumulation of ashfalls is referred to as a **tuff,** which, although of volcanic origin, has many of the characteristics of sedimentary rock.

Ash Flow Tuff. Ash flow tuff is a rock composed of fragments of volcanic glass, broken fragments of crystals, rock fragments, and pieces of solidified lava fused together into a tight, coherent mass. Many fragments typically are flattened or bent out of shape. These unique textures indicate that the rockmass, although composed of ash fragments, at the time of extrusion was hot enough to fuse together. *Figure 5.5* shows these features as they appear under a microscope.

Extrusive Rock Bodies

Statement

Volcanic eruptions are among the most spectacular of all geologic processes, and today more than 500 active volcanoes testify to the dynamics of the earth. Moreover, volcanoes provide a window to the planet's interior, because the lava erupted at the surface provides evidence of the processes operating in the lower crust and the upper mantle.

The two major types of magma, basaltic and granitic, produce contrasting types of eruptions.

1. Basaltic magmas are fluid, and the lava emerges quietly from fractures and fissures to produce a succession of thin flows that can cover large areas.
2. Granitic magmas are thick and viscous and typically erupt with great violence as a result of pressure built up by trapped gases. The explosion of granitic magma ejects pieces of lava, hot ash, and early-formed crystals into the air that move as a thick, dense cloud of hot ash. Extrusions of this type are best thought of as **ash flows.** When granitic magmas do not explode, they are extruded to form thick flows or bulbous domes.

A variety of volcanic phenomena are associated with each type of eruption, many of which are subject to direct observation and study.

Discussion

The type of eruption appears to be related to the viscosity of the magma, which in turn is related to its composition. Magmas rich in silica have a high viscosity, because the silica-oxygen tetrahedra are strongly linked to one another in complex chains and networks, which inhibit flow and make the magma viscous. This retards the escape of gas bubbles that build up pressure and cause the magma to explode. Basaltic magmas have less silica, and this fact allows them to be more fluid and to produce relatively quiet eruptions, because dissolved gases can escape readily.

Basaltic Eruptions.   Basaltic lavas with eruptive temperatures of between 900 °C and 1200 °C have been measured and, upon extrusion, tend to flow freely downslope and spread out into valleys and topographic depressions. The speed at which the fluid basaltic lava flows can reach 30 or 40 km per hour before it cools and congeals, but rates of 20 km per hour are considered unusually rapid. As the flow moves downslope, it loses gas,

Figure 5.7   *The surface of an aa flow at Craters of the Moon, Idaho, showing the jumbled mass of angular blocks formed when the congealed crust was broken as the flow slowly moved. Aa flows are viscous and much thicker than pahoehoe flows.*

Figure 5.8   *Hexagonal columnar joints at Devil's Post Pile, California, formed by contraction when the lava cooled.*

Figure 5.9 *Pressure ridge arched up by trapped gas in recent flow in southern Idaho.*

Figure 5.10 *The surface of a recent pahoehoe lava flow on the flanks of Kilauea, Hawaii, showing the ropy flow structures. Pahoehoe flows are produced from fluid lava and characteristically are very thin.*

cools, and becomes more viscous. Movement then becomes sluggish, and the flow soon comes to rest.

Two types of basaltic flows are common. An **aa flow** (*aa* is a Hawaiian word pronounced ah'-ah') contains relatively little gas and is a slow-moving flow several meters thick. The surface of the flow cools and forms a crust while the interior remains molten. The flow may move only a few meters per hour. As it continues to move, the hardened crust is broken into a jumbled mass of angular blocks and clinkers (*figure 5.7*). Gas within the fluid interior of the flow migrates toward the top but may remain trapped beneath the crust. These "fossil gas bubbles" are referred to as vesicles and form a light, porous rock. The interior of the flow may be massive and nonvesicular. As thick flows cool, they contract and they can develop a system of polygonal cracks called **columnar joints,** which are similar in many ways to mud cracks (*figure 5.8*). In some flows, the sides and top may freeze solid while the interior remains fluid. When pressure is great enough, the fluid interior can break through

the crust and flow out, leaving a long lava tube. More commonly, pressure from the fluid interior causes the crust of the flow to arch up in a **pressure ridge,** a blister characterized by a central crack through which gas and lava can escape (*figure 5.9*).

Pahoehoe flows contain much more gas than aa flows and thus are more fluid. The pahoehoe flows are thinner (usually less than a meter thick) and move much faster than aa flows because of their lower viscosity. As the pahoehoe flow moves, a thin, glassy crust develops that is molded into billowy forms or surfaces that can resemble coils of rope. A variety of flow features (such as those shown in *figure 5.10*) can develop on the surface of pahoehoe flows.

Generally, basaltic lava is extruded from fissures or a central conduit (*figure 5.11*). Locally, it can spew out as lava fountains, which often rise along the fissures, and spatter can collect around the vent to build up small **spatter cones** (*figure 5.12*).

Together, fragments of volcanic rock material blown

Figure 5.11 *Extrusion of basalt along a fissure on the island of Hawaii.*

Figure 5.12 *Spatter cones at Craters of the Moon National Monument, Idaho. Spatter cones are built up by splashes of lava that collect around a vent.*

Figure 5.13 *Tephra—volcanic ash and dust.*

out from a volcanic vent are called **tephra**. Tephra includes fine dust, volcanic ash (*figure 5.13*) (fragments up to about .5 cm) and larger particles called **volcanic bombs** (*figure 5.14*). Commonly, bombs have ellipsoidal shapes twisted as the cooling lava travels through the air. Tephra is ejected vertically from the volcanic vent and generally is cold by the time it is deposited. As it travels through the air, it is sorted according to size, with the large particles accumulating close to the vent to form a **cinder cone** and the finer, dust-size particles being transported afar by the wind. Cinder cones are relatively small features, generally less than 200 m high and 2 km in diameter (*figure 5.15*).

With continued extrusion of large quantities of fluid basaltic lava, a broad cone called a **shield volcano** may be built up around the vent (*figure 5.16*). With each eruption, the fluid basaltic lava flows freely for some distance before congealing and spreading out as a thin sheet, or tongue. Shield volcanoes, therefore, have a wide base (commonly over 100 km in diameter) and gentle slopes (generally less than 10°). The internal structure of the volcano consists of innumerable, thin basalt flows with comparatively little ash. Some of the best examples of shield volcanoes formed the Hawaiian Islands. These volcanoes are enormous mounds of basaltic lava, rising as much as 8 km above the seafloor. The younger volcanoes typically have summit craters up to 3 km wide and several hundred meters deep that resulted from subsidence following eruption of magma from below.

Figure 5.14 *Volcanic bombs—fragments of lava ejected in a liquid or plastic state and formed into spindle-shaped masses during their flight through the air.*

Figure 5.15 *Cinder cones. Volcanic ash and dust accumulate near the vent to form cinder cones. The ejected material is deposited in layers that dip away from the summit crater. The vent beneath the crater commonly is filled with lava or fragmental debris.*

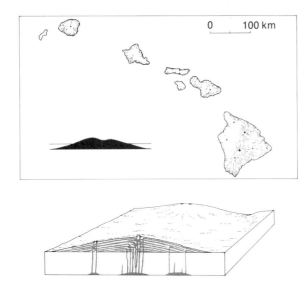

Figure 5.16 *Shield volcanoes of Hawaii. Shield volcanoes are the largest volcanoes on the earth. They are composed almost exclusively of innumerable, thin basalt flows with relatively little ash.*

Figure 5.17 *Pillow basalt, Newfoundland. Pillow basalt is formed when lava is extruded underwater.*

The extrusion of basaltic lava into water produces an ellipsoidal mass referred to as a **pillow lava** (*figure 5.17*). This feature has been observed in the process of formation off the coast of Hawaii, and recent subaqueous photographs show that it is widespread over the seafloor where volcanic activity is common.

Silicic Eruptions. Silica-rich magmas are those with a relatively high silica content that form granitic and intermediate igneous rocks. They are relatively cool, and the mechanics of eruption and flow of their lavas are quite different from those of basaltic composition.

Some silicic magmas are so thick and viscous that small volumes hardly flow at all but form massive plugs or bulbous domes over the volcanic vent (*figure 5.18*). The high viscosity of silicic magmas inhibits the escape of dissolved gas so that tremendous pressure builds up, and, when eruptions occur, they are highly explosive and violent. As a result, they commonly produce large quantities of tephra. The tephra and thick, viscous lava typically produce a high, steep-sided cone around the vent called a **composite volcano** (*figure 5.19*). The composite volcano is probably the most familiar form of continental volcano, with such famous examples as Mounts Shasta, Fuji, Vesuvius, Etna, and Stromboli. At the summit is usually a depression, the **crater,** that marks the position of the vent.

The explosive violence with which silicic volcanoes erupt can blow out large volumes of ash and magma. This can result in the collapse of the summit area to form

Figure 5.18 *Domes of silicic lava, Mono Craters, California. Silica-rich lava is so viscous that it does not flow but piles up over the vent to form large, bulbous domes.*

Figure 5.19 *Mount Fuji, a famous composite volcano in Japan. Composite volcanoes are built up of alternating layers of ash and lava flows, which characteristically form high, steep-sided cones.*

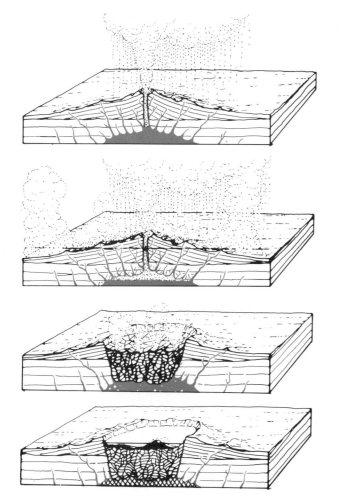

(A) Early eruptions from the prehistoric volcano Mount Mazma.

(B) Great eruptions of vast ash flows empty the magma chamber, causing the top of the volcano to collapse.

(C) Collapse of the summit into the magma chamber forms the caldera.

(D) Lake forms in the caldera, and minor eruptions produce small volcanoes in the lake.

Figure 5.20 *The evolution of the caldera at Crater Lake, Oregon.*

a large, basin-shaped depression known as a **caldera.** A famous example is Crater Lake, Oregon (*figure 5.20*), which formed when a volcano's summit collapsed after its last major eruption and the resulting caldera filled with water. Wizard Island, a small cinder cone in the caldera, resulted from subsequent minor eruptions. Krakatoa is another well-known example. The 1883 eruption caused 80 km³ of rock to disappear. The cone was demolished, and great quantities of volcanic ash were blown high into the atmosphere. The explosion and subsequent subsidence produced a caldera 6 km in diameter that completely altered the configuration of the island (see *figure 19.10*).

A spectacular type of eruption associated with silicic magmas is called an ash flow, the lateral flowing movement of ash and lava particles en masse. It is not a liquid lava flow nor an ashfall, in which particles settle out independently, but a flow phenomenon of suspended fragments and gas. As a magma works its way to the surface, confining pressure is released and the trapped gas bubbles rapidly expand. When it nears the surface, the magma then can literally explode. The ejected material consists of pieces of lava, bits of solid rock, ash, early-formed crystals, and gas. This material is very hot, sometimes incandescent, and, being denser than air, it moves or flows across the surface as a thick, dense cloud of hot ash; hence, the name ash flow. An ash flow can reach velocities in excess of 60 km per hour, because the gas continually expands between the cooling lava particles, forcing them apart. When the ash flow comes to rest, the particles of hot crystal fragments, glass, and ash are fused together to form a rock that has been described as a **welded tuff.** Ash flows can be very large and form flow units more than a hundred meters thick and spread over an area of thousands of square kilometers. As they cool, they can develop columnar jointing as the mass contracts.

Although eruptions of ash flows are rare and catastrophic events, a few geologists have had the unique opportunity to witness them from afar and to make direct observations of this type of extrusion.

An example was the 1951 eruption of Mount Lamington in New Guinea. Up until 1951, Mount Lamington had been considered extinct. It had never been examined by geologists and had not even been considered to be a volcano by the local inhabitants. Volcanic activity began with preliminary emissions of gas and ash, accompanied by earthquakes and landslides near the crater. Then, on Sunday, January 20, 1951, a catastrophic explosion burst from the crater and produced an ash flow that completely devastated an area of about 200 km². The main eruption was observed and photographed at close quarters from a passing aircraft, and a qualified volcanologist began recording events on the spot within 24 hours after the main eruption. Sensitive seismographs soon were installed near the crater to monitor earth movements, and daily aerial photographic records were made. The ash flow descended radially from the summit crater, with the direction of movement being controlled to some degree by the topography. As the ash flow moved down the

Figure 5.21 *Devastation caused by the 1951 ash flow of Mount Lamington, New Guinea.*

slopes, it scoured and eroded the surface. Estimated velocities of 470 km per hour were calculated by the force required to overturn certain objects. Entire buildings were ripped from their foundations, and automobiles were picked up and deposited in the tops of trees (*figure 5.21*).

Intrusive Rock Bodies

Statement

Masses of igneous rock formed by magma cooling beneath the surface are called **intrusions.** These rocks can never be observed in the process of formation, but, when uplift and erosion expose them for study, the following lines of evidence indicate that the rock body was once in a magmatic state.

1. Intrusive rocks are characterized by a suite of high-temperature minerals similar to those that form volcanic rocks, but the crystals are much larger, which suggests slow cooling.
2. The mineral grains typically are smaller close to the contacts with surrounding rock, which suggests that cooling was more rapid near the margins than in the interior of the intrusion.
3. Tongues and stringers of intrusive rock extend out, filling fractures in the surrounding rock, which suggests that the rock was once liquid and was squeezed into surrounding cracks.
4. Fragments of the surrounding rock commonly are included in the intrusive mass, a feature difficult to explain if the pluton were not liquid at one time.
5. The rocks adjacent to an intrusion commonly are recrystallized into new, higher-temperature mineral assemblages requiring an input of heat in order to form.

Magma rises because it is less dense than the sur-

rounding rocks and can be intruded by forceful injection into fractures, or it can melt and assimilate the rock it invades.

Intrusions usually are classified according to their sizes, shapes, and relationships to the older rocks that surround them. These include **batholiths, stocks, dikes, sills,** and **laccoliths.**

Discussion

Batholiths. Batholiths are very large masses of coarsely crystalline rock, generally of granitic composition. They are formed almost exclusively in continents and large island arcs and do not occur in oceanic islands. Batholiths are the largest rock bodies in the earth's crust, and many have an areal extent of several thousand square kilometers. The Idaho batholith, for example, is a huge body of granite exposed over an area of nearly 41,000 km², the British Columbia batholith to the north is over 2000 km long and 290 km wide and at least 3000 m thick. The true, three-dimensional form of batholiths, however, is difficult to determine because of the uncertainty regarding its shape and extension at depth. Evidence showing the layered nature of the crust upon the mantle indicates that batholiths are limited in depth to the thickness of the crust and do not extend down into the mantle. Thus, they must be less than 60 km thick. The nature of the base of the floor of a batholith, however, remains a matter of conjecture. Indirect evidence (such as gravity measurements) suggests that batholiths may increase in size at depth for some distance and then, possibly, taper off at greater depth, somewhat like the root of a tooth. The diagrams in *figure 5.22* give a rough idea of the geometric form of some of the better-known batholiths in North America. From these data, batholiths appear to be huge, slablike bodies whose horizontal extent is much greater than their thickness. Studies of the thickness of strata upwarped along the flanks of batholiths suggest that they were emplaced at depths of more than 7 km below the surface.

A diagrammatic illustration of a batholith and its relationship to associated rock bodies is shown in *figure 5.23.* The mass of granitic rock can be elongate and elliptical or circular. It generally cuts across the surrounding rock, but locally the wall of the batholith can be parallel to the stratification in the sedimentary rocks of foliation in metamorphic rocks. In the simplest cases, batholiths

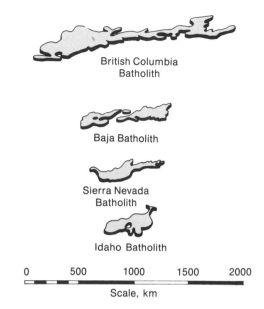

British Columbia
Batholith

Baja Batholith

Sierra Nevada
Batholith

Idaho Batholith

0   500   1000   1500   2000

Scale, km

Figure 5.22   *Diagrams showing the relative sizes and shapes of several well-known batholiths in North America.*

result from single intrusions in which the magma is injected into older rock and cools, but many batholiths are composed of several intrusions of somewhat similar composition.

Batholiths typically are formed in the deeper zones of mountain belts and are exposed only after considerable uplift and erosion. Some of the highest peaks of mountain ranges such as the Sierra Nevada and the Coast ranges of western Canada are carved in the granite of batholiths, a rock that originally cooled thousands of meters below the surface. The trend of the batholith usually parallels the axis of the mountain range, although the intrusion locally can cut across folds within the range.

Extensive exposures of batholiths also are found in the shield areas of the continents (see *plate 4*). These exposures are considered to be the roots of ancient mountain ranges that have long since been eroded to lowlands.

Figure 5.23   *Schematic diagram showing various relationships of sedimentary rocks to igneous intrusions.*

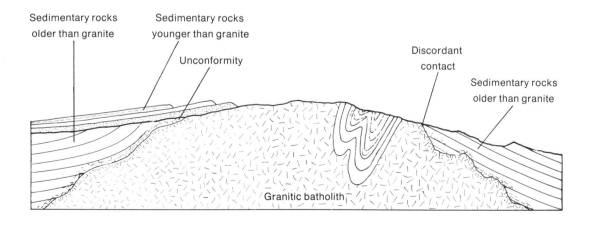

Sedimentary rocks older than granite

Sedimentary rocks younger than granite

Unconformity

Discordant contact

Sedimentary rocks older than granite

Granitic batholith

The young mountain ranges and the ancient shields, therefore, are believed to expose a different part of the batholith. In mountains, remnants of the roof of the magma chamber are seen, whereas, in the shield, only the lower and deeper parts of the batholiths remain.

Stocks.   Smaller plutonic bodies ranging from a few thousand meters to approximately 15 km in plain view are called stocks. Some stocks pass upward into volcanic necks, and others never reach the surface and are exposed only after considerable erosion. It can be demonstrated that some stocks are small protrusions up from the main batholith, but the downward extent of others is unknown. Generally, stocks are composed of granitic rock, and many are porphyritic, with a fine-grained groundmass. Many deposits of silver, gold, lead, zinc, and copper are associated with fractures and veins extending out from a stock into the surrounding rock.

Dikes.   One of the most familiar signs of ancient igneous activity is a narrow, tabular body of igneous rock called a dike (*figure 5.24*). A dike forms where magma is squeezed into fractures of the surrounding rock and cools. The width of a dike can range from a fraction of a centimeter to hundreds of meters. The largest known dike is the Great Dike of Rhodesia, which extends 600 km and has an average width of 10 km.

The implacement of dikes is controlled by fracture systems within the **country rock.** Dikes, therefore, cut across sedimentary and metamorphic layers and are said to be discordant. Dikes radiating from ancient volcanic necks are common, reflecting stresses associated with volcanic activity. In some areas, upward pressure from the magma chamber produces circular or elliptical fracture systems into which magma can be injected to form ring dikes. Large ring dikes can be as much as 25 km in diameter and thousands of meters deep.

After erosion, the surface expression of a dike is usually a narrow, elongate ridge, but dikes can erode as fast, or faster, than the surrounding rock and even can erode down to form long, narrow trenches.

Sills.   When selected bedding planes offer a zone of least resistance, magma can be injected between the layers to form a sill, a tabular intrusive body parallel or concordant to the layering. Sills can range from a few centimeters to hundreds of meters thick and can extend laterally for several kilometers.

Therefore, a sill can resemble a buried lava flow lying within the sequence of sedimentary rock. It is, however, an intrusion forced between layers of older rock, and many features at the contact with adjacent strata should indicate its mode of origin. Intrusion of magma to form a sill alters or recrystallizes the rocks above and below it, and a sill shows no signs of weathering on its upper surface. It also commonly contains **inclusions** of the surrounding rocks. A buried lava flow, in contrast, has an eroded upper surface marked by vesicles, and the younger overlying rock commonly contains fragments of the eroded flow.

Sills can form as local offshoots from dikes, or they can be connected directly to a stock or a batholith. Characteristically, they are formed from basaltic magma that is highly fluid and capable of being squeezed between layers of older rock. The more viscous granitic magma rarely forms sills.

Laccoliths.   When viscous magma is injected between layers of sedimentary strata, it tends to arch up the overlying strata. The resulting intrusive body is lens shaped, with a flat floor and an arched roof. These intrusions are known as laccoliths. They usually occur in

Figure 5.24   *Relationships of dikes and sills to sedimentary rocks.*

Dike   Sill

Dike

(A)

Figure 5.25 *Laccoliths. (A) The Henry Mountain laccolith in the Colorado Plateau. Note the light-colored, tilted, and eroded sedimentary strata along the flanks of the intrusion that forms the central mass of the mountain. (B) Schematic diagram showing the shape of a typical laccolith and its relationship to the sedimentary rocks that it intrudes.*

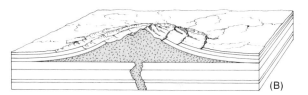

(B)

blisterlike groups in areas of flat-lying strata (*figure 5.25*). They can be several kilometers in diameter and thousands of meters thick and typically are porphyritic.

In the Colorado Plateau, where laccoliths were first described, a series of laccoliths occurs along the margins of a central stock and appears as an inflated sill. Other laccoliths can be fed from a conduit from below.

The Origin of Magma

Statement

The origin of a magma and the mechanics by which it rises through the crust remain fundamental problems that are not yet understood completely. We know from seismic evidence that the crust and the mantle are solid bodies because of the way they transmit earthquake waves, so it is evident that there are no permanent, worldwide reservoirs of magma from the surface to a depth of nearly 2900 km. The core of the earth responds to seismic waves as if it were a liquid, but gravity measurements indicate that the core is made of material far denser than any magma. Thus, the material in the core would produce rocks several times denser than any known igneous rock. Contrary to popular belief, therefore, it is very unlikely that magma comes from the liquid core or even from within the middle and lower parts of the mantle. The most likely place for magma to develop is in local pockets where solid rock can be melted in the upper mantle and lower crust, at depths of possibly 50 to 200 km.

It is important to understand clearly that most rocks are composed of more than one mineral and do not have a specific melting temperature. Each component mineral melts at a different temperature, so the rock melts over a range of temperatures. The major factors influencing melting are:

1. Temperature.
2. Pressure.
3. Amount of water present.
4. Composition.

Therefore, the melting of a rockmass can be initiated by an increase in temperature *or a change in pressure*. As the temperature-pressure combination changes in a direction to induce melting, certain minerals melt first, followed by others in a definite sequence. Thus, at a given temperature-pressure combination, a rock body may be only partly melted and the liquid (or melt) may have a composition quite different from that of the original rock. This process is known as **partial melting.** Also, as a magma cools, certain minerals crystallize first, followed by others in a definite sequence. When partial crystallization occurs and the remaining liquid becomes separated from the crystals, it forms a magma quite different from the parent material. This process is known as **magmatic differentiation.**

The theory of plate tectonics relates the generation of magma to processes operating at active plate margins.

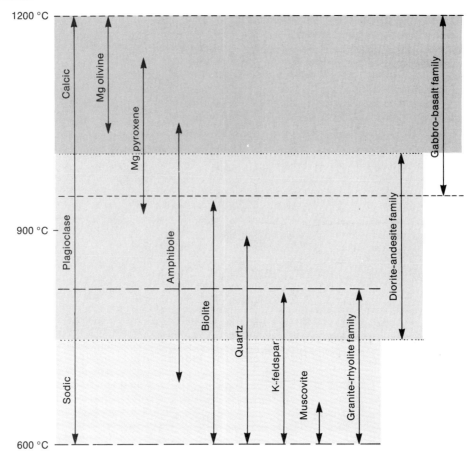

Figure 5.26 *Diagram showing the order of crystallization of the most common rock-forming minerals.*

The two fundamental types of magma are believed to form in separate tectonic settings.

1. Basaltic magma is generated by partial melting of upwelling mantle along spreading centers, where the plates move apart and dominate the igneous activity of the ocean basins.
2. Granitic magmas are generated along the **subduction** zone by partial melting of the oceanic crust and by partial melting associated with metamorphism in orogenic belts.

Discussion

Laboratory studies have shown that the crystallization of a silicate melt is extremely complex and that not only temperature and pressure are important but also composition and amount of water. Early-formed crystals of one mineral can react with the remaining liquid to form new minerals of different composition. The general order of crystallization is summarized in *figure 5.26*. Olivine and Ca-plagioclase are the minerals that crystallize at the highest temperatures (between 1050 °C and 1200 °C). These are followed by pyroxene, amphibole, and Na-plagioclase. Quartz and K-feldspar are lower-temperature minerals crystallizing at temperatures below 900 °C.

Figure 5.27 *The origin of basaltic magma by partial melting of upwelling mantle material at a spreading center. As mantle material moves upward along the spreading zone between two plates, partial melting of hot peridotite (the major rock in the mantle) occurs due to a decrease in pressure. The resulting melt forms a basaltic magma.*

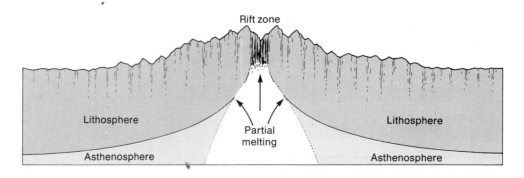

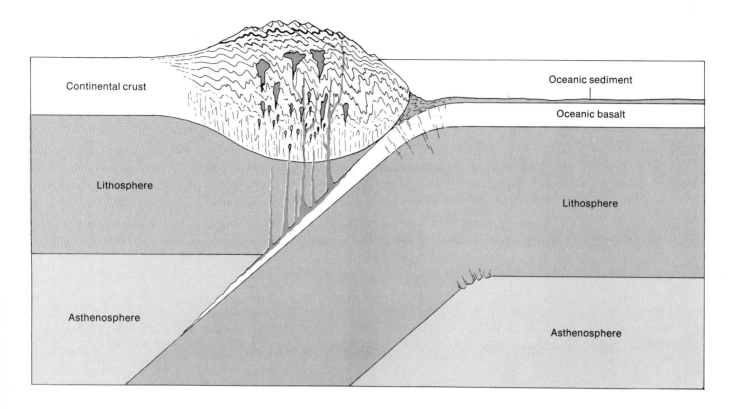

Figure 5.28 *The origin of granitic magma in the subduction zone. Partial melting of the basaltic oceanic crust, which contains oceanic sediments and water-rich clays, produces a silica-rich magma that rises to form granitic and andesitic rocks.*

Generation of Basaltic Magma. As discussed in chapter 1, the asthenosphere is believed to be composed of peridotite, a rock consisting of the minerals olivine and pyroxene. The balance between temperature and pressure in the asthenosphere is just about right for a slight degree of melting to occur, while, at greater depths in the mantle, pressure is too great for partial melting. This explains why the asthenosphere is soft and weak. As the asthenosphere slowly moves upward along a spreading zone between plates, additional partial melting occurs due to the decrease in pressure. Partial melting of peridotite produces a basaltic magma, because the first minerals to melt yield basaltic constituents. Laboratory experiments on melting peridotite at high pressures have indicated that a basaltic magma would be produced if less than 10% to 30% of peridotite were melted. The basaltic magma, being less dense than peridotite, would rise along the oceanic ridge and would be extruded as new crust in the spreading ocean basins (*figure 5.27*).

Generation of Granitic Magma. At the subduction zone, the basaltic oceanic crust and its veneer of marine sediment descend into the mantle, where they are heated (*figure 5.28*). There, melting is much more complex, because, in addition to basalt, the oceanic crust contains oceanic sediments with water—rich clay and silica-rich material derived from erosion of the continent. Partial melting of this material produces a magma richer in silica, which forms andesitic or granitic rock. Where the converging plate margins coincide with a continental border and mountain belts are produced, a greater variety of silica-rich rocks develop from the partial melting of the metamorphic rocks in the roots of mountain systems. The lighter magma rises upward, collecting in larger bodies to produce the great granitic batholiths that characteristically are formed in mountain belts.

In summary, then, the simple model of plate tectonics is able to explain the major facts about most igneous rocks on earth—facts derived from field observations, studies of composition and texture of rock types, and laboratory experiments on synthetic magma.

Summary

Magma is molten rock material that is capable of penetrating into, or through, rocks of the crust. It can include early-formed crystals (solids), silicate melt (liquid), and water and other volatiles (gas). Two major types of magma are recognized: (1) basaltic magma, which typically is very hot (900 °C to 1200 °C), is highly fluid, and contains about 50% SiO_2; and (2) granitic magma, which is relatively cooler (less than 800 °C), is highly viscous, and contains about 60% to 70% SiO_2.

The texture of a rock refers to the size, shape, and arrangement of the constituent mineral grains. The major textures in igneous rocks are (1) glassy, (2) aphanitic, (3) phaneritic, (4) porphyritic, and (5) pyroclastic.

The classification of igneous rocks is based upon

texture and composition. High-silica magmas produce rocks of the granite-rhyolite family characterized by quartz, K-feldspar, and Na-plagioclase. Low-silica magmas produce rocks of the gabbro-basalt family characterized by Ca-plagioclase, amphibole, and pyroxene, with little or no quartz. Magmas with an intermediate composition produce rocks of the diorite-andesite family with a composition intermediate between granite and basalt.

The two major types of magma—basaltic and granitic—produce contrasting types of eruptions. Basaltic magmas are fluid and are extruded as quiet fissure eruptions producing a succession of thin flows that cover a broad area. Basaltic flows commonly develop columnar jointing, cinder cones, pillow lava, and shield volcanoes. Granitic magmas are viscous and contain large amounts of trapped gas. Therefore, they explode violently, and the lava flows as a thick pasty mass or as ash flows. Extrusion of granitic magma commonly produces composite volcanoes and ash flows.

Masses of igneous rock formed by magma cooling beneath the surface are called intrusions, which are classified according to size, shape, and relationships with the older rocks that surround them. The most important types of intrusions include batholiths, stocks, dikes, sills, and laccoliths.

Magma is believed to originate by the partial melting of the lower crust and the upper mantle at depths usually between 50 and 200 km below the surface. It is not generated in the earth's core. Basaltic magma is generated by the partial melting of upwelling mantle along spreading centers when tectonic plates move apart. Granitic magmas are generated along subduction zones by the partial melting of the oceanic crust and by the partial melting associated with metamorphism in the continental crust.

Additional Readings

Ernst, W. F. 1969. Earth Materials. Englewood Cliffs, N.J.: Prentice-Hall.

Schumann, W. 1974. Stones and Minerals. London and Guildford: Lutterworth Press.

Simpson, B. 1966. Rocks and Minerals. New York: Pergamon Press.

Tindall, J. R., and R. Thornhill. 1975. Rock and Mineral Guide. London: Blandford Press.

Williams, H., F. J. Turner, and C. M. Gilbert. 1954. Petrography. San Francisco: W. H. Freeman and Company.

Sedimentary Rocks

The geologic processes that operate upon the earth's surface generally produce only subtle changes in the landscape during a human lifetime, but, over a period of tens of thousands or of millions of years, the effect of these processes is considerable. Given enough time, the erosive power of the hydrologic system is capable of reducing an entire mountain range to a featureless lowland over which the ocean may expand and deposit new sedimentary layers.

The record of erosion through time and the changing landscape it produces are preserved in the sequences of sedimentary rocks. Each bedding plane is a remnant of a former surface of the earth, and each rock layer, together with its internal structure, is the product of a previous period of erosion.

To interpret the sedimentary record correctly, you must first understand something about such modern sedimentary environments as deltas, beaches, and rivers. The study of sediment deposited today in these areas provides an insight into how ancient sedimentary rocks originated. From the sedimentary record, geologists can find the trends of ancient shorelines, map the positions of former mountain ranges, and determine the drainage patterns of ancient river systems. With this information, we are able to make paleographic maps of ancient landscapes and, looking back through time, see the planet as it was millions of years ago.

In this chapter, the various types of sedimentary rocks and the various environments in which they are deposited will be introduced.

Major Concepts

1. Sedimentary rocks form at the earth's surface through the activity of the hydrologic system. Their origin involved erosion of some preexisting rock and transportation and deposition of the eroded material.
2. Two main types of sedimentary rocks are recognized: (a) clastic rocks consisting of rock and mineral fragments and (b) chemical and/or organic rocks consisting of chemical precipitates or organic material.
3. Stratification is the most significant sedimentary structure and results from changes in erosion, transportation, and deposition. Other important structures include crossbedding, graded bedding, ripple marks, and mud cracks.
4. Sedimentary processes result in sedimentary differentiation, in which material is sorted and concentrated according to grain size and composition.
5. The major sedimentary environments include (a) fluvial (stream), (b) alluvial fan, (c) eolian (wind), (d) glacial, (e) delta, (f) shoreline, (g) organic reef, (h) shallow marine, and (i) deep marine.

The Nature of Sedimentary Rocks

Statement

Sedimentary rocks are simply sediments that have been compacted and/or cemented to form solid rock bodies. The original material can be:

1. Fragments of other rocks and minerals, such as gravel in a river channel, sand on a beach, and mud in the ocean.
2. Chemical precipitates, such as salt in a saline lake and calcium carbonate in a shallow sea.
3. Organic materials, such as coral reefs and vegetation in a swamp.

All are sediment, and all can be preserved as future sedimentary rocks.

The single most distinctive feature of sedimentary rocks is that they are stratified; that is, they were laid down in a series of individual beds, one on top of another, with each bedding plane representing part of a preserved segment of a former surface of the earth (such as a river bed, a flood plain, or the seafloor). These layers of sedimentary rock now blanket most of the continents, and major layers or groups of layers, called **formations,** can be traced over areas of thousands of square kilometers and preserve a record of ancient landscapes, climates, and mountain ranges, as well as the erosional history of the earth. In addition, evidences of

Figure 6.1 *The sequence of sedimentary rock formations exposed in the Grand Canyon, Arizona. Each major rock unit erodes into a distinctive landform. The resistant formations (such as sandstone and limestone) erode into vertical cliffs. Nonresistant rocks (such as shale) erode into slopes.*

past life are found in abundance in sedimentary rocks, constituting a record of changing plant and animal communities that extends over a period of 2 to 3 billion years. It is from this record of sediments and fossils that the details of the geologic time scale (from which the relative ages of the major events in later parts of the earth's history were established) were worked out.

Discussion

Sedimentary rocks are probably more familiar to most people than the other major rock types, because they cover approximately 75% of the surface of the continents and, therefore, form most of the landscape. However, few people are aware of the nature and extent of sedimentary rock bodies. The Grand Canyon is an area in which many features of sedimentary rocks are well exposed (*figure 6.1*).

The most obvious characteristic of the sedimentary rocks in the Grand Canyon is that they occur in distinct layers, many of which are more than 100 m thick. The rock types that are resistant to weathering and erosion form cliffs, while the nonresistant rocks erode into gentle slopes. From *figure 6.1,* you should be able to recognize the cross section of the rock formations shown in the diagram in *figure 6.2.*

A high-altitude aerial photograph of the canyon (*figure 6.3*) shows that the major layers can be traced across much of northern Arizona and that they cover an area of more than 256,000 km². A closer view of sedimentary rocks shows that their texture, composition, and internal structure are quite different from the other major rock types. *Figure 6.4* shows that the major layers of a sandstone actually consist of smaller units separated by bedding planes; these bedding planes are marked by some change in composition, grain size, color, or other physical features. Fossil animals and plants are common

in most of the rock units and can be preserved in great detail (*figure 6.5*). The texture of most sedimentary rocks consists of mineral grains or rock fragments that show evidence of abrasion (*figure 6.6*) or interlocking grains of the mineral calcite. In addition, many layers show ripple marks (*figure 6.7*), mud cracks (*figure 6.8*), and other evidence of water deposition preserved on the bedding planes. All of these features support the conclusion that sedimentary rocks form at the earth's surface in environments similar to present-day deltas, streams, beaches, tidal flats, lagoons, and shallow seas.

Types of Sedimentary Rocks

Statement

Sedimentary rocks are classified on the basis of the size, shape, and composition of their constituent particles. Two main groups are recognized.
1. Clastic rocks (such as gravel, sand, and mud), consisting of deposits made from fragments of other rocks.
2. Chemical and/or organic rocks (such as limestone, dolomite, rock salt, gypsum, and coal), which were formed by chemical or biologic processes.

Discussion

Clastic Rocks. Generally, clastic rocks are subdivided on the basis of the grain sizes of the component materials. These components include the accumulation of familiar sedimentary materials such as gravel, sand, silt, and mud, which were eroded from the continents and transported as fragments or grains. When deposited and consolidated into hard rock, they are referred to as conglomerate, sandstone, and shale.

Conglomerates. A conglomerate consists of consolidated deposits of gravel, with variable amounts of sand and mud deposited in the spaces between the larger grains. The cobbles and pebbles usually are well-rounded fragments over 2 mm in diameter. Most conglomerates show a crude stratification and include beds and lenses of sandstone. Conglomerates are accumulating today at the foot of mountains, in stream channels, and on beaches.

Sandstones. Sandstone is probably the most familiar sedimentary rock, since it is well exposed, easily recognized, and generally resistant to weathering. Sand can be composed of almost any material, but quartz grains are usually most abundant because quartz is a common constituent in many other rock types and does not break down easily by abrasion or chemical action. The particles of sand in most sandstones are cemented by calcite, silica, or iron oxide.

The composition of a sandstone provides an important clue to its history. During prolonged transportation, unstable minerals such as olivine, feldspar, and mica and small rock fragments are broken down to finer particles and winnowed out, leaving only the ultrastable quartz. Clean, well-sorted sandstone composed of well-rounded quartz grains indicates prolonged transportation or even several cycles of erosion and deposition.

Shales (Mudstone). Deposits of fine, solidified mud and clay are known as shale. The particles that make up the rock are less than 0.166 mm in diameter and, in many cases, are too small to be clearly seen and identified under a microscope. Shale is the most abundant sedimentary rock, but it usually is soft and weathers into a slope, so that relatively few fresh, unweathered exposures are found. It usually is well stratified, with thin laminae.

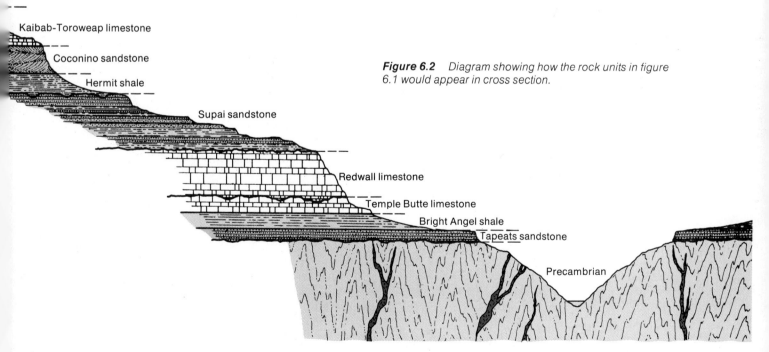

Kaibab-Toroweap limestone
Coconino sandstone
Hermit shale
Supai sandstone
Redwall limestone
Temple Butte limestone
Bright Angel shale
Tapeats sandstone
Precambrian

Figure 6.2 *Diagram showing how the rock units in figure 6.1 would appear in cross section.*

Figure 6.3 *High-altitude photograph of the Grand Canyon, Arizona. The uppermost cliff-forming unit, the Kaibab Limestone, forms the bedrock for the entire plateau shown in this view (see also plate 15).*

Figure 6.4 *Stratification in the Tapeats Sandstone (the lowest horizontal formation shown in figures 6.1 and 6.2) consists of numerous layers 0.5 to 1 m thick, each of which contains crossbedding or other types of layering.*

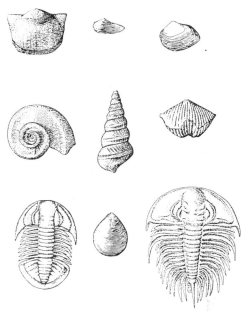

Figure 6.5 *Typical invertebrate fossils found in the sedimentary rocks of the Grand Canyon.*

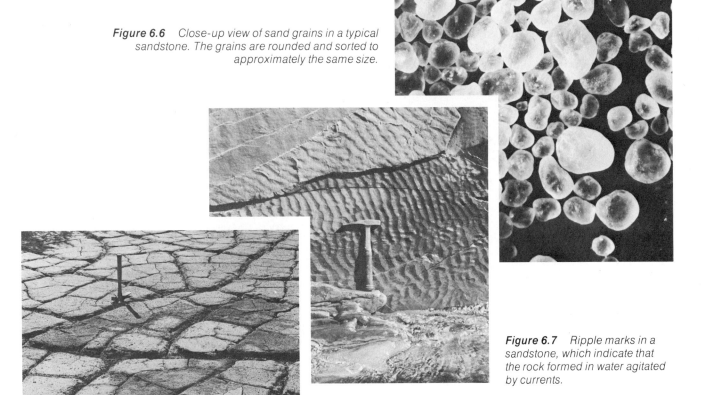

Figure 6.6 *Close-up view of sand grains in a typical sandstone. The grains are rounded and sorted to approximately the same size.*

Figure 6.7 *Ripple marks in a sandstone, which indicate that the rock formed in water agitated by currents.*

Figure 6.8 *Mud cracks formed where sediment was exposed temporarily and dried, such as in a tidal flat or a shallow lakebed.*

Black shales are rich in organic material and accumulate in a variety of quiet-water environments, such as lagoons, restricted shallow seas, and tidal flats. Red shales are colored with iron oxide and indicate oxidizing conditions in the environments in which they accumulate, such as stream channels, flood plains, and tidal flats.

Chemical and Organic Rock. The following rocks are the most common in the group.

Limestone. Limestone is by far the most abundant nonclastic rock. It is composed principally of calcium carbonate ($CaCO_3$) and originates by both chemical and organic processes. Limestones have a great variety of rock textures, and there are many different types of limestone. Some of the major groups are skeletal limestone, oolitic limestone, and microcrystalline limestone.

Many plants and invertebrate animals extract calcium carbonate from water and use it to construct their shells and hard parts. When the organisms die, the shells accumulate on the seafloor and, over a long period of time, these shells build up a deposit of limestone in which the texture consists of shells and shell fragments. This type of limestone, composed mostly of skeletal debris, can be several hundred meters thick and extend over thousands of square kilometers. Chalk, for example, is a skeletal limestone in which the skeletal fragments are remains of microscopic plants and animals.

Some limestones are composed of small, spherical concretions of calcium carbonate called **oolites.** The individual grains are about the size of a grain of sand and can be observed forming at the present time in the shallow waters off the Bahamas, where currents and waves are active. Evaporation and increased temperatures in the seawater increase the concentration of $CaCO_3$ to a point at which it can be precipitated. Small fragments of shells and tiny grains of $CaCO_3$ moved by waves or currents become coated with successive layers of $CaCO_3$ as they roll along the seafloor. The sorting action of the waves and the currents keeps the grain size of oolites quite uniform. Naturally, fragments of shells and oolites accumulate together in many places, forming an oolitic skeletal texture. Oolitic limestone is a special type of nonclastic sedimentary rock in which moving currents, rather than running water or wind, carry the grains to the places where they are deposited.

In quiet water, calcium carbonate is precipitated to form tiny, needlelike crystals that settle to the bottom and accumulate as limy mud. Soon after deposition, the grains commonly are modified by compaction and recrystallization. This modification produces a rock with a dense, very fine grained texture known as microcrystalline limestone. In such a rock, individual crystals can be seen only under high magnification. Microcrystalline limestone also is precipitated from springs and dripping water in caves, but the total amount of **dripstone,** compared to the amount of marine limestone, is negligible.

Dolostone. Dolostone is a rock composed of the mineral dolomite, a calcium-magnesium carbonate [$CaMg(CO_3)_2$] that is similar to limestone ($CaCO_3$) in most textural and structural features and general appearance. It can develop by direct precipitation from seawater, but most dolostones appear to originate by a substitution of magnesium for some calcium in limestones.

Rock Salt and Gypsum. Rock salt is composed of the mineral halite and forms by evaporation in saline lakes (for example, the Dead Sea) or in restricted bays along the shore of the ocean. Gypsum, composed of $CaSO_4 \bullet 2H_2O$, also originates from evaporation and collects in layers as calcium sulfate is precipitated. Since evaporites accumulate only in restricted basins subjected to prolonged evaporation, they are important indicators of ancient climatic and geographic conditions.

Sedimentary Structures

Statement

Most sediment is transported by current systems in streams, along shorelines, and in shallow seas before it finally is deposited as sedimentary accumulations. For this reason, sedimentary rocks commonly show layering and other structures formed as the material is moved, sorted, and deposited by current motion. The layering and related features formed at the time of deposition are called **primary sedimentary structures** and provide key information about the conditions under which the sediment accumulated.

Stratification is probably the most significant sedimentary structure, since it is found in almost all sedimentary deposits. It results from changes and fluctuations in the processes or the conditions of erosion, transportation, and deposition.

Crossbedding is a type of stratification produced by currents that build dunes, sand waves, and ripple marks. They form as the particles migrate up and over the dunes and accumulate as inclined layers on the downcurrent slope. Crossbedding is especially significant, because it reflects the direction in which the currents were moving during sedimentation. From it, geologists can determine patterns of sediment dispersal.

Other important sedimentary structures include **ripple marks, mud cracks,** and tracks or borings of animals that lived within the sedimentary environment.

Discussion

Stratification. Stratification occurs on a wide range of scales and reflects many types of changes that can occur during the formation of a sedimentary rock. In large exposures, the larger scale of stratification is expressed by major changes in rock types, which can be expressed as major cliffs of limestone or sandstone alternating with slopes of weaker shale, as in *figure 6.1.* Within each of these major rock units, bedding occurs on several smaller scales and is expressed by differences in the texture, color, and composition of the rock.

A very important aspect of stratification is that the rock layers do not occur in a random fashion but overlie one another in definite sequences and patterns. One of the more simple and common patterns in a vertical sequence is the cycle of sandstone, shale, and limestone. This pattern can be produced by the transgression and

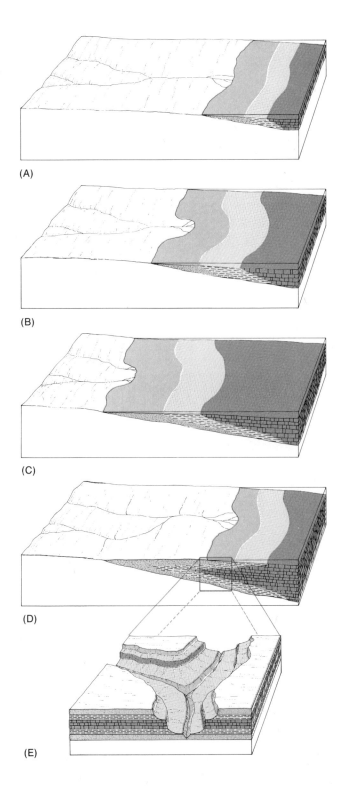

(A)

(B)

(C)

(D)

(E)

Figure 6.9 *Schematic diagrams showing how expansion and contraction of a shallow sea produce a cycle of sandstone-shale-limestone-shale-sandstone.*

ment are deposited simultaneously, each in a different environment. Stream deposits (not shown in the diagram) accumulate on the flood plain of the river system.

As the sea expands over the lowland, each environment shifts landward with each major movement of the shoreline (diagrams *A, B,* and *C*). This results in the beach sands being deposited over the stream sediments, the offshore mud being deposited over the previous beaches, and the lime over the mud. As the sea continues to expand inland, layers of sand, mud, and limestone are deposited.

As the sea withdraws (diagram *D*), the mud is deposited over the limestone and the nearshore sand over the mud. The net result is a wedge of limestone encased in a wedge of shale that, in turn, is encased in a wedge of sandstone. Below and above the marine deposits are fluvial sediments deposited by the river system. Subsequent uplift and erosion of the area will produce a definite pattern of rock (diagram *E*), beginning at the base with a beach sand overlain by shale and limestone deposited offshore that, in turn, is overlain by mud and beach sand. (See the lower rock layers in *figure 6.2.*)

Crossbedding. The formation of crossbedding is shown in *figure 6.10*. As clastic particles are moved by currents, they form ripples or dunes called **sand waves.** Sand waves range in scale from small ripples less than a centimeter high to giant sand dunes several hundred meters high. Typically, they are asymmetrical, with the steep slope facing in the direction of the moving current. As the particles migrate up and over the sand wave, they accumulate on the steep downcurrent face and form inclined layers. The direction of flow of ancient currents that formed sets of cross-strata can be determined by measuring the direction in which the strata are inclined. It is possible, therefore, to determine the patterns of ancient current systems by mapping the direction of crossbedding in sedimentary rocks.

Graded Bedding. A very distinctive type of stratification, called **graded bedding,** is characterized by a progressive decrease in grain size upward through the bed (*figure 6.11*). This type of stratification commonly is produced on the deep-ocean floor by turbidity currents that transport sediment from the continental slope to the adjacent deep ocean. A turbidity current is a current generated by turbid (muddy) water, which, being denser than the surrounding clear water, sinks beneath it and moves out rapidly down the submarine slope (*figure 6.12* and *plate 21*). The principle of turbidity current flow can easily be demonstrated in the laboratory by pouring muddy water down the side of a tank filled with clear water. The mass of muddy water moves down the slope of the tank and across the bottom at a relatively high speed, without mixing with the clear water. Turbidity currents also are observed where streams discharge muddy water into a clear lake or a reservoir. The denser muddy water continues to move out along the bottom of the basin and may flow for a considerable distance. For example, the turbidity currents produced by the Colorado River flowing into Lake Mead continue to flow beneath the clear water of the lake for more than 200 km, and occasionally they boil up into the outlet pipe at Hoover Dam. Turbidity currents also can be generated by an earthquake or a sub-

regression of shallow seas over the continental platform (see *figure 6.9*). In the first diagram, the sea begins to expand over a lowland drained by a river system. Sand accumulates along the shore, mud is transported in suspension offshore, and limestone is precipitated farther offshore beyond the zone of mud. All three types of sedi-

marine landslide, an event during which mud and sand and even gravel can be thrown into suspension.

A turbidity current triggered by an earthquake near Grand Bank off Newfoundland broke a series of transatlantic cables as it swept across the continental slope. The speed at which the current moved, timed through the intervals between cable breaks, was determined to be 80 to 95 km an hour. This mass of muddy water swiftly spread out over a large area on the floor of the Atlantic and formed a graded layer of sediment.

As a turbidity current moves across a given point on the flat floor of a basin, its velocity gradually decreases. The coarser sediment is deposited first, followed by successively smaller and smaller particles. After the turbid water ceases to move, the remaining sediment held in suspension gradually settles out. This one pulse of sedimentation, therefore, deposits a single layer of sediment in which there is a continuous gradation from coarse material at the base to fine material at the top, without a break or a sudden change in grain size. Apparently, deposition is continuous, with no internal lamination. A subsequent turbidity flow would deposit a succeeding layer of graded sediment with a sharp contact between layers. The entire sequence is characterized by a succession of widespread horizontal layers, each of which is a graded unit deposited by a single pulse of sedimentation from a turbidity flow.

Ripple Marks, Mud Cracks, and Other Surface Impressions. In addition to stratification, many sedimentary rocks contain various structures that provide important information concerning the environment of deposition (*figures 6.7* and *6.8*). Ripple marks can be preserved on the bedding surfaces and can show current directions. Mud cracks show that certain sedimentary rocks were formed in an environment that occasionally was exposed to the air during the process of deposition. The presence of this feature in rocks suggests deposition in shallow lakes, tidal flats, and exposed stream banks.

Tracks, trails, and borings of animals commonly are associated with ripple marks and mud cracks and can provide important clues about the environment in which the sediment accumulated. Imprints of raindrops can even be preserved on bedding planes where the succeeding sediment settled out without destroying the bedding surface.

The Origin of Sedimentary Rocks

Statement

Sedimentary rocks originate at the earth's surface through the interactions of the hydrologic system and the crust. Fortunately, we are able to observe many of these processes in operation today, and such research has taken geologists to the rivers, deltas, and oceans of all parts of the earth. These studies indicate that the genesis of sedimentary rocks involves four major processes.
1. Weathering.
2. Transportation.
3. Deposition.
4. Compaction and cementation.

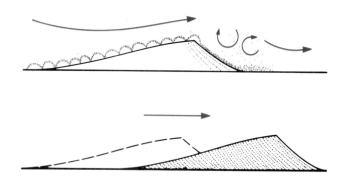

Figure 6.10 *The origin of crossbedding. Crossbedding is formed by the migration of ripple marks, sand waves, and dunes. Particles of sediment are carried by the currents up and over the sand wave and are deposited on the steep, downcurrent face to form inclined layers called crossbedding.*

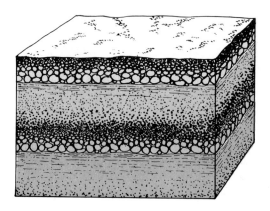

Figure 6.11 *Graded bedding produced by turbidity currents occurs in widespread layers, generally less than a meter thick. The deep-marine environment commonly produces a great thickness of graded units that can be easily distinguished from sediment formed in most other environments.*

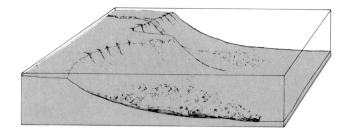

Figure 6.12 *Schematic diagram showing movement of turbidity currents down the slope of the continental shelf. Sediment is moved largely in suspension, and, as the current slows down, the coarse grains are deposited first, followed by successively finer grained sediment.*

Weathering breaks down and decomposes preexisting solid rock exposed at the surface. This material then is transported by various agents, such as streams, wind, and glaciers, and finally is deposited in many different sedimentary environments, such as rivers, deltas, beaches, and shallow seas. Subsequently, the loose, unconsolidated material is buried, compacted, and cemented to form a solid rock. As a result of these selective operations, the original rock material eroded from the land is sorted according to grain size and composition, a process described as **sedimentary differentiation.** Large particles accumulate as gravels, medium-grained materials concentrate as sand, and the fine silt and clay-size particles settle out as mud. The dissolved material is precipitated as limestone, salt, or other chemical deposits.

Since sedimentary rocks are the products of interactions between the hydrologic system and the crust, they bear record of physical events in the history of the earth's surface, including such things as uplift and mountain-building, erosion of the continents, climatic changes, and the constantly changing patterns of land and sea.

Discussion

Probably the most significant factor in the origin of sedimentary rocks is the sedimentary environment (that is, the place where the sediment is deposited) and the physical, chemical, and biologic conditions that exist there. The idealized diagram in *figure 6.13* shows in a general way the regional setting of the major sedimentary environments. Continental environments include those areas of sedimentation that occur exclusively on the land surface, landward from the effects of tides. Most important are major river systems, alluvial fans, desert dunes, lakes, and margins of glaciers. Marine environments include the shallow seas that cover parts of the continental platform, as well as the floors of the deep-ocean basins. Between these two are transitional or mixed environments that occur along the coasts and are influenced by both marine and nonmarine processes. These include deltas, beaches, tidal flats, barrier islands, reefs, and lagoons.

Each of these environments is characterized by certain physical, chemical, and biologic conditions that develop distinctive types of texture, composition, internal structure, and assemblages of fossils in the rocks they produce. Illustrations of modern sedimentary environments, together with examples of the rocks they produce, are shown in *figures 6.14* through *6.24*.

Figure 6.13 *Schematic diagram showing the major environments of sedimentation.*

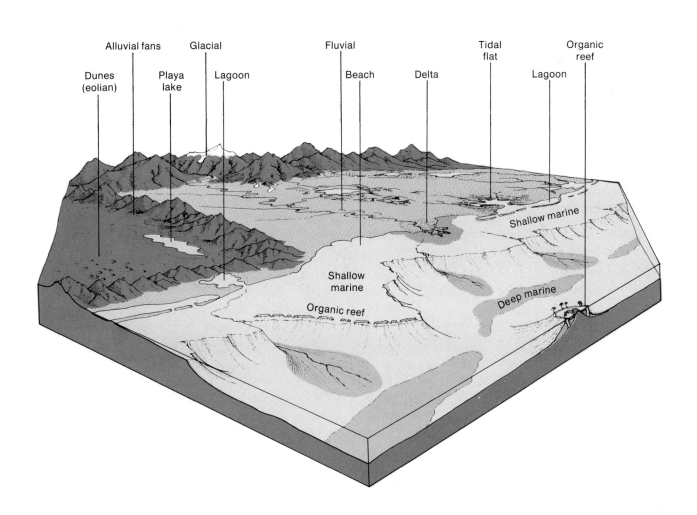

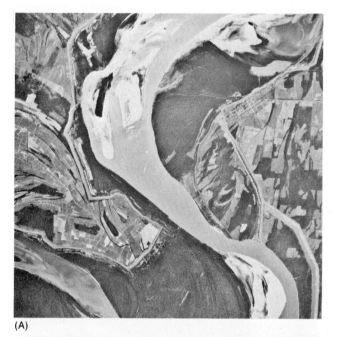

(A)

(A)

(B)

(B)

Figure 6.14 *The fluvial environment (see also plate 14). (A) Point-bar deposits in a modern river. (B) An ancient stream channel in Tertiary sediments in central Utah. The great rivers of the world act as the major channels by which erosional debris is transported from the continents to the ocean. However, before reaching the ocean, most rivers meander across flat alluvial plains and deposit a considerable amount of sediment. Within this environment, sedimentation occurs in the stream channels, in the bars, and on the flood plains. Perhaps the most significant type of sedimentation occurs on the bars on the inside of a meander bend (see figure 11.23). Stream deposits are characterized by channels of sand or gravel cut into horizontal layers of silt and mud.*

Figure 6.15 *The alluvial fan environment. (A) Modern alluvial fans in Death Valley, California. (B) Ancient alluvial fan deposits in central Utah. In many arid regions of the world, thick deposits of sedimentary rock accumulate in a series of alluvial fans at the bases of mountain ranges. Deposition occurs because, in arid regions, streams do not have enough water to transport their sediment load over the flat surface of the basin. Flash floods are an important process in this environment, and torrents from cloudbursts sweep away all the loose debris and deposit it out on the basin floor. The sediment in an alluvial fan characteristically is coarse grained, with conglomerate as the most abundant rock type. In the central part of the basin, fine silt and mud can accumulate in temporary lakes and commonly are associated with the coarser fan deposits.*

(A)

(B)

(A)

(B)

Figure 6.16 *The eolian (wind) environment (see plate 22). (A) Modern sand dunes in the Sahara Desert. (B) Ancient dune deposits in Zion National Park, Utah. Wind is a very effective sorting agent. Silt and dust are lifted high into the air and can be transported thousands of kilometers before being deposited as a blanket of loess. Sand is winnowed out and transported close to the surface and eventually accumulates in dunes, but gravel cannot be moved effectively by the wind. The sand is blown up and over the dunes and accumulates in the wind shadows of the steep dune faces. This forms large-scale cross-strata that dip in a downwind direction (see figure 16.10). Ancient dune deposits consist of well-sorted, well-rounded sand grains deposited in large-scale cross-strata. The most significant ancient wind deposits are sandstones that accumulated in large dune fields comparable to the present Sahara, Arabian, and Australian Deserts. The sandstones form widespread deposits of clean sand that preserve to an unusual degree the large-scale crossbedding developed by migrating dunes.*

Figure 6.17 *The glacial environment (see also plate 17). (A) The margins of the Barnes Ice Cap, Canada. (B) Glacial sediments deposited during the last ice age. A glacier transports large boulders, gravel, sand, and silt suspended all together in the ice and deposits them near the margins of the glacier as the ice melts. The resulting sediment is unsorted and unstratified, with angular individual particles. Typically, the deposits of continental glaciers are widespread sheets of unsorted debris, which can rest on the polished and striated floor of the underlying rock. In many glacial deposits, fine-grained particles dominate, but larger boulders and pebbles are typically present. Streams from the meltwaters of glaciers rework the unsorted glacial debris and redeposit it as stratified, sorted stream deposits beyond the glaciers. Thus, the unsorted glacial deposits commonly are directly associated with well-sorted stream deposits from the meltwaters.*

(A)

(A)

(B)

(B)

Figure 6.18 *The delta environment (see also plate 16). (A) A small delta formed in a lake in Switzerland. (B) Ancient delta deposits in Tertiary rocks of the Colorado Plateau. One of the most significant environments of sedimentation occurs where the major rivers of the world enter the ocean and deposit most of their sediment in large marine deltas. The delta environment can be very large, covering areas of more than 36,000 km²; commonly, it is very complex and involves many distinct subenvironments, such as beaches, bars, lagoons, swamps, stream channels, and lakes. Because deltas are large features and include a number of both marine and nonmarine subenvironments, a large variety of sediment types accumulates in them. Sand, silt, and mud dominate. Therefore, a deltaic deposit is recognized only after considerable study of the sizes and shapes of the various rock bodies and their relationships to each other. Within a delta, both marine and nonmarine fossils may be preserved.*

Figure 6.19 *The beach environment (see also plate 20). (A) A modern gravel beach along the southern coast of France. (B) Ancient beach deposits of Cambrian age in northern Michigan. Much sediment accumulates in the zone where the land meets the ocean. Within this zone, a variety of subenvironments occur as beaches, bars, spits, lagoons, and tidal flats, each with its own characteristic sediment. Where wave action is strong, mud is winnowed out and only sand or gravel accumulates as beaches or bars. Beach gravels accumulate along shorelines where high wave energy is expended along the shore. The gravels are well sorted, well rounded, and commonly stratified in low, dipping cross-strata. Ancient gravel beaches are relatively thin and widespread and commonly are associated with clean, well-sorted sand deposited offshore.*

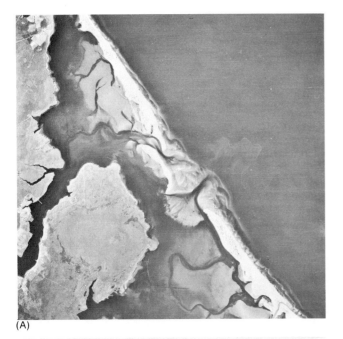

(A)

(B)

(A)

(B)

Figure 6.20 *The lagoon environment (see also plate 20). (A) A lagoon along the central Altantic coast of the United States. (B) Lagoonal deposits of Cretaceous age in central Utah. Offshore bars and reefs commonly seal off part of the coast and develop a lagoon. The lagoon is protected from the high energy of waves, so the water is typically calm and quiet. There, fine-grained sediment, rich in organic matter, accumulates as black mud. Eventually, the lagoon becomes filled with sediment and evolves into a swamp, where the vegetation may provide enough organic matter to form a coal deposit. The rise and fall of sea level will shift the position of the barrier bar so that the organic-rich mud or coal formed in the lagoon or swamp will be interbedded with sand deposited in the barrier island.*

Figure 6.21 *The tidal flat environment. (A) A modern tidal flat in the Gulf of California. (B) Tidal flat deposits of Triassic age in southern Utah. The tidal flat environment is unique in that, alternately, it is covered with a sheet of shallow water and then exposed to the air. Tidal currents are not strong, and they transport fine silt and sand and typically develop ripple marks over a broad area. Mud cracks commonly form during low tide and subsequently are covered and preserved. Ancient tidal flat deposits thus are characterized by accumulations of silt and mud in horizontal layers with an abundance of ripple marks and mud cracks.*

(A)

(A)

(B)

(B)

Figure 6.22 *The organic reef environment (see also plate 9). (A) A modern organic reef in the Bahama Islands. (B) Ancient reefs of Devonian age in Australia. An organic reef is a solid structure of calcium carbonate constructed of shells and secretions of marine organisms. The framework of most reefs, consisting of a mass of colonial corals, forms a wall that slopes steeply seaward. Wave action continually breaks up part of the seaward face, and the blocks and fragments of the reefs accumulate as debris on the seaward slope. Behind the reef, toward the shore or toward the interior of an atoll, a lagoon forms in which lime mud and evaporite salts are deposited. Gradual subsidence of the seafloor permits continuous upward growth of reef material to a thickness of as much as 1210 m. Because of their limited ecological tolerance (they require warm, shallow areas), fossil reefs are excellent indicators of ancient environments. Fossil reefs commonly are found in shallow-marine limestones.*

Figure 6.23 *The shallow-marine environment (see also plates 8 and 20). (A) A modern shallow-marine environment in the Bahamas. (B) Ancient shallow-marine sediments of Pennsylvania age in eastern Kansas. Shallow seas border most of the land areas of the world and extend to the interior of the continent in areas such as Hudson Bay, the Baltic Sea, and the Gulf of Carpentaria (north of Australia). The characteristics of the sediment deposited in a shallow-sea environment depend to a considerable degree upon the supply of sediment from the land and the local conditions of climate, wave energy, circulation of water, and temperature. When there is a large supply of land-derived sediment, sand and mud accumulate to form, ultimately, sandstone and shale, respectively. When sediment from the land is not abundant, limestone generally is precipitated or deposited by biologic means. Ancient shallow-marine deposits are characterized by widespread, interbedded layers of sandstone, shale, and limestone.*

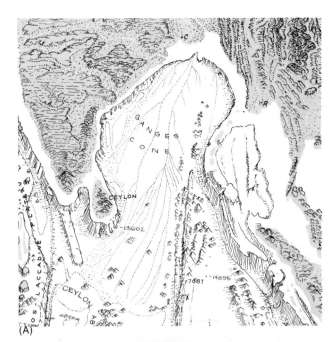

(A)

(B)

Figure 6.24 *The deep-marine environment. (A) A sketch of modern deep-sea fans off the coast of India. (B) Ancient deep-marine deposits near the southern coast of France. The deep-ocean floor adjacent to the continents is, for the most part, extremely smooth because of the deposition of sediment brought down the continental slope by turbidity currents to form deep-sea fans and abyssal plains adjacent to the continental margins. These deposits are characterized by a sequence of graded beds with each layer extending over large areas and, thus, are easily distinguishable from sediment deposited in most other environments. Sediment also accumulates on the floor of the open ocean far from the continents. This material consists of fine particles of mud that are carried in suspension and gradually sink. The most abundant sediment is a fine-grained, brown or red clay. Some sediment in the open ocean apparently crystallized directly from seawater. In large parts of the South Pacific, these deposits form the bulk of the sediment.*

Two main types of sedimentary rocks are recognized: (1) clastic rocks, consisting of rock and mineral fragments (clay, silt, sand, and gravel), and (2) chemical and/or organic rocks.

Important structures in sedimentary rocks include strata, crossbedding, graded bedding, ripple marks, and mud cracks.

The major processes involved in the genesis of sedimentary rocks are (1) weathering, (2) transportation, (3) deposition, and (4) compaction and cementation. During weathering, transportation, and deposition, the sedimentary material is differentiated; that is, material is sorted and concentrated according to grain size and composition.

The sedimentary environment is the place where sediment is deposited and changed under the physical, chemical, and biologic conditions that exist there. The major sedimentary environments include (1) fluvial, (2) alluvial fan, (3) eolian, (4) glacial, (5) delta, (6) shoreline, (7) shallow marine, (8) organic reef, and (9) deep marine. Each major environment is characterized by certain physical, chemical, and biologic conditions and, therefore, develops distinctive rock sequences and fossil assemblages.

Additional Readings

Dunbar, C. O., and J. Rogers. 1963. Principles of Stratigraphy. New York: John Wiley and Sons.

Kuenen, P. H. 1960. "Sand." Sci. Amer. 202(4). (Reprint no. 803.) San Francisco: W. H. Freeman and Company.

Laporte, L. F. 1968. Ancient Environments. Englewood Cliffs, N.J.: Prentice-Hall.

Pettijohn, F. J., and P. E. Potter. 1964. Atlas and Glossary of Primary Sedimentary Structures. New York: Springer-Verlag.

Selly, R. C. 1976. An Introduction to Sedimentology. New York: Academic Press.

Summary

Sedimentary rocks originate from fragments of other rocks and minerals and from chemical precipitation and organic matter. Typically, they occur in layers, or strata, and cover large parts of the continents. They preserve a record of the erosional history of the earth and a history of life on the planet extending back at least 2 or 3 billion years.

Metamorphic Rocks

Most of the rocks exposed in the vast areas of the continental shields and in the cores of mountain belts show evidence that their original texture and composition have been changed. Many have been deformed plastically; this is indicated by contorted parallel bands of minerals resembling swirling layers in a marble cake. Others have recrystallized and developed larger mineral grains, and, in many rocks, the constituent minerals have a strong fabric with a specific orientation. The mineral assemblages in these rocks also are distinctive and are characterized by mineral types that form under high temperature and pressure.

Geologists have interpreted these and other lines of evidence to mean that an original rock was recrystallized in a solid state to produce a new rock type with a distinctive texture and fabric and, in some cases, a new mineral composition. These rocks are called metamorphic rocks. They are important because of their wide distribution in the continental crust and the clear evidence they provide about the dynamics of the earth.

This chapter will consider the nature and the distribution of metamorphic rocks and the processes by which they originate.

Major Concepts

1. Metamorphic rocks result from changes in temperature and pressure and the chemistry of pore fluids. These changes develop new minerals, new textures, and new structures within the rock body.
2. During metamorphism, new crystals grow in the direction of least stress, producing a planar element in the rock called foliation. The three main types of foliation are (a) slaty cleavage, (b) schistosity, and (c) gneissic layering.
3. Rocks with only one mineral (such as limestone) do not develop a strong foliation but instead develop a granular texture with large crystals.
4. The major types of metamorphic rocks are slate, schist, gneiss, quartzite, marble, amphibolite, metaconglomerate, and hornfels.
5. Regional metamorphism develops in the roots of mountain belts as a result of plate collision. Contact metamorphism is a local phenomenon associated with changes close to the contacts of igneous intrusions.

The Nature and Distribution of Metamorphic Rocks

Statement

The earth is a dynamic planet in which physical and chemical conditions are changing continually. As a result, rock bodies can be placed in environments where temperatures and pressures are significantly different from those in which they were formed originally. For example, sedimentary rocks that accumulate at the surface and in the presence of water and atmospheric gases subsequently can be buried deeply, intruded by magma, or compressed by converging tectonic plates. These changes cause modification of the minerals, texture, and structure

of the rocks. New crystals grow that are at equilibrium with high temperature and pressure. Some recrystallization results in new minerals with closer atomic packing. As new crystals grow, new rock textures develop, commonly with preferred grain orientation. Most original structures (such as stratification, fossils, and vesicles) are

Figure 7.1   *Aerial photograph of a portion of the Canadian Shield, showing the nature of metamorphic rocks. The rocks shown here have been compressed and deformed to such an extent that many original features have been obliterated. This occurred at great depths. The area then was eroded to expose the complex rock sequence. Note that, in addition to the tight folding, the rocks are broken by fractures and are intruded by granitic rocks (light tones).*

modified greatly or obliterated completely and new structural features develop in their place. Rocks that have been altered in this way are called **metamorphic rocks** (*meta*, change; *morph*, form).

Most metamorphic rocks are believed to have been formed in the roots of ancient mountain belts as the result of colliding tectonic plates. There, the crust is subjected to tremendous pressure and high temperature capable of modifying rocks along the entire margin of a continent (regional metamorphism). Metamorphism also occurs in local areas from local conditions such as the contact along an igneous intrusion (contact metamorphism) and along the fractures where blocks of the crust slide past each other.

Metamorphic rocks are especially significant in that they constitute a large part of the continental crust below the veneer of sedimentary rocks and indicate that the continents have been mobile and dynamic throughout geologic time, being subjected repeatedly to great horizontal stresses and accompanying changes in pressure and temperature.

Discussion

Many people have some knowledge of various igneous and sedimentary rocks but understand only vaguely the nature of metamorphic rocks. Perhaps the best way to become acquainted with this group of rocks and to appreciate their significance is to study carefully *figures 7.1* through *7.4* (see also *plates 4* and *28*). The vertical aerial photo of a portion of the Canadian Shield (*figure 7.1*) shows that the rocks are deformed extensively into tight folds and have been twisted and compressed. Originally, these were sedimentary and volcanic layers deposited in a horizontal position, but they have been deformed so intensely that it is difficult to determine

Figure 7.3 *Outcrop of metamorphosed conglomerate along the northern shore of Lake Superior, which shows how pressure has stretched and deformed the original spherical boulders.*

the original base or top of the rock sequence. Light-colored granite batholiths and dikes have been injected into the metamorphic series, and the entire mass has been broken, fractured, and displaced by numerous faults.

Figure 7.2 shows a more detailed view of metamorphic rocks. The alteration and deformation of the rock is clearly evident in the contorted layers of dark minerals, and the pattern of the distortions shows that the rock has been subjected to compressive forces while in a plastic or semiplastic state.

The degree of deformation during metamorphism perhaps is easier to comprehend by comparing the shape of pebbles in conglomerate with those in a metaconglomerate. As is evident from the photograph (*figure 7.3*), the original spherical pebbles in the conglomerate have been stretched into long, elliptical blades (the long axis is as much as 30 times their original diameter), all oriented in the same direction.

Even on a microscopic scale, distortion and deformation of the individual grains can be seen. A definite preferred orientation of the grains in *figure 7.4* shows that they either recrystallized under stress or responded to differential pressure and flowed as a plastic.

It is important to note that the typical texture of metamorphic rocks does not show a sequence of formation of the individual minerals like that evident in igneous rocks; all grains in metamorphic rocks apparently recrystallized at roughly the same time and had to compete for space in the already solid rock body. As a result, the new minerals grew in the direction of least stress so that most metamorphic rocks have a layered or planar structure resulting from recrystallization.

Metamorphic rocks constitute a large part of the continental crust. Extensive exposures of metamorphic rocks (such as those shown in *figure 7.1* and *plates 4* and *28*)

Figure 7.2 *Outcrop of highly deformed metamorphic rocks in British Columbia, Canada.*

Figure 7.4 *Microscopic view of a metamorphic rock, showing the degree to which originally spherical or equidimensional grains have been stretched and deformed.*

are found in the vast shield areas of the continents, where the cover of sedimentary rocks has been removed. Deep drilling in the stable platform also indicates that the bulk of the continental crust immediately beneath the sedimentary cover also is made up of metamorphic rock. These facts permit us to estimate that metamorphic rocks, together with associated igneous intrusions, make up approximately 85% of the continental crust, at least to a depth of 20 km. The nature of the earth's mantle is not known from direct observation, but it is difficult to imagine how the mantle, or at least parts of it, could be other than metamorphic rock. In addition to the vast exposures of metamorphic rocks in the shields and beneath the stable platforms of the continents, metamorphic rocks also are found in the cores of eroded mountain ranges.

The widespread distribution of metamorphic rocks in the continental crust, especially in the older rocks, is highly significant, for it indicates that the earth's crust has been mobile and has been subjected to repeated deformation throughout geologic time. The rocks in the continental crust have been compressed almost continually since the earth was formed, a fact that suggests that the tectonic system has operated throughout most of the earth's history.

Metamorphic Textures

Statement

As was emphasized already several times in this chapter, rocks react to higher pressure and temperature by the continuous adjustment and recrystallization of individual grains. New crystals grow in the direction of least stress, so there is a tendency to develop a strong preferred orientation of the mineral grains. This imparts a distinctly planar element to the rock—**foliation.** Three major types of foliation are recognized.

1. **Slaty cleavage**—the tendency for a fine-grained rock to split along parallel, closely spaced surfaces.
2. **Schistosity**—a series of surfaces produced by the parallel arrangement of platy minerals such as mica, chlorite, and talc.
3. **Gneissic layering**—alternate layers of differing mineralogic composition.

Large bodies of limestone and sandstone, which are composed predominantly of one mineral, do not develop good foliation when they are metamorphosed but instead typically develop a granular texture of large crystals, with little or no preferred orientation of the mineral grains.

Discussion

Slaty Cleavage. Slaty cleavage is a type of foliation in which the planar element of the rock is a series of surfaces along which a rock will split (*figure 7.5*). It is produced by the parallel alignment of minute flakes of platy minerals (mica, chlorite, talc). The mineral grains are too small to be obvious without the aid of a microscope, but the parallel arrangement of the perfect cleavage in these small grains develops innumerable parallel planes of weakness that cause the rock to part or split into smooth slabs.

Slaty cleavage is, of course, the type of foliation formed in slates and should not be confused with bedding planes of the parent rock. It is completely independent of bedding and more commonly cuts across the original planes of stratification. Relic bedding can be rather obscure in slates, but it often is expressed by textural changes resulting from interbedded, thin layers of sand or silt (*figure 7.5*). Excellent foliation can be developed in the shale part of the sedimentary sequence, where clay minerals are abundant and are easily altered to mica, but, in thick layers of quartz sandstone, slaty cleavage planes generally will be poorly developed.

Schistosity. When platy mineral grains (such as mica and chlorite) grow large enough to be identified with the unaided eye, they produce an obvious planar structure because of their overlapping parallel arrangement, which is apparent even before the rock is broken. The term *schistosity* comes from the Greek *schistos*, meaning divided or divisible, and, as the name implies, rocks with this type of foliation readily break along the cleavage planes of the parallel platy minerals.

Gneissic Layering. Gneissic layering is a type of foliation consisting of alternating thin layers of light and dark minerals. In contrast to slaty cleavage and schistosity, the planar element in gneissic layering is not a property of splitting or parting of the rock but the result of layers of different mineral compositions. Feldspar is commonly

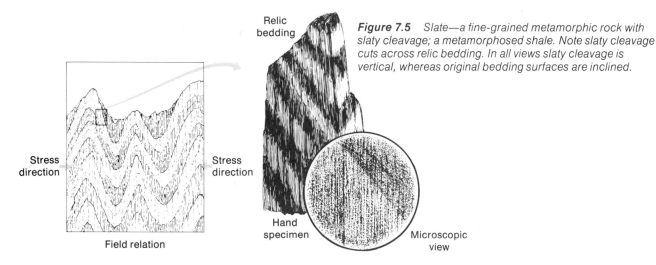

Relic bedding

Stress direction

Stress direction

Field relation

Hand specimen

Microscopic view

Figure 7.5 *Slate—a fine-grained metamorphic rock with slaty cleavage; a metamorphosed shale. Note slaty cleavage cuts across relic bedding. In all views slaty cleavage is vertical, whereas original bedding surfaces are inclined.*

abundant and, together with quartz, constitutes a light-colored (white or pink) layer of interlocking crystals. Mica, amphibole, and other iron-rich minerals form dark layers. Gneissic layering is usually highly contorted and can fracture across the layers or the planes of foliation as easily as along them (see *figure 7.2*).

The Relationship between Foliation and Larger Structures. Foliation is a response to the regional stress that deformed the original rock into folds and other structures as well as having caused recrystallization to occur in the rock. The orientation of foliation, therefore, is closely related to the large folds and structural patterns of the rock seen in the field. This relationship commonly extends from the largest folds to the structures seen only under a microscope. This is clearly illustrated in the cleavage in slate, which generally is oriented parallel to the axial planes of the folds (which can be many kilometers apart). When a slice of the rock is viewed under a microscope, small wrinkles and folds have the same orientation as the larger structures mapped in the field.

Nonfoliated Textures (Granular Texture). Some rocks are composed predominantly of one mineral that crystallizes in an equidimensional form and does not grow in any preferred direction. When these rocks are metamorphosed, the small crystals grow larger and develop a granular, interlocking texture without foliation. The best examples are limestone, which changes to marble, and sandstone, which changes to quartzite. The texture is described as granular or, simply, nonfoliated.

Kinds of Metamorphic Rocks

Statement

A simple classification of metamorphic rocks is based on texture and composition. Following this scheme, two major groups of metamorphic rocks are recognized.
1. Those that are foliated and possess a definite planar structure.
2. Those that lack foliation and have a granular texture.

The foliated rocks are further subdivided on the basis of the type of foliation. The major rock names can be qualified by adjectives describing their chemical and mineralogical composition.

Discussion

Foliated Rocks. The following rocks are common types of foliated rocks.

Slate. Slate is a very fine grained metamorphic rock generally produced by the low-grade metamorphism of shale and other fine-grained rocks. It is characterized by excellent foliation (slaty cleavage), which causes the rock to split into thin sheets (*figure 7.5*). The foliation results from the growth of platy minerals (mica and chlorite) perpendicular to the applied stresses. Individual mineral grains are so small that they rarely can be distinguished without a microscope, and the separation of mineral types into individual layers is not present. It should be emphasized that the foliation in slates (slaty cleavage) is the result of mineral growth under directed stress and is not original stratification.

Schist. Schist is a medium- to coarse-grained, foliated rock. Foliation results from the parallel arrangement of large grains of platy minerals such as mica, chlorite, talc, and hematite. The mineral grains are readily visible and are oriented in an overlapping, parallel arrangement. Thus, the foliation differs from slaty cleavage largely in the size of the crystals. In addition to the platy minerals, significant quantities of quartz, feldspar, garnet, amphibole, and other minerals may occur, providing a basis for subdividing schists into many varieties based on composition (chlorite schist, mica schist, amphibole schist, et cetera).

Schists result from a higher intensity of metamorphism than slates. They have a variety of parent rock types (including basalt, granite, sandstone, and tuff), although shale is the most common. Schists are one of the most abundant metamorphic rock types.

Gneiss. Gneiss is a coarse-grained, granular metamorphic rock in which the foliation results from layers of alternating light and dark minerals (*figure 7.2*). The composition of most gneisses is similar to that of granite; the major minerals are quartz, K-feldspar, and ferromagnesian minerals. Feldspar is especially abundant. The foliation in gneisses generally is less planar than in schists and slates, and the layering often shows tight folds and contortions. Gneiss is formed during high-intensity, regional metamorphism.

Nonfoliated Rocks. Rocks such as sandstone and limestone are composed predominantly of one mineral, which crystallizes in an equidimensional form. Metamorphism of these rocks does not result in a strong foliation, although mica grains scattered through the rock can assume a parallel orientation. The minerals can be flattened, stretched, and elongated and can show a preferred orientation, but the mass of rock does not develop a strong foliation. The resulting texture is best described as granular or, simply, nonfoliated.

Quartzite. Quartzite is a metamorphosed, quartz-rich sandstone. It is nonfoliated because quartz grains, the principal constituents, do not form platy crystals. The individual grains commonly are deformed and fused into a very tight mass, so the rock breaks across the grains as easily as it breaks around them. Pure quartzite is white or light colored, but iron oxide and other minerals may impart various tones of red, brown, green, and other colors.

Marble. Marble is recrystallized limestone or dolomite. Calcite, the major constituent, is equidimensional; therefore, the rock is nonfoliated. Commonly, the crystals are large and compactly interlocked, forming a dense rock. Many marbles show bands or streaks resulting from organic matter or other impurities in the original sedimentary rock.

Amphibolite. Amphibolite is a coarse-grained metamorphic rock composed chiefly of amphibole and plagioclase. Mica, quartz, garnet, and epidote also may be present. Many amphibolites are foliated, but some are not. They result from metamorphism of basalt, gabbro, and other rocks rich in iron and magnesium.

Metaconglomerate. Metaconglomerate is not an abundant metamorphic rock, but it is important locally and illustrates the degree to which a rock can be deformed in the solid state. Under directional stress, the individual pebbles are stretched into a mass that shows a very distinct linear fabric.

Hornfels. A hornfels is a very fine grained, nonfoliated metamorphic rock that is very hard and dense. The grains usually are microscopic in size and are welded into a regular mosaic. Platy minerals, such as mica, have a random orientation and grains of high-temperature metamorphism are present. Hornfelses usually are dark colored and may resemble basalt, dark chert, or dark, fine-grained limestone. They result from metamorphism around igneous intrusions in which partial or complete recrystallization occurs. The parent rock usually is shale, although lava, schists, and other rock types may be baked into a hornfels.

Source Material for Metamorphic Rocks. The origin of metamorphic rocks is a complicated and difficult problem, because a single source rock can be changed into a variety of metamorphic rocks, depending on the intensity or the degree of metamorphism. For example, shale can be changed to slate, schist, or gneiss, and gneiss can form from granite, rhyolite, or shale. A generalized chart (*figure 7.6*), showing the origin of common metamorphic rocks, will help you relate sources of rocks and metamorphic conditions to rock types.

Metamorphic Processes

Statement

Metamorphism is a series of changes in the texture and composition of a rock that result from environmental conditions different from those under which the rock was formed originally. The changes result from an attempt to reestablish equilibrium with the new conditions under which the rock occurs. The principal agents of metamorphism are changes in:
1. Temperature.
2. Pressure.
3. Chemistry.
These agents act in combination, and it is often difficult to distinguish their individual effects in a given rock body.

The changes resulting from metamorphism occur in the solid rock without the constituent minerals passing through a liquid phase. Water generally is present in most rocks as a pore fluid, and it is an important catalyst even when it occurs as only a thin film around the grain boundaries. The pore fluid helps atoms and ions to be exchanged and rearranged into a new crystal structure and permits the development of new textural features, but all this occurs while the rock is in a solid state.

Discussion

Temperature Changes. Heat is one of the most important factors in metamorphism. As the temperature increases, minerals begin to change from the solid to

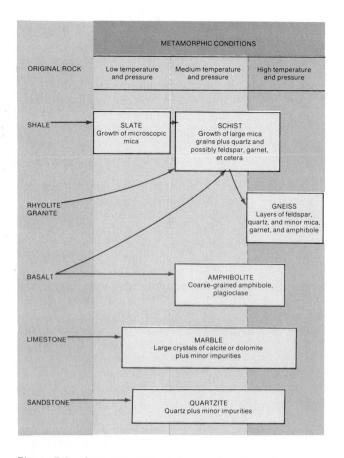

Figure 7.6 *Generalized chart showing the origin of some of the common metamorphic rocks.*

the liquid phase and the amount of pore fluid in the rocks increases. At temperatures below 200 °C, only a small amount of fluid phase exists and most minerals remain unchanged for millions of years. With a rise in temperature, the amount of pore fluid in the rocks increases considerably. Chemical reactions become more vigorous, and new mineral assemblages begin to appear. At temperatures greater than 700 °C, more fusible components of the rock become fluid and there is considerable evidence that a high-temperature fluid phase approaching the magma stage exists in the rock body. In extreme cases, layers of solid material mixed with layers of fluid probably exist and give rise to a rock that is transitional between igneous and metamorphic.

Since different minerals are in equilibrium at successively higher temperatures, the mineral composition of the rock provides a key to the temperatures at which the rock formed. In the field, it has been found that a zonal arrangement of mineral assemblages exists that expresses variations in temperature on a regional scale. This zonation is particularly obvious in metamorphic zones around igneous intrusions.

Pressure Changes. High pressure within the earth's crust causes various changes in the physical properties of many rocks. As was mentioned previously, pressure tends to reduce the space occupied by the mineral components and can produce new minerals with closer atomic packing. An increase in pressure can be brought about by deep burial, but the great bulk of metamorphic rocks results from stress or directed pressure at converging plate margins.

Perhaps the most obvious feature indicating directed pressure is the distinct orientation of minerals in most metamorphic rocks. Uniform crystal orientation of micas and chlorite is very characteristic.

Chemical Changes. Most metamorphic changes take place without chemical constituents being added or removed from the bulk rock. The elements that are present in the minerals simply recombine to form new minerals that are more stable under the new conditions of different temperature and pressure. Low-temperature minerals, which are partly melted, provide the fluid phase in metamorphic rocks and play an important role in the process of metamorphism. Whenever a single atom breaks the bonds that hold it to the crystal structure of a mineral and moves from that mineral to some other place, it is essentially in a fluid state. Although the bulk of the rock remains solid during metamorphism, the rising temperature and pressure cause many atoms to break loose from their crystal lattice and migrate through fluids in pore spaces and along margins of grains. In this manner, there is a constant interchange of atoms. Original crystals break down and new crystal structures, which are stable under the new conditions of temperature and pressure, develop. The metamorphic rock may not have a great amount of pore fluid, but the small amount that does exist provides the medium of transport for material in solution so that it can be diffused through the rock and rearranged into new minerals.

Some types of metamorphism take place in an open chemical system in which elements are added to, or lost from, the existing rock. This process produces a change in the bulk composition of the rock and generally is connected with magmatic intrusions in which new material from the magma passes through the pore fluid of the rock. Volatile solutions from cooling magma can saturate the wall rock and enter into chemical reactions with the surrounding minerals. Some elements can be lost from the preexisting minerals (such as water driven from hydrated minerals).

Metamorphic Rocks and Plate Tectonics

Statement

We can never observe metamorphic processes in action because they occur deep within the crust, but we can study in the laboratory how a mineral reacts to high temperatures and pressures that simulate, to some extent, the metamorphic process. These laboratory studies, together with field observations and studies of texture and composition, provide the rationale for interpreting metamorphic rocks.

In previous sections, it has been shown that contact metamorphism, which is of local extent, occurs around igneous intrusions and that deep burial can alter rock bodies to some extent; but the main concern here is with regional metamorphism, which forms a large part of the continental crust. Two important generalizations can be made about regional metamorphic rocks.

1. The foliation is generally vertical or nearly vertical, which implies strong horizontal stress.
2. Regional metamorphic rocks occur in linear belts and are closely associated with granitic igneous intrusions.

Regional metamorphic rocks are believed to result from horizontal stress associated with mountain-building.

Discussion

Figure 7.7 is a graphic model that summarizes some of the major ideas concerning the origin of metamorphic rocks. According to the theory of plate tectonics, metamorphism occurs in the deep roots of folded mountain belts that form at converging plate margins. The original material is largely sediment derived from erosion of a continent, sediment and volcanic material derived from a volcanic arc, and deep-marine sediments and basalt from the oceanic crust. This material is squeezed at con-

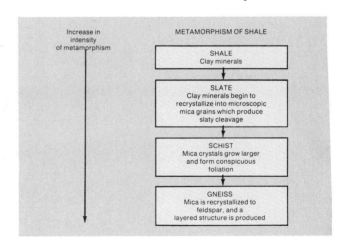

Figure 7.7 *The metamorphism of shale.*

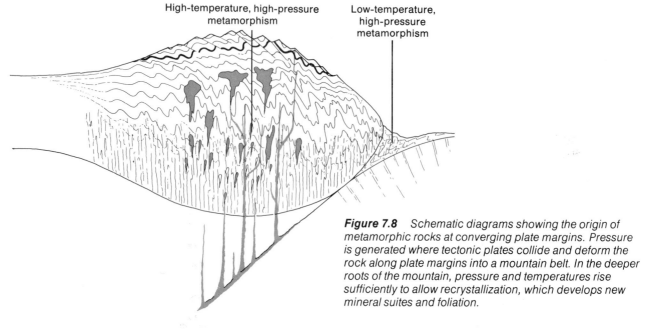

High-temperature, high-pressure metamorphism

Low-temperature, high-pressure metamorphism

Figure 7.8 *Schematic diagrams showing the origin of metamorphic rocks at converging plate margins. Pressure is generated where tectonic plates collide and deform the rock along plate margins into a mountain belt. In the deeper roots of the mountain, pressure and temperatures rise sufficiently to allow recrystallization, which develops new mineral suites and foliation.*

verging plate margins as though it were in a vise. Recrystallization would tend to produce high-angle to vertical foliation in a linear belt parallel to the margins of the converging plates. Metamorphism would be most intense in the deep mountain roots where partial melting may occur to contribute material to the rising magma generated in the subduction zone. Batholiths and dikes, thus, would be associated intimately with the zones of more intense metamorphism. Different groups of metamorphic rocks would be generated from different source materials (sand shale and limestones along continental margins, volcanic sediments and flows along island arcs, and a mixture of deep-marine sediments and oceanic basalt from the oceanic crust).

After the stresses from the converging plates are spent, erosion of the mountain belt would occur and would cause the mountain root to rise upward because of **isostasy.** Ultimately, the deep root and its complex of metamorphic rocks would be exposed to the surface to form a new segment of the continental shield. The entire process would take several hundred million years. Repetition of this process would cause the continents to grow larger with each mountain-building event. Thus, the belts of metamorphic rocks in the shields are considered to be the record of ancient continental collision (see *figure 7.1* and *plates 4* and *28*). More will be said about this topic in chapter 20.

Summary

Metamorphic rocks result from changes in temperature and pressure and the chemistry of pore fluids within a rock body. These changes develop new minerals, new textures, and new structures so that the diagnostic features of the original sedimentary and igneous rocks are greatly modified or completely obliterated. Metamorphic rocks are especially significant in that they constitute a large part of the continental crust (the shield) below the veneer of sedimentary rocks (stable platform) and indi-

cate that the continents have been mobile and dynamic throughout most of the geologic time. Therefore, they constitute the best record of ancient plate movement.

During metamorphism, new minerals grow in directions of least stress and produce in the rocks a planar element called foliation. Three major types of foliation are recognized: (1) slaty cleavage, (2) schistosity, and (3) gneissic layering. Rocks with only one mineral (such as limestone and sandstone) do not develop good foliation but instead develop a granular texture with large crystals.

Metamorphic rocks are classified on the basis of texture and composition. Two major groups are recognized: (1) foliated and (2) nonfoliated. The major types of foliated rocks are slate, schist, and gneiss. The major types of nonfoliated rocks are quartzite, marble, amphibolite, and metaconglomerate.

Heat is one of the most important factors in metamorphism. As temperatures increase, minerals begin to change from solid to liquid. The liquid in the pore spaces then recrystallizes to form new minerals. Directed pressures reduce pore spaces and can produce new minerals with closer atomic packing. Changes in chemistry can result from the partial melting of low-temperature minerals, the migration of the fluid, and subsequent recrystallization.

An increase in heat and pressure can result from deep burial or from igneous intrusions, but most metamorphic rocks of regional extent develop in the deep roots of folded mountain belts as a result of plate collision. The metamorphic mountain roots are welded onto the continent and become a new segment of the continental shield.

Additional Readings

Ernst, W. G. 1969. Earth Materials. Englewood Cliffs, N.J.: Prentice-Hall.

Hyndman, D. W. 1972. Petrology of Igneous and Metamorphic Rocks. New York: McGraw-Hill.

Winkler, H. G. F. 1974. Petrogenesis of Metamorphic Rocks. 3rd ed. New York: Springer-Verlag.

Geologic Time

Some sciences deal with incredibly large numbers, others with great distances, still others are concerned with infinitely small particles. In every field of science, the student must expand his concept of reality, a sometimes difficult, but very rewarding, adjustment to make.

In geology, the student is introduced to new concepts about the duration of time. Because life is so short, man conceives a long time to be 20 or 50 years. A hundred years in most frames of reference is a very long time. Yet, in the study of the earth and the processes that operate on it, you must attempt to comprehend time spans of a million years, 100 million years, and even several billion years.

How do scientists measure such long periods of time? Nature contains many types of time-measuring devices or clocks. The earth itself acts like a clock as it rotates on its axis once every 24 hours. Rocks are records of time and, from their interrelationships, the events of the earth's history can be arranged in their proper chronologic order. Rocks also contain radioactive clocks that permit us to measure with remarkable accuracy the finite number of years that have passed since the minerals that formed the rocks crystallized. Fossils embedded in the rock constitute a separate organic clock from which we can identify synchronous events in the earth's history.

Geologists use all of these methods to date geologic events and with them have devised a geologic calendar, which makes it possible to organize and correlate long periods of geologic time into workable units, thus permitting a systematic study of the earth's history.

Major Concepts

1. The interpretation of past events in the earth's history is based upon the principle that the laws of nature do not change with time.
2. Relative dating (determining the chronologic order of a sequence of events) can be made by applying the principles of (a) superposition, (b) faunal succession, (c) crosscutting relations, (d) inclusion, and (e) succession in landscape development.
3. The standard geologic column was established from studies of the sequences of rocks in Europe and can be used worldwide by correlating rocks on the basis of the fossils they contain.
4. Absolute, or finite, time designates a specific duration in units of hours, days, or years. In geology, long periods of time can be measured by radioactive decay.

Constancy of Natural Laws

Statement

The interpretation of rocks as products and records of past events in the earth's history is based upon one of the few assumptions that make the scientific enterprise possible: the principle of **uniformitarianism**—*that the laws of nature do not change with time.* That is, the chemical and physical laws that operate today are assumed to have operated throughout all time. The physical attraction (gravity) between two bodies operated in the past as it does today. Oxygen and hydrogen that today combine under given conditions to form water did so in the past. In brief, although man's ability to understand and explain how or why physical nature operates has improved and changed over the centuries, the physical laws or processes that scientists study are constant and do not change. All chemical and physical actions and reactions occurring presently have the same causes that operated to produce those actions and reactions 100 years ago or 5 million years ago.

Discussion

The principle of uniformitarianism can best be understood in the historical context in which it originated. In the late sixteenth century, before modern geologic science was established fully, the prevailing concept of the origin and history of the earth throughout much of the world was based upon literal interpretations of the account of the creation given in the Bible. The earth was considered to have been created in 6 days and to be approximately 6000 years old. The creation involving such a short period of time was thought to have involved tremendously violent events surpassing anything experienced in modern times. This type of creation has been referred to as **catastrophism.** The theory was supported by many learned men of the time; foremost among them was Baron Georges Cuvier, a noted French naturalist. Cuvier, an able student of fossils, recognized many extinct forms of life and concluded that each group of fossils was unique to a given sequence of rocks and was the result of a special creation that subsequently was destroyed by a catastrophic event.

This theory was supported generally by theologians until 1785, when it was challenged by James Hutton when he proposed the principle of uniformitarianism—a concept that maintains that the earth has evolved by uniform, gradual processes over an immense span of time. The principle attempts to explain past geologic events in light of natural processes such as erosion by running water, volcanism, and gradual uplift of the earth's crust. Hutton assumed that these processes occurred in the distant past in much the same way as they occur in modern times. The first expression of the concept of uniformitarianism, based on Hutton's observations of the rocks of Great Britain, visualized "no vestige of a beginning—no prospect of an end." Sir Charles Lyell (1797-1875), who used Hutton's uniformitarianism as a basis for his book *Principles of Geology,* accepted Hutton's conclusion. In his writings, Lyell established uniformitarianism as the accepted philosophy for interpret-

ing the geologic and natural history of the earth. Charles Darwin accepted Lyell's uniformitarianism in formulating his theory of the origin of the species and the descent of man. Lyell, however, was perhaps a bit too adamant in not admitting the possibility that the rates at which processes operate could change with time (as more recent studies have indicated).

Modern Views of Uniformitarianism (Naturalism). With the help of modern scientific instruments, geologists have studied much more of the geologic record than did Hutton and Lyell and have observed many subtle details of the rocks, which the earlier scientists could not see. Modern science is making significant advances in understanding the earth, its long history, and how it was formed. By applying discoveries in kinetics, thermodynamics, electromagnetism, chemistry, and related scientific disciplines, we are discovering more specific clues to the nature of the earth's genesis and evolution. Religious thought on this subject has changed also, and today many Christians and Jews believe that the universe was organized (created) over long periods of time.

What then are the modern concepts of uniformitarianism? The assumption on which the idea of uniformitarianism rests is the concept of continuity: natural phenomena (chemical and physical reactions) do not change with time. Our ability to express the relations of natural phenomena may change, but the phenomena do not. As explained earlier, this is not a principle unique to geology; it is basic to the scientific enterprise. Without constancy in natural law, verification of conclusions by repetition of experiments would be impossible. We could not assume that water would freeze at the same temperature tomorrow as it did today, nor could we be sure that oxygen and hydrogen would combine to form water; they might form alcohol or even hydrochloric acid.

However, we must modify the assumption made by Hutton that the geologic past involved only those activities or processes that are observable today and that operate at essentially the same rates as in the past. The earth has evolved throughout its history and has undergone a continuous dissipation of energy; with this dissipation, rates may change. Some processes may have been much more important at times in the past (for example, glaciation and tides), and the rates at which the processes function may have varied with time. Ultimately, the earth will exhaust its energy, and we may anticipate cessation of volcanic activity and earthquakes at some time in the far-distant future.

In summary, the basic assumption endorsed by essentially all geologists today in deciphering the history of the earth is that natural laws do not change with time. The method of the geologist is the inductive (scientific) method (observe, collect data, experiment, make a hypothesis, test the hypothesis, and so on).

The assumption of constancy in natural law is not unique to the interpretations of geologic history, but it constitutes the logic essential in deciphering recorded history as well. We observe only the present and interpret past events on inferences based on present observations. Thus, we conclude that books or other records of history (such as fragments of pottery, cuneiform tablets, flint tools, temples, and pyramids) that were in existence prior to our arrival have all been the works of man despite the fact that postulated past activities have been outside the domain of any possible present-day observations. Having excluded supernaturalism, we draw these conclusions because man is the only known agent capable of producing the effects observed. Similarly, in geology, we conclude that ripple marks in a sandstone formation in the folded Appalachian Mountains were formed by currents or wave action or that coral shells found in limestones exposed in the high Rocky Mountains are indeed the skeletons of corals that lived in a now nonexistent sea.

Many features of rocks serve as records of documents of past events. As discussed in previous chapters, the mineral composition and the texture, as well as the internal structures, of the rock body preserve many clues about how the rock formed.

Concepts of Time

Statement

Everyone is aware of growth and change in the physical and biologic worlds. Were things unchanging and motionless, man would not be aware of time. Time is measured by change. There are many clocks (mechanisms for measuring time), the most fundamental of which utilize some periodic physical phenomenon: the swing of a pendulum, the flow of sand through an hourglass, the revolution of the moon around the earth and the earth around the sun. For most practical and scientific purposes, the earth is the ultimate timepiece, for, as it revolves, distinct changes in the day, the seasons, and the year can be observed and experienced. It is by these changes that man is aware of time. *Efforts to determine the age of the earth are basically efforts to determine how long the earth has been revolving around the sun.*

Discussion

A clock is simply a mechanism for telling time. Generally, people are familiar only with mechanical clocks constructed to measure hours, minutes, and seconds, but there are many kinds of natural clocks that provide measures of time as good as or better than the standard clock on the wall. Natural clocks, however, measure other intervals of time. Vibrations of atoms, for example, provide a means of measuring extremely small intervals of time with great precision. Electromagnetic waves provide another "clock" to measure intervals of time useful to man and his activities. There are also biologic clocks that measure intervals of time between various biologic activities (breathing, heartbeat, hunger, menstruation, and life span or generation). Natural clocks include layers of sediment deposited during specific seasons (layer of sand in the spring and summer, layer of mud in the winter when water freezes), tree rings, and growth marks on corals. For spans of time of a much greater interval, the thickness of sediment in the rock record is a measure of time much like the sand in an hourglass. The most effective clocks for measuring long periods of time, however, are the radioactive clocks, which will be discussed later.

The first serious efforts to estimate the magnitude of geologic time were made in the late nineteenth century. Before Hutton and Lyell, few people even recognized geologic processes or thought about the age of the earth. After Hutton presented his arguments for uniformitarianism and Lyell further developed the concept, much interest was generated in the magnitude of geologic time. Early attempts to estimate the age of the earth were based on (1) salinity of the oceans, (2) thickness of the total sequence of sedimentary rocks, and (3) heat loss from the earth. Each attempt showed evidence of a considerable period of time, but none proved to be accurate.

Relative Dating (to Determine the Order of Events)

Statement

Relative dating is simply determining the chronologic order of a sequence of events. People employ relative dating when they determine that one child is older than another or that some event occurred before another, such as a war or the birth of a famous person. In relative dating, no quantitative or absolute length of time is deduced, only that one event occurred earlier or later than another.

In studying the earth, relative dating is important, because many physical events such as volcanism, canyon cutting, deposition of sediment, and upwarping of the crust can be identified. To establish the relative age of these events is to determine their proper chronologic order. This can be done by applying several principles of remarkable simplicity and universality. The most significant are:

1. The principle of **superposition,** which states that, in a sequence of undeformed sedimentary rocks, the oldest beds are on the bottom and the youngest are on the top.
2. The principle of **faunal succession,** which states that groups of fossil plants and animals occur in the geologic record in a definite and determinable order and that a period of geologic time can be recognized by its respective fossils.
3. The principle of crosscutting relations.
4. The principle of inclusion.
5. The principle of succession in landscape development.

Discussion

The Principle of Superposition and Original Horizontality. The principle of superposition is the most basic guide in determining the relative age of rock bodies. This concept is easily understood, even though a casual observer may have difficulty recognizing rocks as records of events and grasping the magnitude and implications of the rock bodies involved.

In the application of the principle of superposition, two assumptions are made. The first is that the rock layers were originally deposited in an essentially horizontal position; the second, that the rocks have not been deformed to such a degree that the beds are overturned.

The Principle of Faunal Succession. In addition to superposition, the sequence of sedimentary rocks in the earth's crust contains another independent element that can be used to establish the chronologic order of events: the upward succession of fossil assemblages contained in the rocks. Fossils are the remains of ancient organisms, such as bones and shells, or evidence of organisms, such as trails and tracks. Their abundance and diversity are truly amazing. Some rocks (such as coal, chalk, and certain limestones) are composed almost entirely of the remains of former life. Others contain literally millions of specimens. In some areas, fossil shells are so abundant that they have been used as road gravel. Invertebrate marine forms are found most often, but even large vertebrate fossils such as mammals and reptiles are plentiful in many formations. For example, it is estimated that over 50,000 fossil mammoths have been discovered in Siberia, and many more remain covered. A record of ancient life is preserved in the rocks.

The extensively studied fossil record, showing that plants and animals have evolved with time, provides an independent timepiece or a document of the earth's chronology. With it, the relative age of a rock body formed during the last 600 million years can be established independently of superposition (fossils are very rare in rocks older than 600 million years). This was recognized some 150 years ago by William Smith, a British surveyor, even before Darwin developed the theory of organic evolution. Working throughout much of southern England to survey the courses of roads and canals, Smith carefully studied the fresh exposures of rocks in quarries, road cuts, and excavations and collected fossils from the strata. By correlating types of fossils with certain kinds of rock, he developed a practical tool that he could use to predict the location and properties of rocks beneath the surface. In a succession of interbedded sandstone and shale formations, the several shales were very much alike, but the fossils they contained were not. Each shale had its own particular group of fossils.

Soon after Smith announced that the fossil assemblages of England change systematically from the older beds to the younger, other investigations discovered the same thing to be true throughout the world—even in countries separated by oceans.

Today, the principle of faunal succession has been confirmed beyond doubt. It has been used extensively to locate valuable natural resources, such as petroleum and mineral deposits, and is the foundation for the standard **geologic column** (see *figure 8.2*). The geologic record shows that very few species have existed longer than 20 million years, the average being about 5 million.

Fossils provide a means of establishing relative dates in much the same way as artifacts do. Both show evolution and change with time. For example, in a city dump where refuse is buried in succession, one could recognize a period of time prior to the automobile by remains of wagon wheels, saddles, and similar equipment. A layer containing abundant scraps and pieces of Model T Fords would be recognized as being older than one containing remains of Model As. A layer containing newer models such as the Mustang would be recognized as being one of the youngest layers even though it might not rest upon layers containing any of the older materials.

The Principle of Crosscutting Relations. The relative age of certain events is shown by **crosscutting relationships. Faults** and igneous intrusions are younger than the rocks they cut, a fact so obvious that it hardly needs mentioning. However, crosscutting relations can be complex, and careful observation may be required to establish the correct sequence of events in complex areas. The scale of crosscutting features is highly variable, ranging from large faults with displacements of hundreds of kilometers to small fractures less than a millimeter long. The sequence of events in *figure 8.1* can be worked out by applying the principles of superposition and crosscutting relations.

The Principle of Inclusion. The relative age of intrusive igneous rocks (with respect to the surrounding

Figure 8.1 *The sequence of events in a schematic diagram can be determined by the use of the principles of superposition and crosscutting relations.*

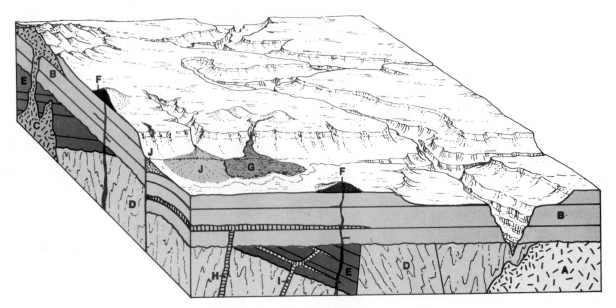

country rock) is commonly apparent from **inclusions,** fragments of older rocks in the younger. As a magma moves upward through the crust, it dislodges and engulfs large fragments of the surrounding material, which remain as unmelted foreign inclusions.

The principle of inclusion is also clear in many conglomerates in which relatively large pebbles and boulders derived from some preexisting rocks have suffered erosion and transportation and have been deposited in a new formation. The conglomerate is obviously younger than formations from which the pebbles and cobbles were derived. In areas where superposition or other methods do not indicate relative ages, a limit to the age of a conglomerate can be determined from the rock formation represented in the pebbles and cobbles.

The Principle of Succession in Landscape Development. Surface features of the earth's crust are being modified continually by erosion and commonly show the effects of successive events through time. Many landforms evolve through a definite series of stages so that the relative age of the feature can be determined from the degree of erosion. This is especially obvious in volcanic features (such as cinder cones and lava flows) created during a period of volcanic activity and then subjected to the forces of erosion until they are completely destroyed or buried by erosional debris.

The Standard Geologic Column

Statement

Using the principles of superposition and faunal succession, geologists have determined the chronologic sequence of rocks on a regional basis and have con-

RELATIVE GEOLOGIC TIME			*ATOMIC TIME
ERA	**PERIOD**	**EPOCH**	
CENOZOIC	Quaternary	Holocene	
		Pleistocene	2-3
	Tertiary	Pliocene	12
		Miocene	26
		Oligocene	37-38
		Eocene	53-54
		Paleocene	65
MESOZOIC	Cretaceous	Late / Early	136
	Jurassic	Late / Middle / Early	190-195
	Triassic	Late / Middle / Early	225
PALEOZOIC	Permian	Late / Early	280
	Carboniferous Systems — Pennsylvanian	Late / Middle / Early	
	Carboniferous Systems — Mississippian	Late / Early	345
	Devonian	Late / Middle / Early	395
	Silurian	Late / Middle / Early	430-440
	Ordovician	Late / Middle / Early	500
	Cambrian	Late / Middle / Early	600
PRECAMBRIAN			3600

Figure 8.2 *The standard geologic column.*   *Estimated ages of time boundaries (millions of years)

structed a standard geologic time scale. Most of the original scale was pieced together from sequences of strata studied in Europe during the mid-nineteenth century. Each major unit of rock (Cambrian, Ordovician, Silurian, et cetera) generally was named after the area where it is well exposed. The rock units are distinguished one from the other by major changes in rock type, unconformities, or abrupt vertical changes in the groups of fossils they contain. In effect, the original subdivision of the geologic column was based simply on the sequence of rock formations in their superposed order as they were found in Europe. Rocks in other parts of the world that contain the same assemblage of fossils as those in a given part of the succession in Europe are considered to be the same age and commonly are referred to by the same names.

Discussion

The standard geologic column, shown in *figure 8.2*, is based upon the principles of superposition and faunal succession and indicates only the relative ages of the rocks. It is a composite of the local succession of rock in many parts of the world. The column by itself tells nothing about the specific duration of time represented by each period or the age (time before the present) of each period. Later radioactive age determinations indicate that the periods are not of equal length (as is illustrated in the right-hand column in *figure 8.2*).

The nomenclature for the geologic column at first may seem nothing more than a collection of meaningless names, but the standard geologic column serves as a type of calendar for the earth's history and is the basic language used to designate large time intervals.

Radiometric Measurements of Absolute Time

Statement

Unlike relative time, which signifies only the chronologic relationships among events, absolute or finite time designates specific durations measured in units of hours, days, or years. Duration of time is measured by any regularly recurring event. One very useful natural clock measures time by the processes of radioactive decay. Radioactive decay is a process in which the atoms of an element lose particles from their nuclei and thereby become atoms of other elements. Because the rate at which they decay is unaffected by conditions such as pressures, temperatures, and chemical binding forces, radioactive decay is a very precise and accurate geological measuring device. The time elapsed since a radioactive element was locked into the crystal of a mineral can be determined when the rate of decay is known and the proportions of the original element and the decayed product can be measured. The most important radioactive clocks for geologic studies are uranium, thorium, rubidium, and potassium.

Discussion

The rate of radioactive decay is measured in terms of **half-life.** In one half-life, half of the original atoms

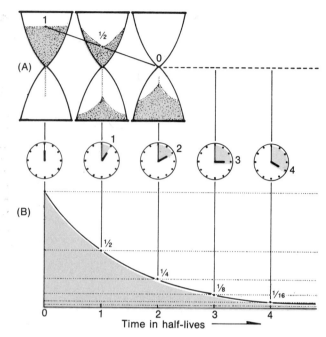

Figure 8.3   *The contrast between linear and exponential rates of depletion. (A) Most processes have a uniform straight-line depletion that is like sand moving through an hourglass. If half of the sand is gone in ½ hour, all will be gone in 1 hour. (B) Radioactive decay, in contrast, is exponential. If half is depleted in 1 hour, half of that remaining, or one-fourth, will be depleted in 2 hours.*

decay; in a second half-life, one-half of the remainder or one-quarter of the original atoms decay; in a third half-life, half of the remaining quarter decay, and so on (see *figure 8.3*). From the rate at which a particular radioactive element decays, the time elapsed since the formation of the crystal containing the element can be calculated comparing the amount of the radioactive element remaining in the crystal (parent) to the amount of disintegration product (daughter).

There are numerous radioactive isotopes, but most have rapid rates of decay (that is, short half-lives) and lose their radioactivity within a few days or years. However, some decay very slowly, with half-lives of hundreds of millions of years, and can be used as atomic clocks for measuring very long periods of time. The parent isotopes and their daughter products most useful for geologic dating are listed in *table 8.1*.

The theory of **radioactive dating** is simple enough, but the laboratory procedures are complex, the principal difficulty being precise measurement of minute amounts of isotopes. The accuracy of the method depends upon the accuracy with which the half-life is determined. (U^{235} to Pb^{207} measurements are considered accurate within 2%.)

At present, the potassium-argon method is of great importance, since it can be used with the micas and amphiboles that are widely distributed in igneous rocks. It also can be used on rocks as young as a few thousand years or on the older known rocks. Whenever possible, two or more radioactive elements from the same specimen are analyzed to verify the results.

Table 8.1 Radioactive Isotopes Useful in Determining Geologic Time

Parent Isotope	Daughter Product	Half-Life
Uranium - 238	Lead - 206	4.5 billion years
Uranium - 235	Lead - 207	713.0 million years
Thorium - 232	Lead - 208	13.9 billion years
Rubidium - 87	Strontium - 87	50.0 billion years
Potassium - 40	Argon - 40	1.5 billion years

Another important radioactive clock is based on the decay of carbon-14 (C^{14}), which has a half-life of 5692 years. C^{14} is produced continually in the earth's atmosphere as a result of the bombardment of nitrogen-14 (N^{14}) by cosmic rays. This newly formed radioactive carbon becomes mixed with ordinary carbon atoms in carbon dioxide gas. Plants absorb the carbon dioxide and animals eat the plants, so that both maintain a fixed proportion of C^{14} while they are alive. After death, no additional C^{14} can replenish the supply in their tissues and radioactive decay begins, causing C^{14} to revert to N^{14}. The time elapsed since the organism died can be determined by measuring the amount of C^{14} remaining. The longer the time elapsed since death, the less C^{14} will remain. Inasmuch as the half-life of C^{14} is 5692 years, the amount of C^{14} remaining in organic matter older than 50,000 years is so small that it cannot be measured accurately. Therefore, this method is useful for dating very young geologic events plus most archaeologic material.

The Radiometric Time Scale

Statement

Absolute dates of numerous geologic events have been determined and are used in combination with the standard geologic column to provide a radiometric time scale from which the absolute age of a geologic event can be estimated. Unfortunately, radiometric dates cannot be determined for any given layer of sedimentary rock, because sediments are composed of eroded debris of preexisting rocks from various sources. We can determine the radiometric dates of minerals in sedimentary rocks, but the dates indicate when the minerals formed, not when the beds of sediment were deposited. Radioactive isotopes can be used to date the time igneous rocks crystallized and the time heat and pressure developed new minerals in metamorphic rocks, because in each case the mineral and rock formed together. The problem in developing a reliable radiometric time scale is accurately placing radiometric dates of igneous and metamorphic rocks in their proper positions in the relative time scale established by sedimentary rock. Layers of volcanic rocks and **bracketed intrusions** are the most suitable rocks for radiometric age determinations that can be used as finite time markers in the standard geologic column. These time benchmarks accurately placed in the geologic column constitute the basis for the radiometric time scale.

Discussion

Layered Volcanics. The best reference points for the radiometric time scale are probably volcanic ashfalls and lava flows. They are deposited instantaneously as far as geologic time is concerned, and, because they commonly are interbedded with fossiliferous sediments, their exact position in the geologic column can be determined, providing the main basis for establishing an absolute time scale within the geologic column.

Bracketed Intrusions. As shown in *figure 8.4,* a molten rock can cool within the earth's crust without ever breaking out onto the surface. Subsequent erosion may expose this rock; later, younger sediments may be deposited on top. In some cases, the entire sequence

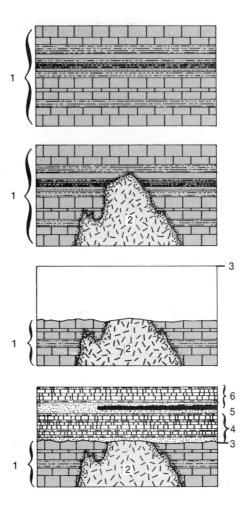

Figure 8.4 *Diagrams showing how radioactive dates of intrusive rocks can be used in developing a radiometric time scale. The sequence of sedimentary rocks (1) is deposited and subsequently intruded by the igneous body (2). Erosion removes part of the sequence (1 and 2). Subsequent deposition of the sediment sequence (4) occurs, followed by lava flows (5) and younger sedimentary rocks (6). A radiometric date of lava flows (5) would provide an excellent age for rocks in that position of the geologic sequence, because the flows occurred as part of the normal sequence of rocks. A date for the intrusive (2) is more difficult to place in the column. We only know that it occurred sometime after 1 and before erosion (3 and 4).*

of events may take only a few million years; in others, the events may require a much longer time. The relative age of the igneous rock falls in the bracket between the older sediments (1) and the younger sediments (4). Unfortunately, the span of time between 1 and 4 is com-

monly too long to permit the date of the intrusion to be useful in detailed geochronology, but radiometric dates on such rocks do establish the time of major igneous events.

The presently accepted geologic time scale is based

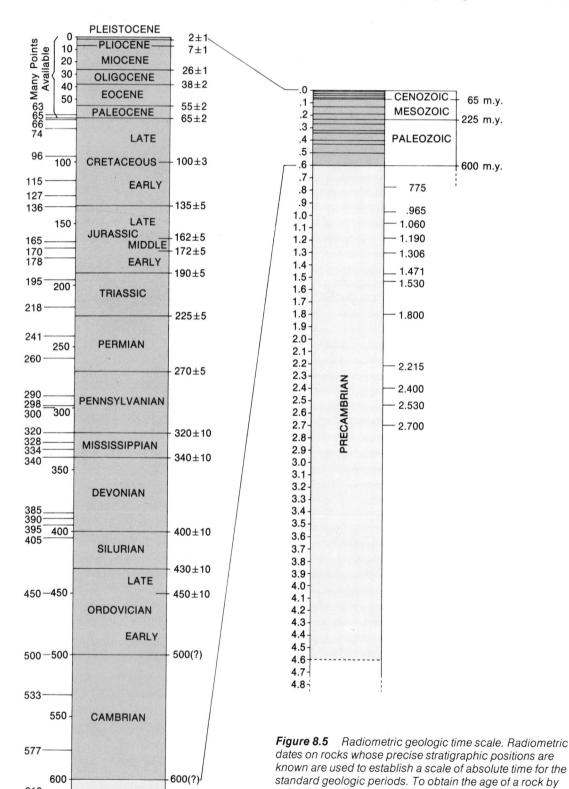

Figure 8.5 *Radiometric geologic time scale. Radiometric dates on rocks whose precise stratigraphic positions are known are used to establish a scale of absolute time for the standard geologic periods. To obtain the age of a rock by using this scale, one needs to know its exact position in the geologic column; from this, an approximate absolute or finite date can be interpolated.*

on both the standard geologic column, established by faunal succession and superposition, plus the finite radioactive dates of rocks that can be placed precisely in that column. Each system of dating provides a cross-check on the other inasmuch as one is based on relative time and the other on absolute time. In most instances, there is remarkable agreement between the two systems, and discrepancies are few. In a sense, the radioactive dates act as the scale on a ruler and provide accurate reference markers between which some interpolation can be made. The geologic column in *figure 8.5* shows the standard geologic column together with radiometric dates of layered volcanics and bracketed intrusions in the sedimentary sequence. Enough dates have been established so that the time span of each geologic period can be estimated with considerable confidence. To

determine the age of a rock, one need only determine its location in the geologic column and interpolate between the nearest radiometric time marks.

From this radiometric time scale, several general conclusions can be made about the history of the earth and geologic time: (1) present evidence indicates that the age of the earth is around 4.5 to 4.8 billion years; (2) Precambrian time constitutes more than 80% of geologic time; (3) Phanerozoic time (Paleozoic and later) began about 570 million years ago (rocks deposited since Precambrian time can be correlated on a worldwide basis by means of fossils, and many important events during this time can be established on the basis of radioactive dates); and (4) some major events in the earth's history may be difficult to place in their relative positions in the geologic column but can be dated by radioactive methods.

Figure 8.6 The length of geologic time compared to a yardstick. On this scale, Precambrian time represents the first 31 in., and all events since the beginning of the Paleozoic are compressed into the last 5 in. Dinosaurs first appeared 2 in. from the top. The glacial epoch occurred in the last fraction of an inch, and historic time is so small that it cannot be represented, even on the enlarged part of the figure.

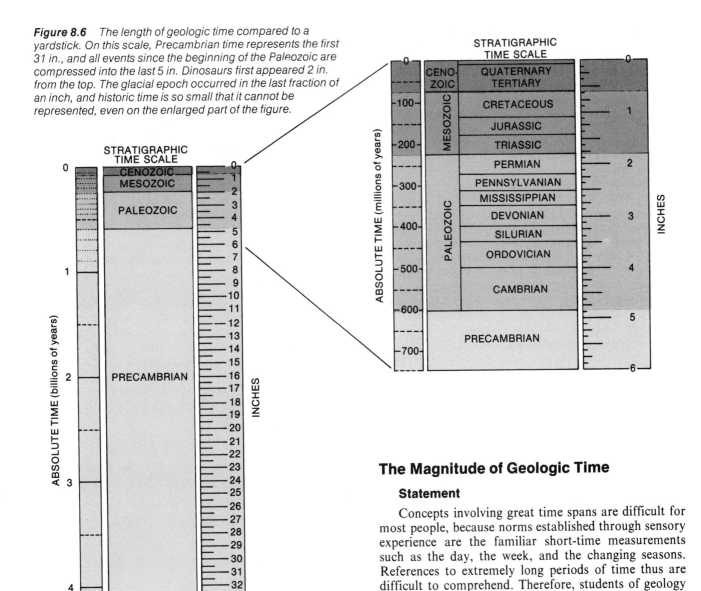

The Magnitude of Geologic Time

Statement

Concepts involving great time spans are difficult for most people, because norms established through sensory experience are the familiar short-time measurements such as the day, the week, and the changing seasons. References to extremely long periods of time thus are difficult to comprehend. Therefore, students of geology must continually attempt to enlarge their time norms to encompass the magnitude of geologic time. Otherwise, extremely slow geologic processes considered in terms of human experience will have little meaning.

Discussion

Perhaps the best way to begin is to consider geologic time in reference to something tangible and familiar rather than in terms of large numbers. In *figure 8.6,* the length of the yardstick represents the length of time from the beginning of the earth to the present. A scale of absolute time is plotted on the right, and the standard geologic periods are shown on the left. This diagram reveals several facts. Subdivisions of geologic time are not equal. Precambrian time constitutes the great bulk of the earth's history, whereas the Paleozoic and younger periods are equivalent to only the last 5 in. on the scale. In order to show events with which most people are familiar, the last few inches on the scale must be enlarged. The first abundant fossils occur 5 in. from the top. The great coal swamps are about 2 in. from the top. The dinosaurs became extinct about 0.5 in. from the top, and the ice age would be shown in the uppermost 0.01 in. Recorded history would be shown only in the upper 0.001 in. of this scale.

Summary

As was seen in previous chapters, mineral types form under very specific environmental conditions and the shape of the mineral grains reveal many things about their history. Rocks, being aggregates of minerals, thus constitute records of events in the changing planet earth.

This chapter has discussed how the events recorded in rocks can be arranged in their proper sequence by using the principles of superposition and faunal succession and how the dates of some events can be determined in terms of a finite number of years. In the next section of this book, the details of the hydrologic system and how the surface features of the earth are formed, modified, and changed with time will be discussed.

Additional Readings

Berry, W. B. N. 1968. Growth of a Prehistoric Time Scale. San Francisco: W. H. Freeman and Company.

Deevey, E. S., Jr. 1952. "Radiocarbon Dating." Sci. Amer. 186(2): 24-28.

Eicher, D. L. 1968. Geologic Time. Englewood Cliffs, N.J.: Prentice-Hall.

Faul, H. 1966. Ages of Rocks, Planets and Stars. New York: McGraw-Hill.

Harbaugh, J. W. 1968. Stratigraphy and Geologic Time. Dubuque, Iowa: William C. Brown and Company.

Hurley, P. M. 1959. How Old Is the Earth? Garden City, N.Y.: Anchor Books (Doubleday and Company).

Toulmin, S., and J. Goodfield. 1965. The Discovery of Time. New York: Harper and Row.

Weathering

A new building gradually deteriorates. The paint chips and peels and is gone in a matter of a few years. Wood soon dries and splits, and even the bricks and cement eventually decay and crumble. Indeed, left by themselves, most buildings decompose into a pile of rubble within a few hundred years. This process of natural decay is called weathering. Solid bedrock is also subject to weathering and eventually decomposes into piles of rubble. In fact, some rocks are even less resistant to weathering processes than the paint on a house.

Weathering occurs because most rocks were formed originally in equilibrium with the high temperatures and pressures deep within the earth. When the rocks are exposed to the much lower temperatures and pressures at the surface, as well as to the gases in the atmosphere and the elements in the water, they become unstable and undergo various chemical reactions and mechanical stresses. As a result, the solid bedrock is changed into loose, decomposed products. Therefore, weathering may be considered as a general term to describe all of the changes that occur in rock materials as a result of their exposure to the atmosphere. The environment of the atmosphere, with abundant water, oxygen, carbon dioxide, and organic activity, may vary widely from day to day, season to season, and over the years in longer cycles so that the rocks must continually adjust to new physical and chemical conditions.

The atmosphere tends to break down rocks in two major ways: (1) by mechanical disintegration, in which the rocks are fragmented by physical forces, and (2) by chemical decomposition, in which decomposition results from a chemical reaction between the elements in the atmosphere and those in the rocks. Although these processes will be considered separately, they operate concurrently, each facilitating and reinforcing the other until the solid, coherent bedrock exposed at the surface is transformed into a layer of loose, decomposed material. It is this weathered rock material that is washed away by river systems in the process of eroding the surface of the land.

From a geologic point of view, the importance of weathering is that it transforms the solid bedrock into small, decomposed fragments and prepares it for removal by the agents of erosion. Without it, there would be no erosion as it is known today and the landscape would be strikingly different. In addition, the products of weathering form a blanket of soil over the solid bedrock, and soil is the basis for most terrestrial life. Therefore, weathering should be considered as a part of the geologic system that has tremendous ecological significance.

Major Concepts

1. The major types of weathering are mechanical disintegration and chemical decomposition.
2. Frost wedging is the most important form of mechanical weathering.
3. The major types of chemical weathering are oxidation, dissolution, and hydrolysis.
4. Joints are important in weathering in that they permit the atmosphere and water to attack a rock body at considerable depth. They also greatly increase the surface area of a rock on which chemical reactions can occur.
5. The major products of weathering are a blanket of soil and regolith and spheroidal rock forms.
6. Climate greatly influences the type and rates of weathering; the major controlling factors are precipitation and temperature and their seasonal variations.

(A) Enlargement of fractures.

(B) Development of columns and pillars.

(C) Piles of rock debris.

(D) Block separation.

(E) Shattering.

(G) Exfoliation.

(F) Granular disintegration.

Figure 9.1 The effects of weathering.

Mechanical Weathering

Statement

Mechanical weathering is the breakdown of rock into smaller fragments by various physical stresses. It is strictly a physical process without a change in chemical composition. No chemical elements are added to, or subtracted from, the rock. The rock simply is broken into small fragments.

The most important types of mechanical weathering are:

1. **Ice Wedging,** a process in which expansion of freezing water in cracks or in bedding planes wedges the rock apart.
2. **Sheeting** or unloading, a series of fractures in a rock body produced by expansion resulting from the removal of underlying material.

Discussion

Ice Wedging. *Figure 9.2* is a simple diagram showing the manner in which ice wedging breaks a rock mass into small fragments. Water from precipitation or melting snow easily penetrates cracks, bedding planes, and other openings in the rock and expands 9%, approximately, when it freezes. With each freeze great pressures (similar to those produced by driving a wedge into a crack) are exerted upon the rock walls. Over a period of time, therefore, the fractured blocks and bedding layers are pried free from the parent material. The force generated by each freeze is approximately 110 kg/cm², which is roughly equivalent to that produced by dropping a shot put from a height of 3 m. With each freeze, the stress is exerted so that, over a period of time, the rock literally is hammered apart.

Conditions necessary for effective ice wedging include (1) an adequate supply of moisture; (2) fractures, cracks, or other voids within the rock into which the water can enter; and (3) temperatures that rise and fall across the freezing point. Temperature is especially important, because pressure is applied with each freeze. Frost action is much more effective in areas where

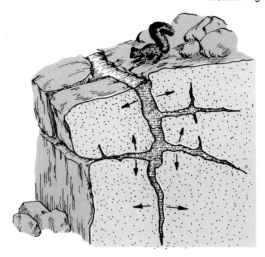

Figure 9.2 *The process of ice or frost wedging. Water seeps into fractures in a rock and expands with great force when it freezes. This wedges the rock apart and produces shattered, angular fragments.*

freezing and thawing occur many times a year. It is less effective in exceptionally cold areas where water is permanently frozen. Therefore, frost action occurs most frequently above the timberline and is especially active in the steep slopes above alpine glaciers, where meltwater produced during the warm summer days seeps into cracks and joints and freezes during the night.

The products of ice wedging commonly accumulate at the bases of cliffs in piles of angular rock fragments called **talus.** Since most cliffs are notched by steep valleys and narrow ravines, the fragments dislodged from the high valley walls are funneled through the ravines to the base of the cliff, where they accumulate in a cone-shaped deposit (*figure 9.3*).

Sheeting. Rocks originally formed deep within the earth's crust are under high confining pressure from the weight of thousands of meters of overlying rocks. When this rock cover is removed by erosion, the confining pressure is released and the rock body tends to expand.

Figure 9.3 *Talus cones, which accumulate at the base of a cliff as a result of rockfalls, produced mostly by frost action.*

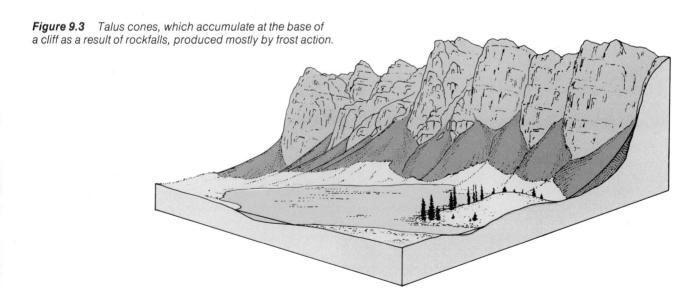

Figure 9.4 *Sheeting in the granite of the Sierra Nevadas. As erosion removes the overlying rock cover, confining pressure is released and the bedrock expands upward. This expansion develops large fractures parallel to the earth's surface, which subsequently may be enlarged by frost wedging.*

The internal stresses set up by expansion can cause large fractures or expansion joints to develop parallel to the earth's surface (*figure 9.4*). This process is observed commonly in quarrying operations when the removal of large blocks is followed by the rapid, almost explosive expansion of the quarry floor. A sheet of rock several centimeters thick bursts up, and, at the same time, numerous new parallel fractures appear deeper in the rock body. The same process occasionally causes rock bursts in mines and tunnels when the confining pressure is released during the tunneling operation.

Other Types of Mechanical Weathering. Animals and plants play a variety of relatively minor roles in mechanical weathering. Burrowing animals such as rodents mechanically mix the soil and loose rock particles, a process that facilitates further breakdown by chemical means. Pressure from growing roots widens cracks and contributes to the rock breakdown. Lichens can live on the surface of bare rock and extract nutrients from the rock minerals by ion exchange. This results in the alteration of minerals both mechanically and chemically. These processes may seem trivial, but the work of innumerable plants and organisms for a long period of time can add significantly to the disintegration of the rock.

Thermal expansion and contraction caused by daily or seasonal temperature changes once were thought to be an effective process of mechanical weathering. The idea is plausible, but experiments show that stresses developed by alternate heating and cooling over long periods of time are significantly smaller than the elastic strength of rock. Even on the moon, where daily temperature changes are much greater than those on the earth, the effects of thermal expansion in rock fragmentation are uncertain.

Chemical Weathering

Statement

Chemical weathering, or decomposition, consists of several important chemical reactions among the elements in the atmosphere and those in the minerals of the earth's crust. In these processes, the internal structures of the original minerals are destroyed, and new minerals with new crystal structures that are stable under atmospheric conditions are created. Thus, there is a significant change in the chemical composition and the physical appearance of the rocks.

Water is of prime importance in chemical weathering, because it takes part directly in the chemical reactions, carries other elements of the atmosphere into contact with the minerals of the rocks, and removes the products of weathering so that fresh rock is exposed. The rates of chemical weathering, therefore, are influenced greatly by the amount of precipitation.

No area of the earth's surface is continually dry; for, even in the most arid deserts, some rain does fall and cause chemical weathering. Therefore, chemical weathering is essentially a worldwide process, which is least effective in deserts and in cold climates where water is frozen the entire year.

The chemical reactions of chemical weathering are very complex, but three main groups are recognized.
1. **Hydrolysis.**
2. **Dissolution.**
3. **Oxidation.**

Discussion

Hydrolysis. The chemical union of water and a mineral is known as hydrolysis, a process involving not merely absorption of water, as in a sponge, but a specific chemical change in which a new mineral is produced from the original material. The reaction is between the H^+ or OH^- ions of the water and the ions of the mineral.

A good example of hydrolysis is found in the chemical weathering of K-feldspar. It can be described chemically as:

$$2KAlSi_3O_8 \quad + \quad H_2CO_3 \quad + \quad nH_2O \quad \longrightarrow \quad K_2CO_3 \quad + \quad Al_2(OH)_2Si_4O_{10} \bullet nH_2O \quad + \quad 2SiO_2$$

| (potassium feldspar) | (carbonic acid) | (water) | (potassium carbonate— readily soluble) | (clay mineral) | (soluble hydrated silica or finely divided quartz) |

The hydrogen ion of the H_2CO_3 displaces the potassium ion of the feldspar and combines with the aluminum silicate radical of the feldspar to form a new clay mineral. The potassium is released as a free positive ion in the soil water. Silica also is released but may remain in solution. The new clay mineral does not contain the potassium present in the original feldspar and has a new crystal structure consisting of sheets of silica tetrahedra that form submicroscopic crystals. The importance of hydrolysis in weathering cannot be overstressed, since hydrolysis acts on the feldspars and ferromagnesian minerals, which are the dominant minerals in most rocks.

Dissolution. Water is one of the most effective and universal solvents known, and practically all minerals are soluble in water to some extent. This is because of the polarity of the water molecule. The structure of the water molecule is such that the two hydrogen atoms are positioned on the same side of a large oxygen atom so that each molecule has a concentration of positive charge on one side between the two hydrogen atoms, balanced by a negative charge on the opposite side. As a result, the water molecule is polar and behaves like a tiny magnet that acts to loosen the bonds of the ions at the surface of minerals in contact with the water. There are a number of good examples of various rock types being completely dissolved and leached away by water. Rock salt is perhaps the best known. It is extremely soluble and survives at the earth's surface only in the most arid regions. Gypsum is less soluble but is easily dissolved by surface water, and there are no outcrops of these rocks in humid regions. Limestone is also soluble in water, especially when the water contains carbon dioxide. It commonly weathers into valleys where water is abundant but forms cliffs in arid regions.

An idea of the effectiveness of solution activity in the weathering of rocks can be gained from analysis of the composition of water in rivers. Fresh rainwater has relatively little dissolved mineral matter, but running water soon dissolves the more soluble minerals in the rock and transports them in solution. The amount of dissolved minerals carried by the rivers of the world to the ocean each year is approximately 3.9 million metric tons. Also, seawater contains 3.5% (by weight) dissolved salts, all of which were dissolved from the continents by pure rainwater.

Oxidation. Oxidation is the combination of atmospheric oxygen with a mineral to produce an oxide. It is especially important in minerals having a high iron content, such as olivine, pyroxene, and amphibole. The iron in silicate minerals unites with oxygen to form hematite $[Fe_2O_3]$ or limonite $[FeO(OH)]$. Hematite is deep red, and, when dispersed in sandstone or shale, it imparts a red color to the entire rock.

The Importance of Joints in Weathering

Statement

Almost all rocks are broken by systems of fractures called **joints,** which result from the strain established by earth movements, the release of confining pressure, and the contraction related to the cooling of lava. Joints greatly influence the weathering of rock bodies by:
1. Effectively cutting large blocks of rocks into smaller ones and thereby greatly increasing the surface area available for chemical reactions.
2. Acting as channelways through which water can penetrate and break down the rock by frost action.

Discussion

The importance of jointing in weathering processes can be appreciated better by considering the amount of new sur-

Figure 9.5 *Diagrams showing how a system of joints cutting a rock body greatly increases the surface area exposed to weathering.*

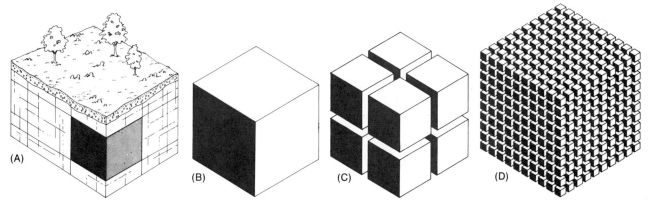

(A) A surface of bedrock, 100 m² with no joints, exposes a total area of 100 m² to weathering processes.
(B) Intersecting joints, 10 m apart, increase the surface area to 500 m².

(C) Two additional joints increase the area to 900 m².
(D) If joints 1 m apart cut the rock, the area exposed to weathering is 5900 m².

face area produced by joints for chemical weathering. For example, consider a cube of rock that measures 10 m on each side (as shown in *figure 9.5*). If only the upper surface of the cube were exposed and the rock were not jointed, weathering could attack only the exposed surface of 100 m². However, if the block were bounded by intersecting joints 10 m apart, the surface area exposed to weathering processes would be 500 m² (the base of the cube is not exposed). If two additional joints cut the cube into eight smaller cubes, the surface area exposed to weathering would be 900 m². If joints 1 m apart cut the rock, 5900 m² of rock surface would be exposed to weathering. Obviously, a highly jointed rock body weathers much more rapidly than a massive one.

In addition to providing a much larger surface area for chemical decomposition, joints act as a system of channels through which water can penetrate below the surface, thus permitting mechanical and chemical weathering processes to attack the rock from several sides hundreds of meters below the surface.

Some of the most striking examples showing the influence of jointing on weathering are found in the Colorado Plateau, where thick sandstone formations have developed prominent joint systems because the rock is so brittle. Weathering proceeds along each joint surface and cuts the rock into large slabs (*figure 9.6*). In other formations, intersecting joints divide the rock into large columns and play an important role in the sculpture of columns and pillars (*figure 9.7*). Similarly, jointing plays a dominant role in the weathering of large masses

Figure 9.6 *Joint systems cutting a sandstone formation near Arches National Park, Utah. Weathering along the joints has produced deep, narrow crevasses, so the joint planes are greatly emphasized.*

Figure 9.7 *Ground view of an intersecting system of joints that divides the rock into rough columns.*

of granite and provides the initial shape for further rock breakup.

The Products of Weathering

Statement

The results of weathering can be seen over the entire surface of the earth from the driest deserts and the frozen wastelands to the warm and humid tropics. The major products include:

1. A blanket of loose, decayed rock debris, known as **regolith,** that forms a discontinuous cover over the solid bedrock.
2. A modification of the shape of the rock body into spherical forms (**spheroidal weathering**).
3. Soluble products that are carried away by streams and ground water.

Discussion

Regolith. The regolith is important to man, because its upper part—the soil—is essential for plant life, the critical link in man's food chain. The thickness of the regolith ranges from a few centimeters to many meters, depending upon the climate, the type of rock, and the length of time that uninterrupted weathering has proceeded. The transition from bedrock to regolith can be seen in road cuts and steep stream valleys. As is shown in *figure 9.8,* the solid bedrock usually is divided by

numerous joints that separate the rock into angular blocks. Above the jointed bedrock is a zone with smaller, highly decomposed blocks. These grade upward into small, sandsize fragments of rocks, clay, and mineral grains. The upper layer commonly is rich in decaying vegetation mixed with fine, decayed mineral matter.

A more regional view of regolith and its relationship to bedrock is shown in *figure 9.9.* It is clear from the photograph that exposures of bedrock are limited to certain areas of resistant limestone and sandstone strata that form discontinuous cliffs along the upper part of the mountain front. In the canyons, very little soil is retained and bedrock is exposed from the base to the top of the canyon walls. The sketch in diagram *B* was made from the photograph and outlines the surface covered with regolith. In this diagram, the outcropping bedrock is not shown, so the regolith appears as a thin, discontinuous blanket over the surface. The holes in the regolith occur where bedrock is exposed. Sediment fills the valley in the foreground, but the regolith developed on it is not shown

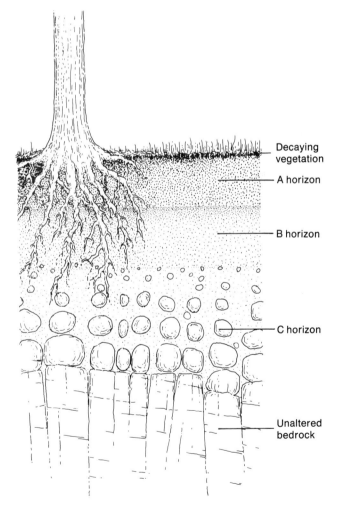

Decaying vegetation

A horizon

B horizon

C horizon

Unaltered bedrock

Figure 9.8 *The major horizons in a soil profile.*

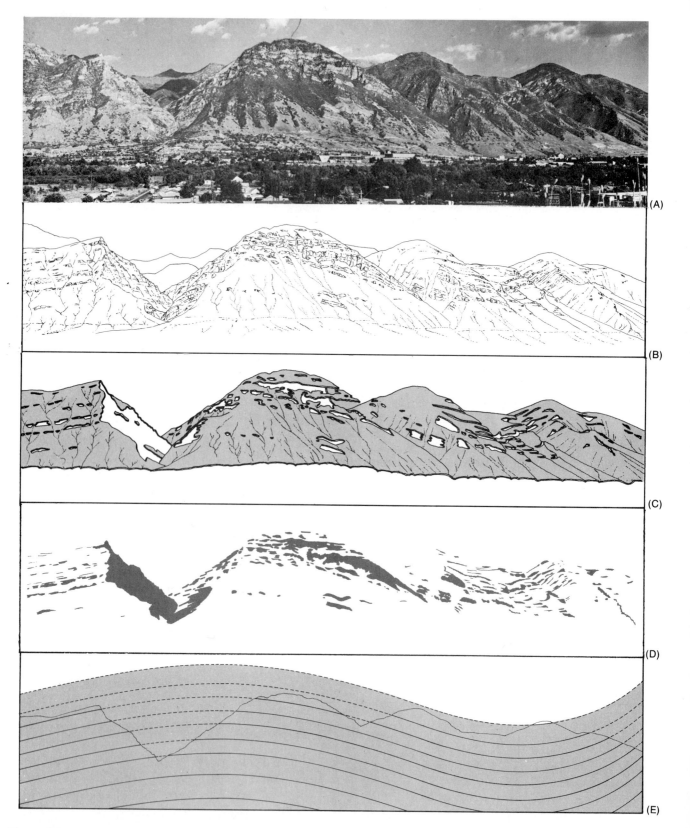

Figure 9.9 *The relationship between bedrock and regolith.*

(A) Photograph of the Wasatch Mountains in central Utah.

(B) A sketch showing outcrops of bedrock in cliffs and canyons, and slopes covered with regolith.

(C) The discontinuous blanket of regolith. Outcrops of bedrock form holes in the regolith cover. Some formations are almost completely covered with regolith, while others are exposed as discontinuous cliffs.

(D) Areas where bedrock is exposed outline the structure of the rocks.

(E) The structure of the bedrock. The rock layers are warped into broad folds, some of which are cut by canyons. Compare with A.

in the diagram. When you carefully study the areas of exposed bedrock in *figure 9.9*, you can see that the strata are warped into broad folds (diagram *D*). These folds form the structure of the mountains. At this point, the important role that regolith plays in the processes of erosion should be emphasized. It is the regolith that is removed by agents of erosion, not the solid bedrock; and were it not for weathering, the style of erosion on the earth would be vastly different.

Soil. Soil is so widely distributed and so economically important to man that it has acquired a variety of definitions. From a geologic point of view, it is the upper part of the regolith and is composed chiefly of small particles of rocks and minerals, plus varying amounts of decomposed organic matter. The mature soil profile shows a rather consistent sequence of zones, distinguished by composition, color, and texture, between the surface and the unbroken bedrock. These are shown in

figure 9.8. **Horizon A** is the topsoil layer often visibly divided into three layers: A_0 is a thin surface layer of leaf mold, especially obvious on forest floors; A_1 is a humus-rich dark layer; and A_2 is a light, bleached layer. **Horizon B** is the subsoil containing fine clays and colloids washed down from the topsoil. It is largely a zone of accumulation and commonly is reddish in color.

The type of soil depends upon a number of factors, the most important of which are climate, parent rock material, and topography. Climate is of major importance, because rainfall, temperature, and seasonal changes all directly affect the development of soil. For example, in deserts, arctic regions, and high mountainous regions, mechanical weathering dominates and organic matter is minimal. The resulting soil is thin and consists largely of broken fragments of bedrock (*figure 9.10*). In

Figure 9.10 *Examples of soil types.*

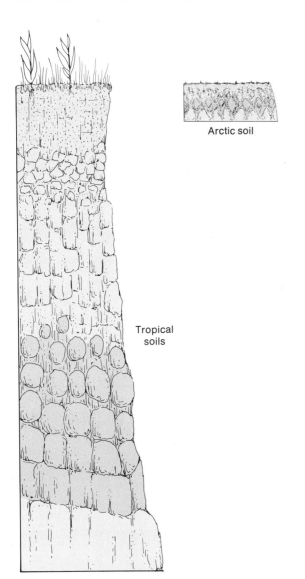

(A) *The influence of climate. Thin soils in arctic and desert regions are characterized by partly decomposed rock fragments. Thick soils of tropical regions have developed from extensive chemical decomposition.*

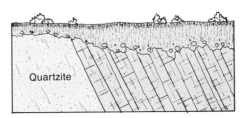

(B) *The influence of rock type. Quartzite resists chemical decomposition, so soil is thin and poorly developed. Limestone is much more susceptible to chemical weathering and develops thicker soil.*

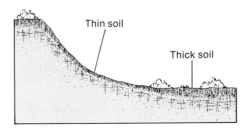

(C) *The influence of topography. Thick soils develop on flat or gently sloping surfaces, while steep slopes permit development of thin soils.*

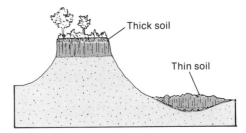

(D) *Influence of time. Thick soils have been developed on old lava flows, contrasted with thin soil development on younger flows.*

equatorial regions, where rainfall is heavy and temperatures are high, chemical processes dominate and thick soils develop rapidly. As a consequence, soil profiles 60 m thick are common in the tropics and subtropics; and, in some areas (such as central Brazil), the zone of decayed rock is more than 150 m thick.

The mineral composition of bedrock exerts a strong influence on the type of soil, because it provides the chemical elements and mineral grains from which the soil develops. A pure quartzite, for example, contains 99% SiO_2 and would develop a thin, infertile soil (*figure 9.10*).

Topography affects soil development, because it influences the amount and rate of erosion and the nature of drainage. Flat, poorly drained lowlands develop a bog-type soil, whereas steep slopes permit rapid removal of regolith and inhibit the accumulation of weathered materials. Well-drained uplands are conducive to thick, well-developed soils.

Spheroidal Weathering. In the weathering process, there is an almost universal tendency to produce rounded (or spherical) surfaces on decaying rock. The rounded shape results because weathering attacks an exposed rock from all sides at once, and, therefore, the depth of decomposition is much greater along the corners and edges of the rock (see *figure 9.11*). As this material falls off, the corners become rounded and the block eventually is reduced to an ellipse or a sphere. The sphere is the geometric form that has the least amount of surface area per volume; once the block attains this shape, it simply becomes smaller.

Although examples of spheroidal weathering can be seen in almost any exposure of rock, it perhaps is appreciated best in rounded blocks in ancient man-made structures, such as the Parthenon of ancient Greece shown in *figure 9.12*. The original blocks had sharp corners and were fitted together with precision, as can be seen in the unweathered blocks used in the restored section. The

Figure 9.12 *Spheroidal weathering in the building blocks of the Parthenon, Athens, Greece. The original blocks were rectangular and closely fit together like the restored section shown in the left part of the photo. Weathering proceeding inward from the sides of each block has produced a spheroidal form.*

edges of the original blocks, in contrast, have decomposed completely and have assumed elliptical or spherical shapes.

In nature, spheroidal weathering is produced both

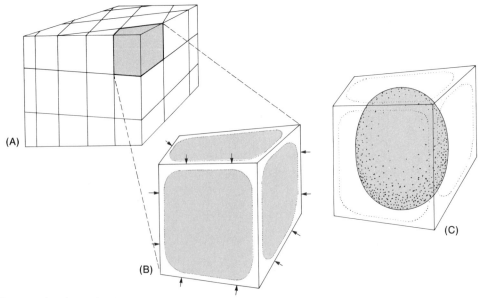

Figure 9.11 *Diagram showing spheroidal weathering.*
(A) Joint systems cut a rock body into angular blocks.
(B) Weathering proceeds inward on each block from the joint face.

(C) The corners of the block soon are completely decomposed so that the unweathered rock assumes a spherical or an elliptical shape.

at the surface and at some depth. The series of sketches in *figure 9.13* shows how an intersecting joint system produces angular blocks, which ultimately decompose into elliptical or spherical boulders.

Exfoliation is a special type of spheroidal weathering in which the rock breaks apart by separation along a

Figure 9.13 *Sketches showing evolution of spheroidal boulders.*

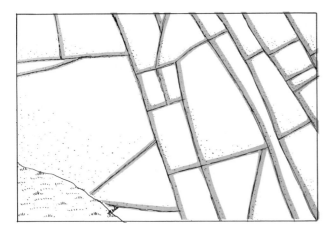

(A) Rock cut by a joint system.

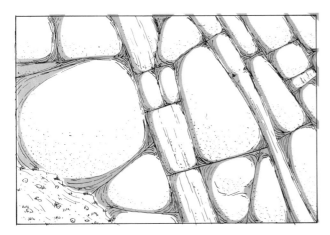

(B) Preliminary stages of spheroidal weathering.

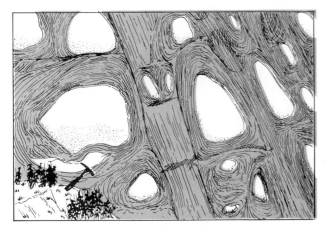

(C) Late stages of spheroidal weathering.

Figure 9.14 *Exfoliation domes in the Sierra Nevada Mountains, California. The Sierras are composed of massive granite cut by joints that separate the rock into large blocks. The exfoliation domes developed as huge slabs of rock spalled off. The joints separating the rock slabs probably resulted from expansion as erosion removed the overlying rock.*

series of concentric shells or layers that look like cabbage leaves (*figure 9.14*). The layers, essentially parallel to each other and to the surface, develop by both chemical and mechanical means. Sheeting can play an important part, because deeply buried rocks like granite have a tendency to expand upward and outward as the overlying rock is removed. In cold climates, frost wedging along the sheeting joints helps in the gradual removal of successive layers. The volume increase associated with the decomposition of feldspar also is thought to function in the development of exfoliation. As a result of exfoliation, massive rocks, such as granite, develop a spherical form on a wide range of scales characterized by a series of concentric layers.

Climate and Weathering

Statement

Climate is probably the single most important factor influencing weathering. It, more than anything else, determines the type of weathering in a given region and the rate at which the process operates. Therefore, the products of weathering usually directly reflect the climate. It has been shown in previous sections that water is the most important agent in almost all types of weathering. The total amount of precipitation is, of course, very significant; but the intensity of rain, seasonal variations, infiltration, runoff, and the rate of evaporation all combine to influence the characteristics of weathering and weathered products in a given region. Temperature is important, because it greatly influences the rate of chemical reaction; this is illustrated by the fact that a 10 °C increase in temperature commonly doubles reaction rates. Other temperature factors include average temperature, temperature ranges, and fluctuations of temperature around the freezing point.

Discussion

Perhaps the best way to appreciate the influence of climate on weathering is to consider variations in the types and thicknesses of soils from the equator to the poles. The diagram in *figure 9.15* summarizes in a general way the relationships between the amount of weathering (chemically altered rock) and variations in precipitation and temperature.

In the humid tropical climates, chemical weathering

Figure 9.15 *Schematic diagram showing variations in weathering from the arctic to the tropics. The graphs at the top of the diagram show variations in precipitation, temperature, and vegetation. The generalized cross section in the lower part of the diagram shows the relative depths of weathering that result from fluctuations of these factors. Weathering is most pronounced in the tropics, where precipitation, temperature, and vegetation reach a maximum; conversely, a minimum of weathering is found in deserts and polar regions, where these factors are also minimal.*

is extreme and rapidly develops thick soils to depths in excess of 70 m. Under such conditions, the feldspars in granites and related rocks are altered completely to clay minerals and all soluble minerals are removed in solution. Only the most insoluble materials (such as silica, aluminum, and iron) remain in the thick, deep soil, with the result that the soil commonly is infertile. High temperatures in tropical zones increase the rate of chemical reaction so that chemical decomposition is very rapid. This is illustrated by the fact that soil develops on new lava flows in Hawaii in only a matter of years. Frost action, of course, is essentially nonexistent in the tropics.

In the low-latitude deserts north and south of the tropical rain forests, chemical weathering decreases to a minimum because of the lack of precipitation. Soil is nonexistent and exposures of fresh, unaltered bedrock are abundant. However, mechanical weathering is evident in the fresh, angular rock debris that litters most slopes.

In the temperate climates that occur in the higher latitudes, the climate is subhumid to subarid and temperatures range from cool to warm. There, both chemical and mechanical processes operate and the soil and regolith develop to depths of several meters. In savanna climates, where alternating wet and dry seasons occur, the type of weathering is somewhat different from regions where precipitation is more evenly distributed throughout the year. An interesting example of weathered products' expressing adjustments to climatic conditions is seen in differences in weathering on north- and south-facing slopes. South-facing slopes are more exposed to the sun and during the winter and spring months experience more melting and a greater number of freezing-thawing cycles than do the north-facing slopes that are more in the shade. The south-facing slopes undergo

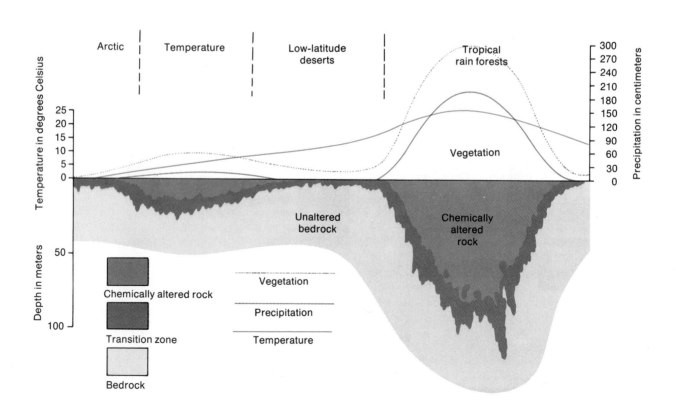

noticeably more mechanical weathering and are typically more rugged; in contrast, the north-facing slopes develop a thicker soil and support more vegetation.

In the polar climates, weathering is largely mechanical. Temperatures are too low for much chemical weathering, so the soil typically is thin and is composed mostly of angular, unaltered rock fragments. In permafrost zones, unique ground patterns result from thermal contractions and differential thawing and freezing in areas of swamps, lakes, and dry ground.

Rates of Weathering

Statement

The rate at which weathering processes are able to decompose and break down a solid rock body is quite variable and depends upon the interrelationship among three main factors.
1. The susceptibility of the constituent minerals to weathering.
2. The climate or the intensity of the weathering processes.
3. The amount of surface exposed to the atmosphere.

Minerals vary widely in their ability to resist weathering, and, although many details are not understood completely, minerals that crystallize at a high temperature are most susceptible to weathering. Thus, olivine and Ca-plagioclase weather most rapidly, followed by pyroxene and amphibole and then by Na-plagioclase and K-feldspar. Quartz is one of the most highly resistant minerals, and it resists alteration by chemical decomposition.

In warm, humid climates, the rate of weathering is very rapid, a fact indicated by some of the recent lava flows on Hawaii, which, within a few years, were decomposed sufficiently to support some vegetation. By contrast, in colder climates, the same rock types have remained fresh and essentially unaltered since the ice age.

Discussion

In spite of the complexity of the weathering process, some idea of the rates of weathering can be gained from measuring the amount of decay on rock surfaces of known age. Tombstones, ancient buildings, and monuments, for example, provide rock surfaces that were

Figure 9.16 *Weathering of the Great Pyramids of Egypt. The weathering process is expressed in two ways: the rectangular building blocks have been modified into an elliptical form and the weathered debris from the blocks has accumulated as talus on each step.*

Figure 9.17 *Pyramid at Saggarah. Weathering of the ancient stepped pyramids. Each step is completely covered with talus, and many individual blocks are weathered to spherical shapes.*

originally fresh and unaltered when constructed and have since suffered weathering over a specific period of time. One interesting example is found in studies of the Egyptian pyramids.

The Great Pyramid of Cheops near Cairo, Egypt, was faced originally with polished, well-fitted blocks of limestone, which protected the inner layers of core rock from weathering until the outer layers were removed about 1000 years ago to build mosques in Cairo. Since then, weathering has attacked all four main rock types used in the construction of the pyramid. The most durable rock (least weathered) in the pyramid is a granite, which remains essentially unweathered. Also resistant is a hard, gray limestone that still retains marks of the quarry tools. However, the shaly limestone used in other blocks has weathered rapidly, and many of the blocks have a zone of decay as deep as 20 cm. Most of the weathered debris has remained as a talus on individual tiers and around the base of the pyramid (*figure 9.16*). The volume of weathered debris produced from the pyramid during the last 1000 years has been calculated to be 50,000 m³, or an average of 50 m³ per year. This is a loss of approximately 3 mm per year over the entire surface of the pyramid. Some of the older stepped pyramids, built 4500 years ago, show much greater weathering, and many large blocks are nearly completely decayed or reduced to small spheroidal boulders. In addition, high piles of debris have accumulated on each major terrace (*figure 9.17*).

Summary

Weathering processes act upon a rock to break it into small pieces and decompose its unstable minerals. The fact that most rocks have been fractured by stresses in the crust greatly affects weathering by increasing the surface area of exposed rock and by providing channels through which weathering agents (air and water) gain access to rock below the surface. As weathering proceeds, there is a universal tendency to produce spheroidal fragments. Climate exerts a profound influence upon weathering and controls the type and rate of weathering processes. One of the main products of weathering is soil.

Weathering processes do not erode and transport the fragmented and decomposed rock materials; they produce them. Other agents (such as running water) pick up the weathered rock fragments and transport them as sediment toward the ocean. It is the process of erosion by running water that will be the subject of the next three chapters.

Additional Readings

Carroll, D. 1970. Rock Weathering. New York: Plenum Press.

Goldich, S. S. 1938. "A Study in Rock Weathering." J. of Geology 46:17-58.

Keller, W. D. 1957. Principles of Chemical Weathering. Columbia, Mo.: Lucas Brothers Publishing Company.

Loughnan, F. C. 1969. Chemical Weathering of the Silicate Minerals. New York: American Elsevier Publishing Company.

Ollier, C. D. 1969. Weathering. New York: American Elsevier Publishing Company.

Reiche, P. 1950. A Survey of Weathering Processes and Products. Albuquerque: University of New Mexico, Publications in Geology, Number 3, pp. 1-95.

River Systems

A river system consists of a network of connecting channels through which water precipitated on the surface is collected and funneled back to the ocean. At any given time, it is estimated, there are about 1300 km³ of water flowing in the world's rivers. As the water flows, it picks up and carries with it the loose, decomposed rock debris produced by weathering and transports it to the ocean. The removal of the weathered regolith is referred to as denudation. Therefore, the river systems continually erode the surfaces of the continents and constitute one of the most important geologic agents on the earth.

By measuring the volume of sediment in motion within a river system, we can measure the rates of denudation. Yet, how can we prove that streams have formed the valleys and canyons through which they flow? How could the Colorado River have cut the Grand Canyon over 2 km deep in solid bedrock? Since the erosion of a canyon cannot be observed from start to finish, geologists must approach the problem of erosion and the development of landforms indirectly. Two methods have been used: we have analyzed the drainage network and found it to have system and order susceptible to mathematical analysis, and we have studied the processes operating in drainage basins today, processes infinitely slow but over a period of several million years capable of eroding the earth's surface to the form in which it is seen today.

In this chapter, the order in river systems and their erosion of the earth's surface will be considered.

Major Concepts

1. Running water is part of the hydrologic system of the earth and is the most important eroding agent. Stream valleys are the most abundant and widespread landforms on this planet.
2. A river system consists of a main channel and all of the tributaries that flow into it. It can be subdivided into three subsystems: (a) tributaries that collect and funnel water and sediment into the main stream, (b) a main trunk that is largely a transporting system, and (c) a dispersing system that is found at the river's mouth.
3. There is a high degree of order among the various elements of a river system.
4. Sediment is transported when the velocity of stream flow exceeds the settling velocity of the sediment.
5. The capacity of running water to erode and transport sediment is largely dependent upon stream velocity, although characteristics of the channel are also significant.
6. There is a universal tendency for a river system to establish equilibrium among the various factors that influence stream flow (velocity, volume of water, gradient of stream, and volume of sediment).
7. Man is able to manipulate river systems by constructing dams, flood-control projects, and cities. Such construction commonly results in many unforeseen, and unwanted, side effects.

The Geologic Importance of Running Water

Statement

Running water is by far the most important agent of erosion and is responsible for the configuration of the landscape of most continental surfaces (see *plates 5, 6, 12, 13* and *15*). Other agents such as ground water, glaciers, and wind are locally dominant but affect only limited parts of the land surface and generally operate in an area during a limited interval of geologic time. Even where other agents are especially active, running water still is likely to be a significant geologic agent.

Discussion

Appreciating the significance of streams and stream valleys in the regional landscape of the earth is a problem of perspective much like the problem of appreciating the abundance of craters on the moon from viewpoints on its surface. To an astronaut, the surface of the moon appears to be an irregular, broken landscape cluttered with rock debris. The crater systems and the patterns of terrain, so striking when viewed from space, are not at all apparent from vantage points on the moon's surface. Indeed, crater rims may appear as rounded hills. Without the aid of maps or aerial photographs, it is difficult to recognize some of the larger craters as circular forms when they are seen only from the Apollo landing sites. When viewed from the ground, the earth's stream valleys may appear only as relatively insignificant irregular depressions between rolling hills, mountain peaks, and broad plains. But, when viewed from space, they dominate the landscape of the earth much as craters dominate the landscape of the moon.

The ubiquitous nature of stream valleys and the importance of running water as the major agent of erosion may be appreciated best by considering a broad regional view of the continents and their major river systems as seen through high-altitude aerial photography. As the photographs of the southwestern United States in *figure 10.1* illustrate, throughout broad regions of the continents, the surface is little more than a complex of valleys created by stream erosion. Even in the desert, where it may not rain for decades, the network of valleys fashioned by streams is commonly the dominant landform. In more humid areas, details of drainage networks can be obscured by vegetation, but stream valleys related to drainage systems dominate most landscapes.

The Major Features of a River System

Statement

A river system consists of a main channel together with all of the tributaries that flow into it. Each river system is bounded by a divide (ridge) beyond which water is drained by another system. The surface of the ground slopes toward the network of tributaries so that the drainage system acts as a funneling mechanism for removing precipitation and weathered rock debris. Essentially, all of the land surface is related to some drainage system.

(A) A Skylab photograph of an area in the arid southwestern United States shows the regional patterns of a river system and its valleys.

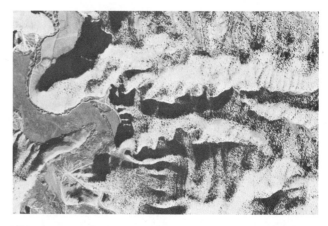

(B) A high-altitude aerial photograph of a portion of the area shown in photo A shows that an intricate network of streams and valleys exists within the tributary regions of the larger streams.

(C) A low-altitude photograph of part of the area shown in B shows many smaller streams and valleys in the drainage system.

Figure 10.1 Stream valleys.

A typical river system can be divided into three subsystems.

1. *A collecting subsystem,* consisting of a network of tributaries in the headwater region that collects and funnels water and sediment to the main stream.

2. *A transporting subsystem,* consisting of the main trunk stream in which the main process is transporting water and sediment out of the drainage area.

3. *A dispersing subsystem,* consisting of a network of distributaries at the mouth of a river where sediment and water enter the ocean, a lake, or a dry basin.

Each of the subsystems and their relationships to one another are illustrated in *figure 10.2.*

Discussion

The collecting subsystem of a river commonly has a dendritic (treelike) pattern with numerous branching

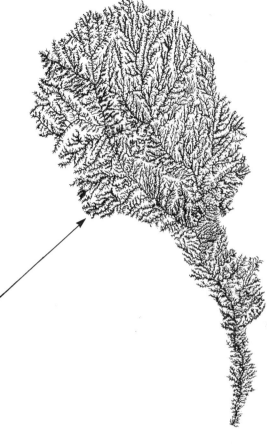

Figure 10.3 *Details traced from aerial photographs show the intricacy of a drainage system (see also plate 12).*

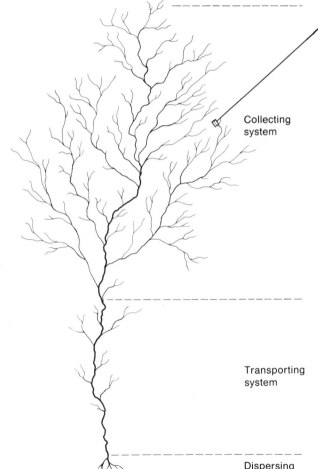

Figure 10.2 *Idealized diagram showing the major parts of a river system. The tributaries in the headwaters constitute a collecting subsystem where water and sediment are collected and funneled into a main trunk stream. Erosion is dominant in this area. The main trunk stream functions as a transporting subsystem, but both erosion and deposition may occur in this area. The lower end of the river is a dispersing subsystem in which most sediment is deposited in a delta or an alluvial fan and water is dispersed into the ocean. Deposition is the dominant process in this part of the river.*

tributaries that extend upslope toward the divide. This subsystem is the primary source area for both water and sediment transported in the system, and it includes the region of most vigorous erosion. On most maps, only the major tributaries are shown, and so one may fail to appreciate the delicately intricate network of the tributary system. When the minor tributaries and the small rills are plotted on a map (as in *figure 10.3*), it soon becomes apparent that all of the land surface is part of a drainage system.

The transporting subsystem is the main trunk stream, which functions as a channelway through which water and sediment move from the collecting area toward the ocean. Although the major process is transportation, additional water and sediment are collected in this subsystem and some deposition occurs as the channel meanders back and forth and at times when a river overflows its bank during its flood stage. Both erosion and deposition, in addition to transportation, occur in the transporting subsystem of a river.

The dispersing subsystem is the small but important part of a river system where the main stream breaks up into a series of small channels, called **distributaries,** as it enters the ocean, a lake, or a dry basin. The major processes active are deposition of the coarse sediment load and dispersal of the fine-grained material and the river waters into the basin.

Order in Stream Systems

Statement

Studies of drainage systems show that where a stream system is able to develop freely on a homogeneous surface there is a definite mathematical relationship (ratio) between the tributaries and the size and gradient of the stream and the valley through which the stream flows. Some of the more important relationships are these.

1. The number of stream segments (tributaries) decreases downstream in a definite geometric progression.
2. The length of tributaries becomes progressively longer downstream.
3. The gradient of tributaries decreases downstream in a constant ratio.
4. The stream channels become progressively deeper and wider downstream.
5. The size of the valley is proportional to the size of the stream and increases downstream.

These relationships among tributaries in a drainage system constitute the basis for the assumption that streams erode the valleys through which they flow.

Discussion

The first meaningful observations about the relationship of stream systems to their valleys were made in 1802 by John Playfair, an English geologist. In a statement, which has become known as Playfair's law, he said:

> Every river appears to consist of a main trunk, fed from a variety of branches, each running in a valley proportional to its size, and all of them together forming a system of valleys connecting with one another, and having such a nice adjustment of their declivities that none of them join the principal valley at either too high or too low

a level; a circumstance which would be infinitely improbable if each of these valleys were not the work of the stream which flows in it.

An appreciation of the order and system in a drainage basin can be gained from careful study of *figure 10.4*, which is a graphic model of a typical stream. The decrease in the gradient of the tributaries is expressed by the curve on the side of the block diagram (a curve that can be expressed by a mathematical formula). The stream channel becomes progressively deeper and wider downstream, as is shown by the series of cross sections. Similarly, the sizes of the valleys increase downstream and are proportional to the sizes of the streams that flow through them. You should note also that the length of the tributaries increases downstream. Numerous statistical studies support these observations and show that there are a number of specific mathematical relationships between a stream and the valley through which it flows.

If valleys were ready-made by some process other than stream erosion, such as faulting or other earth movements, these relationships would be "infinitely improbable" (see *plate 12*). You can easily confirm the high degree of order in streams by carefully studying the aerial photographs in *figure 10.1*. Does each tributary have a steeper gradient than the stream into which it flows? Does each tributary flow smoothly into a larger stream without an abrupt change in gradient? Are the tributary valleys smaller than the valleys into which they drain?

Stream erosion has been studied in great detail over the last 100 years, and we have been able to observe and measure many aspects of stream development and erosion by running water. The origin of valleys by erosion is well established, and it is clear that running water constitutes the most significant process shaping the earth's surface.

Figure 10.4   *Schematic diagram showing the systematic change in characteristics of a river downstream. The gradient decreases, as indicated by the curve of the diagram. In addition, the channel becomes larger, the volume of water increases, and the size of valley increases.*

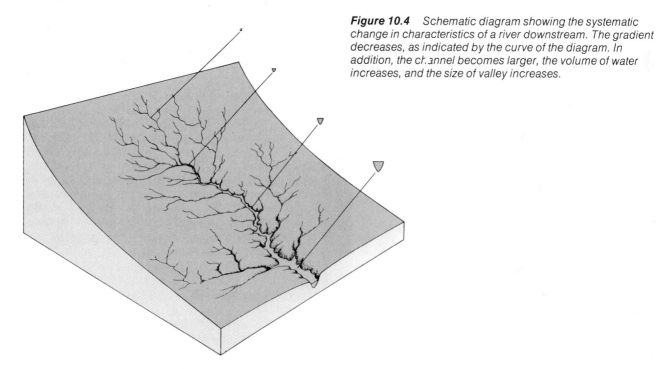

The Flow of Water in Natural Streams

Statement

Water in natural streams moves by **turbulent flow,** which is characterized by numerous secondary eddies and swirls in addition to the main downstream movement. **Laminar flow,** in which water moves in straight lines, is restricted to a thin film a few millimeters thick on the bottom and walls of the stream channel. The flow of water in a stream is very complex and is influenced by a number of variables, the most important of which are:

1. **Discharge** (the amount of water passing a given point during a specific interval of time; discharge is usually measured in cubic meters per second).
2. **Velocity.**
3. Shape and size of channel.
4. **Gradient** (slope of the stream channel).
5. **Base level** (the lowest level to which a stream can erode).
6. **Load** (the material carried by the flowing water).

Discussion

Laminar and Turbulent Flow. The nature of flow in rivers can be studied by injecting dye into a stream and observing the path of movement. In the laboratory, water moving in a glass tube at very low velocities moves in

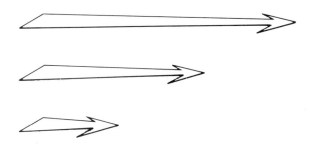

(A) Laminar flow—water particles move in parallel lines.

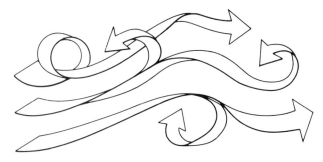

(B) Turbulent flow—many secondary eddies are superposed on the main stream flow.

Figure 10.5 *Diagrams showing types of flow.*

laminar flow, with parallel layers of water shearing over one another. Theoretically, there is no mixing. Laminar flow usually is not found in natural streams except in a very thin layer along the bottom and banks of the channel. However, it is common in groundwater flow.

Turbulent flow is a type of irregular movement, with eddies superposed on the main downstream flow (*figure 10.5*). Dye injected into turbulent flow moves downward, sideways, and upward and shows that the water is capable of lifting loose sediment from the riverbed and transporting it downstream.

Discharge. The discharges of most of the major drainage systems of the world have been monitored by gauging stations for a number of years, and the information collected from such studies provides the basis of planning for the utilization of water resources, flood control, estimates of erosion rates, and so on. An essential point to remember is that the discharge increases downstream because of the addition of water from each tributary (*figure 10.4*). Small tributaries discharge only 5 to 10 cubic meters per second (abbreviated m^3/sec.). Larger streams (such as the Mississippi River) discharge 40,500 to 337,500 m^3/sec. The Amazon River, which has the highest discharge of any river in the world, was measured at 1.5 million m^3/sec. This is because of the great amount of precipitation in the collecting system of the river.

The water for a river system comes from both surface runoff and seepage of ground water into the stream channels. Ground-water seepage is important, because it maintains the flow of water throughout the year and establishes **permanent streams.** When the supply of ground water is depleted seasonally, streams become dry temporarily. Such streams are referred to as **intermittent streams.**

Velocity. The velocity of flowing water is not uniform throughout the stream channel and depends upon channel shape and roughness and stream pattern. The velocity is usually greatest near the center of the channel above the deepest part and away from the frictional drag of the channel walls and floor (*figure 10.6*). In a straight segment of a stream, the maximum velocity is in the central part of the channel; but, as the channel curves around a meander, the zone of maximum velocity shifts to the outside of the bend, with minimum velocity on the inside of the curve (*figure 10.7*). This is important in the lateral erosion of stream channels and the migration of stream patterns.

Obviously, the velocity of flowing streams is related to the gradient of the stream channel. Steep gradients produce very rapid flow as is commonly seen in high mountain streams. Where slopes are very steep, waterfalls and rapids develop and velocity approaches that of free fall. Low gradients produce slow, sluggish flow and, where a stream enters a lake or the ocean, the velocity is soon reduced to zero. The velocity of flowing water also depends upon the volume of water within the channel. The greater the volume in a given channel, the faster the flow.

Channel Shape. The cross-sectional shape of a stream channel greatly influences the nature of the flow and the velocity at which a given volume of water moves

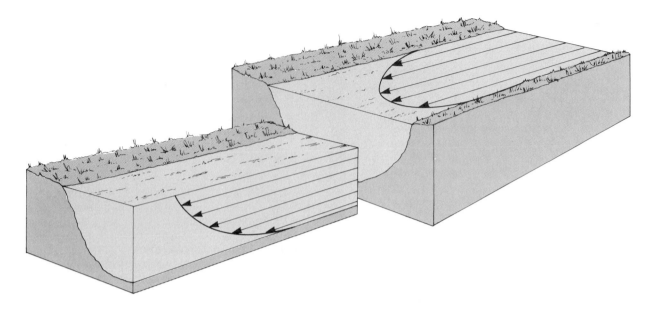

Figure 10.6 *Variations in velocities in natural stream channels. Friction reduces the velocity along the base and the sides of the channels. Maximum velocity in a straight channel is near the top and center of the channel*

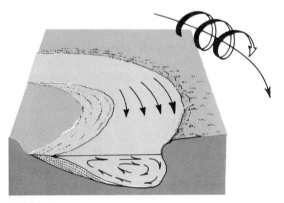

Figure 10.7 *Patterns of flow in a curved channel. Water on the outside of the bend is forced to flow faster than that on the inside of the curve. This, together with normal frictional drag on the channel walls, produces a corkscrewlike pattern of flow. As a result, erosion occurs on the outer bank and deposition occurs on the inside of the bend. This produces an asymmetrical channel, which slowly migrates laterally.*

down a given slope. Velocity is greatest in channels that offer the least resistance to flow. A channel that approaches an ideal semicircular cross section (*figure 10.8C*) has the least surface-area-per-unit volume of water and, hence, offers the least resistance to flow. Wide, shallow channels (*figure 10.8B*) and deep, narrow channels (*figure 10.8A*) both present greater surface-area-per-unit volume of water and retard the flow of water. Streams flowing through narrow channels crowd and erode their banks so that the channel shape approaches that of a semicircle. In flat channels, deposition of sediment

results from the low velocities and the stream produces a more narrow channel. In both cases, changes occur and produce a balance or an equilibrium between the flowing water and the area through which it moves (*figure 10.8*).

A rough channel strewn with boulders and other obstructions retards flow and reduces velocity. A smooth, clay-lined channel offers the least resistance and permits greater velocity. This roughness factor is difficult to measure, but its influence is obvious.

Stream Gradient. The gradient of a stream is the slope of the streambed. It is steepest in the headwaters and decreases downslope so that the **longitudinal profile** (a cross section of a stream from its headwaters to its mouth) is a smooth, concave, upward curve that becomes very flat at the lower end of the stream. The gradient usually is expressed in the number of meters the stream descends during each kilometer of flow. The headwater streams that drain the Rockies can have gradients of over

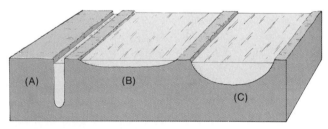

(A) Deep channels present a large surface-area-per-unit volume of water and tend to develop a wider channel.
(B) Flat channels also present a large surface area and tend to develop a deeper channel.
(C) A semicircular channel presents a minimum surface-per-unit volume of water and is the optimum channel shape.

Figure 10.8 *Stream channels show a tendency to adjust to a shape that offers minimum resistance to flow.*

50 m/km, while the lower reaches of the Mississippi River have a gradient of only 1 to 2 cm/km.

Base Level. The base level of a stream is defined as the lowest level to which a stream can erode its channel. It is a key concept in the study of stream activity. The base level is, in effect, the elevation of its mouth where it enters the ocean, a lake, or another stream. A tributary cannot erode lower than the level of the stream into which it flows. Similarly, a lake controls the level of erosion for the entire course of the river that drains into it. The levels of tributary junctions and lakes are temporary base levels, because lakes can be destroyed and major streams can deepen their channels. For all practical purposes, the ultimate base level is sea level, because the energy of a river is quickly reduced to zero as it enters the ocean. Exceptions are fault valleys like Death Valley, which is faulted below sea level. As base level changes (even sea level), the longitudinal profile of the river changes and the stream adjusts to the new condition.

Stream Load. Flowing water in natural streams provides a fluid medium by which the loose, disaggregated regolith is picked up and transported to the ocean. The capacity of a stream to transport sediment increases to a third or fourth power of the velocity; that is, if the velocity is doubled, the stream can move 8 to 16 times as much sediment. The amount of material transported by a stream (its load) is usually less than its **capacity** (the total amount the stream is able to carry under a given set of conditions). The maximum particle size a stream can transport is its **competence**. In many streams, the competence is enormous, especially in those with a relatively steep gradient and a high discharge. Flash floods in the Wasatch Mountains in 1923 carried boulders weighing 80 to 90 metric tons for nearly 2 km. When a dam breaks, the large volume of water is capable of transporting blocks weighing as much as 10,000 metric tons. By far the greatest amount of material carried by a stream is derived from the regolith in the collecting system of the drainage basin. This fact is fundamental to understanding the erosion of the land. Weathering produces a blanket of loose, unconsolidated rock (the regolith), which is easily washed downslope and carried away by the river system. This exposes fresh bedrock, which then is weathered to form new regolith, and the cycle continues. More will be said about transportation of sediment and erosion of the land in subsequent sections.

The Transportation of Sediment by Streams

Statement

Running water is the major agent of erosion not only because of its ability to abrade and erode its channel but because of its enormous power to transport loose sediment produced by weathering. The load, or volume, of erosional debris transported by a river is almost incomprehensibly large. The Mississippi River alone carries 454 million metric tons of sediment a year, a quantity roughly equivalent to a layer of sediment 0.5 m deep spread over an area of 500 km². In some rivers, the sediment carried at a given time even may exceed the volume of water and the stream may become a mud flow. There-

fore, it is probably more accurate to consider a river as a system of moving water and sediment, rather than one of flowing water alone.

The mechanics by which sediment is transported can be understood by taking a handful of soil and dropping it into a container of water. The large, coarse particles sink rapidly, while the smaller particles stay in suspension for hours. Each size of particle has its own **settling velocity,** (the constant velocity at which the particle falls through a still fluid). The settling velocity is determined not only by the size of the particle but also by the specific gravity and the shape of the particle, as well as by the density of the fluid. More time is required for a given particle to settle in muddy water and salt water, because they are denser than clear water. Particles of gold, iron, and lead settle faster than quartz, because they are heavier. This principle is obvious to miners, who have panned for minerals in streams for ages.

Within a stream system, sediment is transported in three ways.
1. The fine particles are moved in suspension (**suspended load**).
2. The coarse particles are moved by traction along the streambed (**bed load**).
3. Dissolved material is moved in solution (**dissolved load**).

Discussion

Suspended Load. The suspended load is the most obvious, and generally the largest, fraction of the material moved by a river. It consists of mud, silt, and fine sand and can be seen in most parts of the transporting segment of a river system. The movement of particles in suspension occurs where the velocity of the upward currents in turbulent flow exceeds the settling velocity of the sediment particles. Thus, in essentially all major streams, most silt and clay-size particles remain in suspension most of the time and move downstream at the same velocity as the flowing water to be deposited in the ocean, the lakes, or on the flood plain.

Samples of the suspended load moved by the Missouri River at Kansas City show that clay-size particles are distributed equally from top to bottom, but sand-size particles occur in larger amounts near the bottom of the channel (*figure 10.9*).

Bed Load. Particles of sediment too large to remain in suspension collect on the bottom of the stream and are moved by sliding, rolling, and saltation (short leaps) (*figure 10.10*). The distinction between the largest particles of the suspended load and the smallest particles of the bed load is not always sharp, because the velocity of a stream fluctuates constantly and part of the bed load suddenly may move in suspension, and vice versa.

Bed load can constitute 50% of total load in some rivers, but in most rivers it ranges from 7% to 10% of the total sediment load. Although the movement of this material is not easily seen in most streams, one can feel and sometimes hear the dull, thundering impact of boulders rolling over the channel bottom in large, rapidly flowing streams. The bed load moves only when there is sufficient velocity to move the large particles and thus differs fundamentally from the suspended load that

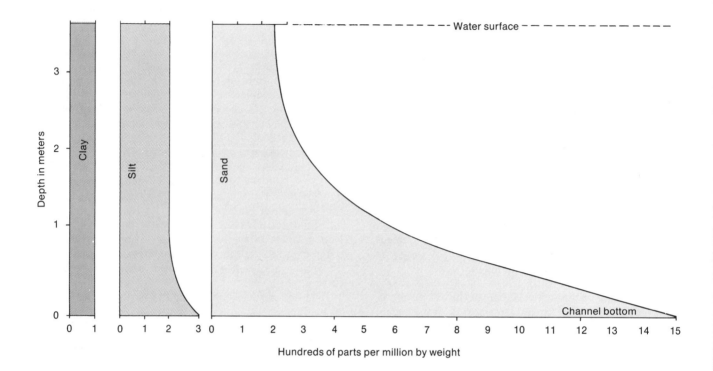

Depth in meters

Clay

Silt

Sand

Water surface

Channel bottom

Hundreds of parts per million by weight

Figure 10.9 *Distribution in the sediment load from the surface to the base of the Missouri River at Kansas City. The vertical scale shows depth of water, and the horizontal scale shows the amount of sediment being transported in hundreds of parts per million (by weight). Clay and silt are transported in suspension and are nearly equally distributed throughout the channel. The number of sand-size particles increases conspicuously toward the bottom of the channel because they are transported mostly by saltation and traction.*

moves constantly. The movement of this bed load is one of the major tools of stream erosion; for, as the sand and gravel move, they abrade (wear away) the sides and bottom of the stream channel.

Dissolved Load. Dissolved matter is essentially invisible and is transported in the form of chemical ions.

All streams carry some dissolved material, which is derived principally from ground water emerging as seeps and springs along the river banks. The most abundant materials in solution are Ca and HCO_3, but sodium, magnesium, chlorine, ferric, and sulfate ions are also common. Various amounts of organic matter are present, and some streams are brown with organic acids derived from the decay of plant material.

The velocity of flow, which is so important to the transportation of the suspended and traction load, has

Figure 10.10 *Movement of the bed load in a stream. Particles too large to remain in suspension are moved by sliding, rolling, and saltation. When discharge increases due to heavy rainfall or spring snowmelt, all loose sand and gravel may be flushed out so that bedrock is eroded by abrasion.*

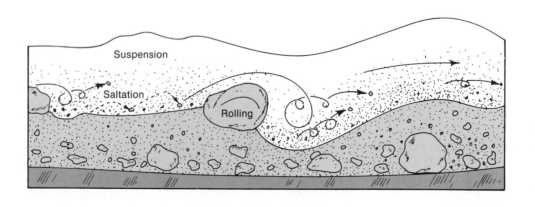

Suspension

Saltation

Rolling

little effect on the capacity of the river to move material in solution. Apparently, once mineral matter is dissolved, it remains in solution, regardless of velocity, and is precipitated and deposited only when the chemistry of the water changes.

Chemical analysis shows that most rivers carry less than a thousand parts per million dissolved load, but some streams in arid regions carry several thousand parts per million and are distinctly salty in taste. Although these amounts of dissolved material seem small, they are far from trivial. Sampling indicates that in some rivers, when rainfall is abundant, more than half of all the material carried to the ocean is in solution. In areas of low relief, such as the Atlantic and Gulf Coast states,

runoff is slow and solution activity lowers the surface at a rate of 52 metric tons per square kilometer, or about 1 m in 25,000 years. The dissolved load is somewhat less in mountainous terrain (1 m in 100,000 to 3 million years), but it is by no means negligible.

In order to appreciate the volume of the sediment load carried by a river, refer again to *plate 16,* which shows silt and mud from the Mississippi River as a large cloud of sediment at the river's mouth in the Gulf of Mexico. In many rivers, the sediment load can be seen in the reservoir behind a dam. A dam across a river effectively traps all the sediment and the reservoir eventually will be filled with mud and silt (*figure 10.11*). Unless this sediment is removed by dredging or other means, it will terminate the useful life of the reservoir.

Relation of Velocity to Erosion, Transportation, and Deposition

Statement

The capacity of running water to erode and transport sediment is largely dependent upon the stream velocity. If the velocity is doubled, the size of the particles that can be transported increases four times. There are two reasons for this exponential increase.

Figure 10.11 *Mono Reservoir, California. This reservoir was built as a debris basin to protect the Gebaltar Reservoir downstream. It was built in 1935 after a severe fire that burned off much of the vegetation in much of the watershed area. Very heavy rains occurred during the following two winter seasons before vegetation could be reestablished, and the accelerated erosion provided enough silt and sand to fill completely the Mono Reservoir and to fill partly the lower Gebaltar Reservoir. This serves as a vivid reminder that a river system is a system of moving water and sediment, not just of water alone.*

1. Each particle of water hits a particle of sediment with twice as much force.
2. Twice as much water strikes the face of a fragment during a given interval of time.

Discussion

The results of experimental studies of water's capacity to erode, transport, and deposit sediments are summarized in *figure 10.12*. The velocity at which a particle of a given size is picked up and moved is shown on the upper curve. On the graph, this is a zone instead of a line, because the value (or erosional velocity) varies according to the characteristics of the water and the shape and the density of the grain to be moved (shape has a definite effect on the suspension of flat mica flakes, for example). Moreover, erosional velocities vary with the depth and the density of the water. The lower curve (dashed line) shows the velocity at which the particle settles out and comes to rest. It is interesting to note that particles such as fine silt and clay require a relatively high velocity to lift them into suspension. Small particles tend to stick together. However, once in suspension, fine particles remain suspended with a minimum velocity.

The graph in *figure 10.12* also shows that sediment is deposited according to size whenever the velocity decreases below a critical settling velocity. Thus, on gentle slopes where the stream's velocity is reduced, a signifi-

cant part of the sediment load is deposited along the channel or on the flood plains of the river as the channels meander and continually develop new courses. Most of the remaining sediment is deposited when velocity is reduced at the point where the river enters a lake or the ocean.

The period of greatest erosion and transportation of sediment is, of course, during floods. The increase in discharge results in an increase in both the maximum size of material to be transported and the total load. Exceptional floods may cause an unusually large amount of erosion and transportation during a brief period of time. In arid regions, many streams carry little or no water throughout the year but, during a cloudburst, great quantities of sediment can be moved.

Equilibrium in River Systems

Statement

In previous sections, it has been emphasized that streams do not occur as separate, independent entities but constitute a drainage system in which each segment is intimately related to the others. Every stream has tributaries and every tributary has smaller tributaries extending down to the smallest gully. The entire river system functions as a unified whole, and any change brought about in one part of the system affects the other parts. The major factors affecting stream flow (discharge, velocity, gradient, base level, and load) constantly change to establish a dynamic balance or a condition of equilibrium.

Therefore, adjustments within a river system are continually being made so that, ultimately, the slope or the gradient of a stream is altered to accommodate the volume of water available, the channel's characteristics, and the velocity necessary to transport the sediment load. A

Figure 10.12 *Graph showing threshold velocities of sediment in stream flow. The upper curve shows the velocity necessary for a stream to pick up and move a particle of a given size. This is a zone on the graph, not a line, because of variations resulting from stream depth, et cetera. The lower curve indicates the velocity at which a particle of a given size will settle out and be deposited. Note that fine particles will stay in suspension at velocities much lower than those required to lift them from the surface of the streambed.*

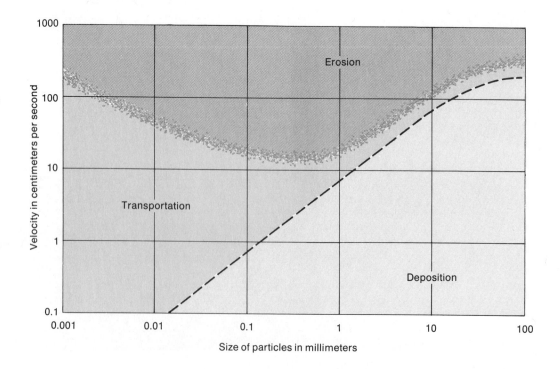

change in any of these factors causes compensating adjustments to restore equilibrium within the entire drainage system.

A profile of complete equilibrium in a river is one in which the channel form and gradient are balanced delicately so that neither erosion nor deposition occurs—an ideal condition toward which rivers are continually adjusting. The concept is important to an understanding of the natural evolution of the landscape. It also has some very practical applications, since man is continually attempting to modify the rivers to better suit his needs and he should know how the river system will respond to artificial modifications.

Discussion

The concept of equilibrium in a river system can be appreciated best by considering the slope of the longitudinal profile of a hypothetical stream in which equilibrium has been established and the changes that occur when the equilibrium is disrupted (*figure 10.13*). In diagram *A*, the variables in the stream system (discharge, velocity, gradient, base level, and load) are in balance so that neither downcutting nor sedimentation occurs along the stream's profile. There is just enough water to transport the available sediment down the existing slope. The energy of the stream system is in balance, and the stream's profile is a smooth curve. In diagram *B*, the stream's profile is displaced by a fault that created a waterfall. The increased gradient across the falls greatly increases the velocity so that rapid erosion occurs and the waterfall (or the rapid) migrates upstream. The eroded sediment added to the stream on the dropped block is more than that segment of the stream can transport, because the stream *was* in equilibrium before faulting occurred. Therefore, the river deposits part of its load and builds up the gradient of the channel (shaded area in diagrams *C, D,* and *E*) until a new profile of equilibrium (like that in diagram *A*) is established (*E*).

An example of the adjustments described above is found in Cabin Creek, a small tributary to the Madison River north of Hebgen Dam in Montana. In 1959, a 3-m fault scarp formed across the creek during the Hebgen Lake earthquake. By June 1960, erosion by Cabin Creek had erased the waterfall at the cliff formed by the fault and developed a small rapid. By 1965, the rapid was removed completely and equilibrium was reestablished.

The tendency for river systems to establish a profile of equilibrium also is illustrated by the results of the construction of dams. Dams are built to store water for industrial and irrigational use, to control floods, and to produce electrical power. Yet, when a dam is built, the balance of the river system established over thousands or millions of years is instantly upset. Many unforeseen, long-term effects occur both upstream and downstream from the new structure and its great reservoir of water. In the reservoir behind the dam, the gradient is reduced to zero; hence, where the stream enters the reservoir, the sediment load carried by the stream is deposited as a delta and as layers of silt and mud over the floor (*figure 10.11*).

Less obvious are the changes in the gradient upstream as the delta builds out into the reservoir. To adjust to the

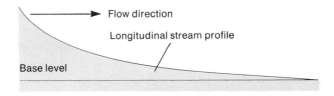

(A) *When the stream profile is at equilibrium, the velocity, load, gradient, and volume of water are in balance so that neither erosion nor deposition occurs.*

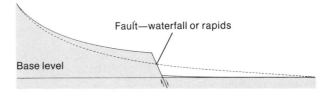

(B) *Faulting disrupts equilibrium by decreasing the gradient downstream and increasing the gradient at the fault line.*

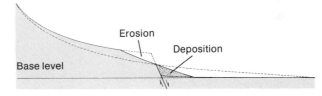

(C) *Erosion proceeds upstream from the fault, and deposition occurs downstream.*

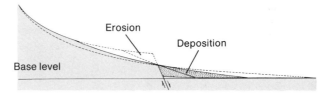

(D) *Erosion and deposition continue to develop a new stream profile.*

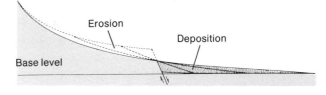

(E) *A new profile of equilibrium, in which neither erosion nor deposition occurs, is established eventually.*

Figure 10.13 *Adjustments of a stream profile to reestablish equilibrium.*

new gradient, deposition occurs over the valley floor upstream from the delta. For example, upstream from Elephant Butte Reservoir in New Mexico, the Rio Grande buried the village of San Marcial with 3 m of sediment,

while at Albuquerque, 160 km farther upstream, 1.2 m of sediment were deposited.

Disturbances of the balance between gradient, sediment load, and discharge are among the consequences downstream of dam building. Since most sediment is trapped in the reservoir, the water released downstream has practically no sediment load. Therefore, it is capable of much more erosion than the previous river, which carried a sediment load adjusted to its gradient, and extensive scour and erosion commonly result downstream from the dam. Another important effect of a dam is that the sediment supply to the ocean is cut off or reduced drastically. The delta built by the river ceases to grow and is vigorously eroded by wave action, which results in significant changes in the shoreline.

Other examples of rivers adjusting toward equilibrium can be found in canal systems where sediment may be deposited along the canal banks as the system adjusts to the new gradient and channel patterns (*figure 10.14*).

Manipulating River Systems

Statement

As population has grown, man increasingly has become involved in changing the natural water systems of the continents, largely through the construction of dams and canals and through urbanization. Dams are built to store water for industrial and irrigational use, to control floods, and to produce electrical power. Yet, when a dam

Figure 10.14 *Sedimentation in a canal, resulting in part from the universal tendency for a stream to develop a meandering pattern. The flow pattern illustrated in figure 10.7 causes variations in velocity that are accentuated on the inside of a bend. Deposition occurs and the stream attempts to develop a meandering pattern.*

is built, the balance of the river system, established over thousands or millions of years, is instantly upset and there are many unforeseen, long-term effects both upstream and downstream from the construction.

Discussion

Examples. The Aswan Dam of the Nile provides a good example of the many consequences of modifying a river system. For centuries, the Nile River has been the sole source of life to Egypt. The principal headwaters of the Nile are located in the high plateaus of Ethiopia. Once a year, for approximately one month, the Nile normally rises to flood stage and covers much of the fertile farmland in the Nile delta area. The Aswan Dam, completed in the summer of 1970, was to provide Egypt with water for irrigating 1 million acres of arid land, generate 10 billion kilowatts of power, double the national income, and permit industrialization. The dam has modified the equilibrium of the Nile drastically, and many adjustments to this major change in the river have resulted.

The Nile is not only the source of water for the delta, it is also the source of sediment. When the dam was finished and began to trap the sediment in Lake Nasser, it destroyed the physical and biological balance in the delta area. Without the annual "gift of the Nile," erosion is eating the delta coastline, which now is exposed to the full force of marine currents. Some parts of the delta are receding several meters a year.

The sediment previously carried by the Nile was an important link in the aquatic food chain nourishing marine life in front of the delta. The lack of Nile sediment has reduced plankton and organic carbon to a third of what they used to be, thus either killing off (or driving away) sardines, mackerel, clams, and crustaceans. The annual harvest of 16,000 metric tons of sardines and a fifth of the fish catch have been lost. The sediment of the Nile also acted as a natural fertilizer; without this annual addition of soil nutrients, Egypt's million cultivated acres need artificial fertilizer. Without its load of sediment, the water discharged from the dam flows downstream much faster and is vigorously eroding the channel bank. This scouring process already has destroyed three old barrier dams and 550 bridges built since 1953. Ten new barrier dams between Aswan and the ocean now must be built at a cost equal to one-fourth the cost of the dam itself.

The annual flood of the Nile was also important to the ecology of the area, because it washed away the salts that formed in the soil. Already, soil salinity has increased not only in the delta but throughout the middle and upper Nile; and, unless corrective measures are taken at a cost of over $1 billion, millions of acres will revert to desert within a decade.

The change in the river system has permitted double cropping, but there are no periods of dryness that previously helped limit the population of bilharzia, a blood parasite carried by snails that infects the intestinal and urinary tracts of humans. One out of every two Egyptians now has the infection, and one out of every ten deaths in the country is caused by it.

Problems also occur in the lake behind the dam. The lake was to have reached a maximum level in 1970, but it may actually take 200 years to fill. More than 15 million m³ of water escape underground into the porous Nubian Sandstone, which lines 480 km of the lake's western bank. It is capable of absorbing an almost unlimited quantity of water. Moreover, the lake is located in one of the hottest and dryest places on the earth, and the rate of evaporation is staggering. This was expected, but additional losses from transpiration by plants growing along the lakeshore and increased evaporation from high wind velocity have nearly doubled the expected loss of water from the lake. This loss equals half the total amount of water that once was "wasted" while flowing unused to the ocean.

The Colorado River, as well as many others in the United States, has been altered by a series of dams that resulted in environmental alterations similar to those experienced on the Nile. The economy and population of the United States are not tied completely to one river, however, so many changes resulting from modifying river systems go unnoticed. In the case of the Colorado River, erosion downstream from Hoover Dam has extended all the way to Yuma, 560 km below the dam, where the channel has been lowered by approximately 3 m. This has eliminated the gravity-flow irrigation system in some

areas and has necessitated pumping to lift the water to the level of irrigation intakes.

Of course, there is another side to the story. In terms of property damage and loss of life, floods are considered to be one of the most disastrous of natural phenomena. Floods are caused primarily by high precipitation during a short period of time in a given drainage basin. Rapid thawing of snow and ice also can supply floodwaters during the spring thaw. In some areas, such as the lower Nile, flooding occurs once a year; in other drainage systems, flooding may occur at irregular intervals, with years passing between floods. However, destructive as they seem to be, floods are natural stream processes that have operated throughout geologic time and are important in maintaining the balance in a river system.

Many tragic floods that destroyed entire towns have been recorded, especially during the early years of the United States. The most extensive hazards are in the flood plains and deltas of the major rivers, areas containing rich soil deposited by the river during previous floods. Deltas are particularly hazardous, because a combination of flooding and storm-induced high water can completely inundate large areas. A combination of such events in 1970 resulted in the deaths of 500,000 people in the lower delta of the Ganges River.

The loss of lives in the United States as a result of flooding is about 80 people a year, which is relatively low compared to preindustrial societies that lack adequate monitoring, warning systems, and disaster relief. For example, in a 20-year period, 154,000 lives were lost in Asia due to floods, compared to 1300 in the United States. The loss of property in the industrialized societies, however, is high, averaging $225 million per year in the United States.

The most common approach to flood control involves manipulations and adjustments in river systems, such as construction of levees to serve as physical barriers and dams to store water that is later released at a safe rate. A long-range problem with most flood-control measures is that they upset the balance of the river system. The gradient is modified, and the sediment is impounded behind dams and levees. Ultimately, the reservoirs are filled with sediment and meandering rivers change courses.

Perhaps the best method of flood control, from an environmental point of view, is the regulation of the use of flood plains. The flood plain is a natural part of the river system, and flooding is a natural phenomenon. Indiscriminant use of the flood plain results in loss of life and tremendous property damage. The rich flood plain is a valuable resource that cannot be abandoned, but man must "design with nature" in its utilization.

Another way in which man has modified river systems is simply by urbanization. The construction of cities at first may seem unrelated to modifying river systems, but a city significantly changes the surface runoff, and the resulting changes in river dynamics are becoming very serious and costly. Water that falls to the earth as precipitation usually follows several paths in the hydrologic system. Generally, 54% to 97% returns into the air directly by evaporation and transpiration; 2% to 27% collects in a stream system as surface runoff; and 1% to 20% infiltrates into the ground and moves slowly through the subsurface toward the ocean. Urbanization disrupts the normal hydrologic system by changing the nature of the terrain, which in turn affects the rates and percentages of runoff and infiltration. Roads, sidewalks, and roofs of buildings render a large percentage of the surface impervious to infiltration. Not only does the volume of surface runoff increase, but runoff is much faster because water is channeled through gutters, storm drains, and sewers. As a result, flooding increases in intensity and frequency.

The problem of urbanization and its effect upon natural runoff can be understood readily by considering the characteristics of stream flow as shown in *figure 10.15*. This hydrograph shows variations in stream flow with time. The shape of the curve depends upon the na-

ture of the stream system, which is influenced by such factors as relief, vegetative cover, permeability of ground, and number of tributaries. The low, flat portion of the curve represents the portion of stream flow attributable to ground water and is referred to as base flow. The shaded part of the graph represents a period of rainfall, and the high part of the curve shows the resulting increase in discharge. An important factor shown in *figure 10.15* is the time lag or interval of time between the storm and the runoff.

The effect urbanization has on runoff is shown on the hydrograph in *figure 10.16*. Lag time is decreased materially because water runs off faster from streets and roofs than from naturally vegetated areas. As the runoff time decreases, the peak rate of runoff increases. As a result, floods are more frequent and more severe.

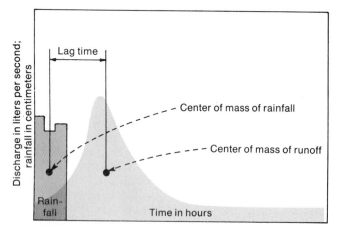

Figure 10.15 *Hydrograph showing variations in stream flow with time under normal terrain conditions. After a flood, there is a considerable lag of time between the periods of maximum precipitation and maximum runoff, because water is held back by vegetation, seepage into the subsurface, and a slowdown of runoff due to the curves and bends in stream channels.*

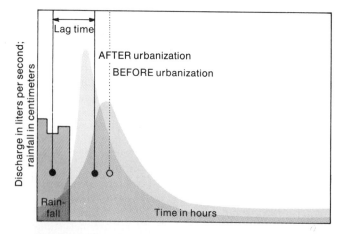

Figure 10.16 *Hydrograph showing variations in stream flow with time in an urban area. There is very little lag time between points of maximum precipitation and maximum runoff, because much of the surface is made impermeable by roads, sidewalks, buildings, and parking lots. In addition, water is channeled directly through sewers and gutters for rapid runoff. As a result, the number and intensity of floods are increased.*

Summary

A river system is a network of connecting channels through which surface water is collected and funneled back to the ocean. It can be divided into three subsystems: (1) tributaries, which collect and transport water and sediment into the main stream; (2) a main trunk stream, which is largely a transporting system; and (3) a dispersing system at the river's mouth.

Studies of river systems show that there is a high degree of order among the tributaries and the size and gradients of their valleys, indicating that streams erode the valleys through which they flow.

The major variables in a river system are (1) discharge, (2) velocity, (3) gradient, (4) base level, and (5) load.

A river system is best thought of as a system of moving water and sediment, because an enormous volume of sediment is constantly being transported to the ocean by running water. The sediment is moved in solution, in suspension, and by traction and is visible in most rivers, especially where it is being deposited as deltas near the mouths of the rivers. The capacity of running water to erode and transport sediment is dependent largely upon the stream's velocity.

A river system functions as a unified whole and adjusts its profile to establish equilibrium among the various factors that influence flow. When one of the factors is changed, adjustments in the river system are made to reestablish equilibrium.

Running water is the most important eroding agent on this planet, and stream valleys are the most characteristic landforms on continental surfaces. The next chapter will consider how streams erode their valleys and canyons and how the resulting sediment is deposited ultimately as deltas, alluvial fans, and other landforms.

Additional Readings

Leopold, L. B., and W. B. Langbein. 1966. "River Meanders." Sci. Amer. 214(6): 60-70. (Reprint no. 869.) San Francisco: W. H. Freeman and Company.

Leopold, L. B., M. G. Wolman, and J. P. Miller. 1964. Fluvial Processes in Geomorphology. San Francisco: W. H. Freeman and Company.

Morisawa, M. 1968. Streams: Their Dynamics and Morphology. New York: McGraw-Hill.

Processes of Stream Erosion and Deposition

Erosion of the land is one of the major effects of the hydrologic system. It has occurred throughout all of recorded geologic time and will continue as long as the system operates and land is exposed above sea level. Evidence for erosion is ubiquitous and varied. It is seen in the development of gullies on farmlands and in the cutting of great canyons. It is seen in the material carried by the rivers and deposited in the ocean. It is seen in the vast thicknesses of sedimentary rocks that cover large parts of the continents and bear witness to erosion during past ages on a magnitude that is difficult to conceive.

There are six major agents of erosion: running water, mass movement, ground water, glaciers, waves and currents, and wind. Running water, however, is by far the most important, and there are few areas of the earth that have not been modified by it.

In order to understand the various processes of stream erosion, you must attempt to analyze them as independent, isolated activities, but you must always remember that each process of erosion is part of a single system.

In the development of a drainage system and the subsequent erosion of a landscape, there is a universal tendency for a river system to (1) remove the regolith produced by weathering, (2) deepen its channel by downcutting when there is sufficient gradient, (3) extend the drainage network upslope by headward erosion, (4) widen its valley by mass movement and slope retreat, and (5) extend its drainage channel downslope as sea level (base level) recedes.

Where stream velocity is reduced, deposition of the sediment load produces a variety of landforms, such as flood plains, deltas, and alluvial fans.

Major Concepts

1. Downcutting of stream channels is accomplished by abrasion, solution, and hydraulic action.
2. Drainage systems grow by headward erosion, which commonly results in stream piracy.
3. Slope retreat causes valleys to grow wider and is associated with downcutting and headward erosion. It results from various types of mass movement under the pull of gravity.
4. As a river develops a low gradient, it releases part of its load by deposition on point bars of meanders and natural levees and across the surface of its flood plain.
5. Where rivers empty into lakes or the ocean, most of the sediment is deposited. This commonly builds a delta at the river's mouth. In arid regions, many streams deposit their loads as alluvial fans at the bases of steep slopes.

Removal of Regolith

Statement

One of the most important processes of erosion is the transport and removal of the regolith produced by weathering. The process is simple but important. Soluble material is carried in solution and insoluble debris is washed downslope and transported by the streams and rivers as sediment load.

Discussion

Evidence of the continuous removal of regolith can be easily observed on any hillside during a rainstorm or during spring runoff as the small particles of silt and mud are washed downslope and into a major stream. The vast amount of material carried by rivers bears witness to the removal of regolith on a magnitude that is difficult to conceive. As the regolith is removed, fresh rock is exposed and is weathered so that the supply of regolith always is being replenished. By this process and by associated downcutting of the stream channel and headward erosion, high mountains and entire continents are reduced to elevations close to sea level.

Downcutting of a Stream Channel

Statement

The basic process of erosion in all streams, from small gullies on hillsides to great canyons of major rivers, consists of the abrasive action of sand, cobbles, and boulders as they are moved by running water along the channel floor. Other erosional processes in stream channels are solution activity and direct hydraulic action.

Discussion

Abrasion. When one clearly realizes that river systems are not only systems of flowing water but systems of moving water and sediment, one can comprehend better the effect of abrasive action on a river channel as gravel and sand are swept along the riverbed. Although the sediment may be deposited temporarily along the way during low water, it is picked up and moved again during high water and floods; and, as it moves, it continually abrades the surface of the stream channel. The process of abrasion in riverbeds is not unlike that of a wire saw used in quarries to cut and shape large blocks of stone. An abrasive such as garnet, corundum, or quartz, when dragged across a rock by a wire, is able to cut through the block with remarkable speed.

One of the most dramatic examples of the power of a stream to cut downward is the steep, nearly vertical gorges in many canyons in the southwestern United States (*figure 11.1*). During spring runoff and periodic flash floods, the load of sand and gravel on the channel floor is moved. This material constitutes a very effective tool of abrasion and is capable of cutting the stream channel to a profile of equilibrium in a short time.

An effective and interesting type of stream abrasion is the drilling action of pebbles and cobbles trapped in

Figure 11.1 *Sand and gravel are tools of stream erosion. Transported by a river, they act as powerful abrasives and cut through the bedrock as they are moved by flowing water. The abrasive action of sand and gravel has cut this vertical gorge through the resistant limestone formation in the Grand Canyon, Arizona.*

a depression and swirled around by the moving water. The rotational movement of the sand, gravel, and boulders spun around by the river currents acts like a drill and cuts deep holes called **potholes.** As the pebbles and cobbles are worn away, new ones take their places and continue to drill into the stream channel, eventually developing holes that can range up to several meters in diameter and more than 5 m deep (*figure 11.2*). Abrasion also effectively undercuts the stream banks, and large blocks of rock and regolith slump off and become part of the stream load (*figure 11.3*).

As they abrade and downcut the stream channel, the pebbles and cobbles themselves are worn down as they strike one another and the channel bottom. The corners and edges are chipped off, and the particles become smaller and smooth and rounded. Thus, large boulders that fall into the stream or are transported only during a flood are continually being broken and worn down to smaller fragments and ultimately are washed away as grains of sand.

Solution. Although solution activity of running water generally is not visible, the chemical content of rivers shows that solution activity is a very effective and important means of lowering the land surface. Estimates based on measurements of the chemical content of the world's major rivers indicate that each year a total of 3.5 billion metric tons of dissolved material is carried from the continents by running water. The effectiveness of chemical erosion is illustrated by the fact that, in humid areas, limestone (which is the most soluble of the major rock types) forms valleys and is considered a weak rock but, in arid regions, it is resistant to erosion by solution and forms cliffs. Most of the dissolved matter probably is derived from ground-water seepage into the stream, but chemical erosion, especially in limestone terrains, also occurs in the stream channel and in the small particles of soluble rock carried by the stream.

Hydraulic Action. The force of running water alone, without the tools of abrasion, quickly can erode soft, unconsolidated material such as soil, clay, sand, and gravel. It is especially effective when water is moving at high velocities. Although directed hydraulic action is unable to erode large quantities of solid bedrock rapidly, pressure exerted by moving water into cracks and bedding planes can be sufficient to remove slabs and blocks from the channel floor and sides.

An important factor in the downcutting of a stream channel is the migration of waterfalls and rapids upstream. Undercutting of weak, nonresistant rocks can cause the stronger resistant layer to collapse, and, where the water is particularly turbulent (such as at the base of a falls), hydraulic action is particularly effective (*figure 11.4*).

Figure 11.2 *Potholes are eroded in the hard rock of a streambed by sand, pebbles, and cobbles whirled around by eddies of the stream.*

Figure 11.3 *Slumping along the banks of a stream.*

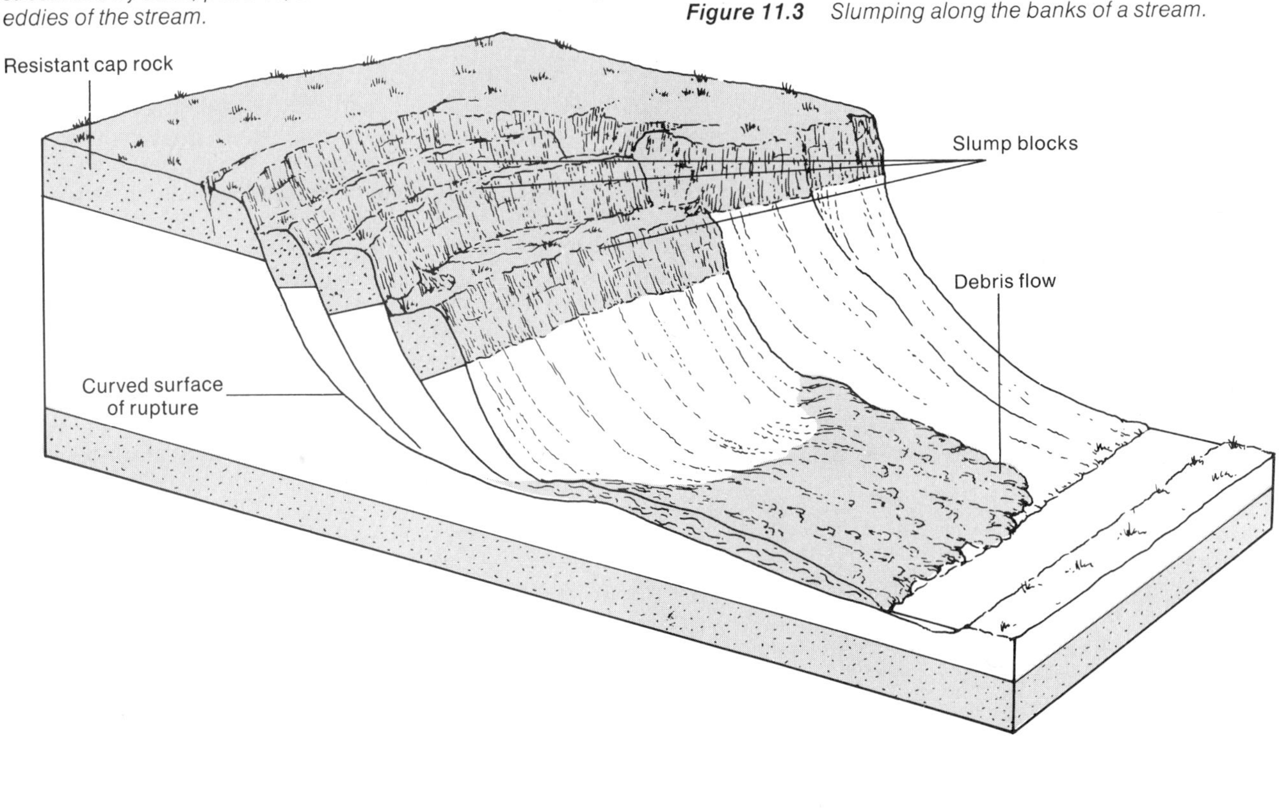

Resistant cap rock

Slump blocks

Debris flow

Curved surface of rupture

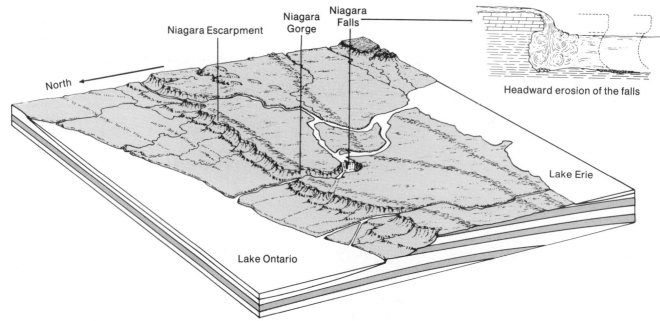

North

Niagara Escarpment

Niagara Gorge

Niagara Falls

Headward erosion of the falls

Lake Erie

Lake Ontario

Figure 11.4 *The retreat of Niagara Falls upstream with the resulting formation of Niagara Gorge. The Niagara River originated as the last glacier receded from the area and water flowed from Lake Erie to Lake Ontario over the Niagaran Cliffs. Erosion caused the waterfalls to migrate upstream at an average rate of 1.3 m per year.*

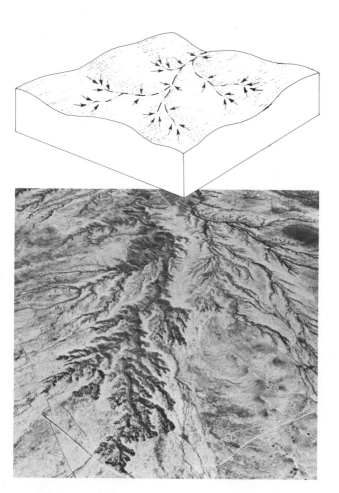

Headward Erosion and Growth of the Drainage System Upslope

Statement

In the process of stream erosion and valley evolution, there is a universal tendency for each stream to erode headward, or upslope, and increase the length of its valley until it reaches the divide. Every tributary is involved in this process until the length of the valley is extended upslope as far as possible.

Discussion

The process of headward erosion can be analyzed by reference to *figure 11.5*. The reason erosion is more vigorous at the head of a valley than on its sides is apparent when you understand the relationships that exist among the valleys and the regional slope. Above the heads of valleys, water flows as a sheet (sheet flow) down the regional slope and converges to a point where a definite stream channel begins. As soon as the water is concentrated into a channel, its velocity and erosive power increase to a point far beyond that of a slower moving sheet of water on the surrounding and ungullied surface. The additional volume and velocity erode the head of the valley much faster than the sheet did on ungullied slopes, so the head of the valley is extended upslope. In addition, ground water moves toward the valley, making the head of the valley favorable for the development of springs and

Figure 11.5 *Headward erosion and extension of the drainage system upslope. Water flows down the regional slope (away from the viewer) as a sheet and converges toward the head of the tributary valleys. The tributary valley thus is eroded headward or up to the regional slope. Headward erosion is caused by the concentration of sheet flow into a channel at the head of the tributary valley. At this point, the velocity of the water is greatly increased, thus increasing its ability to erode. The tributaries, therefore, erode upslope.*

seeps. This, in turn, helps to undercut overlying resistant rock and cause the head of the valley to grow upslope much faster than the retreat of the valley walls.

Careful study of the relationship between the drainage network and regional slope (*figure 11.5*) shows that the position of the tributary is fixed at the point where it joins the next-larger segment of the stream. Continued erosion along this stream segment is restricted to downcutting and valley widening. Thus, it is possible for the drainage system to grow only by extending existing gullies up the regional slope or by developing new tributaries along the walls of the larger stream. Inevitably, the gullies extend up the slope of the undissected surface toward the gathering grounds of the waters that feed the stream. Thus, headward erosion is a basic process in the evolution of drainage systems.

Stream Piracy. With the universal tendency for headward erosion, the tributaries of one stream can extend upslope and intersect the middle course of another, thus diverting the headwater of one stream to another. The process is known as **stream piracy,** or stream capture, and is illustrated in *figure 11.6*.

Stream piracy is most likely to occur in a situation in which the headward erosion of one stream is favored by a steeper gradient or by a course in more easily eroded rocks. Some of the most spectacular examples occur in the folded Appalachian Mountains, where the nonresistant shale and limestone are easily eroded and the hard quartzite formations with which they are interbedded are very resistant. The process of stream capture and the evolution of the drainage system in this region is shown in the series of diagrams in *figure 11.7*. The original streams flowed in a dendritic pattern (a branching, tree-like pattern) on horizontal sediments that covered eroded folds. As uplift occurred, erosion removed the horizontal sediments and the dendritic drainage pattern became

superposed, or placed upon, the folded rocks, cutting across weak and resistant rocks alike. As the major stream cut a valley across the folded rocks, new tributaries rapidly extended themselves headward along the nonresistant formations. In doing so, they progressively captured the superposed tributaries and changed the dendritic drainage to a **trellis pattern.**

Extensive stream piracy and development of a trellis drainage pattern can be seen almost anywhere folded rocks are exposed at the surface. In the folded Appalachian Mountains, the major streams that flow to the Atlantic (such as the Susquehanna and the Potomac) all are superposed across the folded strata but their tributaries flow along the nonresistant rocks parallel to the geologic structure and have captured many superposed streams.

Another example is the Pecos River in New Mexico, which, by extending itself headward along the weak shale and limestone, has captured a series of east-flowing streams that once extended from the Rockies across the Great Plains. The original eastward drainage (shown in *figure 11.8A*) resulted from the uplift of the Rocky Mountains. Now, as a result of headward erosion and stream piracy, the headwaters of the southern rivers flow to the Pecos (*figure 11.8B*).

In summary, headward erosion is one of the most important ways in which a drainage system evolves. It continues to extend the drainage network upslope until the tributaries reach the crest of the divide (the ridge separating two drainage systems [*figure 11.9*]). Through the process of headward erosion, stream piracy can divert large segments of a drainage segment into another as the drainage system adjusts to the structure of the rocks over which it flows. When headward erosion reaches the divide, it is effectively eliminated because no undissected slope remains. Subsequent erosion is restricted to downcutting and valley widening. Therefore, a drainage system grows in a specific way, and, once sheet flow is concentrated into a channel, the future development of the drainage is somewhat predictable.

Figure 11.6 Stream piracy occurs where tributaries with high gradients erode headward and capture the tributaries of another stream.

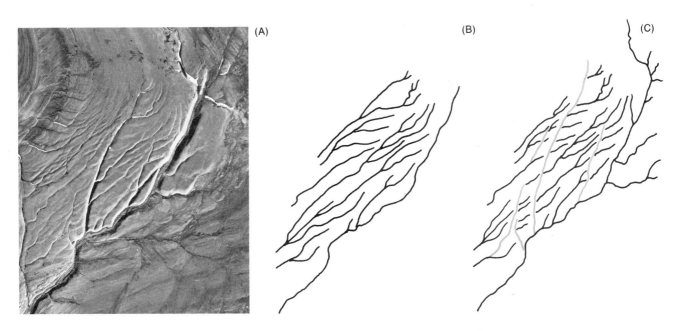

(A) (B) (C)

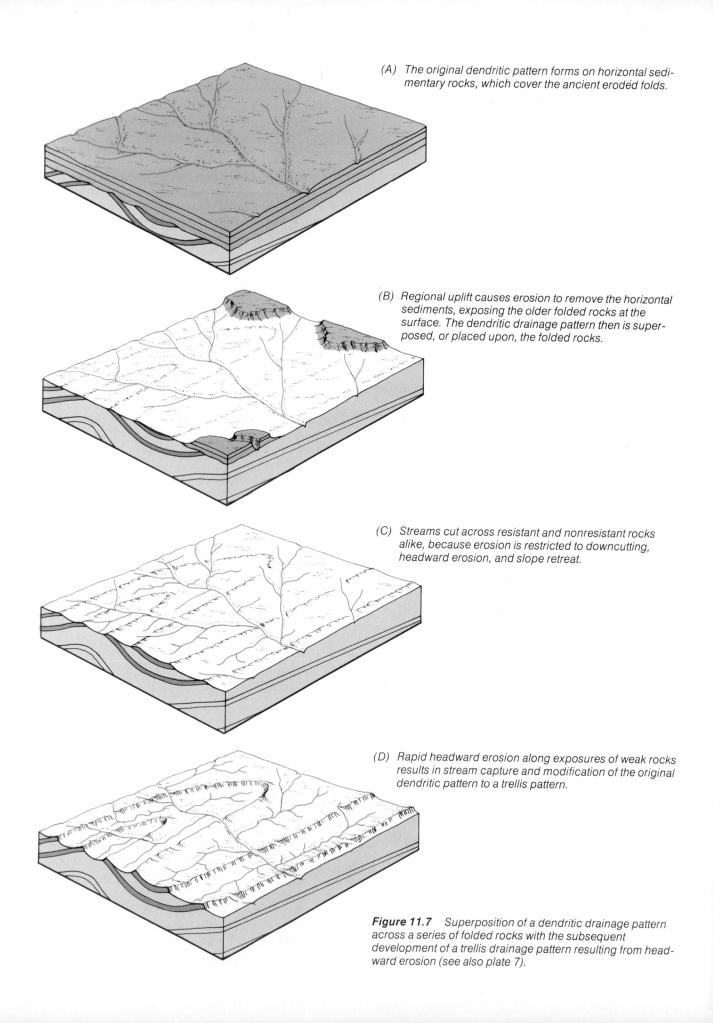

(A) The original dendritic pattern forms on horizontal sedimentary rocks, which cover the ancient eroded folds.

(B) Regional uplift causes erosion to remove the horizontal sediments, exposing the older folded rocks at the surface. The dendritic drainage pattern then is superposed, or placed upon, the folded rocks.

(C) Streams cut across resistant and nonresistant rocks alike, because erosion is restricted to downcutting, headward erosion, and slope retreat.

(D) Rapid headward erosion along exposures of weak rocks results in stream capture and modification of the original dendritic pattern to a trellis pattern.

Figure 11.7 Superposition of a dendritic drainage pattern across a series of folded rocks with the subsequent development of a trellis drainage pattern resulting from headward erosion (see also plate 7).

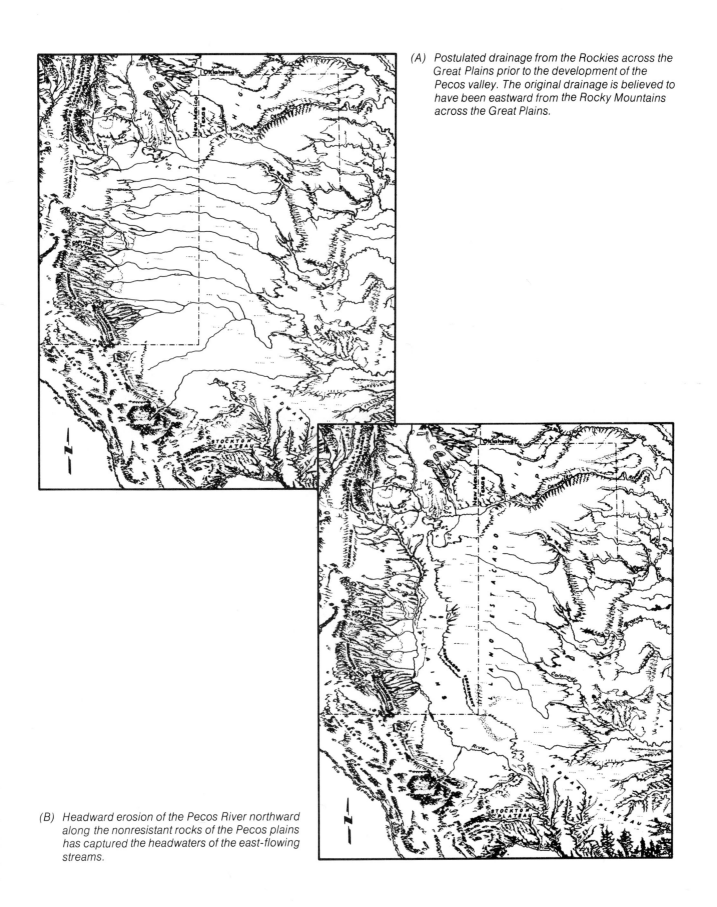

(A) Postulated drainage from the Rockies across the Great Plains prior to the development of the Pecos valley. The original drainage is believed to have been eastward from the Rocky Mountains across the Great Plains.

(B) Headward erosion of the Pecos River northward along the nonresistant rocks of the Pecos plains has captured the headwaters of the east-flowing streams.

Figure 11.8 Evolution of the Pecos River by headward erosion and stream capture.

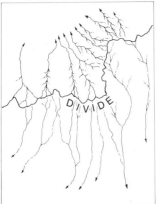

Figure 11.9 *The divide between two drainage systems.*

Extension of Drainage Systems Downslope

Statement

In addition to downcutting and headward erosion, a drainage system can grow in length simply by extending its course downslope as sea level falls or as the landmass rises. This process is probably fundamental in determining the original course of many major streams, especially in the interior lowlands where the ocean covered much of the continent many millions of years ago and then slowly withdrew. As the ocean retreated, drainage lines were extended down the newly exposed slope and were modified later by headward erosion and stream piracy.

Discussion

An example of the beginning of a new segment of a drainage system resulting from ocean withdrawal can be seen in tidal channels along coastal plains. The pattern of land and tidal drainage is shown in *figure 11.10*. Because the material upon which this drainage is established consists of recently deposited horizontal sediments, the

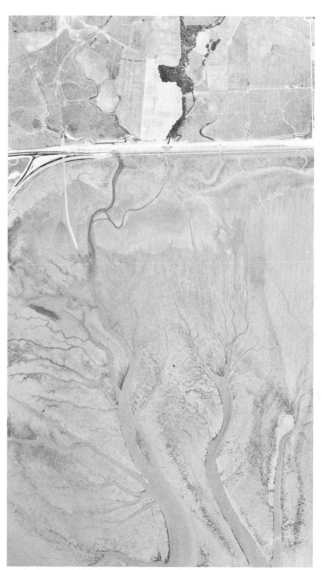

Figure 11.10 *Tidal channels develop a characteristic dendritic drainage pattern between high and low tides.*

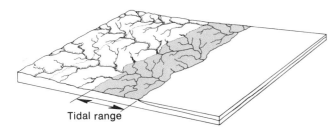

(A) Original position of the shore with tidal channels between high and low tide.

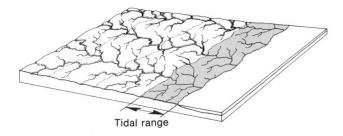

(B) As the sea level recedes, tidal channels become part of the permanent drainage system.

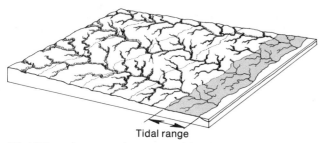

(C) With each successive retreat of the shoreline, new tidal channels develop and drainage again is extended downslope. The drainage pattern typically produced on the homogeneous tidal-flat material is dendritic.

Figure 11.11 *Extension of drainage downslope as a result of shoreline retreat.*

drainage pattern is characteristically dendritic. However, when the slope is pronounced, the tributaries, as well as the major streams, flow parallel for a long distance. If sea level were to drop, the streams would continue to flow downslope, as shown in *figure 11.11.* Major streams would extend their drainage patterns over the deltas that they deposit.

Mass Movement and Slope Retreat

Statement

The valley walls associated with a stream system are subjected to a variety of slope processes that cause them to recede and retreat away from the river channel. These processes, commonly referred to as a **mass movement,** can be rapid and devastating (like a great landslide) or imperceptibly slow (like creep down the gentle slope of a grass-covered field).

Gravity is the driving force behind all slope processes. In waterless areas, as in local regions of Antarctica and on the moon, gravity alone operates to move unconsolidated material downslope; but, in most areas on the earth, water is an important factor, because it lubricates the unconsolidated material and adds weight to the mass of material. Of course, the force of gravity is continuous, but gravity is able to move material only when it is able to overcome inertia. Any factor that tends to reduce resistance to motion aids mass movement. These factors include saturation of the material by water, oversteepening of slopes through undercutting by streams or waves, alternating freezing and thawing, and vibrations from earthquakes.

Mass movement of surface material downslope can be considered the most universal of all erosional processes. It occurs under all geologic conditions on the earth—on the steep slopes of high mountains, the gently rolling plains, the sea cliffs, and the slopes beneath the oceans—and on the moon and Mars and on other planets.

Discussion

The mechanics of mass movement can be very complicated; however, for convenience, examples from two groups will be considered: (1) predominantly rapid downslope movements and (2) predominantly slow movements. Some examples of each are shown in the diagrams in *figure 11.12.*

Rockfalls (free falls of rocks from steep cliffs). Rockfalls are the most rapid types of mass movement. The sizes of rockfalls range from large masses of rock that break loose from the face of a cliff to small fragments loosened by frost action and other weathering processes. In areas where rockfalls are common, the debris accumulates at the base of a cliff as talus.

Rockslides (rapid movements of rock material along a bedding plane or another plane of structural weakness). Rockslides can move as a large block for a short distance, but, more frequently, they break up into smaller blocks and rubble. Most commonly they occur on steep mountain fronts, but they can develop on slopes with gradients as low as 15°. Rockslides are among the most catastrophic of all forms of mass movement, sometimes involving millions of tons of rock that plunge down the side of a mountain in a few seconds (*figure 11.13*).

Debris Slides (rapid movements of soil and loose rock fragments).   Debris slides are composed of dry to moderately wet loose rock fragments that move rapidly over the surface of underlying bedrock. The mass is dry to the surface of underlying bedrock.

Debris Flows (rapid flows of rock fragments and water).   Debris flows consist of a mixture of rock fragments, mud, and water that flows downslope as a thick, viscous fluid. Movement can range from that similar to the flow of freshly mixed concrete to a stream of fluid mud in which rates of flow are equal to those of rivers. Many debris flows begin as slumps and continue as flows near the lower margins of the slump block.

Mudflows (variety of debris flow consisting of a large percentage of silt and clay-size particles).   Mudflows almost invariably result from an unusually heavy rain or a sudden thaw. Water content can be as much as 30%. As a

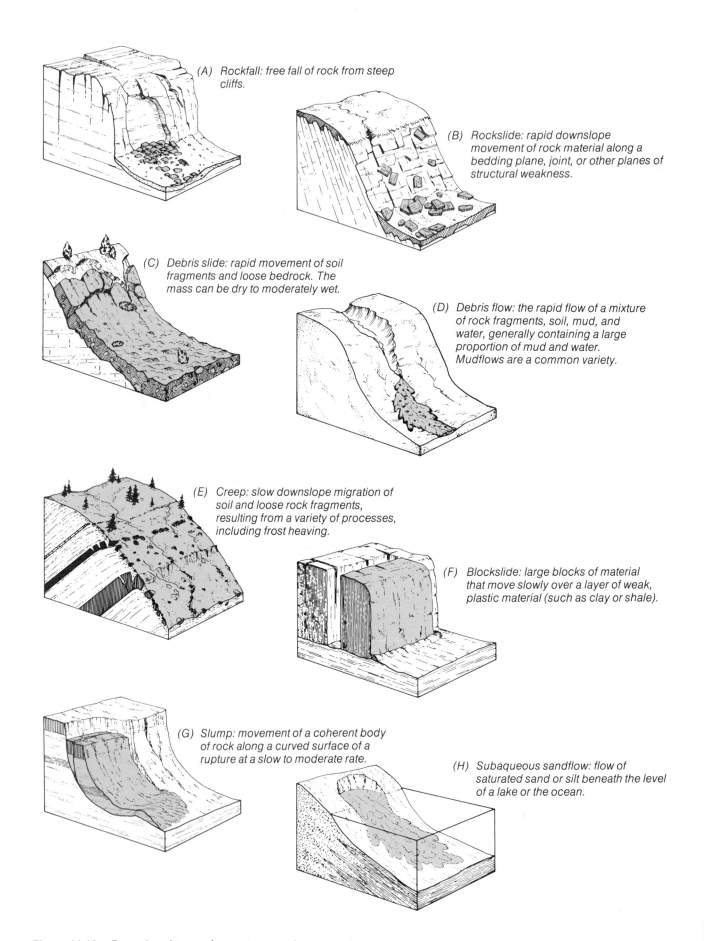

Figure 11.12 *Examples of types of mass movement.*

Figure 11.13 *A recent rockslide in the Swiss Alps.*

result of the predominance of fine-grained particles and the high water content, mudflows follow stream valleys. They are common in arid and semiarid regions and originate typically in steep-sided gullies where there is abundant loose, weathered debris (*figure 11.14*). As they reach the open country at a mountain front, they spread out in the shape of a large lobe or fan. Because of their great density, mudflows can transport large boulders by "floating" them over slopes as gentle as 5° and have been known to move houses and barns from their foundations. Many of the disastrous landslides in southern California are really mudflows that move rapidly down a valley for

considerable distances. Mudflows vary in size and rate of flow, depending upon water content, slope angle, and available debris; but many are over 100 m thick, and some can be as much as 80 km long.

Subaqueous Sand Flows. Mass movement also occurs beneath the water. Commonly, it can include rockfalls and slump blocks, but usually the movement of sediment is in subaqueous sand flows.

Creep (an extremely slow, almost imperceptible downslope movement of soil and rock debris). Creep is so slow that it generally is difficult to observe directly, but it is expressed in a variety of ways. On weakly consolidated, grass-covered slopes, evidence of creep can be seen as bulges or low, wavelike swells. In road cuts and stream banks, creep can be expressed by the bending of

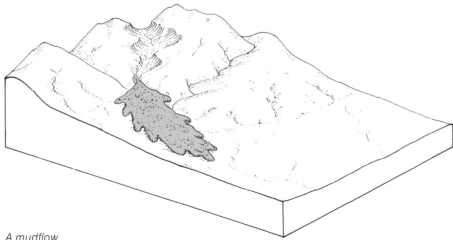

Figure 11.14 *A mudflow.*

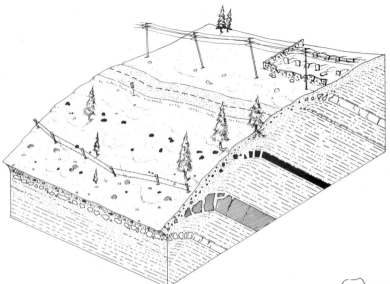

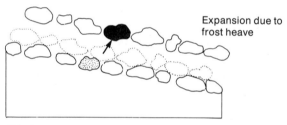

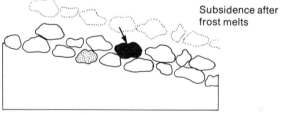

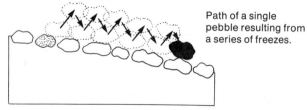

Figure 11.15 *Various expressions of creep. These are shown by downslope displacement of fence lines, roads, telephone poles, headstones in cemeteries, and rock debris from original outcrop.*

steeply dipping strata in a downslope direction or the transporting of blocks of a distinctive rock type downslope from their outcrop. Additional evidence of creep includes tilted trees and posts, displaced monuments, deformed roads and fence lines, and tilted retaining walls (*figure 11.15*). The slow movement of large blocks (blockslides) can be considered a type of creep.

Many factors combine to cause creep. Undoubtedly, soil moisture is important in that it weakens the soil's resistance to movement. In cold regions, creep is produced by frost heaving, a process in which water percolates into the pore spaces of the rocks, freezes, and expands, causing the ground surface to be lifted at right angles to the slope. When thawing occurs, each particle drops vertically, coming to rest slightly downslope from its original position (*figure 11.16*). Repeated freezing and thawing cause the particles on the slope to move downslope in a discontinuous series of zigzags. Wetting and drying can cause a similar motion of loose particles, since moisture causes expansion of clay minerals.

Many other factors contribute to creep, including growing plants that exert a wedgelike pressure between rock particles in the soil and cause them to be displaced downslope. Burrowing organisms also displace particles, permitting the force of gravity to move them.

Landslides (a wide variety of predominantly rapid earth movements, including the subtle slumping of stream banks or sea cliffs as well as the more obvious sliding [slipping] of mountain sides). Movement in landslides can include both the unconsolidated regolith and the solid bedrock. A landslide block moves as a unit (or a series of units) along a definite plane in contrast to debris flows, which move as viscous fluids. The development of a definite fracture or system of fractures in the rock body due to gravitational stresses leads to **slumping.** The material acts as an elastic solid in which a large block moves downward and outward along a curved plane.

Slumping commonly occurs where resistant rock (such as limestone, sandstone, and basalt) caps a much weaker shale formation. Rapid erosion of the weak underlying shale undermines the cap rock to produce an unstable condition. Slump blocks in bedrock can be as

Original position

(A) Water seeps into the pore spaces between loose rock debris.

Expansion due to frost heave

(B) As the water freezes and expands, the soil and rock fragments are lifted perpendicularly to the ground surface.

Subsidence after frost melts

(C) As the ice melts, the particles move down vertically through the pull of gravity and are displaced slightly downhill.

Path of a single pebble resulting from a series of freezes.

(D) Repeated freezing and thawing causes a significant net displacement downslope.

Figure 11.16 *Idealized diagram showing how creep results from repeated frost heaving.*

much as 5 km long and 150 m thick and move in a matter of seconds or gradually slip throughout a period of several weeks.

Solifluction (a downslope movement of regolith saturated with water). Solifluction is a special type of earthflow that commonly occurs in arctic and subarctic regions where the ground is permanently frozen. During the spring and early summer, the ground begins to thaw from the surface downward. Since the meltwater cannot percolate downward into the impermeable permafrost, the upper zone of soil always is saturated and slowly flows down even the most gentle slopes.

Rock Glaciers (glacierlike bodies of rock fragments slowly flowing downslope). In regions of cold temperatures (such as Alaska, the high Rockies, and other alpine areas), tonguelike masses of angular rock debris resembling glaciers move as a body downslope at rates ranging from 3 cm a day to 1 m a year. These bodies are known as **rock glaciers** and represent a type of mass movement involving slow flowage. Evidence of flow includes concentric ridges within the body, its lobate form, and its steep front. Excavations into rock glaciers reveal a considerable amount of ice in the pore spaces between the rock fragments. Presumably, the ice is responsible for much of the movement. With a continuous supply of rock fragments from above, the increased weight causes the interstitial ice to flow. Favorable conditions for development of rock glaciers include a cold climate that keeps ice in the pore spaces frozen and steep cliffs that supply the coarse rock debris that leaves large spaces between fragments for the ice. Some rock glaciers, however, may be debris-covered, formerly active glaciers (no longer active because rate of flow is below the minimum accepted as active).

Slope Systems

Statement

In the previous section, some of the ways in which rock debris is transported downslope through the pull of gravity were described. Now consider the slope as a system in which the effects of weathering, mass movement, and headward erosion of small gully tributaries combine to transport material down the slopes of valleys to the main stream, through which it is then transported to the ocean. The diagram in *figure 11.17* illustrates the major components of the system and how they function. The resistant rock units break up into blocks that accumulate at the base of the cliff in piles of rock fragments. The material on the slope then moves chiefly by rockfalls and rockslides and by rolling. Subsequently, the debris moves slowly downslope by creep. During this process, the rock particles are weathered and broken into smaller fragments, which continue to move downslope toward the stream. The fine-grained rock debris gradually enters the river, where it is carried away by the running water.

Discussion

The talus slope can be thought of as a local open system with an input of coarse rock debris at the base of the cliff and an output of fine rock fragments into the stream. As the system operates, the valley slope gradually retreats away from the stream channel, although the shape of the slope profile may remain essentially constant.

The diagram in *figure 11.17* illustrates how mass movement is involved in slope retreat, but the very important fact that there are numerous tributaries and small gullies operating on essentially all slopes to collect and

Figure 11.17   *Diagram showing the system by which loose debris is moved downslope into a drainage channel. The talus accumulates at the base of the cliff as coarse fragments from rockfalls. Then, it is transformed into smaller and smaller particles by mechanical and chemical weathering and moves downslope by creep. Some of the debris may be collected by minor tributaries and moved to the main stream by running water. The remaining fine-grained debris continues to move downslope by gravity and ultimately is fed into a stream that carries it out through the drainage system.*

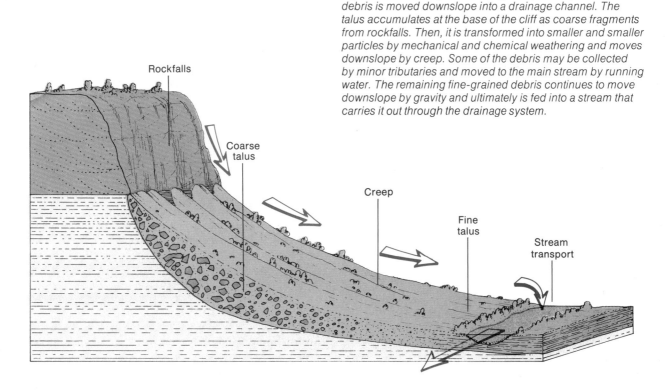

funnel debris to the main stream should be emphasized. The small tributaries work headward and undercut the resistant cliff, making it more susceptible to rockfalls and slumping. During heavy storms, much of the loose rock debris on the slope is moved in the tributary system and debris flows commonly follow the tributary stream channels. Therefore, in the slope system, mass movement and headward erosion of minor tributaries operate together, each facilitating and reinforcing the other in transporting debris and causing slope retreat.

One way to appreciate the importance of mass movement and headward erosion of minor tributaries in the process of valley development is to try to visualize what a valley would look like if these processes were not active. If downcutting of the stream channel were the only process in operation, a deep, vertical, walled canyon would result (*figure 11.18*). This does occur where rocks are strong enough to resist the force of gravity, as in the deep canyons cut in resistant sandstones and limestones in the Colorado Plateau (*figure 11.1*). More commonly, slope retreat occurs contemporaneously with downcutting to produce sloping valley walls.

One of the most important concepts presented in this section is that, although most slopes may appear to be stable and static, they actually are dynamic and involve systems in which surface material constantly is moving downslope in a variety of ways. Large, catastrophic landslides and debris flows are commonly the most hazardous, but imperceptible creep and mass movements of

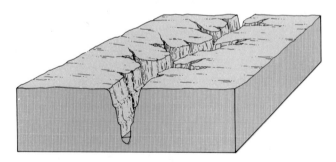

(A) *In resistant rocks where slope processes are minimal, downcutting of the stream channel develops a vertical, walled canyon.*

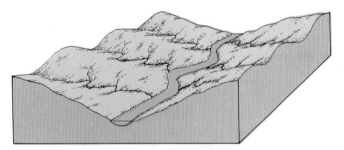

(B) *In most areas, slope processes keep pace with downcutting to produce more gently sloping valley walls.*

Figure 11.18 *The importance of slope processes in development of the landscape.*

moderate magnitude probably move the most material. Downslope movements are natural events that occur everywhere; however, the modification of slopes by man has resulted in both an increase in mass movement and a temporary reduction by the construction of stabilizing structures. Whether catastrophic or slow moving, mass movement of slope material represents a major geologic hazard and a continual challenge for effective land-use planning.

Examples of Man's Interaction with Slope Systems. Many catastrophic landslides are the results of special geologic conditions such as high, steep slopes and weak surface material. These natural processes that occur regardless of man's activity are commonly triggered by exceptionally heavy snowfall or rainfall or by earthquakes. For example, during the earthquake in Peru on May 31, 1970, 4000 m^3 of debris roared down the slope of Mount Huascaran at a speed of more than 300 km per hour, killing approximately 40,000 people as it covered the town of Yungay with up to 14 m of mud and rock.

However, many landslides are caused by man's manipulation of the landscape even though the actual event may have been initiated by excess moisture. An example is the Vaiont Dam slide in Northern Italy. The worst dam disaster in history resulted from a huge landslide into the Vaiont Reservoir on October 9, 1963. The landslide began as a relatively slow creep over a 3-year period. The rate of creep ranged up to 7 cm per week until a month before the catastrophy, when it increased to 25 cm per day. On October 1, animals grazing on the slopes sensed the danger and moved away. Finally, on the day before the slide, the rate of creep was about 40 cm per day. The engineers expected a small landslide but did not realize until the day before the slide that a large area of the mountain slope was moving en masse at a uniform rate. When the slide broke loose, more than 240 million m^3 of rock moved down the hill slope at a speed of 60 m per hour and splashed into the reservoir. This produced a wave of water over 100 m high, which swept over the dam and completely destroyed everything in its path for many kilometers downstream. The entire catastrophic event, including the slide and flood, lasted only 7 minutes, but it took approximately 2600 lives and caused untold property damage. Several adverse geologic conditions contributed to the slide: (1) the rocks consisted of weak limestone with interbedded thin layers of clay, (2) the beds dipped steeply toward the reservoir, (3) the steep slopes created strong driving (gravitational) forces, and (4) the rise of water level in the reservoir reduced the resistance to shear.

The well-known landslide problems in southern California that occur with remarkable frequency are associated with uncontrolled development of hillsides underlain by poorly consolidated rock material. Landslides would be common naturally in the area; but, as man modifies the slopes for building sites and roads, he greatly increases the magnitude and the frequency of mass movement. This results in the loss of millions of dollars' worth of property each year. Careful geologic investigations and proper land-use planning could reduce these losses greatly.

Models of Stream Erosion

Statement

The evolution of a drainage system may require several tens of millions of years, so the origin of stream valleys can be studied only indirectly. One approach is the study of the nature of drainage systems in newly exposed areas, such as coastal plains and former lakebeds. Another is the use of computer models to aid in analyzing the interaction among the major variables of downcutting, headward erosion, slope retreat, and uplift. The results of these studies indicate that a river system evolves through a series of predictable stages and that careful observation in the field reveals the major trends of landscape development in a given area.

Discussion

Drainage Systems on Former Lakebeds.   The development of a drainage system by downcutting, headward erosion, slope retreat, and extension of the drainage net downslope can be seen in the photograph in *figure 11.19.*

This area once was occupied by a prehistoric lake, the remnants of which can be seen in the background. The well-defined terrace midway up the mountain front marks the highest level of the lake. Lower terraces formed as the lake receded.

The lake levels provide important reference lines with respect to the growth of the drainage network. Above the highest lake level, stream erosion has continued without interruption. The drainage was established before the lake was created, and downcutting, headward erosion, and slope retreat have produced the dissected mountain front. Note that the valleys above the lake level are much wider than those below, because they have been subjected to a greater amount of slope retreat.

When the lake came into existence, the drainage system ended at the upper shoreline and sediment carried by the streams was deposited in the lake. These deposits were redistributed by waves to form the prominent terrace in the middle of the photograph. As the lake level receded, the major streams were extended downslope and new, subparallel streams originated on the newly exposed lakebed. Many of these terminated at the lower shore-

Figure 11.19   *The shoreline of ancient Lake Bonneville illustrates all the major processes involved in dissection of the land by running water. Downcutting and slope retreat occur in all segments of the streams. Headward erosion is extending the drainage network upslope in the mountains above the shoreline. As the shore of the ancient lake retreated, the drainage also was extended downslope.*

line, because the lake level remained stationary for some time at this elevation. The lowest zones have been exposed only since the lake completely receded to its present level. This surface is marked by the extension of the major drainage lines downslope.

Within this small area, all the major processes of drainage development can be observed, including headward erosion, downcutting, slope retreat, and extension of the drainage system downslope as the shoreline retreats.

Computer Models of the Erosion of the Grand Canyon. Geologists recently have studied the interaction of downcutting and slope retreat by using a computer model of the Colorado River's erosion of the rock sequence in the Grand Canyon area. Variables such as rates of downcutting and slope retreat affecting various rock formations were analyzed, and a series of profiles showing a series of changes in the canyon with time were calculated and printed by the computer. This study produced a series of hundreds of computer-calculated profiles of the Grand Canyon from the time the Colorado River began cutting through the Colorado Plateau. Several of these profiles are shown in *figure 11.20.* Erosion of the model canyon was not uniformly fast or slow but occurred in a series of pulses. Downcutting of the main stream was extremely rapid and was largely a function of

the rates of uplift. In areas where little uplift occurred, erosion was slow; but, in areas of rapid uplift, erosion was correspondingly fast and a smooth, gently curving longitudinal stream profile was maintained.

On minor intermittent streams, downcutting was slow where resistant rocks were encountered and rapid erosion occurred on nonresistant strata, but the main river was able to cut through resistant and nonresistant rocks with equal ease. The rate of slope retreat was shown to be intimately related to the rate of downcutting.

Although this model cannot be verified directly, you can get a glimpse of the stages of canyon development by studying the nature of the canyon longitudinally. Upstream, near Lee's Ferry (*figure 11.21*), the river flows on the top of the Kaibab Limestone and the entire sequence of strata exposed farther downstream in the Grand Canyon is below the surface. The canyon cut into the resistant formation is a narrow gorge in the Kaibab Limestone. Farther downstream, uplift has permitted the river to cut much deeper into the sequence of rocks and a profile very similar to that developed by the computer model has been formed. Thus, the canyon itself is evidence of the evolution of slope morphology, and it therefore corroborates the findings of the computer model.

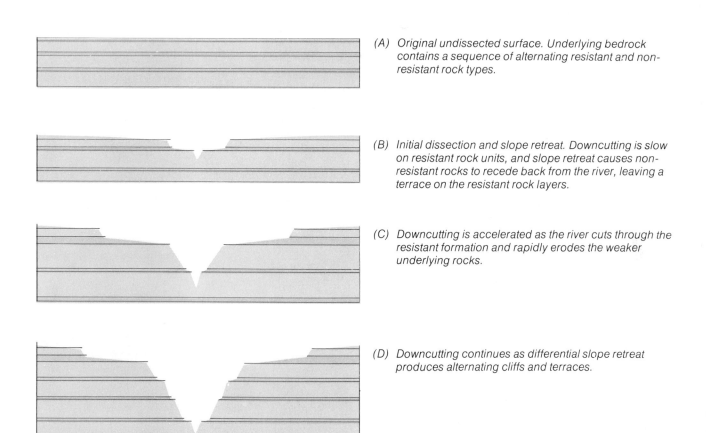

(A) *Original undissected surface. Underlying bedrock contains a sequence of alternating resistant and nonresistant rock types.*

(B) *Initial dissection and slope retreat. Downcutting is slow on resistant rock units, and slope retreat causes nonresistant rocks to recede back from the river, leaving a terrace on the resistant rock layers.*

(C) *Downcutting is accelerated as the river cuts through the resistant formation and rapidly erodes the weaker underlying rocks.*

(D) *Downcutting continues as differential slope retreat produces alternating cliffs and terraces.*

Figure 11.20 *Series of profiles showing the evolution of the Grand Canyon as determined by a computer model.*

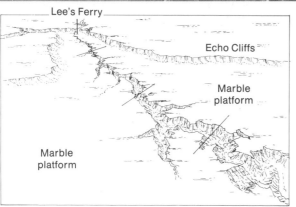

Lee's Ferry

Echo Cliffs

Marble
platform

Marble
platform

Figure 11.21 *Erosion in the eastern Grand Canyon. At Lee's Ferry in the eastern Grand Canyon, the river is just beginning to cut through the rock sequence and has produced a profile like the one in figure 11.20B. Downstream, uplift has permitted the river to cut deeper, producing profiles similar to the ones in 11.20C and 11.20D.*

Flood-Plain Deposits

Statement

Erosion is the dominant process in the high headwaters of a drainage system where both water and sediment are collected by the tributaries and funneled into the main trunk stream. In the transporting subsystem of a river system (which commonly occurs across the stable platform or the shield), the surface of the land slopes gently toward the ocean and the stream gradient is very low. There, the river system commonly is unable to transport all of its sediment load and some deposition occurs. As a result of the low gradient in the transporting segment, a number of landforms are developed that are not found in other parts of the system where stream gradients are high. Foremost among these are:

1. **A flood plain.**
2. **Meanders.**
3. **Point bars.**
4. Natural **levees.**
5. **Backswamps.**

Discussion

A schematic diagram showing the features commonly developed on the flood plain of a river is shown in

Figure 11.22 *Schematic diagram showing the major features of a flood plain. As a stream flows around a meander bend, it erodes on the outside curve and deposits sediment on the inside to form a point bar. The meander bend migrates and ultimately is cut off and forms an oxbow lake. Natural levees build up the banks of the stream, and backswamps develop on the lower surfaces of the flood plain. Yazoo tributaries find it difficult to enter the main stream and flow parallel to it for considerable distances. Slope retreat continues to widen the low valley, which is partly filled with river sediment.*

figure 11.22. The diagram serves as a simple graphic model of flood-plain sedimentation.

Meanders and Point Bars. There is a strong natural tendency for all rivers to flow in a curved, or sinuous, pattern even when the slope is relatively steep. This is because water flow is turbulent and any bend or irregularity in the channel deflects the flow of water to the opposite bank, where the force of the water erodes and causes undercutting. This initiates a small bend in the river channel. As the current continues to impinge on the side of the channel, the bend grows larger and is accentuated with time, ultimately growing into a large meander bend (*figure 11.23*; see also *plate 14*). On the inside of the meander, velocity is at a minimum, so deposition occurs. Therefore, two major processes are involved with water flowing in a meandering channel: (1) erosion on the outside of the meander and (2) deposition on the inside. As these processes operate, the meander loop migrates laterally. Because the valley surface slopes toward the ocean, erosion is more effective on the downstream side of the meander bend and the meander also migrates slowly down the valley (*figure 11.23*). As the meander bend becomes accentuated, it develops an almost complete circle, and eventually the river channel cuts across the meander loop and follows a more direct course to the ocean. The meander cutoff forms a short, but sharp, increase in stream gradient so that the river completely abandons the old meander loop, which remains as a crescent-shaped lake called an *oxbow lake* (*figure 11.24*).

Natural levees. Another key process operating on a flood plain is the development of natural levees. When a river overflows its banks during flood stage, the water is no longer confined to a channel but flows over the surface as a broad sheet. This significantly reduces the velocity of flow, and the coarsest material is deposited close to the channel, building up an embankment known as a natural levee on either side of the river. As natural

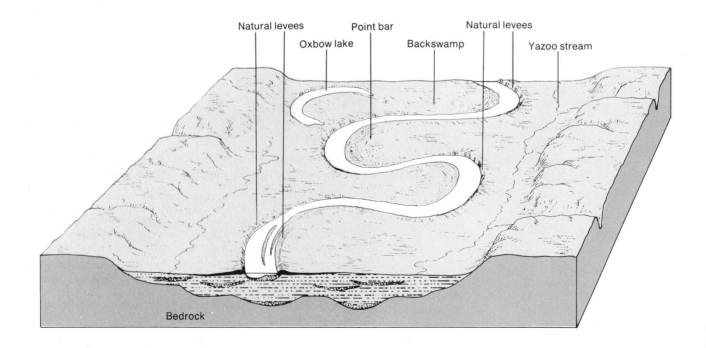

Natural levees   Point bar   Natural levees
Oxbow lake   Backswamp   Yazoo stream

Bedrock

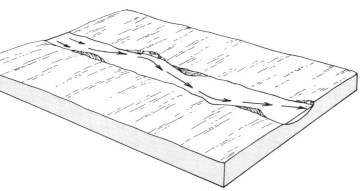

(A) *Patterns of stream flow are deflected by any irregularity and move to the opposite bank, where erosion begins.*

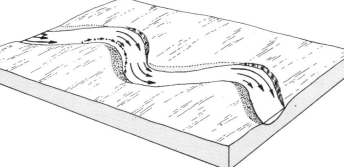

(B) *Once the bend begins, the flow of water continues to impinge upon the outside of the bend, developing a meander loop. At the same time, deposition occurs on the inside of the bend as a result of the lower stream velocities in that area.*

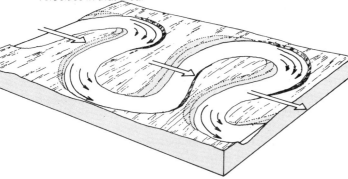

(C) *The meander is enlarged and migrates laterally with contemporaneous growth of the point bar. There is a general downslope migration of the meanders as they grow larger and ultimately cut themselves off to form an oxbow lake.*

Figure 11.23 *Development of meander bends and point bars. The process of meandering involves erosion on the outside of a curve in the stream channel, where velocity is greatest, and deposition on the inside of the curve, where velocity is at a minimum.*

Figure 11.24 *Flood-plain features seen from an altitude of approximately 6000 m. Note meanders, cutoffs, oxbow lakes, and old meander scars (see also plate 14).*

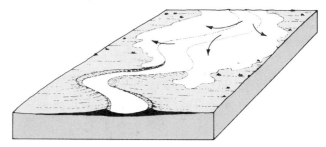

Figure 11.25 *Natural levees are wedge-shaped deposits of silt that taper away from the stream banks toward the backswamp. They form during flood stages because, as the water overflows its banks, the velocity is reduced, causing the silt to be deposited. As the levees grow higher, the stream channel also rises, so the river can be higher than the surrounding area.*

levees grow higher with each flood, they soon build up high embankments and the river actually can build its channel higher than the surrounding area (*figure 11.25*).

Backswamps. As a result of the growth and development of natural levees, a considerable part of the flood plain is below the level of the river. This area, known as the backswamp, is poorly drained and commonly is the site of marshes and swamps. Tributary streams are unable to flow up the slope of the natural levees, so they are forced either to empty into the backswamp or to flow parallel to the main stream for many kilometers before connecting with the main stream. Strangely enough, then, the highest surface on the flood plain may be along the natural levees immediately adjacent to the river.

Figure 11.26 *Flood-plain features of the Mississippi River valley include meander bends (a), point bars (b), natural levees (c), oxbow lakes (d), and backswamps (e).*

There are many excellent examples of flood plain deposits, because most streams that flow across the lowland of the continental interior have a low gradient and deposit part of their sediment load. The lower Mississippi River is a well-known example, and much of the sediment it carries is deposited along a broad flood plain between Cairo, Illinois, and the Gulf of Mexico. An aerial photograph of the river (*figure 11.26*) displays most of the features illustrated in the graphic model shown in *figure 11.22*. The dynamics of the river and the changes that it can bring about by deposition are illustrated by the fact that, during the period from 1765 until 1932, the river abandoned 19 meanders between Cairo, Illinois, and Baton Rouge, Louisiana. Now, the river is controlled by dams and artificial levees, which have modified its hydrology in a manner similar to that of the Nile and Colorado Rivers.

Braided Rivers. Where streams are supplied with more sediment than they can carry, the excess materials are deposited as sand and gravel bars on the channel floors. This deposition forces a stream to split into two or more channels and to form an interlacing network of channels and islands (*figure 11.27*). The term **braided stream** is used to describe the pattern of a low-gradient stream in which deposition is the dominant process.

Braided rivers have wide, shallow channels and are best developed where a river is heavily laden with erosional debris from a high mountain range. When the river emerges from the mountain front, its velocity is reduced rapidly by the abrupt change in gradient and part of the load, including all of the coarser debris, is dropped.

Melting ice caps and glaciers also produce favorable conditions for braided streams, because the low-gradient streams in front of the ice caps and glaciers cannot trans-

port the exceptionally large load of reworked glacial debris.

Stream Terraces. This section has examined how a stream can deposit much of its sediment load across a flood plain and build up a deposit of alluvium. The general process of deposition can be initiated by any change that reduces the capacity of a stream to transport sediment. These changes include (1) a reduction in discharge (climatic changes, stream piracy), (2) a change in gradient (rise of sea level, regional tilting), and (3) an increase in sediment load. During the last ice age, most rivers experienced significant changes in their hydrology. Meltwater from the glaciers carried large quantities of sediment deposited by the glaciers, and stream runoff was increased greatly by the melting ice. In addition, climatic changes accompanied the ice age, with a general increase in precipitation. As a result, many streams show evidence of filling part of their valleys with sediment during the ice age and now subsequently are cutting through the sediment fill to form stream terraces. The basic steps involved in the evolution of stream terraces are shown in the series of diagrams in *figure 11.28*. In block *A*, a stream cuts a valley by downcutting and slope retreat. In block *B*, changes such as those described above cause the stream to deposit part of its sediment load and build up a flood plain that forms a broad, flat valley floor. In block *C*, subsequent changes (uplift, increased runoff, et cetera) result in the renewed downcutting into the easily eroded flood plain deposits and the development of a single set of terraces on both sides of the river. Further erosion can produce additional terraces (block *D*).

Figure 11.27 *A braided river. This pattern commonly results when a river is supplied with more sediment than it can carry. Deposition occurs, causing the river to develop new channels repeatedly.*

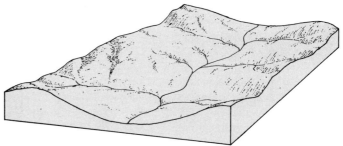

(A) A valley eroded by normal processes.

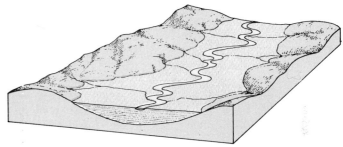

(B) Changes in climate, base level, and other factors that would reduce energy cause the stream to fill its valley partly with sediments, forming a broad, flat floor.

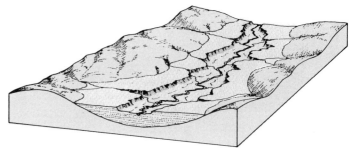

(C) An increase in energy causes the stream to erode through the previously deposited alluvium, leaving a pair of terraces that are remnants of the former flood plain.

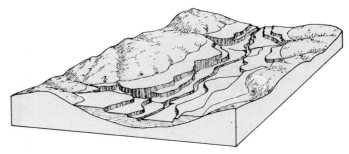

(D) The stream shifts laterally and forms lower terraces as it erodes through the valley fill.

Figure 11.28 *The evolution of stream terraces.*

Deltas

Statement

As a river enters a lake or the ocean, its velocity is suddenly diminished and most of its sediment load is deposited to form a **delta.** The growth of a delta can be very complex, especially for the larger rivers that deposit a huge body of sediment, which extends several hundred kilometers out into the ocean. And there are many distinct subenvironments of deposition in large deltas, such as tidal flats, beaches, lakes, swamps, and flood plains. However, two major processes are fundamental to the formation and growth of a delta; they are:

1. The splitting of a stream into a system of distributary channels, which extend out into the open water in a branching pattern.
2. The development of local breaks or crevasses in natural levees through which sediment is diverted and deposited in the area between the tributaries.

The manner in which a delta grows depends upon a number of factors, including the rate of sediment supply, the rate of subsidence, and the removal of sediment by waves and tides. If stream deposition dominates, the delta is extended seaward. If waves dominate, the sediment delivered by the river is transported along the coast and deposited as beaches and bars. If tides are strong, sediment is transported up and down the distributary channels to form linear sedimentary bodies parallel to the flow of tidal currents.

Discussion

Deltaic Processes. The diagrams in *figure 11.29* illustrate one of the major processes operating in delta construction, the development of distributaries. Distributary channels are formed and extend themselves

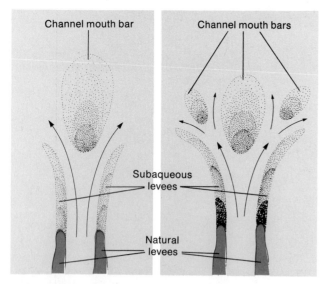

Figure 11.29 *Development of distributaries. A bar develops at the mouth of a river channel when the water previously confined to the channel loses velocity as it enters the ocean. The bar then diverts the water coming from the main stream into two distributary channels, which grow seaward. This process is repeated to form branching tributaries.*

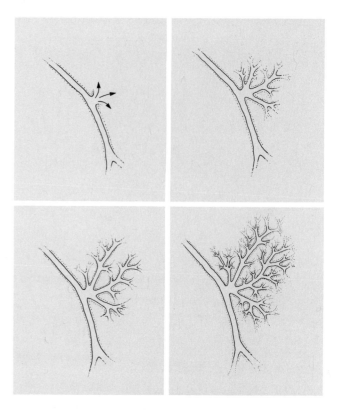

Figure 11.30 *Development of splays. A break in the natural levee permits part of the stream to be diverted to the backswamp, where a reduction in velocity causes deposition of sediment in a fan-shaped splay. Like the main river, a splay has distributaries and a series of smaller subsplays.*

seaward, because sediment tends to accumulate at the channel mouth due to the sudden loss of velocity where the river enters a standing body of water. As the river enters the ocean (or a lake), the flowing water is no longer confined to the channel and the currents flare out, resulting in a rapid decrease in velocity and flow energy. Deposition of the coarse material carried by the streams occurs in two specific areas: (1) along the margin of quiet water adjacent to current flow (this builds up subaqueous natural levees) and (2) in the channel at the mouth of the river where there is a sudden loss of velocity (this builds a bar at the mouth of the channel). These two deposits effectively create two smaller channels (distributary channels), which are extended seaward for some distance. The processes of deposition then are repeated, and each new distributary is divided into two smaller distributaries. In this manner, a system of branching distributaries builds seaward in a fan-shaped pattern.

The manner in which the area between distributaries is filled with sediment is shown in *figure 11.30*. A local break in the levee, called a **crevasse,** forms during high runoffs, and a significant volume of water and sediment is diverted from the main stream through the crevasse. There it spreads out to form a **splay,** which is in essence a small delta itself, with small distributaries and systems of subsplays. The growth of splays in the Mississippi delta can be documented by a series of historical maps dating back to 1838 (*figure 11.31*).

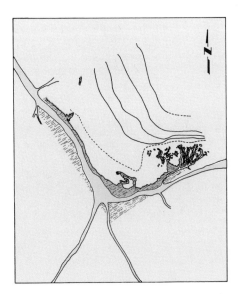

(A) *1838. Configuration of main channel prior to the development of a crevasse.*

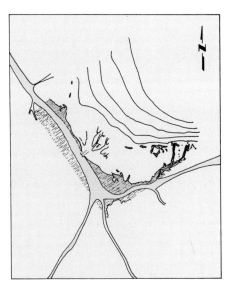

(B) *1870. About 10 years after the opening of a crevasse in the natural levees, a splay began to grow as sediment was deposited adjacent to the levees.*

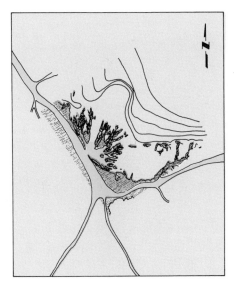

(C) *1877. About 17 years after the crevasse opened. Note the extent of minor splays and distributaries.*

Figure 11.31 *The evolution of a splay on the Mississippi delta.*

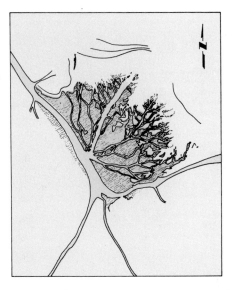

(D) *1903. About 43 years after the crevasse opened. The extensive development of subsplays and minor distributaries caused the splay to fill much of the area adjacent to the channel. The splay has continued to grow by subsplays and distributary channels.*

The sediment deposited by the distributaries and the splays is vulnerable to erosion and transportation by marine waves and tides, so the growth of a delta is influenced greatly by the balance between the rates of sediment input by the river and the rate of erosion by marine processes.

A major phenomenon in delta construction is the shifting of river courses into successive major distributary channels. The extension of the distributaries out into the ocean cannot continue indefinitely, because the gradient and the capacity of the river gradually decrease. The river, therefore, is diverted eventually to a new course, which has a higher gradient. It is generally during a flood that the river breaks through its natural levee far inland from the delta and develops a new course to the ocean. The new river course shifts the site of sedimentation to a new area, and the abandoned or inactive delta is quickly attacked by the ocean since it is no longer fed with sediment from the river. The new active delta builds seaward by developing distributaries and splays until it also is abandoned and another site of active sedimentation is formed.

Examples of Modern Deltas. Several types of deltas are shown in *figure 11.32*. In the Mississippi delta, processes of river deposition dominate. The delta is fed by the extensive Mississippi River system, which drains a large part of North America and discharges an annual sediment load of approximately 454 million metric tons per year. The river is confined to its channel throughout most of its course, except during high floods. Therefore, most of the sediment reaches the ocean through two or three main distributary channels and rapidly has extended the river channels far out into the Gulf of Mexico (see *plate 16*). This extension is known as a **birdfoot delta.** Five major deltas have been constructed during the last 5000 years. These are shown in *figure 11.33*. The oldest lobe (1) was abandoned approximately 4000 years ago and since then has been eroded back and inundated. Only small remnants remain exposed today. The successive lobes, or subdeltas (2, 3, 4, and 5), have been modified to varying degrees, but the abandoned channels of the Mississippi River generally are well preserved and can be recognized on satellite photography. The present active delta lobe (5) has been constructed during the last 500 years.

It is apparent from river studies that the present bird-foot extension is as long as the balance of natural forces will permit and that without continued intervention by man, the Mississippi River will shift to the present course of the Atchafalaya River.

The Nile delta differs from the Mississippi delta in several important ways. Instead of being confined to one channel, the river begins to split up into a series of distributaries at Cairo, Egypt, more than 160 km inland, and fans out over the entire delta (*figure 11.32B*). Before construction of the Aswan Dam, the annual flood of the Nile covered much of the delta during a brief period each year and deposited a thin layer of silty mud. In the past 3000 years, 3 m of sediment has accumulated near Memphis.

Two of the larger distributaries have built major lobes extending beyond the general front of the delta, but

Figure 11.32 *Varieties of deltas formed by different relationships between fluvial and marine processes.*

(A) Mississippi delta. Fluvial processes dominate, with development of a long birdfoot extension.

(B) Nile delta. Wave action dominates, with development of an arcuate delta front.

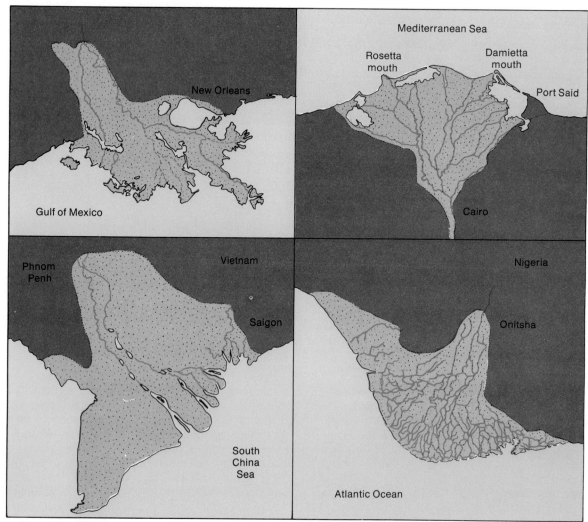

(C) Mekong delta. Tidal forces dominate, with development of wide distributary channels.

(D) Niger delta. Wave action and tidal forces are about equal with arcurate delta front and wide distributary channels.

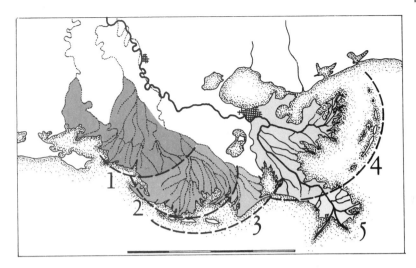

Figure 11.33 *The Mississippi delta and its most recent subdelta (see also plate 16). The active distributary system has built a major birdfoot delta during the last 500 years. The water is confined to the main distributary, and deposition is concentrated in a small sector of the delta front. After one sector is built far out into the ocean, the entire flow is shifted to some other sector and the process is repeated. Wave action then erodes back the birdfoot deltas. Previous subdeltas are indicated by numbers (1 through 4) according to age. Number 5 is the present subdelta.*

strong wave action in the Mediterranean redistributes the sediment at the delta front. The reworked sediment forms a series of arcuate barrier bars, which close off segments of the ocean to form lagoons. The lagoons form a subenvironment and soon become filled with fine sediment.

The difference between the Nile and the Mississippi deltas is due largely to the balance between influx of sediment, which builds birdfoot deltas, and the strength of wave action, which redistributes the sediment to form barrier bars.

The Mekong delta along the southern coast of Vietnam is a large delta in which redistribution of sediment by tidal currents dominates (*figure 11.32C*). The distributaries branch near Phnom Penh, about 500 km inland, into two main courses. Sediment carried by the river is reworked by tidal currents, which form broad distributary mouth channels and a series of elongate shoals.

The Niger delta is symmetrical and is dominated by the forces of tides and waves.

Alluvial Fans

Statement

Alluvial fans are stream deposits that accumulate in dry basins at the bases of mountain fronts. Deposition results from the sudden decrease in velocity as a stream emerges from the steep slopes of the upland and flows on to the gentle gradient of the adjacent basin.

Figure 11.34 *Alluvial fans in the arid southwestern United States. A fan develops as a stream enters a dry basin and deposits its sedimentary load.*

Figure 11.35 *Alluvial slopes in Nevada. As alluvial fans grow, they merge into a broad alluvial slope that covers much of the adjacent basin.*

Alluvial fans form mostly in arid regions, where streams flow only intermittently. In such areas, there is usually a large quantity of loose, weathered rock debris on the surface so that, when rain falls, the streams have a huge load of sediment to transport. Where a river emerges from a mountain front, the gradient is sharply reduced and deposition occurs. The stream channel soon becomes clogged with sediment and is forced to seek a new course. In this manner, the stream shifts from side to side and builds up an arcuate, fan-shaped deposit (*figure 11.34*). As several fans build basinward at the mouths of adjacent canyons, they eventually merge to form a broad slope of alluvium at the base of the mountain range (*figure 11.35*).

Discussion

The Basin and Range Province of the western United States is an arid region with large areas of internal drainage, so most of the sediment eroded from the ranges accumulates as alluvial fans in the basins. The fans range from very small deposits covering an area of less than 1 km² to large alluvial slopes on which a series of fans merge into a continuous alluvial plain along the base of the mountain front (*figure 11.35*).

Although alluvial fans and deltas are somewhat similar, they are different in mode of origin and internal structure. In deltas, stream flow is checked by standing water and sediment is deposited largely in an aqueous environment. The level of the ocean or the lake effectively forms the upper limit to which the delta can be built. In contrast, a fan is deposited in a dry basin, and the upper surface is not limited by water level. The coarse, unweathered sands and gravels of an alluvial fan also contrast with the sand, silt, and mud that predominate in a delta.

Summary

In this chapter, it has been demonstrated that a river system is changing continually. It erodes primarily in the headwater regions by downcutting, slope retreat, and headward erosion. Some sediment is deposited in the transporting segment of the river system and forms a variety of flood-plain features. Most of the sediment load, however, is deposited near the mouth of the river as a delta or an alluvial fan. In any segment of a river, one or more of these processes can be observed.

In the entire river system, the net result of these processes is erosion, transportation, and deposition of sediment. Erosion occurs in a systematic way, and the resulting landscape evolves in a predictable manner through a sequence of stages. Although we cannot observe the evolution of a landscape from start to finish, we can construct several conceptual models that explain landscape development. In the next chapter, some of these models will be considered in order to give you some insight into the changing surface of the earth.

Additional Readings

Crandell, D. R., and H. H. Waldron. 1956. "A Recent Volcanic Mudflow of Exceptional Dimensions from Mt. Rainier, Washington." Amer. J. Sci. 254:359-62.

Morisawa, M. 1968. Streams: Their Dynamics and Morphology. New York: McGraw-Hill.

Sharpe, C. F. S. 1938. Landslides and Related Phenomena. New York: Columbia University Press.

Young, A. 1972. Slopes. Edinburgh: Oliver and Boyd.

Evolution of Landforms 12

River systems are not simply channels through which water and sediment flow from the continents to the ocean; they are the major agents by which the earth's surface is sculptured into an infinite variety of erosional and depositional landforms. Almost every landform on a continent is related in some way to a drainage system, and, as the rivers flow endlessly to the ocean, the landscapes with which they are associated continually change and evolve. In the preceding chapter, it was shown that a stream valley evolves by three main processes (1) downcutting of the stream channel, (2) headward erosion and extension of the drainage net upslope, and (3) slope retreat (or lateral migration) of the valley walls away from the stream channel. As a result of these processes' operation as a system, a landscape evolves through a series of predictable stages from the time erosion begins until it is modified or stopped. No two regions are ever exactly alike, but, through the study of regions in various stages of development, geologists have constructed conceptual models of landscape evolution that permit a better understanding of the importance of river systems in the formation of the surface features of the earth and a fuller recognition of the sequence of steps in landscape development.

In this chapter, first, several general models of landscape development will be considered and each will be illustrated by a series of block diagrams. Second, how the structure and nature of the bedrock cause erosion to occur in different ways and at different rates to produce distinctive landscape types will be dealt with. Third, there will be a discussion of the problem of rates of erosion: how they are measured and what the measurements signify in terms of the earth's history.

Major Concepts

1. A landscape eroded by running water evolves through a series of predictable stages.
2. Climate and rock structure greatly influence the nature of the landscape.
3. Evolution of landscapes by running water can be interrupted by (a) earth movements, (b) glaciation, and (c) volcanism.
4. Rock formations erode at different rates so that resistant and nonresistant units are etched out into relief. This produces alternating cliffs and slopes on horizontal rocks and alternating ridges and valleys on folded rocks.
5. Rates of erosion show that the continents are eroded rapidly and must have been uplifted repeatedly in the geologic past to remain above sea level.

Model of Erosion in a Humid Climate

Statement

Theoretically, a newly uplifted area evolves through a series of stages until it is eroded to a surface of low relief near sea level. In the early stages, headward erosion and downcutting begin to dissect the area. Divides are broad and wide, but headward erosion extends the drainage net upslope, and with time the entire area becomes completely dissected into rolling hills. Slope retreat then becomes a major process, and the valley walls recede from the stream channel. Ultimately, the area is reduced to a surface of low relief near sea level.

Discussion

The block diagrams in *figure 12.1* illustrate the sequence of changes resulting from a newly uplifted area in a humid climate. In this model, it is assumed that uplift was rapid and that the surface was elevated more than 500 m above sea level.

Initial Stage. The newly exposed surface shown in block *A* at first is poorly drained, but eventually an integrated drainage system develops by headward erosion. Streams begin to erode V-shaped valleys into the uplifted landmass; downcutting and headward erosion are the dominant erosional processes. The valley walls extend right down to the stream bank. Tributary development proceeds rapidly as the drainage net is extended up the regional slope by headward erosion. New tributaries develop on the slopes of the valley walls of the major rivers. Stream capture is common in the headwater region. The stream gradients are steep, and the channels are marked by waterfalls and rapids. But broad areas of the original surface remain uncut, and marshes and lakes occupy original depressions in the newly uplifted surface.

As erosion continues, the local relief of the area (difference in elevation between valley floor and divide summit) increases and valleys are cut deeper and wider. Tributary valleys branch out from the larger streams and further dissect the initial surface. Slope retreat becomes more evident as an important process.

Intermediate Stage. As tributaries increase in number and length, the original surface is completely destroyed and the landscape becomes marked by steep slopes leading down to the deeply incised stream channels (block *B*) (see also *plate 12*). The drainage network becomes so well developed that the entire area is well drained. Local relief steadily increases. Streams also begin to erode laterally during this stage and develop flood plains. Slope retreat becomes more prominent. Ultimately, the process of downcutting ceases to produce greater local relief. Thereafter, erosion and deposition tend to reduce the local relief and the general elevation of the area. With the land reduced to a lower elevation, the rate of erosion proceeds more slowly.

Late Stages. Ultimately, the landscape is reduced to a low, nearly featureless erosional surface near sea level (block *C*). This surface is a **peneplain** over which the streams meander slowly, depositing almost as much sediment as they erode. The peneplain is the ultimate result of erosion in this model; it is near sea level and no further effective stream erosion can occur upon it. Subsequent modifications, when conditions remain stable,

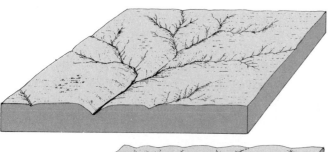

(A) Initial stage. Large areas are undissected. Downcutting and headward erosion are the dominant processes. The streams have steep gradients, with rapids and waterfalls. Only a small percentage of the area consists of valley slopes. The streams have not developed flood plains or an extensive meandering pattern.

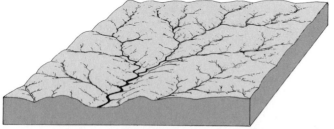

(B) Intermediate stage. The area is dissected completely so that most of the surface consists of valley slopes. Relief is at a maximum, and a well-integrated drainage system is established. The main streams meander, and a flood plain begins to develop. The topography is characterized by smooth, rolling hills.

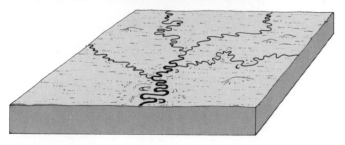

(C) Late stage. The landscape has been eroded to a nearly featureless surface near sea level. Rivers meander over a broad flood plain. Some isolated erosional remnants remain, but deposition is nearly as important as erosion.

Figure 12.1 *Evolution of a landscape in a humid climate.*

are deep weathering and deposition of sediment on the flood plain. Oxbow lakes, meander scars, and natural levees are common, and locally an isolated erosional remnant of resistant rock, a **monadnock,** may protrude above the peneplain surface.

Examples. Erosion of a landscape is much more complex than the model described above. It is doubtful that a large area evolves through all of these stages, but there are numerous examples of regions in various stages of dissection. The coastal plains of the Atlantic and Gulf regions of the United States only recently have emerged from below sea level, so vigorous erosion has not occurred.

Figure 12.2 *An area in the intermediate stage of dissection. Valley slopes make up a large percentage of the surface; relief is at a maximum; and the drainage system is well integrated, with a large number of streams per unit area (see plate 12).*

Figure 12.3 *The Canadian Shield, a low, flat, erosional surface eroded close to sea level.*

The Allegheny Plateau (Pennsylvania), with its adjacent areas, is an example of a region near the intermediate stages of erosion in a humid climate; most of the area is dissected to some extent and, with continued erosion, relief will be diminished (*figure 12.2*). The best example of a surface eroded to the late stage is the broad, low region of eastern Canada (*figure 12.3*). This area is eroded to within a few hundred meters of sea level and has remained low for many millions of years. Its erosional history is much more complex than the model described above, because it has been covered partly by the ocean at various times and has been subjected to recent glaciation.

Model of Erosion in an Arid Climate with Through-flowing Streams

Statement

The evolution of a landscape in an arid or semiarid climate in which there is an integrated drainage system with major streams flowing to the ocean is similar to that which occurs in a humid climate: the surface is dissected systematically through a series of stages until it is worn down to a peneplain. However, there are several significant differences in the details of the landscape. In an arid climate, the topography is much more angular, since resistant limestones and sandstones form steep cliffs and nonresistant shales form broad slopes. Slope retreat is more evident in an arid region, and, in some areas, wind deposits significantly alter the landscape. Many tributary streams are dry except during major rains or spring runoff.

Discussion

The series of diagrams shown in *figure 12.4* illustrates the evolution of an idealized landscape in an arid region with through-flowing streams.

Initial Stage. The profile of canyons developed in the early stages of dissection (block *A*) is characterized by alternating cliffs and slopes, whereas the topography in humid regions characteristically is rounded. Slope retreat is a more obvious process, and, with a minimum of precipitation, an intricate network of tributary stream channels is slow to develop. Cliffs retreat laterally by undercutting of the nonresistant shales below them. It is common to find large areas of resistant rock stripped clean of the overlying formation, even though the surface may be over a thousand meters above sea level. As in the initial stages in a humid climate, large regions of the original surface remain undissected.

Intermediate Stage. As erosion continues (block *B*), the original surface becomes dissected by a network of deep canyons. Local relief reaches a maximum, and alternating hard and soft layers are etched out into cliffs and slopes. Resistant rock layers play an important role in landscape development, since they form resistant cap rock for a variety of landforms. A large upland surface called a **plateau** typically develops where resistant sandstone, limestone, or basalt maintains a nearly flat surface. Continued erosion may dissect the plateau into smaller, tablelike mountains called **mesas,** which ulti-

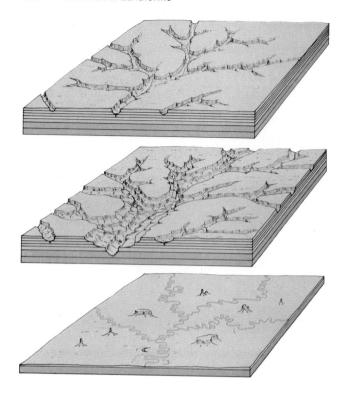

(A) *Initial stage. Alternating cliffs and slopes characterize the typically angular topography. A network of stream channels begins to develop, and cliffs retreat laterally. Much of the original surface is undissected.*

(B) *Intermediate stage. Erosion continues to dissect the area into a network of deep canyons. Local relief reaches a maximum, and resistant rock layers form cap rock for plateaus, mesas, and buttes.*

(C) *Late stage. Erosion shrinks the mesas and buttes and forms a peneplain near sea level. Only a few resistant remnants punctuate the nearly flat region.*

Figure 12.4 *Evolution of a landscape in an arid climate with through-flowing streams.*

mately erode down to small, isolated, flat-topped hills called **buttes.**

Late Stage. In the final stages of erosion (block *C*), the mesas and buttes continue to shrink as slopes retreat from the drainage channels.

Example. The Colorado Plateau of the western United States can be considered an example of this type of landscape development, although the region has been subjected to recent earth movements and is somewhat more complicated. In *figure 12.5*, the spectacular landscape of the Canyonlands National Park can be seen

to be the result of the dissection of an uplifted plateau, with differential slope retreat being responsible for the plateaus and mesas eroded from the resistant rock formations. The patterns of erosion of this region are controlled by dissection of the region by the Colorado River. On a regional basis, all major cliffs recede away from the river and its major tributaries.

Figure 12.5 *The canyonlands of the Colorado Plateau, an area in the intermediate stage of landscape evolution in an arid climate.*

Model of Erosion in an Arid Climate with Internal Drainage and Block Faulting

Statement

Because of limited rainfall in arid regions, many streams are shorter than the slopes down which they flow. That is, the water in the stream either seeps into the ground or evaporates before it reaches a lake or the ocean. Instead of a well-integrated drainage system with a long, continuous stream system flowing to the ocean, streams in desert regions are typically shorter and terminate in alluvial fans or shallow **playa lakes.** This is especially true of regions where rift systems begin to split a continent and block faulting has produced structural depressions of internal drainage. The major landforms in such areas are linear mountain ranges that result from erosion of fault blocks and basins that become filled with sediment.

Discussion

The idealized stages in landscape development in an arid region with internal drainage and block faulting are illustrated in *figure 12.6*.

Initial Stage. The initial stage (block *A*) develops by block faulting. Maximum relief usually occurs in the initial stage as a result of crustal deformation, not erosion. Relief then continues to diminish throughout subsequent stages, unless major uplift recurs to interrupt the evolutionary trend by producing greater relief during the later stages. Depressions between mountain ranges generally do not fill completely with water to form large lakes because of the low rainfall and excessive evaporation. Instead, shallow, temporary lakes called playa lakes form in the central part of the basin and fluctuate considerably in size. They may be completely dry for many years and then expand and cover a large part of the valley floor during years of high rainfall. Having no outlets, these lakes commonly are saline.

More sediment is produced by weathering in the uplifted mountain mass than can be carried away by the intermittent streams, which may flow only during spring runoff. Thus, the streams commonly are overloaded with sediment and deposit much of their load where they emerge from the mountain front. The sediment deposited by the river accumulates in a broad alluvial fan. Throughout the history of an intermontane basin, mountain ranges are eroded and the debris is deposited in the adjacent basin.

Intermediate Stage. As geologic processes continue (block *B*), the mountain mass becomes dissected into an intricate network of canyons and is worn down to

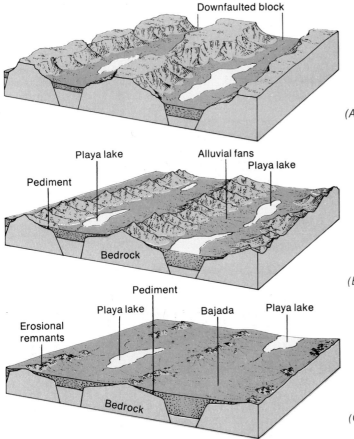

(A) Initial stage. Faulting produces maximum relief. Some areas are mountainous and undissected. Playa lakes develop in the central part of the basins.

(B) Intermediate stage. The mountain range is dissected completely, and the mountain front retreats, developing a pediment. Alluvial fans spread out into the valley.

(C) Late stage. The basins become filled with sediment. Erosion wears down the mountain ranges to small, isolated remnants. The pediments expand, and the alluvial fans merge to form bajadas. Most of the surface is an alluvial slope.

Figure 12.6 *Evolution of a landscape in an arid climate with block faulting and internal drainage.*

a lower level. At the same time, the mountain shrinks in size as the front recedes through the process of slope retreat. The fans along the mountain front grow and merge to form a large alluvial slope called a **bajada.** As the mountain front retreats, an erosional surface called a **pediment** develops on the underlying bedrock and continues to expand as the mountain shrinks. Pediments are generally covered by a thin veneer of alluvium as they form.

Late Stage. The final stage in this model of arid landscape development (block *C*) is characterized by small, islandlike remnants of the mountains surrounded by an extensive erosional surface, the pediment. The pediment shows little relief and is covered largely with erosional debris.

Example. This model, or variations of it, can be used effectively in explaining the landscape of the Basin and Range province of the western United States (*figure 12.7*). There, block faulting has occurred over a large area to produce alternating mountain ranges and intervening fault-block basins. To the north, throughout much of western Utah and all of Nevada, the area is in the initial stages of development. The basins occupy about half of the total area, and pediments are small. The relief of the mountain ranges is high, with most fans just beginning to coalesce to form a broad alluvial slope. In Arizona and Mexico, erosion in the Basin and Range has proceeded much farther (*figure 12.8*) than in Utah and Nevada, and the area could be considered to be in the later stages of development. The ranges are eroded down to small remnants of their original size, and extensive bajadas, extending over approximately four-fifths of the area, cover wide pediments through which isolated remnants of bedrock protrude.

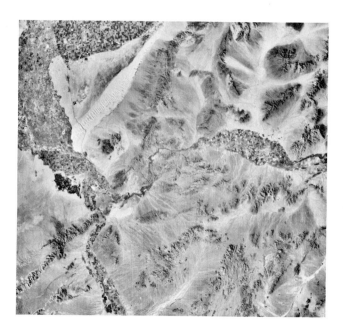

Figure 12.8 *The Basin and Range in Arizona, an area of block faulting and internal drainage in the intermediate to late stages of arid landscape evolution.*

Models of Erosion in Volcanic Terrains

Statement

The extrusion of lava provides a new surface that is attacked by erosion and is modified through a series of stages until the flows are destroyed completely or buried by sediment or younger lava. Lava flows generally are very resistant to erosion. They flow down previously established drainage systems and displace the river channel. In areas of small, local volcanic eruptions, the lava flows downslope and is confined to the stream valley. Thus, the stream channel is displaced and erosion proceeds along the margin of the flow to develop an **inverted valley.**

Where large volumes of lava are extruded, the flows may bury the landscape completely. New drainage established across, or along, the margins of the lava plain ultimately dissects it into a plateau.

Discussion

Local Volcanism. The sequence of landforms resulting from erosion of an area where minor volcanic activity has occurred is illustrated in *figure 12.9.* In the initial stage (block *A*), volcanic flows enter a drainage system, following the river channel and partly filling the valley. The lava flow disrupts the drainage in several ways. Lakes are impounded upstream, and the river is displaced and forced to flow along the margins of the lava. Subsequent stream erosion then is concentrated along the margins of the lava flow.

As erosion proceeds in the displaced drainage (block *B*), new valleys are cut along the margins of the flow, becoming deeper and wider with time. The cinder cones, formed during the initial volcanic activity, soon are obliterated because the unconsolidated ash is easily

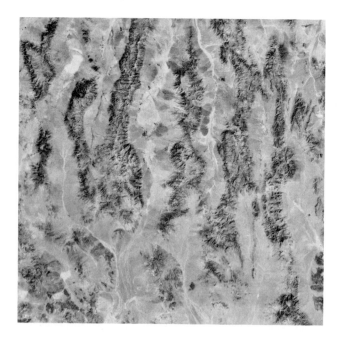

Figure 12.7 *The Basin and Range in central Nevada, an area of block faulting and internal drainage in an early to intermediate stage of development.*

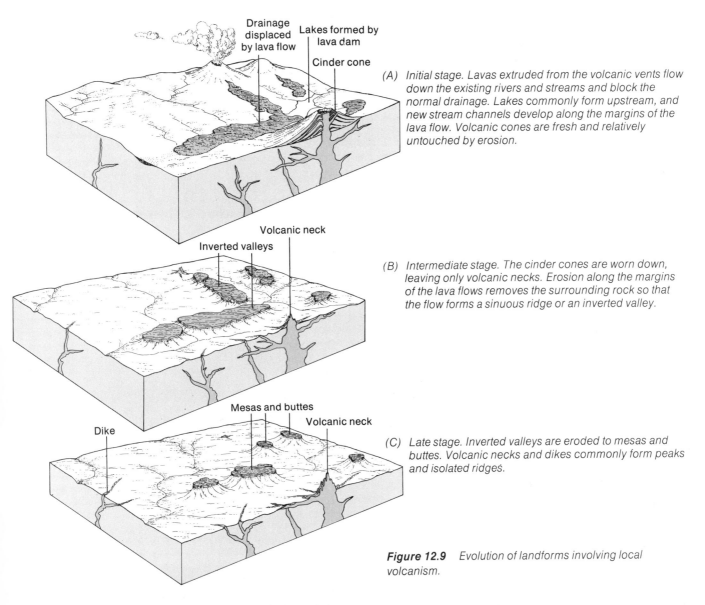

Drainage displaced by lava flow · Lakes formed by lava dam · Cinder cone

(A) *Initial stage. Lavas extruded from the volcanic vents flow down the existing rivers and streams and block the normal drainage. Lakes commonly form upstream, and new stream channels develop along the margins of the lava flow. Volcanic cones are fresh and relatively untouched by erosion.*

Inverted valleys · Volcanic neck

(B) *Intermediate stage. The cinder cones are worn down, leaving only volcanic necks. Erosion along the margins of the lava flows removes the surrounding rock so that the flow forms a sinuous ridge or an inverted valley.*

Dike · Mesas and buttes · Volcanic neck

(C) *Late stage. Inverted valleys are eroded to mesas and buttes. Volcanic necks and dikes commonly form peaks and isolated ridges.*

Figure 12.9 *Evolution of landforms involving local volcanism.*

eroded. Only the conduit through which the lava was extruded remains as a resistant volcanic neck. With time, the lava flow is eroded into a long, sinuous ridge, commonly referred to as an inverted valley, because the previous stream valley beneath the lava is now higher than the surrounding area.

In the final stages of erosion (block *C*), the inverted valley is reduced in size and ultimately becomes isolated mesas and buttes.

Example. Inverted valleys are common along the western and southern margins of the Colorado Plateau (*figure 12.10*), where recent volcanism has occurred during the erosion of the landscape.

Figure 12.10 *Inverted valleys in southwestern Utah, a region with local volcanic activity in an intermediate stage of development. The long, narrow ridge is an ancient valley that became partly filled with lava extruded from vents in the far background. The drainage was displaced to the flow margins, where it cut new valleys, leaving the old valley (filled with lava) stranded high above the present surface.*

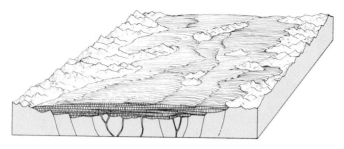

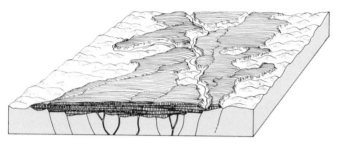

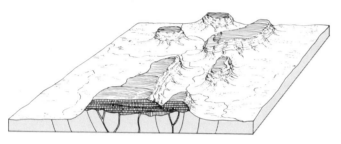

(A) Initial stage. Extensive lava flows fill the preexisting drainage system and flood the surrounding area. New drainage develops along the margins of the volcanic field, as well as across the new surface of the lava.

(B) Intermediate stage. The lava field begins to be dissected by the new drainage system. The igneous rock usually is very resistant, so the surrounding area is eroded faster, leaving the lava field as a plateau.

(C) Late stage. The area surrounding the volcanic flow is reduced to lowland. Remnants of lava form cap rock on isolated high mesas and buttes.

Figure 12.11 Evolution of a landscape involved with regional volcanism.

Regional Volcanism. In regions of extensive volcanic activity, the entire area may be buried completely by lava flows (*figure 12.11*). In the initial stage (block *A*), drainage usually is displaced near the margins of the lava plain, but some rivers may migrate across it. When extrusion ceases (block *B*), stream erosion begins to dissect the lava plain and ultimately cuts it into isolated plateaus and mesas (block *C*). These, in turn, eventually are consumed.

Example. The Snake River plain in Idaho is a classic example of an area of extensive basaltic volcanism. It presently is being dissected by the Snake River (*figure 12.12*).

Interruptions in Landscape Development

Statement

The models just described assume rapid uplift (or rapid fall of sea level) followed by stabilization so that erosional processes continue uninterrupted. In reality, this exact sequence rarely, if ever, occurs, for various forces can interrupt the process at any stage of development. Among the more significant changes that can interrupt the development of a landscape are:

1. Rejuvenation—recurrent uplift can occur and increase the elevation. This rejuvenates the stream or increases its capacity to downcut its channel.
2. Glaciation—the presence of continental glaciers completely interrupts the evolution of a landscape by

Figure 12.12 The Snake River plain of southern Idaho, an area of recent regional volcanic activity in the early stage of erosion.

changing the drainage system and modifying the surface with products of glacial erosion and deposition.
3. Volcanic activity—the extrusion of lava and ash over a terrain previously evolving by stream erosion modifies the surface and displaces drainage systems.
4. Climatic changes—changes in climate can produce different styles of erosion.

5. Changes in sea level—changes in sea level either raise or lower the base level and affect the evolutionary trends in landscape development.

Discussion

Rejuvenation. The most significant interruptions in landscape development usually are due to earth movements: broad regional uplift, subsidence, and folding and faulting associated with mountain-building processes. The result is an increase in elevation and a corresponding increase in stream energy. The evolutionary trends are, therefore, interrupted, and many aspects of landscape development return to the condition of the initial stage and begin development again. When uplift occurs in a late stage of development, the gradient of a meandering stream pattern is increased and the stream begins to downcut its channel. As a result, the meandering pattern characteristic of the late stage in erosion can become entrenched and form a deep, meandering canyon, but the river has many characteristics of the initial stage of erosion, such as steep gradient and rapids. Classical examples of entrenched meanders are found throughout the Colorado Plateau, as shown in *figure 12.13.*

Glaciation. The glacial epoch disrupted the erosional processes not only in the area covered with ice but also in other areas throughout most of the continents. The ice sheets completely obliterated the drainage system over which they moved, and, along the margins of the great continental ice sheets, deposits of glacial debris hundreds of meters thick completely covered the preexisting drainage channels. Thus, much of the surface in Canada and northern Europe shows a strong imprint of glaciation, rather than erosion by streams and rivers.

Figure 12.13 *Entrenched meanders in the drainage system of the Colorado Plateau illustrate rejuvenation of stream erosion. The meandering pattern of the river was formed during an earlier period when the river had a low gradient. Recent uplift has increased the gradient and the river's power to erode. Downcutting became the dominant process, and the meanders became entrenched.*

In addition, the melting ice deposited a huge load of debris along the ice margins so that the remaining streams became overloaded with sediment. Instead of eroding, the streams began to fill their valleys with sediment. For example, in the lower Mississippi valley, more than 60 m of sediment is believed to have accumulated largely from deposition of the great load of glacial sediment. The river now flows over a broad, flat-floored valley built up by fluvial sedimentation.

Volcanism. Floods of lava could cover completely a large area like the Snake River plain and the Columbia River Plateau in the northwestern United States. As illustrated in *figure 12.11,* erosion in such an area has to begin from scratch. Other volcanic terrain can produce a series of high composite volcanoes with associated ash and lava flows, which block drainage, cause lakes to form, and build up a distinctive constructional landscape. The Cascade Mountains are the best example of this phenomenon in the United States.

Differential Erosion

Statement

Different rock types erode at vastly different rates. Factors such as the climate, the degree of compaction and cementation, and the nature of the dominant minerals are all important in causing some rocks to be hard and resistant and others soft and weak. The net result is that, when a sequence of rocks is exposed to the surface, the rocks erode at different rates and to different degrees; this process commonly is called **differential erosion.** Differential erosion occurs on all scales ranging from steep cliffs and gentle slopes formed on alternating hard and soft rock bodies to delicate laminae within a rock etched out in delicate relief. It is this differential erosion occurring on such a vast range of scales that is responsible for much of the beauty and the spectacular scenery of the earth.

Discussion

Differential erosion perhaps is best developed in arid regions. In these areas, not only is the type of rock important, but jointing and availability of surface and ground water combine to produce fascinating details of the landscape.

Cliffs and Slopes. Probably the most widespread example of differential erosion is the alternating cliffs and slopes developed on a sequence of alternating hard and soft sedimentary rocks. Soft shales typically form slopes, and the more resistant sandstones and limestones produce cliffs (*figure 12.14*). The height of the cliff and the width of the slope are largely functions of the thickness of the formations involved. Most of the topography of the Colorado Plateau is in some way a product of differential weathering on sequences of alternating shales and sandstones or limestones.

Columns and Pillars. Buttes, pinnacles, and columns are simply details of differential erosion on receding cliffs. When a cliff is a massive unit with well-developed joints, vertical columns like those shown in *figure 12.15* develop. When stratification produces alter-

Figure 12.14 *In the Colorado Plateau, slopes form on nonresistant shale formations, which erode rapidly and quickly retreat from the river channel. Cliffs form on hard sandstone and limestone, which erode very slowly.*

Figure 12.15 *Eroded columns in a resistant sandstone formation.*

Figure 12.16 *The eroded columns of Bryce Canyon, Utah. Rapid erosion along joint systems separates the columns from the main cliff. Differential erosion accentuates the difference between rock layers to produce the fluted columns.*

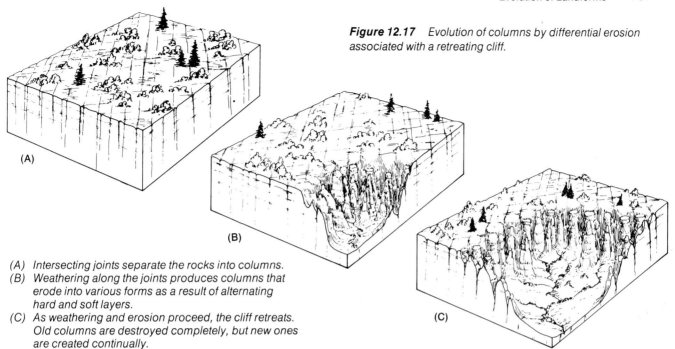

Figure 12.17 *Evolution of columns by differential erosion associated with a retreating cliff.*

(A)

(B)

(C)

(A) *Intersecting joints separate the rocks into columns.*

(B) *Weathering along the joints produces columns that erode into various forms as a result of alternating hard and soft layers.*

(C) *As weathering and erosion proceed, the cliff retreats. Old columns are destroyed completely, but new ones are created continually.*

nating layers of hard and soft rock, additional detail may be etched out (*figure 12.16*). Jointing commonly plays an important role, for it permits weathering to attack a rock body from many sides at once. The columns and pillars famous in Bryce Canyon National Park, for example, result from differential weathering along a set of intersecting joints. The joints separate the rock into columns. Differential erosion of the nonresistant shales that separate the more resistant sandstone and limestone forms deep recesses in the columns to produce the fascinating slopes and forms (*figure 12.17*).

Natural Arches. An interesting form of differential erosion is the natural arch formed in massive sandstones in the Colorado Plateau of the western United

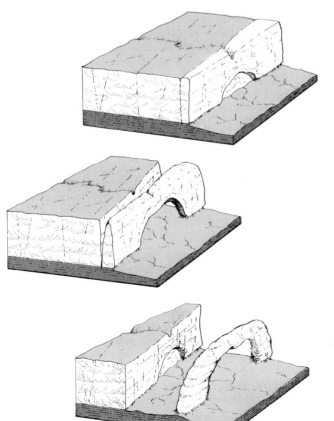

(A) *In arid regions, much water seeps into the subsurface below a stream channel. This water moves laterally above an impermeable layer and eventually emerges as a seep near the base of a cliff. The cement that holds the sand grains together soon is dissolved in this area of greatest moisture and the sand grains fall away, forming a recess or alcove beneath the dry falls from the stream above.*

(B) *If a joint system in the sandstone is roughly parallel to the cliff face, it may be enlarged by weathering, which separates a slab from the main cliff.*

(C) *An arch is produced as the alcove enlarges. Weathering then proceeds inward from all surfaces until the arch collapses.*

Figure 12.18 *The formation of natural arches in massive sandstones.*

States. Most of the arches result from solution activity, which dissolves the calcite cement that holds the sand grains together. Once the cement is dissolved, the loose sand is carried away by gravity, wind, or water, producing an alcove or cave that may enlarge to form an arch. Solution activity to dissolve the cement is most effective where there are concentrations of surface and ground water, so many natural arches are associated with drainage systems. The cement is distributed unevenly throughout the rock, however, so arches are formed in places that contain the least amount of cement.

A model illustrating the process by which natural arches may be formed is shown in *figure 12.18*. In the arid western parts of the United States, there is little precipitation and much of the surface water seeps into the porous sandstone. The subsurface water is most abundant beneath the dry stream channels and the movement of ground water follows the general surface drainage lines. The ground water emerges as a seep in a cliff beneath a dry waterfall and dissolves the cement in that area. The loose sand grains are washed or blown away so that an alcove soon develops at the base of a dry waterfall. When the sandstone is cut by a series of joints, a large block can be separated from the cliff and the alcove and joint surface enlarge to produce an isolated arch (*figure 12.19*). Weathering then proceeds inward from all surfaces on the arch until the arch is destroyed and leaves only standing columns.

Figure 12.19   A natural arch being produced by the method illustrated in figure 12.18.

Erosional Landforms and Structures

Statement

Erosional landforms are influenced greatly by the structure and nature of the rock into which they are cut. As was discussed in the previous section, different rock types erode at different rates, causing the more resistant units to stand out as ridges above the more readily eroded material adjacent to them. Thus, the structure of the rock commonly is expressed in the landforms and exerts a significant influence upon their shapes and patterns.

Discussion

Inclined Strata.   When a sequence of alternating resistant and nonresistant rock is tilted, the resistant formations are eroded into long, asymmetrical ridges and the interbedded soft units form elongate valleys parallel to the ridges (*figure 12.20*). Ridges formed on gently inclined strata are known as **cuestas;** sharp ridges formed on steeply inclined rocks are referred to as **hogbacks.**

Usually, inclined strata are parts of larger flexures, such as domes and basins. Major streams are superposed and cut across resistant and nonresistant rocks alike, but new tributaries begin to develop on the weak rock and rapidly extend themselves headward, eroding a subsequent valley along the nonresistant unit and leaving the inclined resistant beds protruding as a hogback ridge.

Domes and Basins.   Sedimentary rocks commonly are warped into large domes and basins (see *plates 5* and

6). When erosion is initiated on these structures, there may be little more than local modifications of the drainage, but, with differential erosion, a sequence of alternating hard and soft rocks is expressed by a series of circular hogbacks or cuestas with intervening valleys cut on the less resistant rocks (*figure 12.21*). The valleys commonly are referred to as **strike valleys,** because they are developed along the strike or trend of the inclined strata.

If the older rocks in the center of the dome are non-resistant, the center of the uplift may be eroded down into a topographic lowland bordered by inward-facing escarpments formed on the younger resistant units. If the older rocks are more resistant, the center of the dome remains high and forms a dome-shaped hill or ridge. Large domal structures have inward-facing cliffs, whereas large basins have outward-facing escarpments.

Folded Rocks.   As erosion proceeds on a sequence of folded rocks, resistant formations are etched out and form ridges that follow the outcrop pattern of the limbs of the folds. The Jura Mountains of Switzerland and the Appalachian Mountains of the eastern United States illustrate well the variations in topography of folded rocks (see *plates 7* and *26*). The rocks of the Jura Mountains, folded very recently in late Tertiary time, are so young that erosion is just beginning to break through the crests of the folds (*figure 12.22*). **Anticlines** (uparched rocks) form most of the mountain ridges; and **synclines** (downfolded rocks), the valleys. In places, a resistant limestone formation is stripped off the crest of some anticlines and forms a hogback along the flanks of the fold.

By contrast, in the Appalachian Mountains, where the folding occurred much earlier (late Paleozoic), the topography bears no relation to the structural relief; ridges are carved on resistant sandstone and conglomerate formations and the valleys are developed on weak shales and limestones. These mountains have suffered a much greater amount of erosion; therefore, the resistant units form ridges corresponding to the trend of the limbs of the folds. Long strike valleys have developed by headward erosion along the weak formations, and the drainage network has assumed a trellis pattern (*figure 12.23*). (See also *plate 7*.)

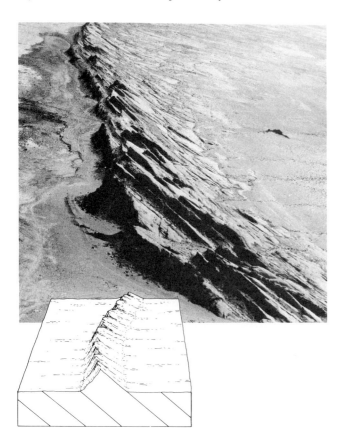

Figure 12.20   *A hogback formed by differential erosion on a sequence of tilted sedimentary strata. A resistant sandstone forms the hogback ridge, while nonresistant shales erode into valleys parallel to the strike of the beds.*

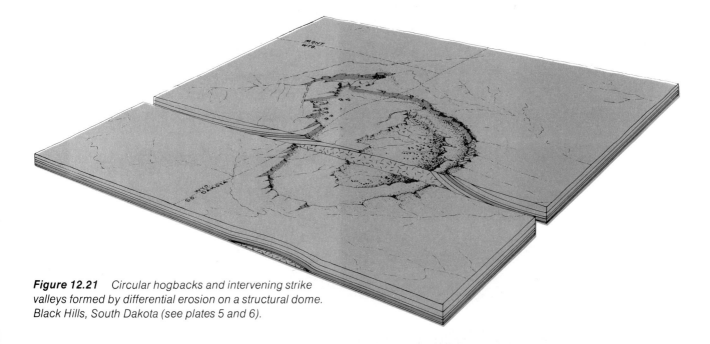

Figure 12.21   *Circular hogbacks and intervening strike valleys formed by differential erosion on a structural dome. Black Hills, South Dakota (see plates 5 and 6).*

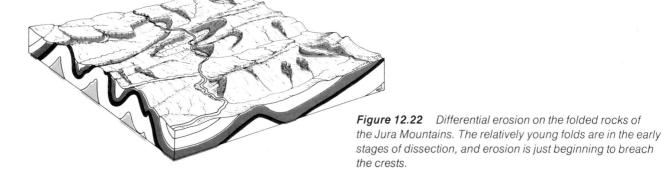

Figure 12.22 *Differential erosion on the folded rocks of the Jura Mountains. The relatively young folds are in the early stages of dissection, and erosion is just beginning to breach the crests.*

Figure 12.23 *Differential erosion on the resistant sandstone formations in the Appalachian Mountains. Erosion has proceeded to a much later stage than in the Juras. The tops of the folds are eroded off, and only the resistant rock units remain as ridges (see plate 7).*

Figure 12.24 *The Hurricane fault in southern Utah. The fault is located at the base of the 400-m cliff and has displaced the rocks on the left downward relative to those on the right. Total displacement is 1500 m.*

Faults. Faults are fractures in the earth's crust along which there has been displacement. Normal faults are those in which displacement is largely vertical and results from tensional stresses. Many of the major normal faults are expressed topographically as cliffs like the one shown in *figure 12.24*. Erosion soon dissects the cliff to produce a series of triangular faces called **faceted spurs** on the cliff front (*figure 12.25*). Erosion forms gullies along the blunt face of the faceted spur, so the cliff pro-

duced by the fault is modified considerably. Recurrent movement along the fault can produce a fresh cliff at the base of the older modified faceted spurs, but it also soon is modified by gullying (*figure 12.25C*). When movement on the fault ceases, the cliff erodes down and back from the fault line.

Strike-slip faults are fractures in which displacement has been horizontal rather than upward or downward and one block has moved past the other parallel to the

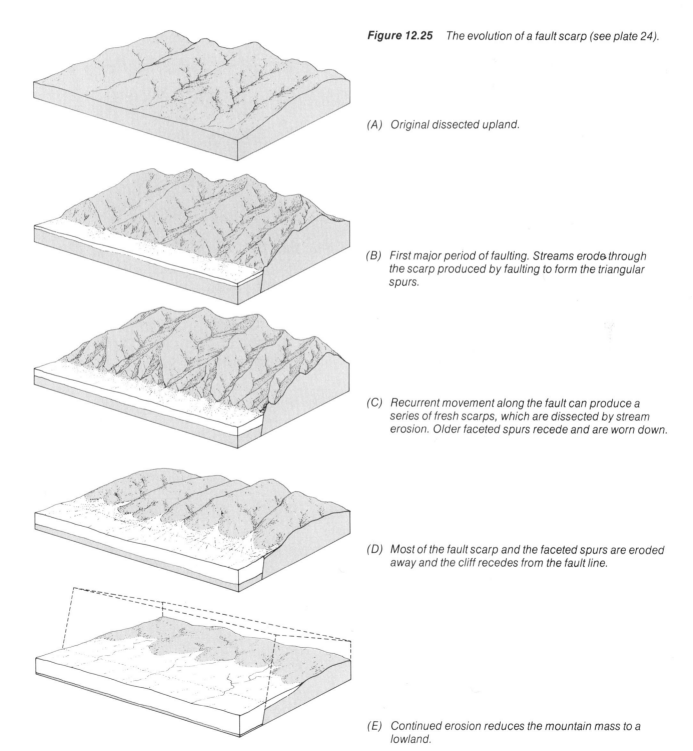

Figure 12.25 *The evolution of a fault scarp (see plate 24).*

(A) *Original dissected upland.*

(B) *First major period of faulting. Streams erode through the scarp produced by faulting to form the triangular spurs.*

(C) *Recurrent movement along the fault can produce a series of fresh scarps, which are dissected by stream erosion. Older faceted spurs recede and are worn down.*

(D) *Most of the fault scarp and the faceted spurs are eroded away and the cliff recedes from the fault line.*

(E) *Continued erosion reduces the mountain mass to a lowland.*

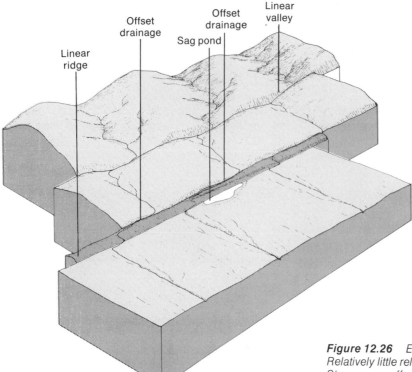

Figure 12.26 *Erosional features along a strike-slip fault. Relatively little relief is produced by strike-slip displacement. Streams are offset, linear ridges and valleys form in sliver blocks, and local sag ponds may form in depressions along the fault line.*

strike or trend of the fault plane. Since there is little vertical movement, high cliffs do not form from strike-slip faults; instead, the fracture is expressed at the surface by a low ridge or offset drainage (*figure 12.26* and *plate 25*).

Rates of Erosion

Statement

In the study of landforms, critical questions about erosion include: How fast does erosion occur? How long has it been going on? Can it continue indefinitely into the future? Answers to these questions are not only important to the origin of surface features but provide fundamental information about the dynamics of the earth as a whole.

In many ways, slow processes like erosion are more difficult to measure than extremely rapid events, but various approaches to the problem provide some very reliable estimates of the rates at which the continents are being worn away. One approach is to measure the amount of sediment transported each year to the ocean by the major river systems. In essence, this is a measure of the rate at which the regolith is being removed (the rate of **denudation**). Another approach has been to study inverted valleys and measure the rates at which streams have been able to downcut their channels.

Discussion

Rates of Denudation. Many significant estimates of the rate at which the cover of regolith is being removed

by a drainage system have been made by carefully measuring the amount of sediment carried out of the drainage basin each year. Most of the major rivers of the world have gauging stations, and it is a simple matter to sample the sediment load of a river and calculate the volume of material transported from a continent during a given period of time. Of course, rates of erosion vary greatly, depending on climate, topography, rock composition, and stage of stream evolution; but this method provides a good estimate of the rate at which the general surface of a continent is being lowered.

On the basis of the volume of sediment transported by the major rivers, the average rate of erosion in the United States is estimated to be approximately 6 cm per 1000 years. It should be emphasized that the estimate of 6 cm per 1000 years is an *average*. The surface is *not* lowered uniformly by this amount; indeed, some areas within the drainage basin may be unaltered, or even built up, by local accumulations of sediment. Others may be lowered more than 25 cm in only a few hours by the erosive power of a flash flood. Measurement of the amount of sediment removed from an area indicates only how much erosion occurs, not where it occurs. On a regional basis, the most rapid rates of erosion occur where the relief is greatest and where the total potential energy available from rivers for erosion and transportation is highest.

Rates of Downcutting. Preliminary estimates of the rate at which a stream is able to downcut its channel have been made by studying stream profiles preserved beneath lava flows that can be dated by radiometric methods and by measuring the amount of downcutting that has occurred since the lava was extruded. An example is shown in *figure 12.27*. A basalt flow was extruded near the Hurricane fault, flooded the Virgin River, and ponded in a depression at the base of the Hurricane Cliffs. Subsequent displacement along the fault offset the basalt 134 m. The age of the flow, as determined by radiometric dating methods, is 250,000 years. Thus, in the last 250,000 years, the Virgin River has been able to cut a gorge 134 m deep through the resistant limestones that form the Hurricane Cliffs and reestablish a smooth profile. On the downthrown side of the fault, the Virgin River has cut only through the basalt down to its original profile. Along with other examples, this suggests that a perennial stream has a tremendous capacity for downcutting but it does so only when uplift occurs and the stream's profile is increased significantly.

Another measurement of very rapid downcutting has been made in the Grand Canyon. Recent spectacular lava cascades on the rim of the inner gorge and a sequence of older lava dams with an aggregate thickness of 1190 m have been eroded during the last 1.2 million years (*figure 12.28*). This is an average rate of 1 m per 1000 years. In another example, modern measurements on the erosion

of Niagara Falls indicate a retreat of nearly 1 m per year, or 1000 km in a million years.

The important point that these measurements emphasize is that a perennial stream has the capacity to erode through practically any barrier placed in its way, with remarkable speed in a geologic time frame, and that the amount of downcutting is determined largely by the amount of uplift of continental crust (or lowering of sea level).

Summary

The evolution of a landscape is complex and is influenced by many factors, such as climate, rock type, structure, and elevation. Various graphic models can be constructed to show how, under ideal conditions, a landscape evolves through various stages in (1) a humid climate, (2) an arid climate with external drainage, (3) an arid climate with internal drainage, and (4) volcanic terrains. It should be emphasized that these are ideal models and that the real world is much more complex. At any time, the evolution of a landscape can be interrupted or changed by such things as earth movements, glaciation, and volcanism. Much of the earth's scenic landscape is the product of differential erosion in which rock units of varying resistance are etched out into cliffs, slopes, columns, and natural arches. Rates of erosion are slow by human standards, but measurements of the amount of sediment transported by rivers indicate that, on an average, the continental surface of North America is being lowered at a rate of 6 cm per 1000 years.

Figure 12.27   *Basalt flows displaced by recurrent movement along the Hurricane fault near the tower of Hurricane, Utah. Downcutting by the Virgin River has cut a gorge 134 m deep on the upthrown block, while stream erosion on the downthrown block has cut only through the basalts. This indicates that perennial streams have the capacity for very rapid downcutting when uplift occurs.*

River profile

Figure 12.28 *Lava cascades in the western Grand Canyon. The remnants of basalts at the base of the cliff are 1.2 million years old. These formed a lava dam across the river 350 m high. Erosion by the Colorado River has been able to cut down through this dam plus an additional 15 m since it was formed. The average rate of downcutting is approximately 1 m per 100 years.*

Although running water is by far the most effective agent of erosion, water also seeps into the ground and percolates through the pore spaces of the rock. As it moves, it dissolves soluble minerals and erodes through subsurface solution activity. In the next chapter, ground water as part of the hydrologic system will be considered.

Additional Readings

Bloom, A. L. 1969. The Surface of the Earth. Englewood Cliffs, N.J.: Prentice-Hall.

Easterbrook, D. J. 1969. Principles of Geomorphology. New York: McGraw-Hill.

Garner, H. F. 1974. The Origin of the Landscapes. New York: Oxford University Press.

Gordon, R. B. 1972. Physics of the Earth. New York: Holt, Rinehart and Winston.

Hunt, C. B. 1973. Natural Regions of the United States and Canada. San Francisco: W. H. Freeman and Company.

Thornbury, W. D. 1969. Principles of Geomorphology. New York: John Wiley and Sons.

Tuttle, S. D. 1970. Landforms and Landscapes. Dubuque, Iowa: William C. Brown and Company.

Ground-Water Systems 13

Water moving in the pore spaces of rocks beneath the earth's surface is a geologic process not easily observed and, therefore, not readily appreciated. Yet, ground water is an integral part of the hydrologic system, and it is one of the most important natural resources. Ground water is not a rare or unusual phenomenon; it is distributed everywhere beneath the surface. It occurs not only in humid areas but beneath the desert regions, as well as under the frozen arctic and the high mountain ranges. In many areas, the amount of water seeping into the ground may be equal to, or greater than, the surface runoff.

Ground water rarely flows as distinct underground streams but slowly percolates through the pore spaces of the rocks and eventually is discharged back to the surface. As it moves, it dissolves the more soluble minerals, producing caves and caverns, and is responsible for a special type of topography unlike that formed by streams and rivers.

This chapter will consider the role of ground water as part of the hydrologic system.

Major Concepts

1. The movement of ground water is controlled largely by the porosity and permeability of the rocks through which it flows.
2. The water table is the surface below which all pore spaces in the rock are saturated with water.
3. Ground water moves slowly (percolates) through the pore spaces in rocks by the pull of gravity, while, in artesian systems, it is moved by hydrostatic pressure.
4. Natural discharge of ground water is generally into streams, marshes, and lakes.
5. Artesian water is water confined under pressure, like water in a pipe. It occurs in permeable beds bounded by impermeable formations.
6. Erosion by ground water produces karst topography, which is characterized by sinkholes, caves, and disappearing streams.

Porosity and Permeability

Statement

Water is able to infiltrate into the subsurface because solid bedrock, as well as loose soil, sand, and gravel, contain **pore spaces.** The pores or voids within a rock can be the spaces between grains, vesicles, cracks, and solution cavities. Two important physical properties of a rock largely control the amount and movement of ground water. One is **porosity,** that is, the percentage of the total volume of the rock consisting of voids. Porosity determines how much water a rock body can hold. The second factor is **permeability,** the capacity of a rock to transmit fluids. Permeability depends upon such things as the size of the voids and the degree to which they are interconnected. Porosity and permeability are not synonymous; some rocks (such as shale) have a high porosity, but the pore spaces are so small that it is difficult for water to move through them, so the rock, even though it has high porosity, is impermeable. Contrary to popular belief, the only underground rivers occur in cavernous limestone and in some lava tunnels in volcanic terrain.

Discussion

Porosity. There are four main types of porosity: (1) spaces between mineral grains, (2) fractures, (3) solution cavities, and (4) vesicles. These are illustrated in *figure 13.1.* In sand and gravel deposits, pore space (*figure 13.1A*) can occupy 12% to 45% of the total volume. When several sizes of grain are abundant or there is a significant amount of cementing material, the pore space is reduced greatly. All rocks are cut by fractures that constitute a significant type of porosity, and, in some dense rocks (such as granite), fractures can constitute the only significant pore spaces. Solution activity, especially in limestone, commonly dissolves the most soluble material to form pits and holes so that limestones have a high porosity. As water moves along joints and bedding planes in limestone, it commonly enlarges the fracture by solution activity and develops passageways that may enlarge to become caves. One important type of porosity in basalts and other volcanic rocks is the vesicles formed by trapped gas bubbles (*figure 13.1D*). These commonly are concentrated near the top of a lava flow and form a zone of exceptionally high porosity, which may be interconnected by columnar joints and cinders and rubble at the top and base of the flow.

Permeability. Permeability is the ability to transmit fluids. It varies with the viscosity of the fluid, the **hydrostatic pressure,** the size of openings, and, particularly, the degree to which the openings are interconnected. A rock can have high porosity but low permeability.

Rocks commonly having high permeability are conglomerates, sandstones, basalt, and certain limestones. Permeability in sandstones and conglomerates is high because of the relatively large, interconnected pore spaces between the grains. Basalt is permeable, because it often is extensively fractured with columnar jointing and the tops of most flows are vesicular. Fractured limestones are permeable, as are limestones in which solution activity has created many small solution cavities. Rocks

(A) *Porosity resulting from space between grains.*

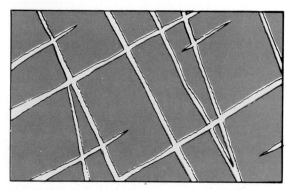

(B) *Porosity due to fractures.*

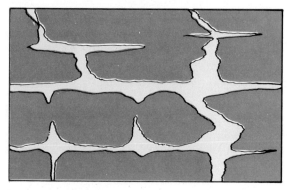

(C) *Porosity resulting from solution activity.*

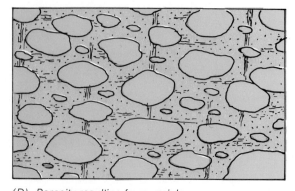

(D) *Porosity resulting from vesicles.*

Figure 13.1 *Types of porosity in rocks.*

that have low permeability are shale, unfractured granite, quartzite, and other dense, crystalline metamorphic rocks.

Water moves through the available pore spaces, twisting and turning through the tiny voids. Regardless of the degree of permeability, the rate at which ground water moves is extremely slow when compared to the turbulent flow of rivers. Whereas the velocity of water in rivers is measured in kilometers per hour, the velocity of ground water flow commonly ranges from 1 m per day to 1 m per year. The highest rate of percolating movement measured in the United States in exceptionally permeable material is only 250 m per day. Only in special cases, such as the flow of water in caves, does the movement of ground water approach the velocity of slow-moving surface streams.

The Water Table

Statement

As water seeps into the ground, it migrates downward by the pull of gravity through two zones in the soil and rock. In the upper zone, the pore spaces in the rocks are only partly saturated and the water occurs as a thin film clinging to grains by surface tension. This zone, in which pore space is filled partly with air and partly with water, is called the **zone of aeration.** Below a certain level, all the openings of the rock are completely filled with water (*figure 13.2*). This area is called the **zone of saturation.** The **water table** is the upper surface of the zone of saturation and is a very important element in the ground-water system. It can be only a meter or so deep in humid regions, but, in the desert, it can be hundreds of meters

below the surface. In swamps and lakes, the water table is, in essence, at the surface.

Discussion

The Water Table.   Although the water table cannot be observed directly, it has been studied and mapped from data collected about wells, springs, and surface drainage. In addition, the movement of ground water has been studied by means of dyes and other tracers, so our knowledge of the invisible body of ground water is quite extensive.

Several important generalizations can be made about the water table and its relation to the surface topography and surface drainage. These are shown diagrammatically in *figure 13.3*. In general, the water table roughly parallels the topographic surface, being highest beneath hills and lowest in the valleys. In flat country, the water table is flat; in areas of rolling hills, it rises and falls with the surface of the land. The reason for such configurations is that ground water moves very slowly, so a difference in the level of the water table is built up and maintained in areas of high elevation. During periods of heavy rain-

Figure 13.2   *The water table. (A) As water seeps into the ground through the pore spaces in the rock and soil, it passes first through the zone of aeration, in which the pore spaces are occupied by both air and water, and then into the zone of saturation, in which all the pore spaces are filled with water. The water table marks the top of the zone of saturation. The depth of the water table varies with the climate and amount of precipitation. Essentially, it lies at the surface in lakes and swamps, but it can be hundreds of meters deep in desert regions. (B) The zones of aeration and saturation and the water table in microscopic view.*

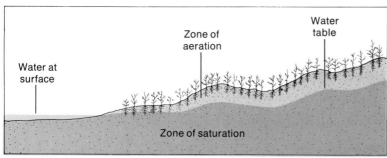

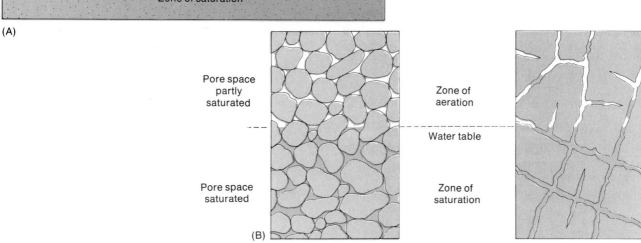

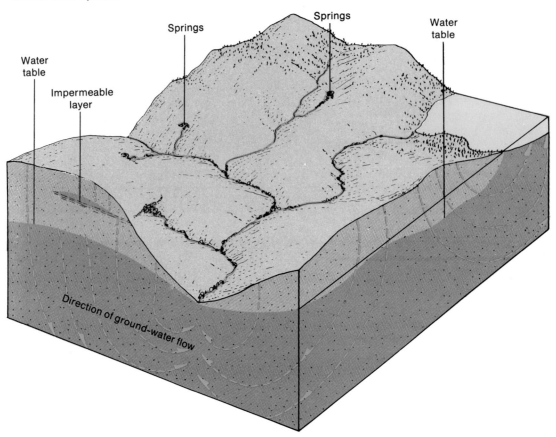

Water table

Springs

Springs

Water table

Water table

Impermeable layer

Direction of ground-water flow

Figure 13.3 *The movement of ground water. The configuration of the water table is a subdued replica of the topography. In cross section, the water table has hills and valleys much like those on the surface. Differences in the height of the water table cause differences in the pressure on the water in the saturation zone at any particular point. Thus, water moves down beneath the high areas of the water table because of the higher pressure and upward beneath the low areas where pressure is less. It commonly seeps into streams, lakes, and swamps where the water table is at the surface. A line of springs and seeps occurs where an impermeable layer of rock that has caused a perched water table to form is exposed at the surface.*

fall, the water table rises in the areas beneath the hills, and, during drought it tends to flatten out.

Where impermeable layers (such as shale) occur within the zone of aeration, the ground water is trapped and cannot migrate down to the main water table and a local **perched water table** is formed within the zone of aeration. When the perched water table extends to the side of a valley, springs and seeps occur.

The water table is at the surface in lakes, swamps, and most streams, and water moves from the subsurface toward these areas. Most streams in arid regions are above the water table and lose much of their surface water through seepage into the subsurface.

Movement of Ground Water. The difference in elevation of the water table is called the **hydraulic head** and causes the water to follow the paths illustrated in *figure 13.3*. If the path of a particle of water could be traced it would be found that gravity pulls it slowly downward

through the zone of aeration to the water table. When it passes through the zone of aeration and encounters the water table, it continues to move downward with the pull of gravity along curved paths from areas where the water table is high toward areas where it is low, that is, toward lakes, streams, and swamps. The path of ground-water movement is *not* down the slope of the water table as one first might suspect. The explanation for this seemingly indirect flow is that the water table is not a solid surface like the ground surface. It is a surface of a liquid that in some ways resembles the surface of a wave. Water on any given point below the water table is under greater pressure from the higher areas of the water table beneath a hill and, therefore, moves directly downward and toward points of lesser pressure.

Although the paths of ground-water movement may seem indirect, they conform to the laws of fluid physics and have been mapped in many areas by tracing the movement of dye injected into the system. The movement of the dye shows that there is a continual slow circulation of ground water from infiltration at the surface to seepage into streams, rivers, and lakes. Thus, the ground-water body is not stagnant and motionless; rather, it is an important part of the hydrologic system and is intimately related to surface drainage. At considerable depths, all pore spaces in the rocks are closed by high pressure and there is no free water. This, then, is the lower limit, or base, of the ground-water system. Like other parts of the hydrologic system (rivers and glaciers), the ground-water system is open, with an input, transport, and discharge of water.

Natural and Artificial Discharge

Statement

Natural discharge of ground-water reservoirs occurs wherever the water table intersects the surface of the ground. Generally, such places are in the channels of streams and on the floors and banks of marshes and lakes. Springs and seeps occur wherever the water table intersects the ground surface.

Natural discharge of the ground-water reservoir into the drainage system introduces a significant volume of water into the surface streams and is the major link between the ground-water reservoir and other parts of the hydrologic system. Indeed, were it not for ground-water discharge, many permanent streams would be dry during

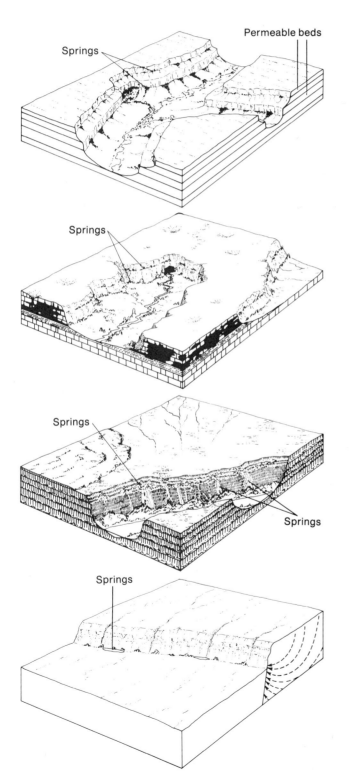

(A) Permeable beds separated by impermeable beds. A spring line develops where permeable beds are exposed in valley walls.

(B) Area of cavernous limestone. Springs form where cavernous limestone is exposed in valley walls.

(C) Porous basalt. Surface water readily seeps into vesicular and jointed basalt flows. It then migrates laterally and forms springs where basalt units are exposed in canyon walls.

(D) Fault lines. Many faults displace rocks so that impermeable beds are placed next to permeable beds. A spring line commonly results as ground water migrates along a fault line.

Figure 13.4 Geologic conditions conducive to the formation of springs.

parts of the year. Most natural discharge is near, or below, the surface of streams and lakes and, therefore, usually goes unnoticed. It is detected and measured by comparing the volume of precipitation to surface runoff. Water wells are a form of artificial discharge and are made simply by drilling holes down into the water table.

Discussion

Natural Discharge. The most obvious natural discharge consists of seeps and springs. Springs result from a variety of geologic structures and by variations in the rock sequence, some of which are shown in *figure 13.4*. In block *A*, permeable beds alternate with impermeable layers and the ground water is forced to move laterally to the outcrop of the bed. Conditions such as this usually are found in mesas and plateaus where interbedded sandstone and shale occur. The spring line commonly is marked by a line of vegetation. Block *B* shows a limestone terrain in which springs occur where the base of the cavernous limestone outcrops. The Mammoth Cave area of Kentucky is a good example. In block *C*, lava formations outcrop along the sides of a canyon and develop important springs, as ground water migrates readily through the vesicular and jointed basalt. An excellent example of this type of discharge is the Thousand Springs area of Idaho, where there are copious flows along the side of the Snake River valley. Block *D* shows springs developed along a fault that produce an avenue of greater permeability. Faults frequently displace strata so that impermeable beds block the flow from the permeable layers.

Wells. Ordinary wells are made simply by digging or drilling holes through the zone of aeration into the zone of saturation, as shown in *figure 13.5*. Water then flows out of the pores into the well and fills it to the level of the water table. When a well is pumped, the water table is drawn down around the well in the shape of a

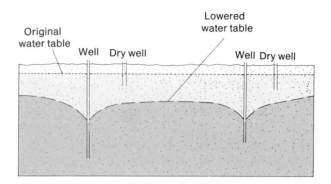

Figure 13.6 *When water is pumped from a deep well, the cone of depression can extend outward for hundreds of meters and effectively lower the water table over a large area. Shallow wells in the area then run dry because they are above the water table.*

cone, called the **cone of depression.** If water is withdrawn faster than it can be replenished, the cone of depression continues to grow, and, ultimately, the well can go dry. In cases of large wells such as those used by cities and industrial plants, the cone of depression can be many hundreds of meters in diameter. All wells within the cone of depression are affected (*figure 13.6*). This undesirable condition has been the cause of "water wars" fought physically as well as in the courts. Inasmuch as ground water moves and is not fixed in a stationary place as mineral deposits are, it is difficult to state who owns it. Many disputes now are being arbitrated by using computers that simulate actual subsurface conditions such as permeability, direction of flow, and level of the water table. Computers can predict what changes will occur in the ground-water system when given amounts of water are drawn out of a well over specified periods of time.

Extensive pumping lowers the general surface of the water table. This has been done with some serious consequences in some metropolitan areas of the southwestern United States, where the water table has been lowered hundreds of meters. There is a limited supply of ground water and, although the ground-water reservoir is continually being replenished by precipitation, migration of ground water is so slow that it may take hundreds of years to raise a water table to its former position of balance with the hydrologic system.

Artesian Water

Statement

Artesian water is ground water that is confined in some way so that it builds up an abnormally high hydrostatic pressure. In sedimentary rocks, an artesian system commonly results where permeable beds such as sandstone lie between impermeable beds such as shale. The water confined in the sandstone bed behaves much as water in a pipe (*figure 13.7*). Hydrostatic pressure builds up so that, when a well or fracture intersects the bed, water rises in the opening and can produce a flowing well or artesian spring.

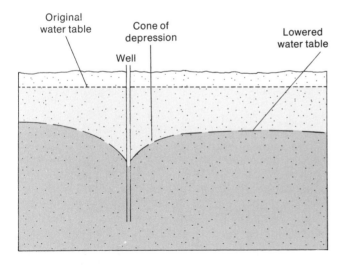

Figure 13.5 *Diagram showing the cone of depression formed around a pumping well.*

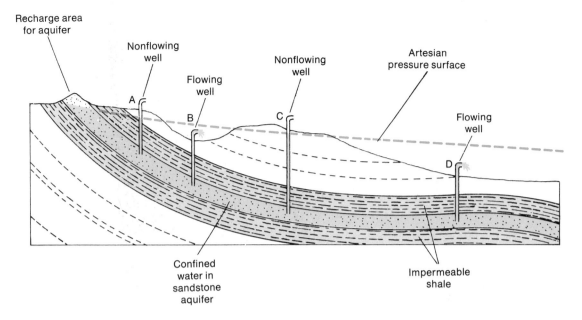

Recharge area
for aquifer

Nonflowing
well

Flowing
well

Nonflowing
well

Artesian
pressure surface

Flowing
well

Confined
water in
sandstone
aquifer

Impermeable
shale

Figure 13.7 *Geologic conditions necessary for an artesian system: (1) a permeable bed (aquifer) must be confined between impermeable layers, (2) rocks must be tilted so the aquifer can receive infiltration from surface waters, and (3) there must be adequate infiltration to fill the aquifer and create a hydrostatic pressure.*

Discussion

Artesian Systems. The geologic conditions necessary for artesian water are illustrated in *figure 13.7* and include the following.

1. The rock sequence must contain interbedded permeable and impermeable strata. The sequence occurs commonly in nature as interbedded sandstone and shale. Permeable beds usually are referred to as **aquifers**.
2. The rocks must be tilted and exposed in some elevated area where infiltration into the aquifer can occur.
3. Sufficient rainfall must exist in the outcrop area to furnish an adequate supply of water.

The height to which artesian water rises above the aquifer is indicated by the dashed colored line in *figure 13.7*. The surface indicated by this line is called the **artesian pressure surface**. It might be expected to be a horizontal surface at the elevation of the water surface in the aquifer in the source area, but it slopes away from the recharge area because of resistance to flow through the pores in the aquifer and loss of pressure from fractures which causes leaks in the underground plumbing system. If a well were drilled at location A or C, water would rise in the well but it would not flow to the surface, because the level of the artesian pressure surface is below the ground surface. At location B or D, however, a flowing well would result, because the level of the artesian pressure surface is above the ground surface. All of these wells that tap the aquifer are artesian wells, that is, the water is under artesian pressure and rises above the top of the aquifer.

Examples of Artesian Systems. In most areas underlain by sedimentary rocks, artesian water is a commonplace occurrence, because the geologic conditions necessary for an artesian system can be produced in sedimentary rocks in a variety of ways. Examples of well-known areas in the United States should help you appreciate how artesian systems result from different geologic settings.

One of the best-known artesian systems underlies the Great Plains states (*figure 13.8A*). The sequence of interbedded sandstones, shales, and limestones is nearly horizontal throughout most of Kansas, Nebraska, and the Dakotas but is warped up along the eastern front of the Rockies and the margins of the Black Hills. The sandstone formations constitute several important aquifers. Water confined in these formations is under hydrostatic pressure and gives rise to an extensive artesian system in the Great Plains states, with the recharge area along the foothills of the Rockies.

Figure 13.8B illustrates a regional artesian system in the inclined strata of the Atlantic and Gulf Coast plains. The rock sequence consists of permeable sandstone and limestone beds alternating with impermeable clay. Surface water flowing toward the coast seeps into the beds where they are exposed at the surface. It then slowly moves down the dip of the permeable strata.

A third example is from the western states, where the arid climate makes artesian water a very important resource (*figure 13.8C*). There the subsurface rocks in the intermontane basin consist of sand and gravel deposited in ancient alluvial fans, which grade basinward into playa clay and silt. The playa deposits act as confining layers between the permeable sand and gravel. Water seeping into the fan deposits becomes confined as it moves away from the mountain front.

Artesian systems also underlie some of the great desert regions of the world, and natural discharge from them

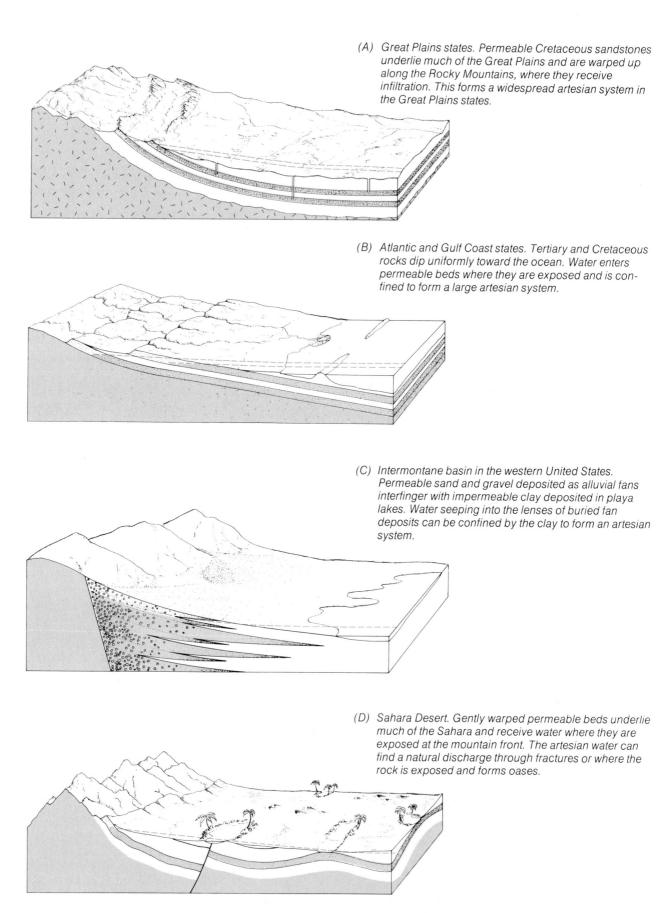

(A) Great Plains states. Permeable Cretaceous sandstones underlie much of the Great Plains and are warped up along the Rocky Mountains, where they receive infiltration. This forms a widespread artesian system in the Great Plains states.

(B) Atlantic and Gulf Coast states. Tertiary and Cretaceous rocks dip uniformly toward the ocean. Water enters permeable beds where they are exposed and is confined to form a large artesian system.

(C) Intermontane basin in the western United States. Permeable sand and gravel deposited as alluvial fans interfinger with impermeable clay deposited in playa lakes. Water seeping into the lenses of buried fan deposits can be confined by the clay to form an artesian system.

(D) Sahara Desert. Gently warped permeable beds underlie much of the Sahara and receive water where they are exposed at the mountain front. The artesian water can find a natural discharge through fractures or where the rock is exposed and forms oases.

Figure 13.8   Examples of artesian systems.

to the surface is largely responsible for the oases. An example of the system in the Sahara is shown in *figure 13.8D*. Oases occur where artesian water is brought to the surface by fractures or folds or where the desert floor is eroded down to the top of the aquifer.

Thermal Springs and Geysers

Statement

In areas of recent igneous activity, rocks in buried lava flows and old magma chambers can remain hot for hundreds of thousands of years. Ground water migrating through these geothermal areas becomes heated and, when discharged to the surface, produces thermal springs and geysers.

Discussion

The three most famous regions of hot springs and geysers are Yellowstone National Park, Iceland, and New Zealand. These areas are all regions of recent volcanic activity, so the temperature of the rocks just below the surface is quite high. Although no two geysers are alike, several conditions must occur in order for them to develop: (1) there must be a body of hot rocks relatively close to the surface, (2) a system of irregular fractures must extend downward from the surface, and (3) there must be a relatively large and constant supply of ground water. The eruption of geysers is caused by pressure building up to a critical point in ground water contained in fractures, caverns, or layers of porous rock (*figure 13.9*). Since the water at the base of the fracture is under greater pressure than that above, it must be heated to a higher temperature than the water above before it will boil. Eventually, a slight increase in temperature or a decrease in pressure resulting from liberation of dissolved gases causes the deeper water to boil. As steam is produced, it expands, throwing the water from the underground chambers and fractures high into the air. After the pressure is released, the fractures and caverns are refilled with water and the process is repeated. This process accounts for the periodic eruption of many geysers, the interval between eruptions being that amount of time required for water to percolate into the fracture and be heated to a critical temperature. Geysers like Old Faithful in Yellowstone National Park erupt at definite intervals, because the rocks are permeable and the "plumbing system" can be refilled rapidly. Other geysers require more time for water to percolate into the chambers and, as a result, erupt at more irregular intervals.

Geothermal Energy. The thermal energy involved in ground water offers an attractive source of usable energy for man. Presently, it is utilized in a variety of ways in local areas of Italy, Japan, and Iceland. Present estimates show that 1% to 2% of man's current needs for energy could come from geothermal sources.

In Iceland, geothermal energy has been used successfully since 1928. The plan is simple. Wells are drilled into geothermal areas, and the steam and hot water is piped to storage tanks and then pumped to homes and municipal buildings where it is used for heating and hot-water supply. The cost of this direct heating is only

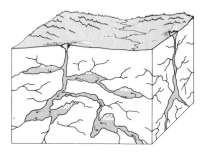

(A) Ground water circulating through the hot rocks in an area of recent volcanic activity collects in caverns and fractures. As steam bubbles rise, they grow in size and number and tend to clog in restricted parts of the geyser tube.

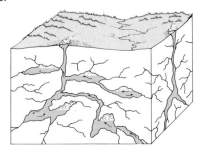

(B) When this happens, the expanding steam forces water upward and it is discharged at the surface vent. The deeper part of the geyser system then becomes ready for the major eruption.

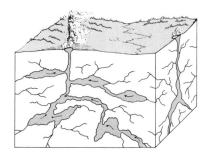

(C) The preliminary discharge of water reduces the pressure on the water at depth. Water from the side chambers and pore spaces begins to flash into steam, forcing the water in the geyser system to erupt.

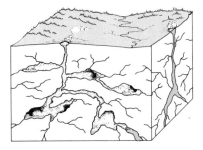

(D) When the pressure from the steam is spent and the geyser tubes are empty, eruption ceases. The system then begins to fill with water again, and the eruption cycle starts anew.

Figure 13.9   *The origin of geysers.*

about 60% of that of fuel-oil heating and about 75% of the cheapest method of electrical heating. Steam from geothermal energy is used also to run electric generators, producing a form of energy somewhat easier to transport. Corrosion is a problem, however, because most thermal waters are acid and contain undesirable dissolved salts.

Erosion by Ground Water

Statement

Slow-moving ground water cannot erode by abrasion in the same manner as a surface stream, but ground water is capable of dissolving great quantities of soluble rock and moving it in solution. In some areas, it is the dominant agent of erosion and is responsible for the development of the major features of the landscape. The process of ground-water erosion starts as ground water, percolating through joints, faults, and bedding planes, dissolves the soluble rock. In time, the fracture enlarges to form a cavern system, some of which extends as a subterranean network for many kilometers. The caves grow larger and ultimately the roof collapses to produce a craterlike depression called a **sinkhole.** Solution activity then enlarges the sinkhole to form a solution valley, which continues to grow until its soluble rock is removed completely.

Discussion

Rock salt and gypsum are the most soluble rocks and are eroded rapidly by solution activity, but they are relatively rare and are not widely distributed throughout any of the continents. Limestone is also fairly soluble and, inasmuch as it is a common rock type, solution

activity plays an important role in eroding limestone terrains in humid regions.

Perhaps the best way to appreciate the significance of solution activity is to consider the nature of a **cave** system and the amount of rock removed by solution. A map of a cave system may show long, winding corridors, with branched openings that enlarge into chambers, or a maze

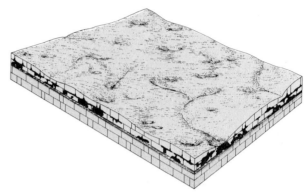

(A) Initial stage. Scattered sinkholes dot the landscape and grow in size and number as caverns enlarge and their roofs collapse.

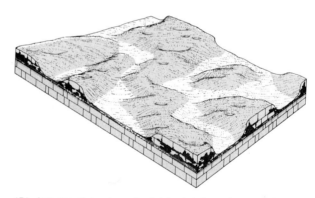

(B) Intermediate stage. Individual sinks enlarge and merge with those in adjacent areas to form solution valleys. Much of the original surface is destroyed. Disappearing streams and springs are common.

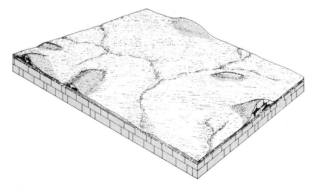

(C) Late stage. Solution activity has removed most of the limestone formation. Only isolated knolls remain as remnants of the former surface.

Figure 13.11 Evolution of karst topography (see plate 19).

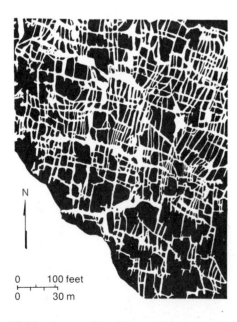

0 100 feet
0 30 m

Figure 13.10 A map of Anvil Cave, Alabama, shows the extent to which caverns are controlled by a fracture system. The joints occur in two intersecting sets, one trending nearly north-south and the other east-west. Solution activity along the joints has produced the network of caverns.

Figure 13.12 *Karst topography in the limestone regions of Kentucky.*

of interlacing passageways and channels, controlled by intersecting joint systems (*figure 13.10*). Where a vertical sequence of limestone formations occurs, several levels of cave networks may exist. Mammoth Cave, Kentucky, for example, has over 50 km of continuous subterranean passages. In humid areas where limestone is exposed, collapse of caves develops numerous sinkholes and the sinks grow and increase in number to produce a unique landscape known as **karst topography.** Instead of a well-integrated surface drainage system with principal valleys and tributaries, karst topography is characterized by a surface pitted with sinks, large closed depressions known as **solution valleys,** and **disappearing streams.**

A karst topography evolves through a series of stages, which are shown in the block diagrams in *figure 13.11*. In the initial stages of development, water follows surface drainage until a large river cuts a deep valley below the limestone layers. Ground water then moves through the joints and bedding surfaces in the limestone and emerges at the river. As time goes on, the passageways become larger and caverns develop. Surface waters disappear into solution depressions. The roofs of caves collapse, producing numerous sinkholes (block *A*). Springs commonly occur along the margins of valleys of major streams. Sinkholes increase in number and continue to grow in size as solution activity dissolves the limestone terrain. The cavernous terrain of central Kentucky, for example, has over 60,000 holes. Sinks ultimately coalesce to form solution valleys (block *B*). Most of the original surface finally is dissolved (block *C*), leaving only scattered mesas and small buttes. When the soluble bedrock has been removed by ground-water solution, surface drainage then reappears.

An example of a karst topography in an intermediate stage of development is shown in *figure 13.12*. Sinkholes in this area have been enlarged to such an extent that the dominant features are rounded mounds.

Deposition by Ground Water

Statement

The mineral matter dissolved by ground water can be deposited in a variety of ways. The most spectacular deposits are formed commonly in caves as **dripstone.** Less obvious are the deposits in permeable rocks such as sandstone and conglomerates, where ground water commonly deposits mineral matter to form a cement between grains. The precipitation of minerals by ground water also is responsible for the formation of certain mineral deposits (such as the uranium found in the Colorado Plateau).

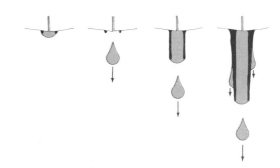

Figure 13.13 *Formation of dripstone. Dripstone originates on the ceiling of a cave when a drop of water seeps through a crack and deposits a small ring of calcite as it evaporates. The ring grows into a tube, which commonly acquires a tapering shape, as water seeps from adjacent areas and flows down its outer surface.*

Discussion

Cave deposits are familiar to almost everyone, and a great variety of forms have been named. However, most originate in a similar way and are referred to collectively as dripstone. The process of dripstone formation is shown in *figure 13.13*. As the water enters the cave (usually from a fracture in the ceiling), part of it evaporates so that a

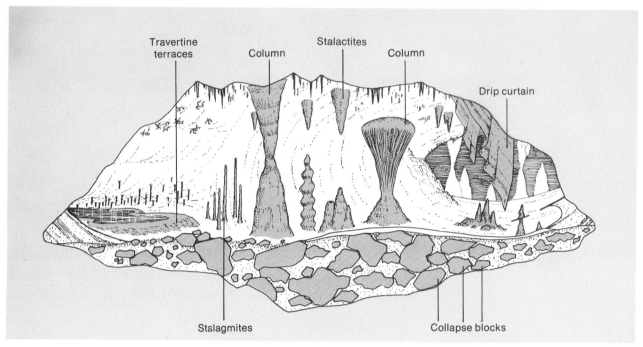

Figure 13.14 *Varieties of cave deposits. The many interesting forms of cave deposits all are composed of calcite and originate when water seeping into the cave evaporates and deposits the dissolved mineral.*

Figure 13.15 *Onadaga Cave, Crawford County, Missouri, shows many of the forms of dripstone illustrated in figure 13.14.*

small amount of calcium carbonate is left behind. The next drop adds more calcium carbonate and, eventually, a cylindrical or cone-shaped projection is built downward from the ceiling. Many beautiful and strange forms result, some of which are shown in *figures 13.14* and *13.15.* Iciclelike forms growing down from the ceiling are called **stalactites.** These commonly are matched by columns growing up from the floor (**stalagmites**), as the water dripping from the stalactite precipitates additional calcium carbonate on the floor directly below. Many stalactites and stalagmites unite to form columns. Water percolating from a fracture in the roof may form **drip curtains,** and pools of water on the cave floor flow from one place to another, making **travertine terraces.**

Alteration of a Ground-Water System

Statement

In previous sections of this chapter, it has been shown that ground water is part of the hydrologic system and is intimately related to precipitation, surface drainage, and discharge. As in many other natural physical systems, there is a tendency for a balance or a condition of equilibrium to be established. For example, the level of the water table represents an equilibrium between the amount of precipitation and infiltration, surface drainage, porosity, and permeability of the rock and the volume of ground water discharged back to the surface. When one of these factors is modified, the others respond to reestablish equilibrium.

A variety of ground-water problems result from man's activities, because ground water constitutes a valuable resource and is being exploited at an ever-increasing rate. Some of the more important problems resulting from altering the ground-water system are:
1. Changes in chemical composition of ground water.
2. Saltwater encroachment.
3. Changes in position of the water table.
4. Subsidence.

Discussion

Changes in Composition. The composition of ground water can be changed by increasing the concentration of dissolved solids in surface water. When the soil mantle of the earth is regarded as an infiltration system through which ground water moves, it is evident that any concentration of chemicals or waste creates local pockets that potentially could contaminate the ground-water reservoir. Material **leached** (dissolved by percolating ground water) from waste-disposal sites includes both chemical and biologic contaminants. Upon entering the ground-water flow system, the contaminants move according to the hydraulics of that system. The character and strength of the **leachates** depends partly upon the length of time the infiltrated water is in contact with the waste deposit and partly upon the volume of infiltrated water. In humid areas where the water table is shallow and in constant contact with refuse, leaching constantly produces maximum potential for pollution.

Figure 13.16 illustrates four geologic environments in which waste disposal affects the ground-water system. In

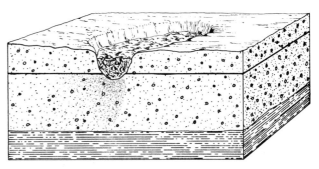

(A) A permeable layer of sand and gravel overlying an impermeable shale creates a potential pollution problem, because the contaminants are free to move with ground water.

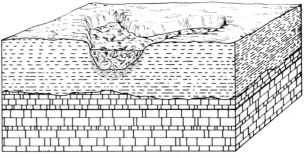

(B) An impermeable shale (or a clay) confines the pollutants and prevents significant infiltration into the ground-water system in the limestone below.

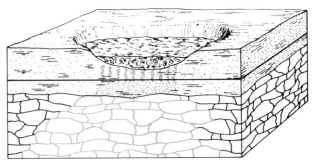

(C) A fractured rock body provides a zone where pollutants can move readily in the general direction of ground-water flow.

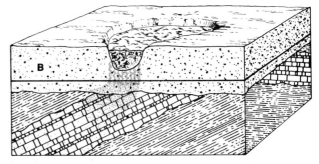

(D) An inclined, permeable aquifer below a disposal site permits pollutants to enter an artesian system and move down the dip of the beds, contaminating the artesian system.

Figure 13.16 *The effects of solid-waste-disposal sites on a ground-water system.*

block *A,* the geological setting is one in which the near-surface material is permeable and essentially homogeneous. Leachate percolates downward through the zone of aeration and, upon reaching the water table, becomes part of the ground-water flow system. As part of the flow system, the leachate moves in the direction of the slope of the water table and ultimately becomes part of the surface drainage system. In block *B,* an impermeable layer of shale confines pollutants and prevents their free movement in the ground-water system. Block *C* illustrates a disposal site above a fractured rock body. Upon reaching the fractured rock, the contaminants can move more readily in the general direction of ground-water flow. Also, dispersion of the contaminants is limited because of the restriction of flow to the fractures. Block *D* illustrates a critical condition in which a waste-disposal site is located in permeable sand and gravel above an inclined aquifer. Here, leachate moves down past the water table and enters the aquifer as recharge. If the waste-disposal site were located directly above the aquifer, as shown in the diagram, most of the leachate would enter the aquifer but some would move down the gradient of the impermeable shale. If the site were located at position B, very little leachate would penetrate the aquifer but it would move downgradient over the impermeable shale.

Saltwater Encroachment. Where permeable rocks are in contact with the ocean, as on an island or a peninsula, a lens-shaped body of fresh ground water is buoyed up by the denser salt water below, in the manner illustrated in *figure 13.17A.* In a very real sense, the fresh water floats on the salt water and is in a state of balance with it. If pumping is excessive and develops a large cone of depression in the water table, the pressure of the fresh water on the salt water below the well is decreased and a large cone of saltwater encroachment develops below the well, as shown in *figure 13.17B.* With excessive pumping, the cone of salt water would extend up the well and contaminate the fresh water. It then would be necessary to stop pumping for a long period of time to allow the water table to rise to its former position and depress the cone of salt water. Restoration of the balance between the freshwater lens and the underlying salt water could be hastened if fresh water were pumped down into an adjacent well (*figure 13.17C*).

Changes in the Position of the Water Table. The water table is intimately related to the surface runoff, the configuration of the landscape, and the ecological conditions at the surface. The balance between the water table and surface conditions, established over thousands or millions of years, can be upset completely by causing the water table to change its position. Two examples will serve to illustrate some of the many potential ecological problems.

In southern Florida, water from Lake Okeechobee flowed for the past 5000 years as an almost imperceptible river, only a few centimeters deep and 64 km wide, to create the Everglades. The movement of the water was not confined to channels. It flowed as a sheet in a great curving swath for more than 160 km (*figure 13.18*). The surface slope of the Everglades is only 2 cm per km to the south, but it was enough to make the water move slowly to the coast, preventing salt water from invading the

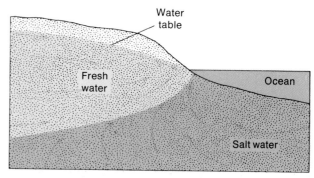

(A) *A lens of fresh ground water beneath the land is buoyed up by denser salt water.*

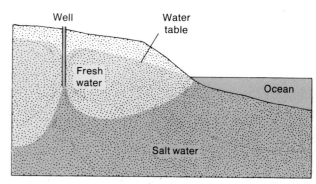

(B) *Excessive pumping causes a cone of depression in the water table and a cone of saltwater encroachment at the base of the freshwater lens.*

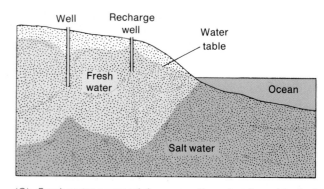

(C) *Fresh water pumped down an adjacent well would raise the water table around the well and lower the interface between the fresh and salt water.*

Figure 13.17 *The relationship between fresh and salt water on an island or a peninsula.*

Everglades and subsurface aquifers along the coast. In effect, the water table in the swamp was at the surface, and the ecology of the Everglades was in balance with the water table.

Today, many canals have been constructed to drain swamp areas for farmland, to help control flooding, and to supply fresh water to the coastal megalopolis (*figure 13.19*). The canals have diverted the natural flow of water across the swamps and, in effect, have lowered the water table, in some places as much as half a meter below sea

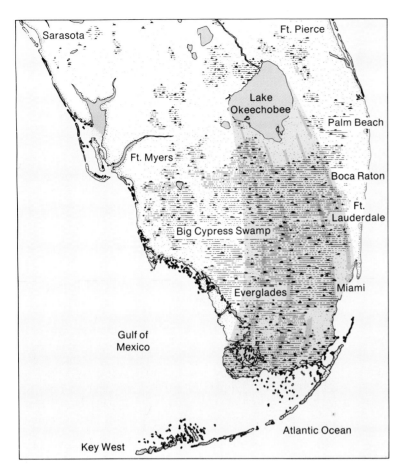

Figure 13.18 Natural drainage of southern Florida in 1871. The surface water spread southward from Lake Okeechobee in a broad sheet only a few centimeters deep. This maintained a swamp condition in the Everglades and established a water table very close to the surface.

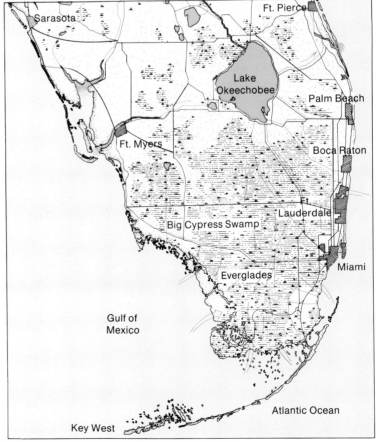

Figure 13.19 Modification of the natural drainage in southern Florida. The system of canals diverts the natural flow of surface water across swamps, the water table is lowered, swamp conditions are destroyed, and salt water encroaches in the wells along the coast.

level. This has produced many unforeseen, and often unfortunate, results. As the water table lowered, saltwater encroachment occurred all along the coast, intruding upon public and private wells. This has forced some cities to move their water wells far inland in order to obtain fresh water.

The most visible effects, however, are in the ecology of the swamp. In the past during periods of natural drought, the high water table was sufficient to maintain a marsh condition. Now during droughts, the surface is dry. Forest fires ignite the dry organic muck, which burns like peat and smolders for long periods after the surface fires are out. This effectively destroys the ecology of the swamp. Lowering the water table also causes the muck to compact and subside, in places as much as 2 m. In addition, when the muck is exposed to the air, it oxidizes and disappears at a rate of about 2.5 cm per year. Once the muck is gone from the swamp, it can be replaced only by nature.

In contrast to problems produced by lowering the water table, raising the water table can modify many surface processes. An example is found in the environmental changes caused by irrigation in the Pasco basin of Washington. This area lies in the rain shadow of the Cascade Mountains and receives only 15 to 25 cm of precipitation a year. The surface conditions of the basin have developed in response to an overall increase in aridity over the last several million years, and the surface material, slope, and vegetation are in balance with an arid climate and a low water table.

In recent years, extensive irrigation has caused the water table to rise, introducing many changes in surface conditions. Today, 100 to 150 cm of water is applied to the ground by irrigation, which produces the effect of a simulated climatic change of considerable magnitude. The higher water table has rapidly developed large springs along the sides of river valleys. The springs now are permanent, reflecting saturation of much of the ground. Erosion is accelerated, and many farms and roads have been damaged severely. Landslides present the most serious problems, as the slopes that were stable under arid conditions are now unstable because they are partly saturated from the greatly raised water table and from the formation of perched water bodies.

In many areas, it is imperative that man modify his environment by reclaiming land or by irrigation. But unless people are careful, the detrimental effects may outweigh the advantages. Before an environment is modified seriously, people must attempt to understand as many consequences of altering the natural systems as possible.

Subsidence. **Subsidence** of the surface related to ground water can be the result of natural earth processes, such as the development of sinks in a karst area, or it can be artificial as a result of fluid withdrawal.

Collapse into sinkholes is an ever-present hazard in limestone terrains. Numerous examples exist where buildings and roads have been damaged by sudden collapse into previously undiscovered caverns below. In the United States, important karst regions occur in central Tennessee, Kentucky, southern Indiana, Alabama, Florida, and Texas (*figure 13.20*). The problem of collapse is difficult to solve. Important construction in karst regions should be preceded by test borings to determine the possible presence of subterranean cavernous zones. Concrete slurries can be pumped into solution cavities, but such remedies can be very expensive.

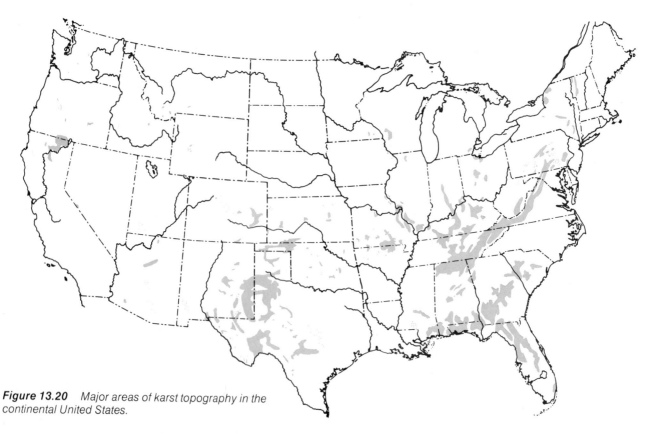

Figure 13.20 *Major areas of karst topography in the continental United States.*

Compaction and subsidence present serious problems in many areas of newly deposited sediments. In New Orleans, for example, large areas of the city are now 4 m below sea level, due largely to pumping ground water. As a result, the Mississippi River flows some 5 m above the city, and rainwater must be pumped out of the city at considerable cost. In addition, as the earth subsides, water lines and sewers are damaged.

When ground water, oil, or gas is withdrawn from the subsurface, significant subsidence also can occur, causing damage to construction, water supply lines, sewers, and roads. Long Beach, California, has subsided 9 m as the result of 34 years of oil production from the Wilmington Oil Field. This has resulted in almost $100 million damage to wells, pipelines, transportation facilities, and harbor installations. Houston, Texas, has subsided as much as 1.5 m from withdrawal of ground water.

Figure 13.21 *Subsidence of buildings in Mexico City resulted from compaction after ground water was pumped excessively from unconsolidated beds beneath the city. Subsidence has tilted this building more than 2 m so that the windows in the background of the photograph are now at the level of the sidewalk.*

Probably the most spectacular example is Mexico City, which is built on a former lakebed. The subsurface formations consist of water-saturated clay, sand, and volcanic ash. As ground water is pumped for domestic and industrial use, the sediment compacts and slow subsidence is widespread. Large structures such as the Opera House (weighing 54,000 metric tons) have settled more than 3 m, and half of the first floor is now below ground level. Other buildings are noticeably tilted, as illustrated in *figure 13.21.*

Summary

Ground water is an integral part of the hydrologic system and is intimately related to surface water. The movement of ground water is very slow and controlled largely by the porosity and permeability of the rock. At some depth below the surface, all pore spaces are filled with water. The upper surface of this saturated zone is called the water table. Movement of ground water below the water table is closely associated with surface runoff, and the ground-water reservoir discharges into streams, lakes, and swamps.

Artesian water is water that is confined between impermeable beds and is under hydrostatic pressure.

Ground water erodes by solution activity in soluble rocks, such as limestone, salt, and gypsum, and produces karst topography characterized by sinkholes, solution valleys, and disappearing streams.

Under normal conditions, ground water is in some degree of balance with surface runoff, surface topography, and salt water in the ocean. Alteration of the ground-water system may produce many unforeseen problems, as does alteration of any aspect of the hydrologic system.

In the next chapter, a natural alteration, in fact a complete disruption, of the hydrologic system will be considered. When the earth's climate changes sufficiently to produce widespread and prolonged cold, the water in the hydrologic system is frozen and glaciers form. While an ice age is rare in the earth's history, this phenomenon significantly modifies and interrupts the geologic processes on the earth's surface.

Additional Readings

Davis, S. N., and R. J. M. deWiest. 1966. Hydrogeology. New York: John Wiley and Sons.

Sayre, A. N. 1950. "Ground Water." Sci. Amer. 183(5): 14-19. (Reprint no. 818.) San Francisco: W. H. Freeman and Company.

Glacial Systems

The earth is just emerging from an ice age—a rare event in its history, for, throughout most of geologic time, its climate has been relatively mild. When glaciation does occur, it completely disrupts the hydrologic system, and many geologic processes are interrupted or modified significantly. During an ice age, much precipitation becomes trapped in the glaciers and does not flow immediately back to the ocean. Consequently, sea level drops. As the great ice sheets advance over the continents, they disrupt and obliterate the preexisting drainage networks, and the hydrology of streams is altered greatly. The moving ice scours and erodes the landscape and deposits the debris over the preexisting topography. The crust of the earth is pushed down by the weight of the ice, and meltwater commonly collects along the ice margins and establishes lakes. New drainage systems carrying large volumes of water and lakes are established along the ice front. Far beyond the margins of the glaciers, many stream systems are modified by the heavy runoff produced by the melting ice. Even in arid regions, changes in climatic conditions associated with glaciation are reflected in the development of large lakes that form in closed basins.

The area beneath the ice is completely depopulated as both plants and animals are forced to migrate in front of the advancing ice. This causes great stress on animal populations, and, because many species fail to adapt, they become extinct.

The last ice age also had a profound effect upon man. The history of his early migrations is largely an account of his response to the changing environment of the ice age, and the rigorous glacial climate probably stimulated many amazing adaptations that otherwise might not have occurred.

Even today in mountainous areas, tongues of ice hundreds of meters thick move down high stream valleys. As they move, they sculpture broad, U-shaped valleys, form sharp glacial peaks, and deposit eroded debris as moraines at the front of the ice. In Antarctica and Greenland, continental glaciers approximately 4 km thick cover most of the landmass so that these areas are, in effect, still in the ice age.

The causes of an ice age remain a tantalizing, unanswered question. Many hypotheses explaining worldwide temperature changes have been proposed, but recently attention has been focused on continental drift and the shifting ocean currents as major controlling factors.

Major Concepts

1. Glacial ice forms where more snow accumulates each year than melts. When ice is of sufficient thickness, it flows under its own weight and forms a glacier.
2. As ice flows, it erodes the surface by abrasion and plucking. Sediment is transported by the glacier and deposited where the ice melts.
3. The two major types of glaciers—continental and valley—produce distinctive erosional and depositional landforms.
4. The major effects of the ice age include the rise and fall of sea level, isostatic adjustments of land, modification of drainage systems, creation of numerous lakes, and migration and selected extinction of plants and animals.
5. The cause of glaciation is not clearly understood, but it may be related to oceanic currents and continental drift.

213

Glacial Systems

Statement

A **glacier** is a system of flowing ice that originates on land through the process of accumulation and recrystallization of snow. The conditions necessary for the development of a glacier are simple: more snow must accumulate each year than is lost by melting and evaporation. When this happens, a layer of new snow is added annually to that which has already accumulated and, over a period of many years, the mass of ice becomes thick enough to flow under its own weight. Perennial snowfields that do not move are not considered to be glaciers nor is pack ice formed from seawater in polar latitudes.

Glaciers are open systems and have much in common with other gravity-flow systems such as rivers and ground water. Water enters the system primarily in the upper parts of the glacier, where snow accumulates and is transformed into ice. The ice then flows out of the zone of accumulation at rates generally in the range of a few centimeters per day. At the lower end (or terminus) of the glacier, the ice leaves the system by a combination of melting and evaporating.

There are two principal types of glaciers.

1. **Valley (or alpine) glaciers,** which are streams of ice that originate in the snowfields of high mountain ranges.

2. **Continental glaciers,** which are enormous sheets of ice thousands of meters thick that spread out and cover large parts of continents.

Discussion

Glacial Processes.   *Figure 14.1* is a highly idealized diagram showing the major parts of a glacial system and comparing the size and form of a continental glacier with that of a valley glacier. The continental glacier is roughly circular or elliptical and spreads out radially from the zone of accumulation. The ice is several thousand meters thick and covers thousands of square kilometers. The valley glacier is much smaller and is confined largely to the valleys of a stream system. It has branching tributaries, and flow direction is controlled by the stream

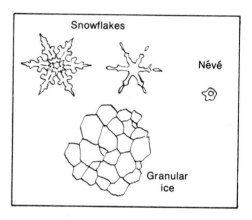

Figure 14.2   *Transformation of a snowflake into granular ice. The process involves melting the arms of the snow crystal and refreezing the water near the center to produce a more compact ice grain.*

Figure 14.1   *Glacial systems.*

(A) *Continental glacier. A continental glacier constitutes an open system of ice flowing under the pull of gravity. Snow enters the system by precipitation and is transformed into ice. The ice flows outward from the zone of accumulation under the pressure of its own weight. The ice leaves the system by evaporating and melting in the zone of ablation. The balance between the rate of accumulation and the rate of melting determines the size of the glacial systems.*

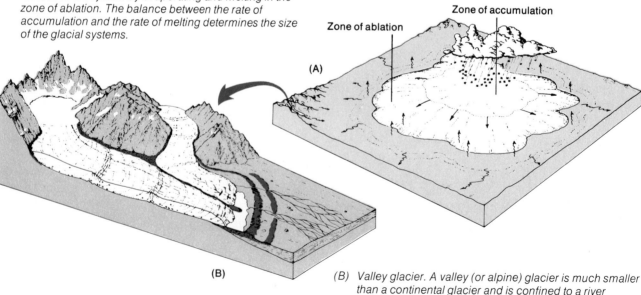

(B) *Valley glacier. A valley (or alpine) glacier is much smaller than a continental glacier and is confined to a river valley with a sloping floor. The movement of ice, therefore, is controlled by the topography. Snow enters the system in the headwaters, is transformed into ice, and flows down the valley to the end of the system, where it melts or evaporates.*

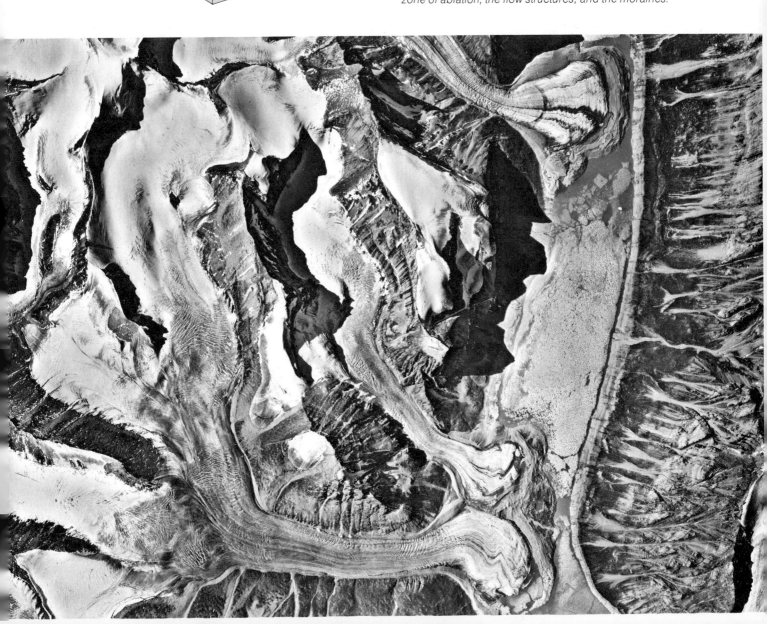

Figure 14.3 Structure and flowage of valley glaciers (see also plate 17).

(A) Sketch showing the major parts of a simple valley glacier and some of the important landforms resulting from valley glaciation.

Zone of accumulation

Snow line

Crevasses

Zones of melting and evaporation

Moraines

Outwash plain

(B) Valley glaciers on Baffin Island, Canada, showing many of the features illustrated in diagram A. Note the snow-covered zone of accumulation, the rough ice in the zone of ablation, the flow structures, and the moraines.

valley that the glacier occupies. The essential parts of both glacial systems are (1) the zone of accumulation, where there is a net gain of ice, and (2) the zone of **ablation,** where ice leaves the system by melting and evaporating.

In the zone of accumulation, snow is transformed into glacial ice through a process identical to rock metamorphism. Freshly fallen snow consists of delicate, hexagonal ice crystals, or needles, with as much as 90% void space (*figure 14.2*). As snow accumulates, the ice at the points of the snowflake melts and migrates toward the center, eventually forming an elliptical granule of recrystallized ice approximately 1 mm thick. The accumulation of these particles packed together is called **firn** or **névé.** With repeated annual deposits, the loosely packed névé granules are compressed by the weight of the overlying snow. Meltwater, which results from daily temperature fluctuations and the pressure of the overlying snow, seeps through the pore spaces between the grains and freezes, aiding in the process of recrystallization. Air in the pore spaces is driven out and, when the ice reaches approximately 30 to 40 m in thickness, it can no longer support its own weight and yields to plastic flow. The upper part of a glacier is rigid and fractures as the ice beneath moves by plastic flow, so the surface of a glacier is characterized by numerous crevasses.

At the lower end of the glacier, loss of ice by combined melting and evaporating exceeds the rate of accumulation. This area of net loss is known as the zone of ablation and is the exit boundary of the system.

It is important to understand that the margins of a glacier constitute the boundaries of a system of flowing ice in much the same way as the banks and mouth of a river constitute the boundaries of a river system. When more snow is added in the zone of accumulation than is lost by melting or evaporating at the end of the glacier, the mass of ice increases and the form of the glacier expands. When the accumulation of ice is less than ablation, there is a net loss of mass and the size of the glacial system is reduced; that is, the form of the glacier recedes.

When there is a balance between accumulation and ablation, the mass of ice remains constant, the size of the system remains constant, and the front of the ice remains stationary. However, *ice within the glacier flows continually toward the terminal margins* regardless of whether the end of the glacier is advancing, retreating, or stationary. The behavior of a glacial system is determined by the balance of the rate of input of ice versus the rate of output. The two major variables in this balance are temperature and precipitation. A glacier can grow or shrink at a given temperature simply by varying the amount of precipitation (rate of input). Also, it can grow or shrink with a given rate of precipitation through temperature variations, which would increase or decrease the rate of melting (rate of output). The length of the glacier in no way represents the amount of ice that has moved through the system, just as the length of a river does not represent the volume of water that has flowed through it. It simply shows the amount of ice that is currently in the system.

Two examples will illustrate this point. A glacial valley 20 km long can be eroded 600 m deeper than the original stream valley. This large amount of erosion was not accomplished by movement of 20 km of ice down the valley. It was the result of many thousands of kilometers of ice flowing through the valley. If the ice occupied the valley during each glacial epoch and moved 0.3 m per day, a total of approximately 72,000 km of ice would move down the valley in the glacier. Abrasion

Figure 14.4   *Schematic diagram of a continental glacier and some of the major features it produces. The weight of the ice depresses the continent, so the land slopes toward the glacier. This produces glacial lakes along the ice margin or permits an arm of the ocean to invade the depression. The original drainage is greatly modified as some streams flow toward the ice margins and form lakes. The glacier advances more rapidly into lowlands, so the margins are typically lobate. As the system expands and contracts, lobate moraines are deposited along the margins and a variety of erosional and depositional landforms are formed beneath the ice.*

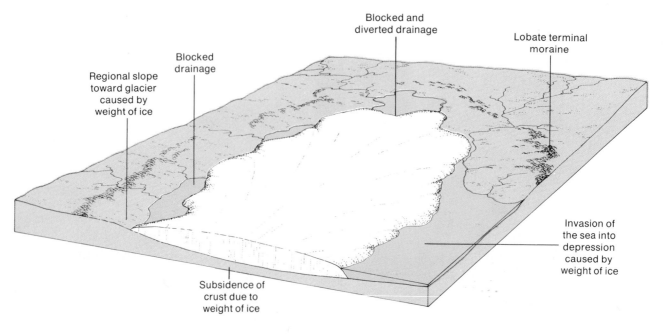

Regional slope toward glacier caused by weight of ice

Blocked drainage

Blocked and diverted drainage

Lobate terminal moraine

Subsidence of crust due to weight of ice

Invasion of the sea into depression caused by weight of ice

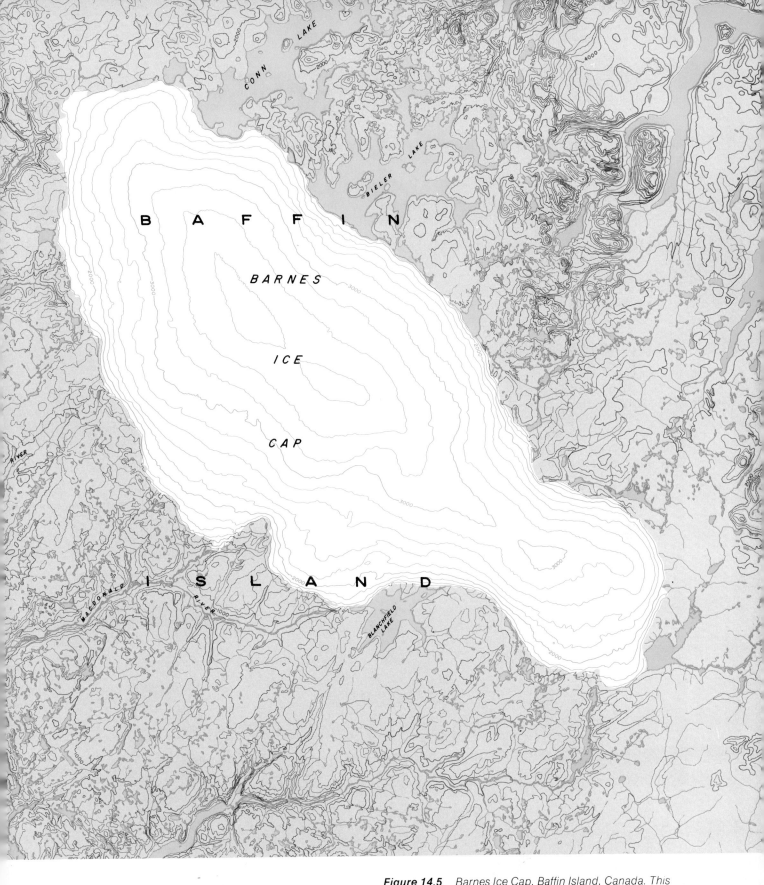

Figure 14.5 *Barnes Ice Cap, Baffin Island, Canada. This small ice sheet shows many features produced by continental glaciation. The drainage coming from the north is blocked by the ice and forms lakes near the ice margin. Erosional debris is carried to the margins of the glacier and is deposited as moraine. Irregularities in the surface over which the ice flows cause the ice margins to be irregular or lobate.*

caused by such a long stream of ice would be enormous and adequate to wear down the valley 600 m deep. Similarly, the continental glacier moving southward from the Hudson Bay area did not simply move a distance of 600 to 800 km. The system expanded out that distance from the center of accumulation and, over the total period of 800,000 years, roughly 72,000 km of ice moved through the system.

Valley Glaciers.   The nature of valley glaciers is well known, because they are relatively accessible and occur in some of the most scenic mountain ranges of the world. The diagram and photograph in *figure 14.3* show some of the features typical of most valley glaciers (see also *plate 17*). As was mentioned before, valley glaciers are long, narrow rivers of ice that originate in the snow-fields of the high mountain ranges and flow down pre-existing stream valleys. They range from a few hundred meters to more than a hundred kilometers long. In many ways, they resemble a river system. They receive a water input in the higher reaches of the mountains, they flow downslope through a system of the tributaries leading to a main trunk stream, and they erode and transport considerable amounts of rock debris, much of which comes from mass movement down the sides of the valleys through which they flow.

As can be seen in the diagram and photograph, the boundary between the zone of accumulation and the zone of ablation is approximated by the snow line. Above the snow line, the surface of the glacier is smooth and white, because more snow accumulates than is lost by melting and any irregularities are soon covered and filled with snow. Below the snow line, melting and evaporating exceed snowfall. There, the surface of the ice is rough and pitted and commonly is broken by open crevasses. Rock debris transported down the valley walls by mass movement adds to the roughness of this part of the glacier, since the debris is concentrated as the ice melts. At the end of the glacier, sediment transported by the ice simply is dropped as the ice melts, and it accumulates in a pile along the glacier's margins. The concentration of sediment at the terminus of the ice may be so great that the end of the glacier may be covered completely with rock debris that is capable of supporting the growth of a forest.

Continental Glaciers.   In terms of their effect upon the landscape and hydrologic system of the earth, continental glaciers (ice sheets) are by far the more important type of glacier. Studies of the Antarctic ice sheet by teams of scientists from various countries during the

Figure 14.6   *The margins of Barnes Ice Cap. The upper surface of the glacier is gently arched, and meltwater has formed meandering streams on the glacier's surface. Erosional debris carried by the moving ice is deposited at the glacier's margins to form a terminal end moraine.*

last 20 years have provided much new information that permits us to construct a reasonably accurate model of the continental glacial system. The diagram in *figure 14.4* illustrates the major features of a continental glacier and some of the important effects it produces.

A continental glacier is a roughly circular or elliptical plate of ice rarely more than 3000 m thick. (Ice does not have the strength to support the weight of a thicker accumulation. If appreciably more ice is added, the glacier simply will flow away from the center of accumulation more quickly.) The ice is not confined to a system of valleys but is capable of moving over relatively large topographic obstructions. However, continental glaciers can move more rapidly down broad preexisting valleys or lowlands, and, as a result, the margins of the glaciers are commonly lobate.

An important result of continental glaciation is the effect the weight of the ice has upon the crust. The huge mass of ice depresses the continental crust so that the surface of the land commonly slopes down toward the center of the ice. This depression acts as a trap for meltwater and, during the retreat of the glacier, large lakes form along the ice margins; in some areas, an arm of the ocean may invade this depression (*figure 14.4*). Previous drainage systems beneath the ice are obliterated completely, and rivers flowing toward the ice margins are impounded to form lakes that may overflow and develop a river channel parallel to the ice margins.

Three examples of existing glaciers will serve to illustrate the way in which a continental glacial system operates.

The Barnes Ice Cap of Baffin Island (*figure 14.5*) is the last remnant of the glacier that recently covered much of Canada and parts of the northern United States. The importance of this example is that it illustrates the relationship between a continental glacier and the regional landform. As can be seen on the map, the glacier has an elliptical shape and irregular or lobate margins. The ice is thickest in the central part of the glacier and thins toward the edges. The presence of the glacier has disrupted the former drainage system, and meltwater has accumulated to form a group of lakes along the ice margin. A photograph of the southern margins of the ice cap (*figure 14.6*) shows the large, gently arched surface of the glacier, the sediment deposited along the ice contact, and the stream channels formed by meltwater on the glacial surface.

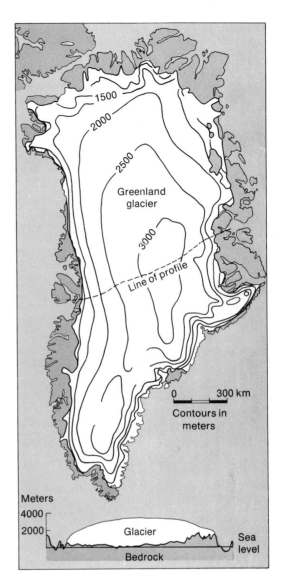

Figure 14.7 *Map showing surface elevations of the Greenland ice sheet. Note from the cross section that part of central Greenland is below sea level. It has been depressed by the weight of the ice.*

Figure 14.8 *Schematic diagram showing outlet glaciers advancing through a mountain pass as the continental glacier moves upon a mountain range.*

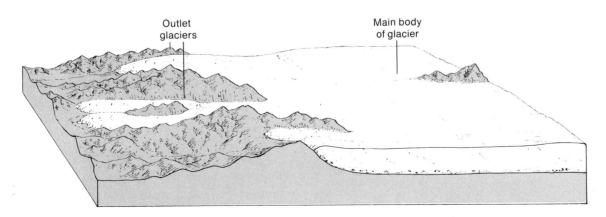

The ice cap of Greenland is much larger than the remnant on Baffin Island, and the ice covers nearly 80% of the island of Greenland. In cross section, the glacier is shaped like a lens (*figure 14.7*). The upper surface is a broad, almost flat-topped arch and is typically smooth and featureless. The base of the glacier also is curved but is considerably more irregular than the top, because the shape of the ice is governed partially by the topography of the land over which it flows. The Greenland ice cap is over 3000 m thick in the central part of Greenland, but it thins toward the margins. The zone of accumulation is in the central part of the island, where the ice sheet is nourished by snow from storms moving from west to east. The snow line varies from 50 to 250 km inland, so the area of ablation constitutes only a narrow belt along the margins. In rugged terrain, especially close to the margins of the glacier, the direction of ice movement is influenced greatly by the mountain ranges, and the ice moves through the mountain passes in large streams called **outlet glaciers** (*figure 14.8*). The outlet glaciers closely resemble valley glaciers in that they are confined by the topography. Measurements in Greenland show that the main ice mass moves forward at approximately 10 to 30 cm per day, but the speed of the outlet glaciers through the mountain passes may be extreme, with velocities as great as 1 m per hour. In some places, the ice can even be seen to move.

The glacier of Antarctica is similar to that of Greenland in that it covers essentially the entire continent (*figure 14.9*). The Antarctic glacier is much larger than Greenland's, and it contains more than 90% of the earth's ice. Much of the glacier is over 3000 meters thick and has depressed large parts of the surface of the continent below sea level. The topography of parts of Antarctica (mostly near the continental margins) is mountainous, with higher peaks and ranges protruding above the ice. In the mountains, the ice moves through outlet glaciers that funnel the ice from the interior to the coast. Two large ice shelves and numerous small ones occur along the coast and break up to form large icebergs in the South Atlantic Ocean.

Erosion, Transportation, and Deposition by Glaciers

Statement

As glacial ice flows over the surface of the land, it erodes, transports, and deposits rock material. Glaciers erode bedrock in two ways: by glacial **plucking** and by grinding (**abrasion**). Beneath the glacier, meltwater seeps into joints or fractures, freezes, and expands, wedging blocks of rock loose. The loosened blocks freeze to the bottom of the glacier and are plucked or quarried from the bedrock below, becoming incorporated into the moving ice (*figure 14.10*). The angular blocks are gripped firmly by the glacier as they become frozen to the ice and act as tools that grind and scrape the bedrock like a file. Aided by the pressure of the overlying ice, the angular blocks become very effective agents of erosion capable of wearing away large quantities of bedrock. The rock fragments incorporated into the glacial ice also are abraded and worn down as they grind against the

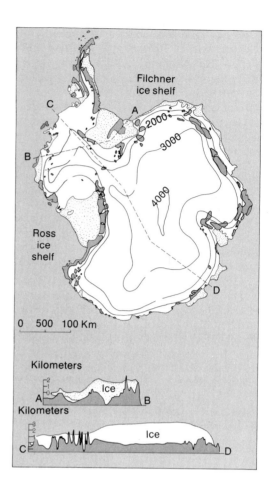

Figure 14.9   *Map of Antarctica, showing elevations of the ice sheet.*

walls and floor of the valley and usually develop flat surfaces that are deeply scratched. The bedrock surface eroded by a glacier commonly shows grooves and striations from abrasion.

The rock material incorporated in the ice moves in suspension or is pushed like soil in front of a bulldozer.

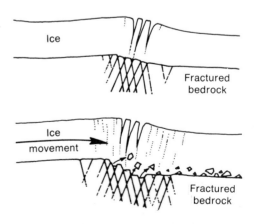

Figure 14.10   *Erosion by glacial plucking. Blocks of rock separated by joints freeze to the base or sides of the glacier and are lifted from the outcrop. They then act as abrasives and wear down the surface over which they move.*

Figure 14.11 *Roche moutonnée. Ice moving over bedrock commonly erodes the surface into streamlined shapes, with glacial plucking producing a ragged ridge on the downcurrent side.*

Figure 14.12 *Glacial striations in the Andes Mountains of South America.*

Deposition occurs as the ice melts, so **till** (glacial sediment) is characteristically unsorted and unstratified.

Discussion

Evidence of the distinctive abrasive and quarrying action of glaciers can be seen on most bedrock surfaces

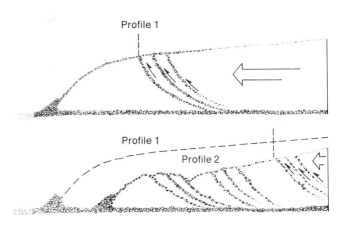

Figure 14.13 *Movement of sediment near the margins of a glacier. The ice near the ends of a glacier commonly is stagnant so that much of the sediment load is carried upward along shear planes and concentrates on the surface of the ice. Melting causes the form of the glacier to recede from profile 1 to profile 2, but the ice continues to move forward, transporting more sediment, which is deposited as end moraines.*

over which glacial ice has moved. Small hills of bedrock commonly are streamlined by glacial abrasion, and their upstream side typically is rounded off while the downstream side is made steep and rugged by glacial plucking (*figure 14.11*). Glacial striations such as those illustrated in *figure 14.12* may range from a hairline scratch to large furrows over a kilometer long.

The inclusions of rock fragments in the glacier are collectively referred to as the **load** of the glacier. The load is concentrated near the contact between the ice and the bedrock from which it was derived. Near the terminus, the ice becomes stagnant, forcing the ice upstream to move up and over it. Therefore, much of the load is carried upward along shear planes and concentrated on the surface of the ice (*figure 14.13*).

A part of the glacial load in valley glaciers consists of rock fragments that avalanche down the steep valley sides and accumulate along the glaciers' margins. Frost action is especially active in the cold climate of valley glaciers and produces large quantities of angular rock

Figure 14.14   *Lateral and medial moraines on a valley glacier. Rock debris moving down the valley slopes by mass movement accumulates along the margins of the glacier as lateral moraines and is transported downstream by the moving glacier. Where tributary glaciers merge into a main stream, the lateral moraines also merge to form a medial moraine (see plate 17).*

fragments. This material is transported along the surface of the margins of the glacier, forming conspicuous dark bands called **lateral moraines** (*figure 14.14*). Where a tributary glacier enters the main valley, the lateral moraine of the tributary glacier merges with a lateral moraine of the main glacier to form a **medial moraine** in the center of the main glacier. Thus, in addition to transporting the load near its base, a valley glacier acts as a conveyor belt and transports a large quantity of surface sediment to the terminus. At the terminus of a glacier, the ice leaves the system through melting and evaporating, and the load is deposited as an **end moraine** (*figure 14.15*).

Rates of Flow and Glacial Erosion.   The rate of ice flow in a glacier may be considered extremely slow when compared to the flow of water in rivers, but the movement is continuous and, over the years, vast quantities of ice can move over the landscape. Measurements show that some of the large valley glaciers of Switzerland move as much as 180 m per year; smaller glaciers move 90 to 150 m per year. Some of the most rapid rates have been measured on the outlet glaciers of Greenland, where ice is funneled through mountain passes at the rate of 8 km per year. From these and other measurements, it appears that flow rates of a few centimeters per day are common and that velocities of 3 m per day are exceptional. The most rapid movement occurs in outlet glaciers where pressure builds up behind a mountain range and helps to force the ice through a mountain pass.

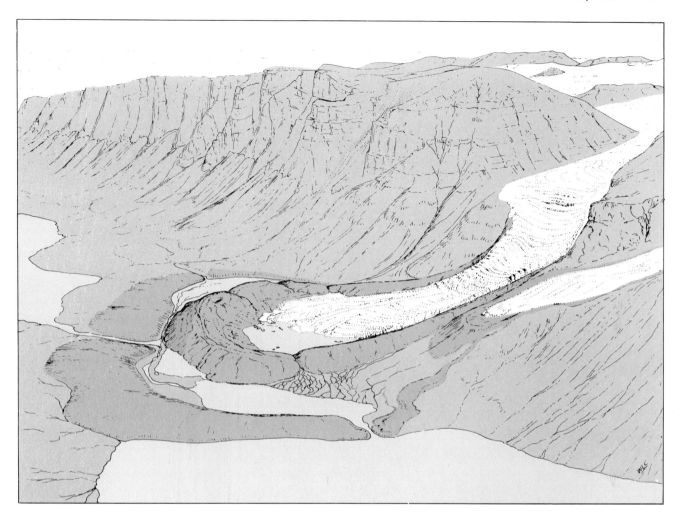

Figure 14.15 *Sketch of end moraines formed by a glacier on Baffin Island, Canada. The moraine accumulates as the sediment load is dropped where the ice melts and leaves the glacial system.*

Recent close observation of glacial movement shows that, at times, the entire glacier may surge forward more than several hundred meters per day. These surges are known to be fairly common, since the movement of ice now can be monitored by satellite photography. In the past, only a few surges were seen because the flow is so short-lived. Some of the best-known examples include a glacier in the Himalayas, which moved 11 km in three months, and the Steele glacier in the Yukon Territory, where a surge in 1966 had a maximum surface displacement of approximately 8 km. Glacial surges apparently are the results of sudden slippages along the bases of the glaciers caused by the buildup of extreme stress upstream. Stagnant or slow-moving ice near the end of a glacier can act as a dam for the faster-moving ice upstream. When this happens, stress builds up behind the slow-moving ice and can cause a sudden surge in flow when a critical point is reached. Surges also can be caused by a sudden addition of mass to the glacier, such as a large avalanche or a landslide onto the glacier's surface.

In considering the flow of ice in a glacier and the erosion it can cause, it is important to remember that a glacier is an open system with material entering the glacier in the zone of accumulation, flowing through, and then leaving the system at the distal margins. The ice within the glacier continually flows through the system regardless of whether the front of the glacier is advancing, retreating, or stationary. The length of a valley glacier is, therefore, no indication of the amount of erosion it can accomplish or has accomplished. *Erosion by a glacier is a function of the time the glacier exists and the velocity of the flowing ice within the system.*

Landforms Developed by Valley Glaciers

Statement

Valley glaciers are responsible for some of the most rugged and scenic mountainous terrain of the world. The Alps, Sierra Nevadas, Rockies, and Himalayas were all greatly modified by glaciers during the last ice age, and the shape and form of their valleys, peaks, and divides retain the unmistakable imprint of erosion by ice. Ice in a valley glacier commonly fills more than half of the valley, and, as it moves, it modifies the former V-shaped stream valley to a broad U-shaped, or trough-like, form. The head of the valley is sculptured into a

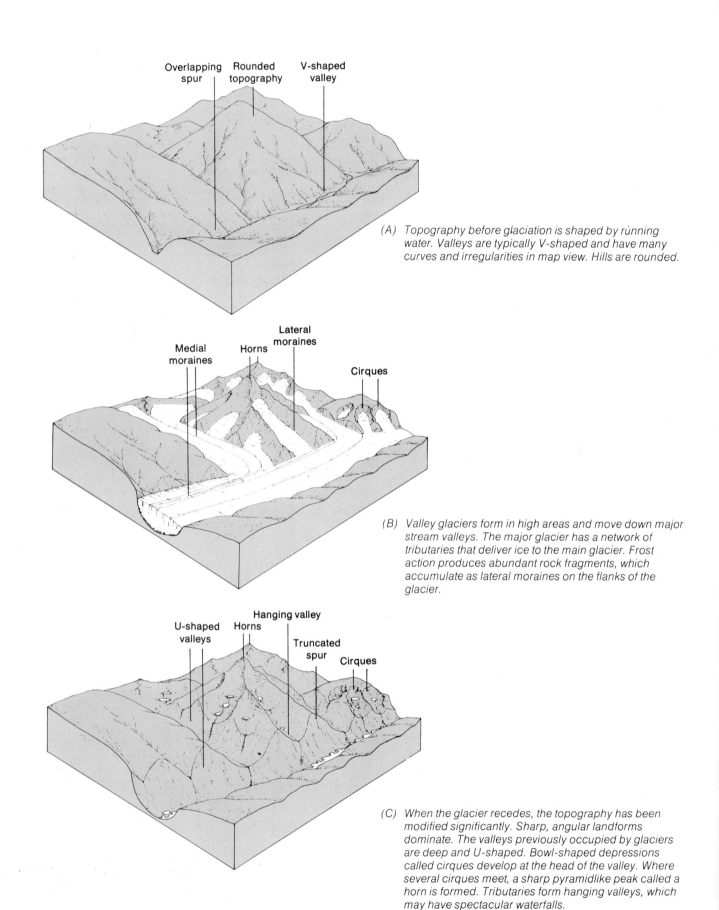

Overlapping spur · Rounded topography · V-shaped valley

(A) Topography before glaciation is shaped by running water. Valleys are typically V-shaped and have many curves and irregularities in map view. Hills are rounded.

Medial moraines · Horns · Lateral moraines · Cirques

(B) Valley glaciers form in high areas and move down major stream valleys. The major glacier has a network of tributaries that deliver ice to the main glacier. Frost action produces abundant rock fragments, which accumulate as lateral moraines on the flanks of the glacier.

U-shaped valleys · Horns · Hanging valley · Truncated spur · Cirques

(C) When the glacier recedes, the topography has been modified significantly. Sharp, angular landforms dominate. The valleys previously occupied by glaciers are deep and U-shaped. Bowl-shaped depressions called cirques develop at the head of the valley. Where several cirques meet, a sharp pyramidlike peak called a horn is formed. Tributaries form hanging valleys, which may have spectacular waterfalls.

Figure 14.16 Landforms developed by valley glaciation.

large amphitheater called a **cirque,** and, where several cirques approach a summit from different directions, a sharp, pyramid-shaped peak called a **horn** is formed. The projecting ridges and divides between glacial valleys are subjected to rigorous ice wedging, abrasion, and mass movement, which produces sharp, angular crests and divides (called **arêtes**) instead of the rounded topography developed by stream erosion. The rock debris transported by valley glaciers is deposited as end moraines, which commonly block the ends of the valleys so that meltwater from the ice accumulates and forms ponds and lakes. Lateral moraines develop along the valley walls and merge downstream to form medial moraines in the main glacier. Downstream from the glacier, meltwater reworks the glacial sediments and redeposits them to form an **outwash plain.**

Discussion

The series of idealized diagrams in *figure 14.16* provides an illustration of some of the major landforms resulting from valley glaciation and permits a comparison and contrast of landscape formed only by running water with that which has been modified by valley glaciers (*see also plate 17*). Block *A* shows the topography typical of a mountain region being eroded by streams. A relatively thick mantle of soil and weathered rock debris covers the slopes. The valleys are V-shaped in cross section and have many bends at tributary junctions so that ridges and divides between tributaries appear to overlap when one looks up the valley. In block *B*, the area is shown occupied by glaciers. The glaciers grow and flow down the tributary valleys and merge to form a major glacier. During epoch B, several thousand kilometers of ice might flow down the valley. This enormous quantity of moving ice erodes the valley as much as 600 m below the original level of the stream valley.

Frost action is a major process that effectively sharpens the mountain summits. Cirques are enlarged by glacial plucking and grow headward toward the mountain crest to form a horn. It should be noted that, where tributaries enter the main glacier, the surface of each glacier is at an even level. However, the main glacier is much thicker and therefore erodes its valley to greater depths than the tributaries. Thus, when the glaciers recede from the area, the floors of the tributary valleys are left higher than that of the main valley and **hanging valleys** result.

Figure 14.16C shows the region after the glaciers have disappeared. The most conspicuous and magnificent landforms developed by valley glaciers are the long, straight, U-shaped valleys or troughs. Many are several hundred meters deep and tens of kilometers long. The heads of glacial valleys terminate in large amphitheater-like or bowllike cirques, which commonly contain small lakes. The general topography of the peaks and ridges adjacent to the glacier is rugged and sharp because of the severe effects of frost action.

The block diagrams in *figure 14.16* do not show the landforms developed at the terminus of a valley glacier. These are illustrated in the photographs in *figure 14.17*.

Figure 14.17 *Valley glaciers and their landforms.*
(A) *Regional view of glacial systems on Baffin Island, Canada. Note the snowfield in the zone of accumulation, lateral moraines that merge into medial moraines, end moraines, and the outwash plain.*
(B) *Details of the terminus of the central glacier shown in photograph A.*

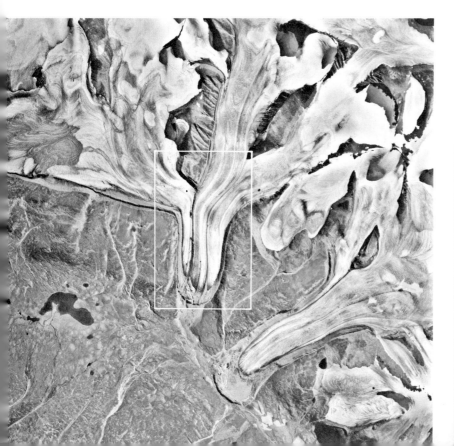

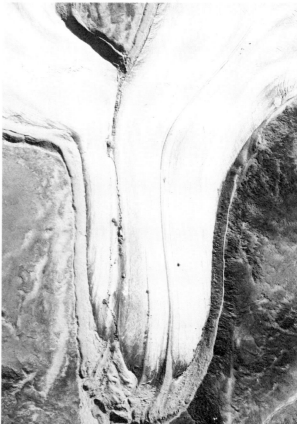

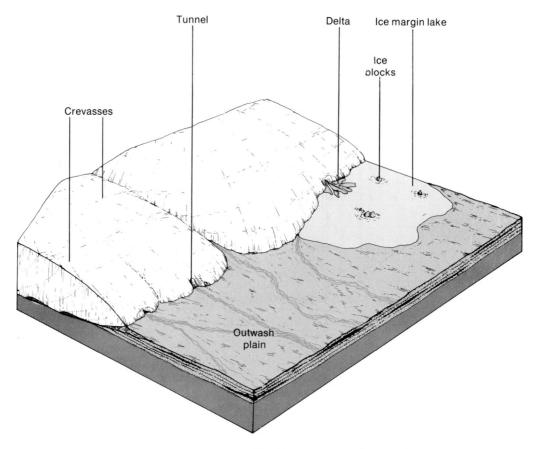

Crevasses

Tunnel

Delta

Ice margin lake

Ice blocks

Outwash plain

(A) Margins of a continental glacier.

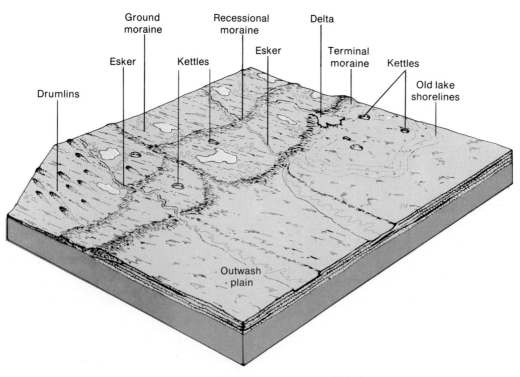

Drumlins

Esker

Ground moraine

Kettles

Recessional moraine

Esker

Delta

Terminal moraine

Kettles

Old lake shorelines

Outwash plain

(B) Area after ice front has retreated.

Figure 14.18 Features formed by continental glaciers.

In *figure 14.17A,* three types of moraine are immediately obvious: (1) lateral moraines along the margins of the glacier, (2) medial moraines formed where two lateral moraines join, and (3) an end moraine. The end moraine characteristically conforms to the shape of the terminus of the ice and forms a broad arc, which commonly traps meltwater and forms a temporary lake (*figure 14.17B*). If during the recession of the ice there are periods of stabilization, recessional end moraines may form behind the terminal end moraine.

The great volume of meltwater released at the terminus of a glacier reworks large amounts of the previously deposited moraine and redeposits the material in an outwash plain beyond the glacier. Outwash sediments have all the characteristics of stream deposits and typically are rounded, sorted, and stratified.

Landforms Developed by Continental Glaciers

Statement

From viewpoints on the ground, the landforms developed by continental glaciers may be relatively inconspicuous and not nearly as spectacular as those produced by valley glaciers; but, on a regional basis, continental glaciation greatly modified the entire landscape and produced many important and distinctive surface features. The great, thick continental ice sheets moved out over the flat lowlands of the shield and the stable platform, removing vast quantities of existing soil and eroding down into the bedrock several meters. Many thousands of square kilometers of North America and northern Europe have little or no soil cover and the effects of glaciation are seen everywhere in the polished and grooved fresh bedrock. Ice was thickest in lowlands such as those now occupied by the Great Lakes, and its erosive action was of greater magnitude, resembling that of valley glaciers.

The vast quantities of regolith picked up by continental glaciers were transported long distances and deposited near the ice margins, producing a variety of depositional landforms such as moraines, **kettles, drumlins, eskers,**

Figure 14.19 *Varves formed in a glacial lake. A layer of relatively coarse grained, light sediment accumulates during the spring and summer runoff. During the winter when the lake is frozen over, fine, dark mud settles to form a dark layer. Each set of light and dark layers represents a year's accumulation.*

(A) Eskers—Canadian Shield.

(B) Drumlins—Canadian Shield.

(C) Terminal end moraine—Alberta, Canada.

(D) Till deposited on bedrock—central Iowa.

Figure 14.20 Features formed by continental glaciers
(see plate 18).

lake sediments, and outwash deposits. These features are very widespread in the Great Lakes area and in the Dakotas and, when accurately mapped, their surface patterns reveal the various forms of the previous ice margins. Sediments deposited by continental glaciers may be more than 300 m thick, and so they blanket most of the preglacial topography on which they rest. One of the most significant topographic effects of continental glaciation, however, is not due so much to the action of ice itself upon the land but to the impact the glacier had in modifying, disrupting, and obliterating the previously established drainage system.

Discussion

Landforms produced by continental glaciers are usually regional in extent and can be explained best by reference to the block diagrams shown in *figure 14.18* and the photographs in *figure 14.20.* Block *A* in *figure 14.18* shows the margins of an ice sheet. Debris transported by the glacier accumulates at the glacier margin as a **terminal end moraine.** Beneath the ice is a variable thickness of debris being dragged forward by the glacier. This material may be reshaped by subsequent advances of ice to produce streamlined hills called drumlins. Meltwater from the ice forms braided streams that flow over the outwash plain and deposit glacial sediment. During retreat of the glaciers, meltwater forms subglacial channels and tunnels, which open into the outwash plain. Temporary lakes may develop where meltwater is trapped along edges of the glacier, and deltas and other shoreline features develop along the lake margins. Deposits on the lake bottom typically are stratified into a series of layers (**varves**) (*figure 14.19*). The coarse, light-colored material accumulates during spring and summer runoff, and the fine mud that forms the dark layers settles out during the winter when the lake is frozen over. Ice blocks left behind by the retreating ice front may be partly or completely buried in the outwash plain or in the moraines. When an isolated block of debris-covered ice melts, a depression called a kettle is formed.

Figure 14.18B shows the area after the glacier has disappeared completely. The end moraine appears as a belt of hummocky hills that mark the former position of the ice. The topography of the end moraine may be several kilometers wide, with a local relief of 100 to 200 m. From the ground, the moraine probably would not be recognized by the untrained observer as anything more than a series of hills; but, when mapped, it is accurate in pattern, conforming to the lobate margin of the glacier. Many small depressions occur throughout the moraine, some of which may be filled with water and form small lakes and ponds.

Pleistocene Glaciation

Statement

The glacial and interglacial periods that occurred during the last 2 to 3 million years constitute one of the most significant events in recent history of the earth. During this time, the normal hydrologic system was interrupted completely throughout large areas of the world and considerably modified elsewhere. Evidence for such a recent event is overwhelmingly abundant, and, over the last hundred years, field observations have provided incontestable evidence that continental glaciers covered large parts of Europe, North America, and Siberia. These ice sheets disappeared only within the last 10,000 to 15,000 years (*figure 14.21*). The general extent of glaciation in the Northern Hemisphere is shown in *figure 14.22,* and a more detailed map showing glacial features in the eastern United States is given in *figure 14.23.* These maps were compiled after many years of work by hundreds of geologists. It took that much effort to map the location and orientation of drumlins, eskers, moraines, striations, and glacial stream channels.

Four major periods of Pleistocene glaciation are recorded in the United States by deposits of broad sheets of till and complex moraines separated by ancient soils and layers of windblown silt. Striations, drumlins, eskers, and other glacial features show that all of Canada, the mountain areas of Alaska, and eastern and central United States down to the Missouri and Ohio Rivers were covered with ice (*figures 14.21, 14.22,* and *14.23*). There were three main centers of accumulation. The largest was centered over Hudson Bay. Ice from it advanced radially northward to the Arctic islands and southward into the Great Lakes area. A smaller center was located in the Labrador peninsula. Ice spread southward from this center into what is now the New England states. In the Canadian Rockies in the west, valley glaciers coalesced into ice caps. These ice caps merged into a single ice sheet and then moved westward to the Pacific shores and eastward down the Rocky Mountain foothills.

Throughout much of central Canada, the glaciers eroded the regolith and solid bedrock to an average depth of 15 to 25 m. This material was transported to the margins of the glaciers and accumulated as ground moraine, end moraines, and outwash in a broad belt from Ohio to Montana. In places, the glacial debris is over 300 m thick, but the average thickness is about 15 m. Meltwater from the ice transported much debris down the Mississippi River and a large amount of the fine-grained debris was transported and redeposited by wind.

The presence of so much ice upon the continent had a profound effect upon the entire hydrologic system of the earth. Sea level dropped more than 100 m. River patterns were obliterated or modified. Glacial processes of erosion and deposition replaced running water in many areas as the dominant geologic processes. Even areas far removed from the ice were affected in many unsuspected ways.

Discussion

Rhythms of Climatic Changes (Substages of the Ice Age). Even before the theory of worldwide glaciation was generally accepted, many observers recognized that the Pleistocene epoch consisted of more than a single advance and retreat of the ice. Extensive evidence now shows that there were a number of periods of growth

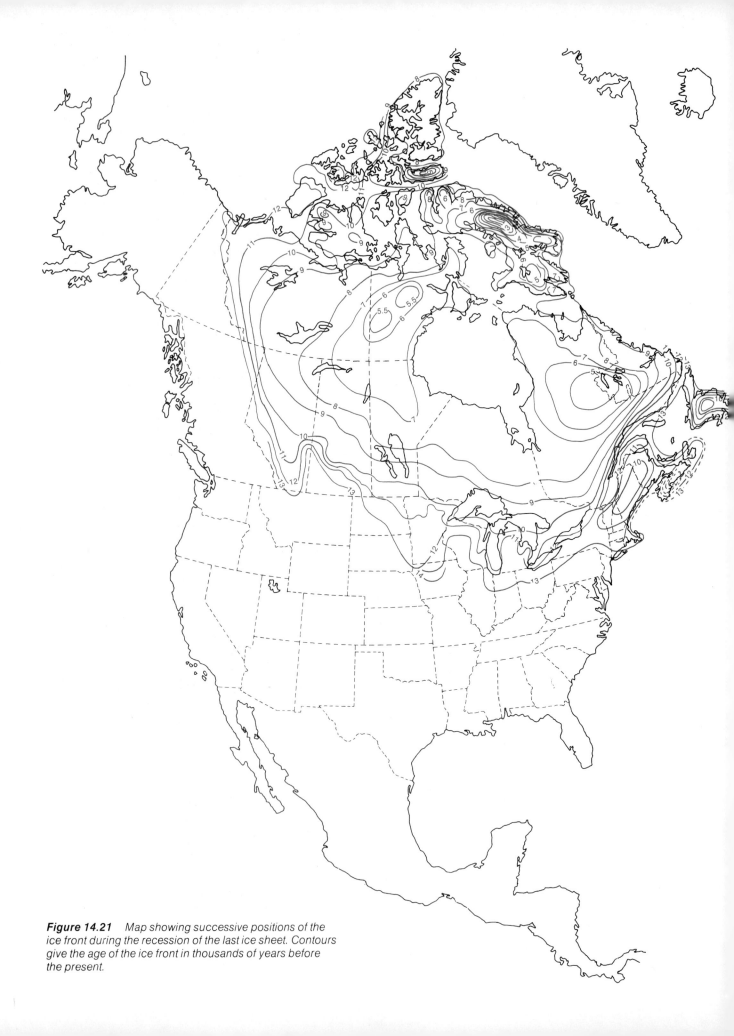

Figure 14.21 Map showing successive positions of the ice front during the recession of the last ice sheet. Contours give the age of the ice front in thousands of years before the present.

and disappearance of continental glaciers during the ice age. The interglacial periods of warm climates are represented by buried soil profiles, peat beds, and lake and stream deposits that separate deposits of unsorted, unstratified glacial debris.

Potassium-argon dating methods show that the first advance of the ice began between 2 and 3 million years ago, and radiocarbon dating indicates the last glaciers began to retreat 10,000 to 15,000 years ago. Remnants of these last glaciers, occupying about 10% of the world's land surface, still exist in Greenland and Antarctica. Thus, man may be living in an as yet unfinished ice age.

Studies of the details of the advance and retreat of the ice are significant, because they provide a surprising amount of climatic data helpful in determining whether the earth is now in a minor interglacial substage or in the beginning of an ice-free period typical of most of geologic history. Although we are unable to predict whether the climate will be warmer or colder, based on the established pattern of climatic change in the past, we can say that significant climatic changes are likely to occur. The earth is near the end of a glacial cycle. A return to normal (warmer) climates would lead to the melting of existing ice caps and the submergence of all major seaports and lowland cities of the world. A return to an ice age with another major advance of continental glaciers would drive mankind from much presently habitable land. Whatever does happen, it is not likely that stability will be established with the present conditions of ice covering approximately 10% of the earth's land surface. If the past is indicative of normal events, this ice will disappear before climatic conditions can be considered normal.

Isostatic Adjustment. Major isostatic adjustment of the earth's crust resulted when the weight of the ice depressed the continents. In Canada, a large area around Hudson Bay was depressed below sea level, as was the area around the Baltic Sea in Europe. A rebound from this depression has been in effect ever since the ice melted, and the former seafloor has been elevated almost 300 m. The rate of rising in this area is still about 2 cm per year. It is calculated that an additional rise of 80 m will occur before the land regains its preglacial level and before isostatic balance is reestablished.

Tilting of the earth's crust as it rebounds from the weight of the ice can be measured by mapping the elevation of shorelines of ancient lakes (*figure 14.24*). The shorelines were level when they were formed but were tilted as the crust rebounded from the unloading of the ice. In the Great Lakes region, old shorelines slope downward to the south, away from the centers of maximum accumulation of ice, indicating a rebound of 400 m or more.

Changes of Sea Level. One of the most important effects of Pleistocene glaciation was the repeated rise and fall of sea level, a phenomenon that corresponded to the retreat and advance of the glaciers. During a glacial period, water that normally returned to the ocean by runoff became locked upon the land as ice, and sea level was lowered. As the glaciers melted, sea level rose. The amount of change in sea level can be calculated because the area of maximum ice coverage is known in consider-

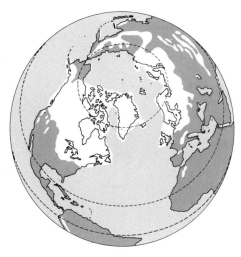

Figure 14.22 *The extent of Pleistocene glaciation in the Northern Hemisphere.*

able detail, and the thickness of ice can be estimated by comparison with known volumes of ice in the glaciers of Antarctica and Greenland. The Antarctic ice sheet alone contains enough water to raise sea level throughout the world by about 70 m.

The dates of sea-level changes are well documented by radiocarbon samples of terrestrial organic matter and nearshore marine organisms obtained from drilling and dredging off the continental shelf. These dates show that about 35,000 years ago the sea level was near its present position. Gradually, it receded; by 18,000 years ago, it had dropped nearly 137 m (*figure 14.25*). It then rose rather rapidly to within 6 m of its present level. These fluctuations caused the Atlantic shoreline to recede 100 to 200 km, exposing vast areas of the shelf. Early man probably inhabited large parts of the continental shelf now more than a hundred meters below sea level.

The glaciers extended far out across the exposed shelf of the New England coast, as evidenced by unsorted morainal debris and even remains of mastodons dredged from these areas. In the central and southern Atlantic states, depth soundings reveal drainage systems which extended across the shelf and eroded stream valleys.

Modification of Drainage Systems. Prior to glaciation, the landscape of North America was carved by running water, and well-integrated drainage systems were established that collected runoff and transported it to the ocean. Much of North America was drained by rivers flowing northward into Canada, because the regional slope throughout the north-central part of the continent was to the northeast. The preglacial drainage patterns are not known in detail, but various features of the present drainage, together with segments of ancient stream channels now buried by glacial sediments, suggest a pattern similar to that shown in *figure 14.26*. Prior to glaciation, the major tributaries of the upper Missouri and Ohio Rivers were part of a northward-flowing drainage system that included the major rivers that drained the Canadian Rockies, such as the Saskatchewan, Athabasca, Peace, and Liard Rivers. This drainage system emptied into the Arctic Ocean, probably through Lancaster Sound.

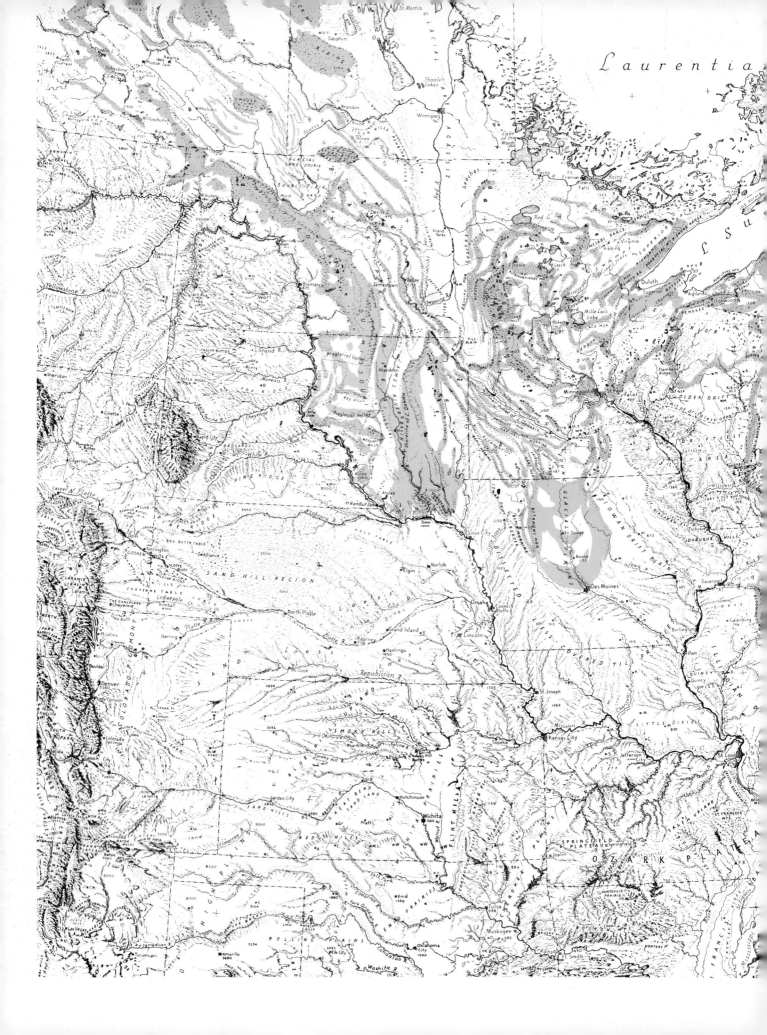

Figure 14.23 *Physiographic map showing major moraines in the central and eastern United States.*

(A)   A lake develops along the glacier's margins. Shoreline features such as beaches and bars are horizontal.

(B)   As the ice recedes, isostatic rebound occurs. Shoreline features formed during phase A are tilted away from the ice. Younger horizontal shoreline features are formed in lake B.

(C)   Continued retreat of ice permits further isostatic rebound and tilting of both shorelines A and B, which converge away from the glacier.

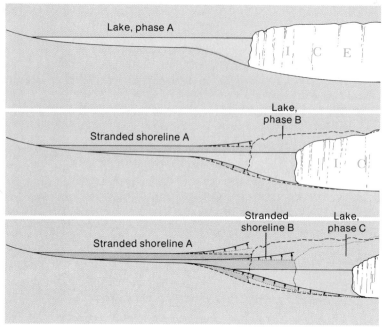

Figure 14.24   Tilted shorelines produced by isostatic adjustment of the crust after a glacier recedes.

As the glaciers overrode the northern part of the continent, they effectively buried the trunk stream of the major drainage systems and caused the north-flowing tributaries to be dammed along the ice front. This created a series of lakes along the margins of the glacier. As the lakes overflowed, the water drained along the ice front and established the present course of the Missouri and Ohio Rivers. A similar situation created Lake Athabasca, Great Slave Lake, and Great Bear Lake and their drainage through the Mackenzie River (*figure 14.27*). Compare this diagram with *figure 14.5,* which shows a drainage system undergoing similar modifications at the present time as a result of the Barnes Ice Cap.

Extensive and convincing evidence for these changes is most clearly seen in South Dakota, where the Missouri River flows in a deep, trenchlike valley roughly parallel to the regional contours and is cut at right angles to the regional slope. All important tributaries enter from the west. East of the Missouri River, abandoned pre-

glacial valleys now are filled with glacial debris and mark the remnants of preglacial drainage. This hypothesis also is supported strongly by recent discoveries of huge, thick, deltaic deposits in the mouth of Lancaster Sound, deposits difficult or impossible to explain by the present drainage pattern.

Beyond the margins of the ice, the hydrology of many streams and rivers was affected profoundly by the increased flow resulting from meltwater or by the greater precipitation associated with the glacial epoch. With the appearance of the modern Ohio and Missouri Rivers, significant drainage that formerly emptied into the Arctic and Atlantic Oceans was diverted to the Gulf of Mexico through the Mississippi River.

Many other streams became overloaded and partly filled their valleys with sediment. Some became more effective agents of downcutting and deepened their valleys. Although the history of each river is complex, the general effect of glaciation on many rivers was to produce thick alluvial fill in their valleys, now being eroded to form stream terraces.

Lakes.   The ice age created more lakes than those produced by all other geologic processes combined. The reason for this becomes obvious in light of the vast amount of water released during the retreat of the glaciers and the complete disruption of the preglacial drainage system. The surface of the shield was scoured and eroded by the ice, which left a myriad of closed, undrained depressions in the bedrock. These depressions filled with water and became lakes (*figure 14.28*).

Southward in the north-central United States, the drainage system was modified in a different manner. There, the glaciers did not erode the surface but deposited ground and end moraines over extensive areas. These deposits formed many closed depressions throughout Michigan, Wisconsin, and Minnesota that soon were filled with water to form tens of thousands of lakes.

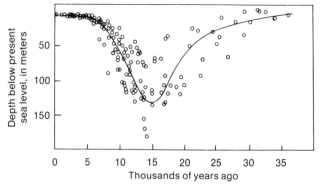

Figure 14.25   Changes in sea level based on radiocarbon ages of nearshore marine organisms obtained from samples from the Atlantic and Gulf coasts.

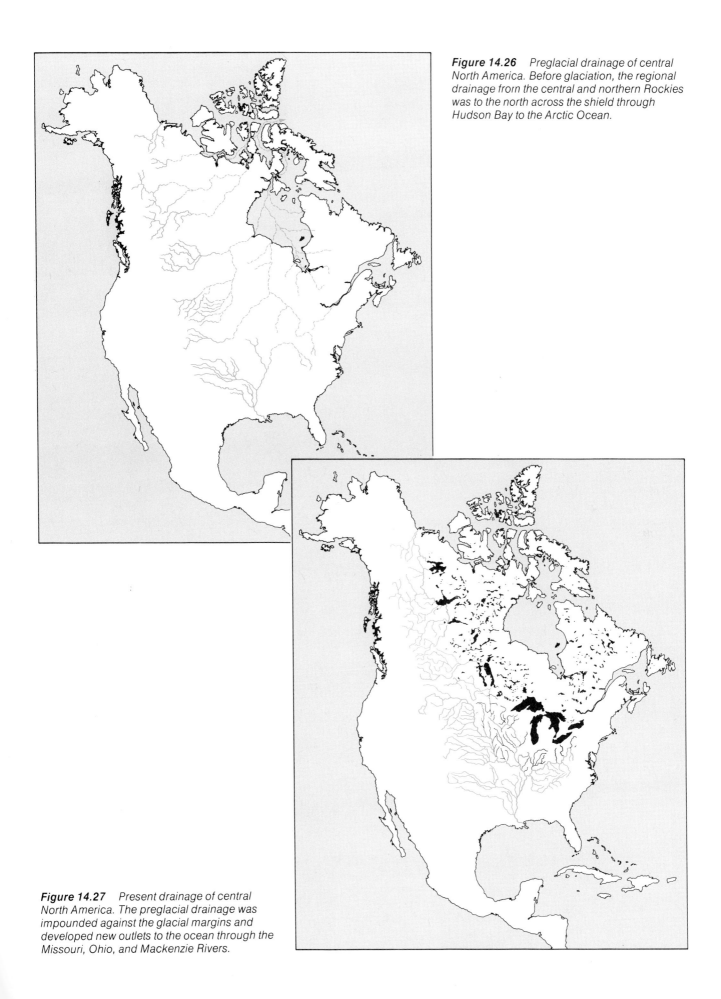

Figure 14.26 Preglacial drainage of central North America. Before glaciation, the regional drainage from the central and northern Rockies was to the north across the shield through Hudson Bay to the Arctic Ocean.

Figure 14.27 Present drainage of central North America. The preglacial drainage was impounded against the glacial margins and developed new outlets to the ocean through the Missouri, Ohio, and Mackenzie Rivers.

Figure 14.28 *Lakes created by continental glaciation in the shield area of North America.*

Many of the lakes still exist; many others have been drained or filled with sediment, leaving a record of their existence in peat bogs, lake silts, and abandoned shorelines.

During the retreat of the ice sheets of both Europe and North America, several conditions combined to create large lakes adjacent to the glaciers. Their formation can be envisioned with the help of the basic model of continental glaciation shown in *figure 14.4*. The ice on both the American and European continents was more than 2400 m thick near the centers of maximum accumulation, but it tapered toward its margins. Subsidence of the crust was greatest beneath the thickest accumulation of ice. In parts of Canada and Scandinavia, the crust was depressed over 600 m. As the ice melted, rebound of the crust naturally lagged behind, producing a regional slope toward the ice. This slope formed basins that lasted for thousands of years and became lakes or were invaded by the ocean. The Great Lakes of North America and the Baltic Sea of northern Europe were formed basically in this way.

Although the origins of the Great Lakes are extremely complex, the major elements of their history are illustrated in the four diagrams of *figure 14.29*. The preglacial topography of the Great Lakes region was influenced greatly by the structure and character of the rocks exposed at the surface. The geologic map of this region (*figure 14.30*) shows that the structural feature is the Michigan basin, which exposed a broad circular belt of weak Devonian shale surrounded by a more resistant Silurian limestone. Preglacial erosion undoubtedly formed a wide valley or lowland along the shale, while escarpments developed on the resistant limestone. As the glaciers moved southward into this area, large lobes of ice advanced down the great valleys to erode them into broad, deep basins. Lakes Michigan, Huron, and Erie all lie along this belt of weak Devonian shale scoured by large lobes of the glaciers. As the glaciers receded, meltwater accumulated in the basins to form the ancestors of the Great Lakes. Drainage was initially to the south through an outlet near present-day Chicago as ice dammed the previous northward-flowing drainage. The continued recession of the ice permitted the lakes to expand; and, eventually, the basin of Lake Superior was exposed, which ultimately developed drainage south from Duluth to the Mississippi. When Lake Superior was again overrun by ice, Lake Michigan drained through the Chicago outlet and the eastern lakes found an outlet through the Mohawk Valley and down the Hudson River. As the ice receded farther, a new outlet was found through the St. Lawrence estuary. The opening of this channel drained much of the lakes' water and caused a low level in the lakes, but crustal rebound closed the Duluth outlet and caused the lake basins to fill again. The three upper lakes were fused into a single large body of water leaving three outlets: one into Lake Erie, another to the south past Chicago, and a third by way of the Ottawa River. With additional crustal rebound, the present drainage system was established.

Niagara Falls came into existence when drainage was established from Lake Erie to Lake Ontario across the Niagara Escarpment. Undercutting of the weak shale below the dolomite caused the falls to retreat upstream about 1.2 m per year. Since their inception, the falls have retreated 11.2 km upstream (see *figure 11.4*).

To the northwest, another group of lakes formed in much the same way, but they have since been reduced to only a small remnant of their former selves (*figure 14.31*). The largest of these marginal lakes, known as Lake Agassiz, covered the broad flat region of Manitoba, northwestern Minnesota, and the eastern part of North Dakota. It drained first into the Mississippi River and at lower stages developed outlets into Lake Superior. Later, when the ice dam retreated, it drained into Hudson Bay. Remnants of this vast lake include Lakes Winnipeg and Manitoba and Lake of the Woods. The vast expanse of lake sediments deposited on the floor of Lake Agassiz provides much of the rich soil for the wheat lands of North Dakota and Manitoba. Even now, ancient shorelines of Lake Agassiz remain clearly etched along its former margins.

Northward, along the margin of the Canadian Shield, Lake Athabasca, Great Slave Lake, and Great Bear Lake are remnants of other great ice-margin lakes.

In northern Europe, the recession of the Scandinavian ice sheet caused similar depressions along the ice margins, and large lakes were formed that ultimately connected to the ocean to form the Baltic Sea.

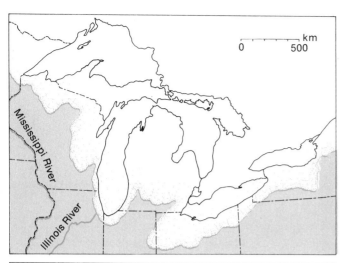

Figure 14.29   The evolution of the Great Lakes.

(A) Approximate position of the ice front 16,000 years ago. The ice advanced into lowlands surrounding the Michigan basin, with large lobes extending down from the present sites of Lake Erie and Lake Michigan.

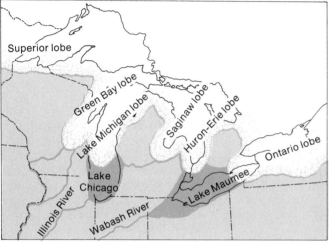

(B) The ancestral Great Lakes appeared about 14,000 years ago as the ice receded. The northern margins of the lakes were against the retreating ice. Drainage was to the south, to the Mississippi River.

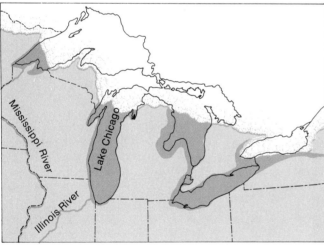

(C) As the ice front continued to retreat, an eastern outlet developed to the Hudson River, but the western lakes still drained into the Mississippi River. The lakes began to assume their present outlines about 10,500 years ago.

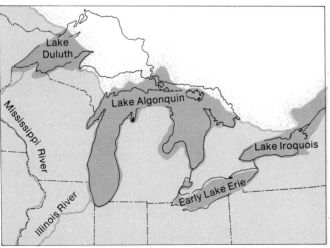

(D) Niagara Falls originated about 8000 years ago when the glacier receded back past the Lake Ontario basin and water from Lake Erie flowed over the Niagara Escarpment into Lake Ontario.

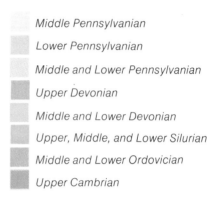

Figure 14.30 *Generalized geologic map of Michigan. The basins of Lake Michigan and Lake Huron were originally wide valleys formed by streams eroding the nonresistant Devonian shales. These lowlands then were enlarged by large lobes of the advancing glaciers.*

Middle Pennsylvanian

Lower Pennsylvanian

Middle and Lower Pennsylvanian

Upper Devonian

Middle and Lower Devonian

Upper, Middle, and Lower Silurian

Middle and Lower Ordovician

Upper Cambrian

Figure 14.31 *Large lakes that originated at, or near, the margins of the glacier in North America.*

Pluvial Lakes. The climatic conditions that brought about the growth and development of glaciers also had profound effects in arid and semiarid regions far removed from the large ice sheets. The increase in precipitation that fed the glaciers also increased the runoff of major rivers and intermittent streams, resulting in the growth and development of large **pluvial** (Latin: *pluvia,* rain) **lakes** in numerous isolated basins throughout the world. Most pluvial lakes were restricted to the relatively arid regions where, prior to the glacial epoch, there was insufficient rain to establish an inte-

grated, through-flowing drainage system. Instead, stream runoff flowed into closed basins and formed playa lakes. With increased rainfall, the playa lakes enlarged and sometimes overflowed. These lakes developed a

Figure 14.32 *The increase in precipitation associated with the glacial periods created numerous, large pluvial lakes in the closed basins of the western United States. Lake Bonneville, in western Utah, was the largest, with present remnants being the Great Salt Lake, Utah Lake, and Sevier Lake.*

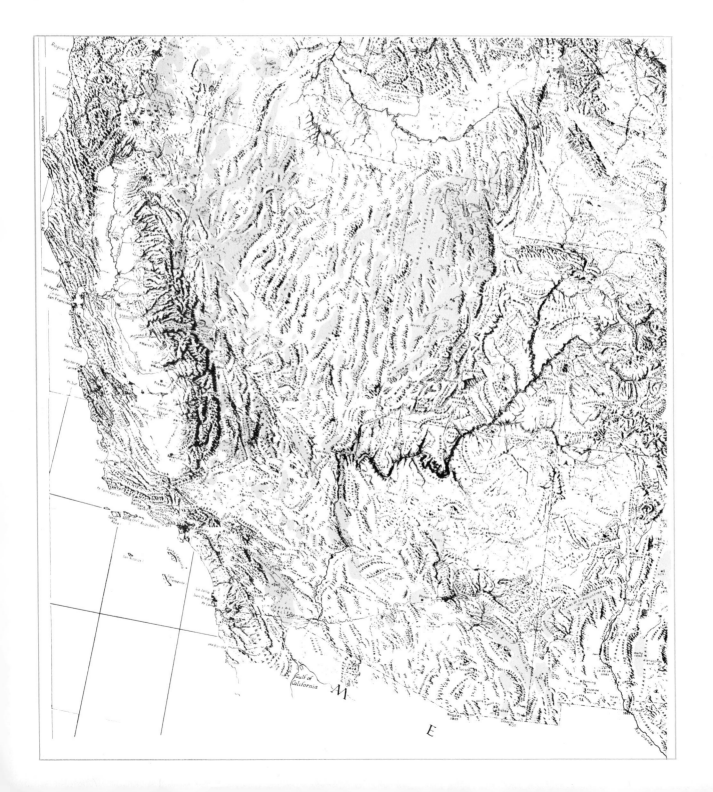

variety of shoreline features, such as wave-built terraces, bars, spits, and deltas, now recognized as high-water marks in many desert basins. Pluvial intervals during which the lakes covered their greatest area usually corresponded with glacial intervals. During interglacial stages when precipitation was less, the pluvial lakes shrank to small salt flats or dry, dusty playas.

The greatest concentration of pluvial lakes in North America was in the northern part of the Basin and Range Province of western Utah and Nevada, where the fault block structure has produced more than 140 closed basins, many of which show evidence of former lakes or former high-water levels of existing lakes. The distribution of the former lakes is shown in *figure 14.32.* Lake Bonneville was the largest by far and occupied a number of coalescent intermontane basins. Remnants of this great body of fresh water are the Great Salt Lake, Utah Lake, and Sevier Lake. At its maximum extent, Lake Bonneville was approximately the size of Lake Michigan, covering an area of 50,000 km², and was 300 m deep. The principal rivers entered the lake from the high Wasatch Mountains to the east and built large deltas, shoreline terraces, and other coastal features that are now high above the valley floors (*figure 14.33*). The most conspicuous feature is a horizontal terrace high on the Wasatch Mountain front.

As the level of the lake rose to 300 m above the floor of the valley, it overflowed to the north into the Snake

Figure 14.33 *Shoreline features of Lake Bonneville at the southern margins of Salt Lake Valley. The horizontal terrace along the base of the range marks the high-water mark of the lake.*

River and then to the ocean. The outlet was established on unconsolidated alluvium and rapidly eroded down to bedrock, 100 m below the original pass. The level of the lake was then stabilized and fluctuated with the pluvial epochs associated with glaciation. Some valley glaciers from the Wasatch Mountains extended down to the shorelines of the old lake, and some of their moraines have been carved by wave action. This indicates clearly that glaciation was contemporaneous with the high level of the lake. As the climate became dryer, the lakes evaporated, leaving faint shorelines at lower levels.

The Channeled Scablands. The continental glacier in western North America moved southward from Canada only a short distance into what is now Washington, but it played an important role in producing a strange complex of interlaced deep channels, a type of topography that is perhaps unique. This area, known as the Channeled Scablands, covers much of eastern Washington and consists of a network of braided channels 15 to 30 m deep. The term *scabland* is appropriately descriptive because, when viewed from the air, the surface of the area has the appearance of great wounds or scars (*figure 14.34*). Many of the channels have steep walls, "dry" waterfalls, and cataracts. In addition, there are deposits with giant ripple marks and huge bars of sand and gravel (*figure 14.35*). These features attest to the erosion by running water of exceptional degree that would be considered catastrophic by normal standards; yet today, there is not enough rainfall to maintain a single permanent stream across the area.

The process by which the Scablands were eroded is outlined as follows. A large lobe of ice advanced southward across the Columbia Plateau and temporarily

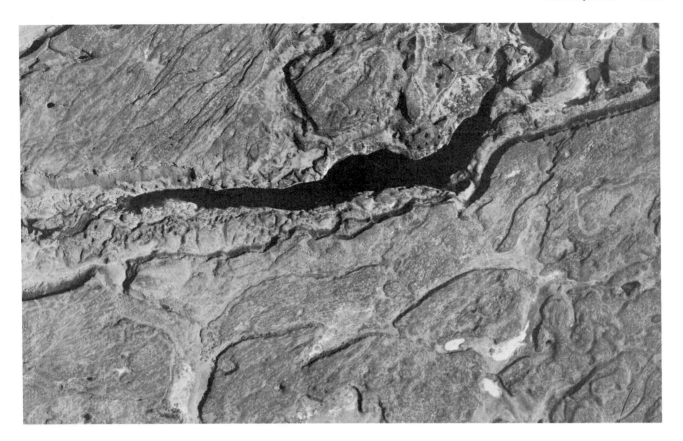

Figure 14.34 *The Channeled Scablands of Washington consist of a complex of deep channels cut into the basalt bedrock. The Scabland topography is completely different from that produced by a normal drainage system and is believed to have been produced by catastrophic flooding.*

blocked the Clark Fork River, one of the major north-flowing tributaries of the Columbia River (*figure 14.36*). The impounded water backed up to form glacial Lake Missoula, a long, narrow lake extending diagonally across part of western Montana. Sediments deposited in this lake now partly fill the long, narrow valley. As the glacier receded, the ice dam failed, releasing a tremendous flood over the southwestward-sloping Columbia Plateau. The enormous discharge was but little diverted by the preexisting shallow valleys and spread over the basalt surface, scouring out channels and forming giant ripple marks, bars, and other sediment deposits. Estimates suggest that, during the flood, as much as 40 km³ of water per hour might have discharged from glacial Lake Missoula. Since the glaciers advanced several times into the region, catastrophic flooding probably occurred many times. Glacial Lake Missoula formed each time the ice front advanced past the Clark Fork River and then flooded the Scablands with each ice recession and subsequent dam failure.

Effect of the Ice Age on Life. The severe climatic changes occurring during the ice age had a drastic impact on most life forms. With each advance of the ice, large areas of the continents (the area beneath the

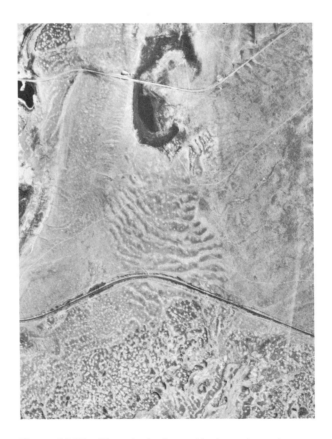

Figure 14.35 *Giant ripples formed in the surface of a gravel bar by catastrophic flooding in the Channeled Scablands, Washington.*

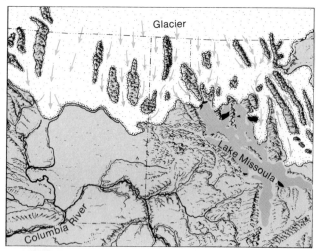

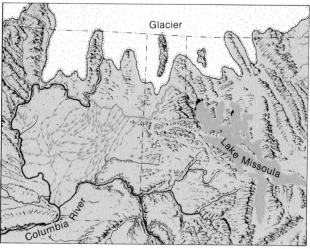

Figure 14.36 *Postulated origin of the Channeled Scablands.*

(A) *The ice sheet in northern Washington blocked the drainage of the north-flowing Clark Fork River to form a long, deep lake in northern Idaho and western Montana.*

(B) *As the glacier receded, the ice dam that formed Lake Missoula failed catastrophically, and water from the lake quickly drained across the Scablands, eroding deep channels. Repeated advance and retreat of the glacier probably produced several ice dams that failed as the ice melted, each causing catastrophic flooding.*

ice) became totally depopulated. In addition, a tremendous stress was put upon plants and animals retreating southward in front of the advancing glacier. The most severe stresses resulted from drastic climatic changes, reduction of living space, and curtailment of total food supply. Most species were displaced, along with their environment, across distances of approximately 3200 km as the glacier advanced. As the ice retreated, new living space became available in the deglaciated areas, whereas the formerly exposed continental shelves were inundated by the rising sea level. During the major glacial advances, when sea level was lower, new migration routes opened from Asia to North America (see *figure 14.22;* much of Alaska and Siberia were not glaciated) and from Southeast Asia to the islands of Indonesia. Land plants, of course, were forced to migrate with the climatic zones in front of the glacier. Displaced storm tracts and changes in precipitation affected even the tropics as the glaciers pushed cold-weather belts southward.

The repeated and overwhelming changes in the environment brought about by the cycles of advancing and retreating ice were too great for many life forms, and a large number of species, particularly giant mammals, became extinct. The Imperial Mammoth, 4.2 m high at the shoulders, once roamed much of North America. The

sabertooth tiger became extinct about 14,000 years ago. Fossils of the giant beaver, as large as a black bear, and the giant ground sloth, which measured 6 m tall when standing on its hind legs, have been found in Pleistocene sediments. In Africa, fossil sheep 2 m tall have been found, in addition to pigs as big as the present-day rhinoceros. In Australia, there were even giant kangaroos and other marsupials during the Pleistocene.

Effects of Winds. The presence of ice over so much of the continents clearly modified patterns of atmospheric circulation. Winds near the glacial margins were strong and unusually persistent because of the abundance of dense cold air coming off the glacier fields. These winds picked up and transported large quantities of loose, fine-grained sediment brought down by the glaciers. The dust deposited by the winds accumulated as **loess** (windblown silt) hundreds of meters thick, forming an irregular blanket over much of the Missouri River valley, central Europe, and northern China.

Sand dunes were much more widespread and active in many areas during the Pleistocene. A good example is the Sand Hills region of western Nebraska, which covers an area of about 60,000 km². This region was a large, active dune field during the Pleistocene, but today it is stabilized by a cover of grass.

The Oceans. The effect of Pleistocene glaciation

was felt in the waters of the oceans. In addition to the changing sea level that altered shorelines and exposed much of the continental shelves, the glacial periods caused the ocean waters to cool. The cold water undoubtedly affected the kind and distribution of marine life, as well as the chemistry of the seawater. In addition, patterns and strengths of oceanic currents were changed significantly. Circulation was restricted significantly by features such as the Bering Strait, extensive pack ice, and exposed shelves.

Even the deep-ocean basins did not escape the influence of glaciation. Where glaciers entered the ocean, icebergs broke off and rafted their enclosed load of sediment out into the ocean. As the ice melted, the debris, ranging in size from huge boulders to fine clay, settled out on the deep-ocean floor, resulting in an unusual accumulation of coarse glacial boulders in fine oceanic mud. Ice-rafted sediment is most common in the Arctic, Antarctic, North Atlantic, and northeastern Pacific.

In the warmer reaches of the ocean, the glacial and interglacial periods are recorded by alternating layers of red clay and small calcareous shells of microscopic organisms. The red mud accumulated during the cold periods when fewer organisms inhabited the cold water. During the warm interglacial periods, life flourished and layers of shells mixed with mud were deposited.

Records of Pre-Pleistocene Glaciation

Statement

Glaciation has been a very rare event in the history of the earth. Prior to the last ice age, which began 2 to 3 million years ago, the climate of the earth was typically mild and uniform throughout long periods of time. This is indicated by the type of fossil plants and animals preserved in the stratigraphic record as well as by the characteristics of sediments themselves. However, there are widespread glacial deposits consisting of unsorted, unstratified debris containing striated and faceted cobbles and boulders, which clearly record several major periods of ancient glaciation prior to the last ice age. Moreover, the glacial deposits commonly rest upon striated and polished bedrock and commonly are associated with varved shales and deposits of sandstone and conglomerate typical of former outwash deposits. Such evidence implies several periods of glaciation in the remote history of the earth.

Discussion

The best-documented record of pre-Pleistocene glaciation is found in late Paleozoic rocks of South Africa, India, South America, Antarctica, and Australia. The evidence indicates widespread continental glaciation during the late Paleozoic era (200 to 300 million years ago). In these areas, numerous exposures of ancient glacial deposits are found, many of which rest upon a striated surface of older rock (*figure 14.37*).

Other deposits of still older glacial sediment that are found on every continent but South America indicate that two other very widespread periods of glaciation occurred during late Precambrian time.

Small bodies of glacial sediment of other geologic periods have been found in local areas, but they are not nearly so well documented or widespread. Glaciation,

Figure 14.37 *Areas of late Paleozoic glaciation in the southern continents.*

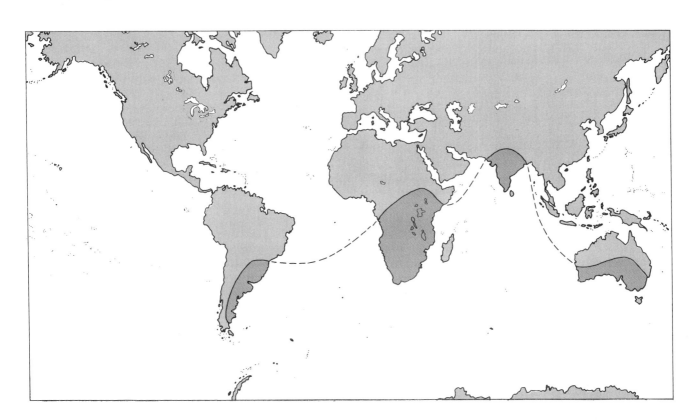

therefore, has been a relatively rare phenomenon and appears not to have occurred in any regular cycles. The only feature that seems to be common to all periods of glaciation is that the continents appear to have been relatively high and were undergoing periods of mountain-building. Mountain-building has occurred many times in the geologic past but often without accompanying glaciation, a fact that suggests that an elevated landmass is not the primary cause of glaciation, only one of several prerequisites. A period of glaciation must require a special combination of conditions that has occurred only several times in the last 4 billion years.

Causes of Glaciation

Statement

Although the history of Pleistocene glaciation is well established and the many and varied effects of glaciation are clearly recognized, it is difficult to explain why glaciation takes place. For over a century, geologists and climatologists have struggled with the problem, but it still remains unsolved.

Any theory that adequately explains glaciation must explain the following facts.

1. During the last ice age, which lasted 2 to 3 million years, there were repeated advances of the ice, separated by interglacial periods of warm climate. Therefore, glaciation is not related to a slow process involving long-term cooling.

2. Glaciation is a rare and unusual event in the earth's history. There was widespread glaciation 200 to 300 million years ago and during late Precambrian time, approximately 700 million years ago. The few other episodes of glaciation were apparently not widespread. Therefore, glaciation does not appear to be a result of some normal cyclic event; rather, it seems to stem from a rare combination of conditions on the earth.

3. Throughout most of the earth's history, the climate has been milder and more uniform than it is now. A period of glaciation would require a lowering of the earth's average surface temperature by about 5 °C and perhaps an increase in precipitation.

4. Precipitation is critical for the growth of glaciers. In order for continental glaciers to grow, an elevated landmass or a polar landmass must be situated in such a position that storms would bring moist cold air to them. Glaciers can move into lowlands in lower latitudes, but they originate in highlands or in high latitudes. Greenland and Antarctica provide favorable topographic conditions today, as do the Labrador peninsula, the northern Rocky Mountains, Scandinavia, and the Andes.

5. Adequate precipitation is required for glacial development. There are a number of areas sufficiently cold to produce glaciers even today, but these areas are too dry and have insufficient snowfall for a glacier system to develop.

Discussion

Astronomical Hypotheses. A number of hypotheses have been advanced to explain glacial periods by extra-terrestrial or astronomical influences. These may be grouped into categories involving (1) variations in solar radiation and (2) variations in the earth's orbital path.

Variations in solar radiation with short-term fluctuations of as much as 3% have been observed, and some meteorologists believe larger changes may be possible. If minimum output in radiation occurred simultaneously with favorable topographic conditions on the earth, a glacial period could result. Unfortunately, there is presently no way to determine variations in solar radiation in the past, so this hypothesis cannot be verified.

Variations in the earth's orbital path are measurable. They alter the length of seasons, perhaps to a degree that could produce alternating colder and warmer cycles. When the cooling produced by rotational wobble is synchronous with the cooling of orbital deviation and minimal radiation from the sun, the earth (at least one hemisphere) could be cooled sufficiently to initiate glaciation. These astronomical variations occur in cycles repeated every few thousand years and apparently have existed throughout geologic time. However, there seems to have been no corresponding cyclic repetition of glaciation in the geologic record. Unless astronomical variations are supported by geologic accidents on the earth's surface, they fail to explain the erratic occurrence of glaciation throughout geologic time.

Atmospheric Changes. Changes in the composition of the earth's atmosphere sufficient to insulate the earth from the sun's radiation have been suggested to explain the glacial periods. The most widely supported hypothesis of this kind attributes the temperature fluctuation necessary for growth of ice sheets to a decrease in the carbon dioxide content of the atmosphere. Carbon dioxide and water vapor produce an important greenhouse effect. With a high carbon dioxide content in the atmosphere, shortwave solar radiation can reach the earth but the longwave thermal radiation cannot escape, so the earth's climate remains moderate. A drop in air temperature of 4 °C could be sufficient to initiate an ice age. Extensive growth of vegetation could remove carbon dioxide from the atmosphere and bring about such a drop in temperature. However, the drop in temperature would retard plant growth and rebalance the carbon dioxide content!

Another proposal is that fine volcanic dust injected into the atmosphere would insulate the earth by reflecting back into space part of the solar radiation, thus causing a drop in temperature. However, periods of exceptional volcanic activity do not correspond with glaciation, so this hypothesis is generally discarded.

Oceanic Controls. Oceanic currents are intimately involved in the temperature and moisture balance in climate and in the past may have played an important role in the growth and decay of ice sheets.

One hypothesis is based on circulation of oceanic currents in the Atlantic and Pacific. Prior to the Pleistocene, North and South America were not connected by the Isthmus of Panama. The warm equatorial currents of the Atlantic moved westward into the Pacific instead of northward in the Gulf Stream. The Arctic Ocean, being frozen, could not supply the moisture needed to feed the glaciers, although the temperature was sufficiently low.

As the Isthmus of Panama grew and connected the Americas, the equatorial currents were deflected northward, providing enough heat to melt the Arctic Ocean's ice. The Arctic waters were then free to evaporate and provide the moisture for the glaciers. As the ice sheets grew, sea level dropped and the effectiveness of the Gulf Stream was reduced. The Arctic waters then froze, and the glaciers decayed and disappeared from lack of nourishment. As the glaciers receded, the Gulf Stream became more effective and melt from the Arctic Ocean's ice provided the precipitation necessary for another advance of the ice sheets. An important aspect of this hypothesis is that it explains the repeated advance and retreat of the glaciers during the Pleistocene epoch under an essentially stable heat budget.

Continental Drift. The theory of continental drift recently has been utilized to help explain the growth and decay of ice sheets during the Pleistocene and earlier glacial periods as well. Throughout most of geologic time, the polar regions appear to have been occupied by a broad, open ocean that allowed major ocean currents to move unrestricted to the polar regions. Equatorial waters were spread over the polar regions, warming them with water from more temperate latitudes. This unrestricted circulation produced mild, uniform climates that persisted throughout most of geologic time.

Movement of the large American continental plate westward from the Eurasian plate throughout Tertiary time culminated in the development of the north-south-trending Atlantic Ocean with the north pole in the small, nearly landlocked basin of the Arctic Ocean. Meanwhile, the Antarctic continent had drifted over the south pole by late Miocene time. Continental glaciation probably began in the Antarctic as the high landmass moved over the cold polar region where a supply of precipitation from the surrounding oceans sufficient to nourish the glaciers was available. Evidence from deep-sea cores in the southern oceans strongly suggests that glaciation in the Antarctic began long before the Pleistocene epoch and has continued there ever since. The presence of a continental glacier on Antarctica probably caused further global cooling of both hemispheres. By the beginning of Pleistocene time, the present location and configuration of the continents and ocean basins had been established. With the present belts of climatic zones, areas near the polar regions were at the glacial threshold.

According to a theory proposed by Donn and Ewing, precipitation derived from the ice-free Arctic Ocean initiated continental glaciation in the high latitudes of the Northern Hemisphere. Channels connecting the Arctic Ocean with the Atlantic occurred in the shallow seaways on either side of Greenland. The cold polar air originating over the snow-covered continents had a very low percentage of water vapor, but the air masses from the open oceans contained substantial amounts of water. Ice caps grew into ice sheets, which spread over the northern parts of North America and Eurasia. Sea level dropped, restricting the circulation through the shallow channels connecting the Atlantic and Arctic Oceans. This restriction reduced the amount of warm water carried to the Arctic, thus lowering the temperature of the water. The glaciers themselves also lowered the average atmospheric temperature, and the ice-free Arctic Ocean became frozen over, probably by the middle of the glacial stage. Evaporation from the Arctic (then essentially nil) cut off the supply of snow to the major centers of the ice sheets. The southern extent of glaciation was determined by the amount of moisture available from the North Atlantic and North Pacific Oceans.

An interglacial period was initiated when the temperatures of the ocean surface water, particularly the North Atlantic, were lowered sufficiently to decrease the rate of evaporation and, therefore, of precipitation over the ice sheets. The glaciers, then starved for moisture, began to decay and retreated toward their centers of spreading.

As the ice sheets receded, the ocean surface waters became warmer and the Arctic ice eventually began to melt. This initiated another ice age, but a time lag of about 5000 years between the melting of the ice and the warming of the surface of oceanic waters permitted the ice to disappear completely before the beginning of the new glacial cycle.

Greenland and Antarctica sustain continental glaciers because they are surrounded by the open ocean, which supplies the moisture necessary for glaciers to exist.

This hypothesis accounts for the complete short-term cycle of glacial and interglacial stages that recurred throughout the Pleistocene. It also agrees with the fact that glaciation is a rare event requiring special geographic conditions involving continental plates, polar regions, and ocean currents. In addition, it explains the glacial period in the Southern Hemisphere during late Paleozoic time; the supercontinent at that time was located in the area of the south pole and was surrounded by oceans that supplied the necessary moisture. A major ice age thus developed but terminated as the continents split and drifted toward the equator. Moreover, this theory emphasizes precipitation as the major controlling factor, whereas other theories are based on almost inexplicable changes in temperature. However, preliminary exploration of the sediment on the Arctic seafloor does not show traces of alternating cold- and warm-water fauna that should be found if the Arctic was alternately frozen and unfrozen.

Summary

Several times in the geologic past, systems of flowing ice called glaciers have interrupted completely the normal hydrologic system over much of the world. Water enters a glacier system as snow, is recrystallized into solid ice, flows through the glacier, and ultimately leaves the system near its outer margins by melting or evaporating. Two main types of glaciers are recognized: (1) valley glaciers and (2) continental glaciers. Each type produces distinctive landforms.

The last ice age began to terminate 10,000 to 15,000 years ago. The period of glaciation lasted for more than a million years and had a profound effect on many aspects of the physical environment. The most significant results of glaciation include (1) rise and fall of sea level, (2) isostatic adjustment of the crust, (3) modification of drainage systems, (4) creation of numerous lakes, and (5) migration and selective extinction of many plants and animals.

The causes of glaciation are not completely clear, but a glacial period may be related to continental drift and modification of oceanic currents.

Melting of the glaciers raised the level of the oceans and inundated many coastlines. Today, these shores are continually modified by movement of the water in the oceans. In the next chapter, these shoreline processes will be discussed.

Additional Readings

Flint, R. F. 1971. Glacial and Quaternary Geology. New York: John Wiley and Sons.

Paterson, W. J. B. 1969. The Physics of Glaciers. London: Pergamon Press.

Post, A., and E. R. LaChapelle. 1971. Glacier Ice. Seattle: University of Washington Press.

Price, R. J. 1973. Glacial and Fluvioglacial Landforms. New York: Hagner Publishing Company.

Sharp, R. P. 1960. Glaciers. Eugene, Ore.: University of Oregon Press.

Turekian, K. K., ed. 1971. The Late Cenozoic Glacial Ages. New Haven, Conn.: Yale University Press.

Shoreline Systems

Water in the ocean is in constant motion. It is moved by wind-generated waves, tides, tsunamis (seismic sea waves), and a variety of density currents. As it moves, it constantly modifies the shores of all the continents and islands of the world, reshaping coastlines with the ceaseless erosive activity of waves and currents. The intensity of shoreline processes can change from day to day and from season to season, but the processes never stop.

However, the present shorelines of the world are not the result of modern shoreline processes alone. Nearly all coasts have been affected profoundly by the rise in sea level associated with the melting of the glaciers, which began 10,000 to 15,000 years ago. With a rise in sea level, large parts of the continents were flooded, and the shorelines moved inland against a landscape resulting from continental processes. Therefore, the configuration and the other characteristics of a given shoreline may be largely the results of stream erosion or deposition, glaciation, volcanism, earth movements, marine erosion or deposition, or even the growth of organisms.

The great concentration of population on, or near, shorelines indicates that they are of major importance to man. To live properly in this environment, man must understand its processes and work within the framework of shoreline dynamics. In this chapter, the origin of wind-generated waves, tides, tsunamis, and longshore currents will be described. Then, the formation of erosional and depositional features along the coasts and how the coasts evolve with time will be considered. A brief analysis of shorelines, based on the geologic agents responsible for their development, concludes the chapter.

Major Concepts

1. Waves generated by the wind provide most of the energy for shoreline processes.
2. Wave refraction concentrates energy on headlands and disperses it in bays.
3. Longshore currents are generated by waves advancing obliquely toward the shore and are one of the most important processes operating on coasts.
4. Erosion along a coast tends to develop a sea cliff by the undercutting action of waves and longshore currents. As the cliff recedes, a wave-cut platform develops until equilibrium is established between the wave energy and the configuration of the coast.
5. Sediment transported by waves and longshore currents is deposited in areas of low energy, forming beaches, spits, and barrier islands.
6. Processes of erosion and deposition along a coast tend to develop a straight or gently curving shoreline along which energy is expended equally.
7. Reefs grow in a very special environment and form coasts that can evolve into atolls.
8. The worldwide rise in sea level associated with the melting of the glaciers has drowned many coasts. As a result, many coasts have been modified only slightly by modern marine processes.
9. Tides are produced by gravitational attraction of the moon (and, to a much smaller extent, the sun) and exert a major local influence on shorelines.
10. Tsunamis are waves generated by earth movements on the seafloor. They become high waves as they near the shore and can be very destructive.

Waves

Statement

Most shoreline processes are either directly or indirectly the result of wave action; therefore, an understanding of wave phenomena is fundamental to the study of shoreline processes. All waves are a means of moving some form of energy from one place to another. This is true of sound waves, radio waves, and water waves. All waves must be started by some force or energy. By far the most important types of waves in the ocean are those generated by the wind.

As wind moves over the open ocean, the turbulent moving air distorts the surface of the water. Gusts of wind depress the surface where they move downward, and, as they move upward, they cause a decrease in pressure and elevate the surface. This produces an irregular, wavy surface on the ocean, and part of the energy of the wind is transferred to wave energy in the water. Waves tend to be choppy and irregular and often are capped by sheets of spray in a storm area. A number of wave systems can be superposed on each other in the area where they are generated. The size and orientation of the various wave systems are different; but, as the waves move out from their place of origin, the shorter waves move more slowly and are left behind, and the wave patterns develop some measure of order. Large waves called swells can travel great distances, some as much as 10,000 km, before breaking against the shore.

Wind-generated waves are oscillatory; that is, a given water particle in the wave moves in a circular orbit and returns approximately to its original position. The net forward motion of the water itself in a wave is negligible; only the form of the wave moves forward. As the wave approaches the shore, however, important changes occur. The height of the wave increases as it approaches shallow water, and, ultimately, the wave form collapses forward or breaks into a surf. The water then rushes forward onto the shore and returns as a backwash. It is this energy that causes erosion, transportation, and deposition along the shores of all the continents and islands of the world.

Discussion

Water Motion in Waves. The description of ocean waves utilizes the same terms as those applied to other wave phenomena. These are illustrated in *figure 15.1.* The **wave length** is the horizontal distance between adjacent crests or adjacent troughs. **Wave height** is the verti-

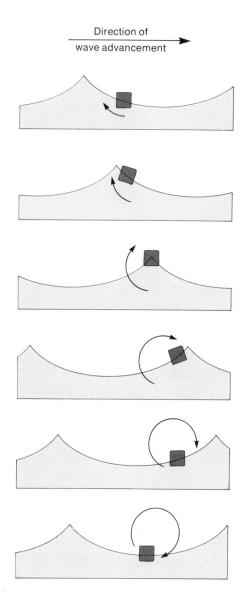

Figure 15.2 *Particle motion as a wave advances from left to right is shown by the movement of a floating object. As the wave advances, the object is lifted up to the crest and returns to the trough. The wave form advances, but the water moves in an orbit, returning to its original position.*

cal distance between **wave crest** and **wave trough.** The time between the passage of two successive crests is referred to as the **wave period.**

The nature of wave motion can be observed easily by watching a floating object move forward as the crest of the wave approaches and then sink back into the following trough. Viewed from the side, the object moves in a circular orbit with a diameter equal to the wave height (*figure 15.2*).

Beneath the surface, this motion in the wave dies out rapidly and becomes negligible at a depth equal to about one-half the wave length. This level is known as the **wave base** (*figure 15.3*). Water motion in waves, therefore, is distinctly different from water motion in current systems, where it moves in a given direction and does not return to its original position.

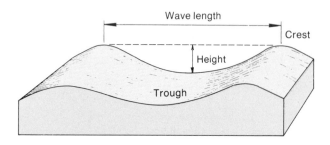

Figure 15.1 *Nomenclature of wave morphology.*

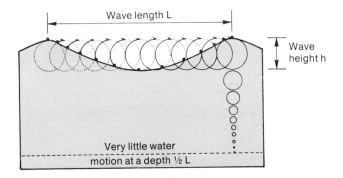

Figure 15.3 *The orbital motion of water in a wave decreases with depth and dies out at a depth equal to about half the wave length.*

Breakers. The energy of a wave depends upon its length and height. The greater the height of a wave, the greater the size of the orbit in which the water moves. The total energy of a wave can be represented by a column of water in orbital motion. As a wave approaches shallow water, some very important changes occur (*figure 15.4*). (1) The wave base encounters the bottom; friction causes the wave to slow down gradually so that each wave progresses more slowly than the one behind. Therefore, the waves are crowded together. (2) The wave height increases as the column of water affected by wave motion encounters the seafloor. Thus, energy from a deep column of water is concentrated into a shallower one. Each wave, therefore, becomes higher as it passes into shallow water. As the wave form becomes progressively higher and the velocity decreases, a critical point is reached where the forward velocity of the orbit distorts the wave form. The wave crest then extends beyond the support of the underlying column of water, and the wave collapses or breaks. (3) At this point, all the water moves forward. The wave form is largely lost, and the energy is released as a wall of moving, turbulent water. After the breaker

collapses, the **swash** (turbulent sheet of water) flows up the beach slope. The swash is a powerful surge that causes a landward movement of sand and gravel on the beach. As the force of the swash is dissipated against the slope of the beach, the backwash flows down the beach slope, but much of it seeps into the permeable sand and gravel of the beach.

In summary, then, waves are generated by the wind on the open ocean. The wave form moves out from the storm area, but water moves in a circular orbit with little or no forward motion. As the wave approaches the shore, it breaks and the energy in the forward-moving surf is expended on the shore and causes erosion, transportation, and deposition of sediment.

Wave Refraction

Statement

As waves approach the shore, they commonly are bent (or refracted) so that the crest line of the waves tends to become parallel to the shore. This is a key factor in shoreline processes, because it has considerable control over the distribution of energy along the shore, as well as the direction in which coastal water and sediment move. **Wave refraction** occurs because part of the wave begins to drag bottom and is slowed down while the remainder of it in deeper water moves forward at normal

Figure 15.4 *As a wave approaches the shore, several significant changes result as the orbital motion of the water interacts with the seafloor. (A) The wave length decreases due to frictional drag, and the waves become crowded together closer to shore. (B) The wave height increases as the column of water moving in an orbit stacks up on the shallow seafloor. (C) The wave becomes asymmetrical due to increasing height and frictional drag on the seafloor and ultimately breaks. The water then ceases to move in an orbit and rushes forward to the shore.*

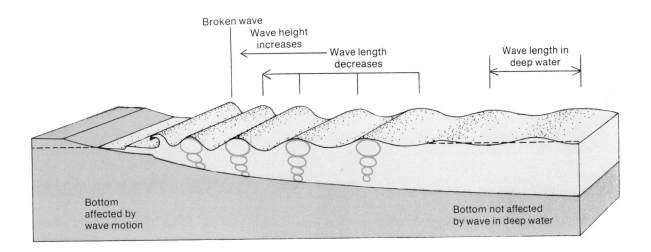

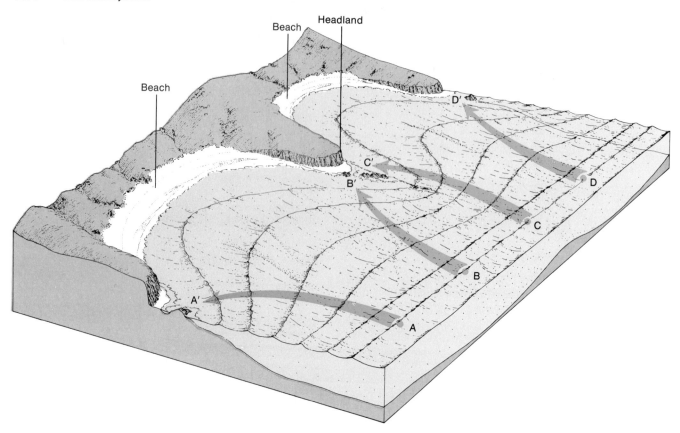

Figure 15.5 *Wave refraction concentrates energy on the headlands. Each segment of the wave, A-B, B-C. and C-D, has the same amount of energy. The segment B-C encounters the seafloor sooner than A-B or C-D and moves more slowly. This causes the wave to bend so that the energy contained in segment B-C is concentrated on the headland, while that contained in A-B and C-D is dispersed along the beach.*

velocity. Therefore, the wave actually changes its direction and the crest line is curved. As a result, wave refraction concentrates energy on the headlands and disperses it in the bays.

Discussion

The effect that wave refraction has on concentrating and dispersing energy can be appreciated by carefully analyzing the energy in a single wave. In *figure 15.5,* the unrefracted wave is divided into several equal parts, each of which has an equal amount of energy. As the wave moves toward the shore, the segment of the wave in front of the headland (B-C) meets the shallow floor first and is slowed down. Meanwhile, the rest of the wave (segments A-B and C-D) moves forward at normal velocity. This causes the crest line of the wave to bend as it advances shoreward. The energy of the wave between points B and C is concentrated throughout a relatively short distance (B'-C') on the headland, whereas an equal amount of energy between A-B and C-D is distributed throughout a much greater distance (A'-B' and C'-D'). As a result, breaking waves are powerful erosional agents on the headlands but are relatively weak in bays, where they commonly deposit sediment to form beaches. Patterns of wave refraction where major wave fronts are

refracted around islands and headlands can be seen easily from the air (*figure 15.6*).

Longshore Drift

Statement

One of the most important processes operating along the coast is the **longshore drift** generated as waves advance obliquely to the shore. The basic elements of this process are shown in *figure 15.7.* As a wave strikes the shore at an angle, water and sediment moved by the breaker are transported obliquely up the beach in the direction the wave is advancing. When the energy of the wave is spent, the water and sediment return with the backwash directly down the beach, perpendicular to the shore. The next wave moves the material obliquely up the shore again, and the backwash returns it directly down the beach slope. If one follows the path of a single grain of sand, one sees that it is moved in an endless series of small steps, with a resulting net transport parallel to the shore. This process is known as **beach drift.** As the waves approach the shore obliquely, they also set up a current in the breaker zone known as a **longshore current.**

Thus, there are two zones in which longshore movement occurs: one is along the upper limits of wave action and is related to the surge and backwash of the waves described above; the other is in the surf and breaker zone in which material is transported in suspension and by saltation. Both processes work together, and their combined action is referred to as longshore drift.

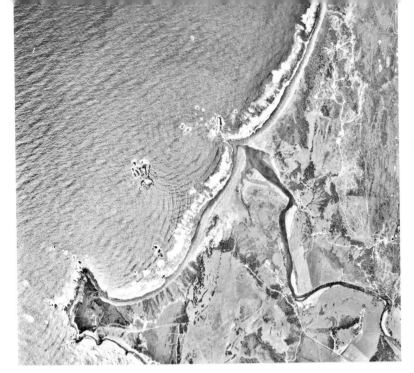

Figure 15.6 *Aerial photograph showing wave refraction along a coastline.*

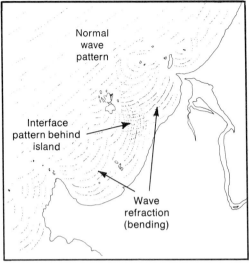

Discussion

Longshore drift results in the movement of an enormous volume of sediment, and, in a sense, the beach can be thought of as a river of sand. When wave direction is constant, longshore drift is in one direction. When the waves approach the shore at different angles during different seasons, longshore drift can be reversed. Longshore currents build up with increasing distance downshore and ultimately return seaward through the breaker zone as a narrow **rip current,** which is strong enough to be dangerous to swimmers.

One of the best ways to appreciate the process of longshore drift is to consider examples in which the process has influenced the affairs of men. A good example is in the area of Santa Barbara along the southern coast of California, where data has been collected over a considerable period of time.

Santa Barbara is a picturesque coastal town at the base of the Santa Ynez Mountains. It is an important educational, agricultural, and recreational area, and the local people wanted a harbor capable of accommodating deep-water vessels (*figure 15.8*). Studies by the United States Army Corps of Engineers indicated that the project was unfavorable because of the strong longshore currents that transported large volumes of sand to the south. But, in 1925, $750,000 was raised locally to construct a breakwater 460 m long. In this area, sand supplied by rivers draining the mountains is transported southward by longshore drift. Sediment on the beach may appear to be stable, but, in the continual motion of longshore drift, sand is moved southward from beach to beach. Ultimately, the sand transported by the longshore drift is delivered to the head of a submarine canyon and then moves down the canyon to the deep-sea floor. It has been estimated that new sand reaches this area at the rate of 592 m³ per day, or about 214,000 m³ per year. The currents are so strong that boulders 0.6 m in diameter can be transported.

In spite of reports advising against the project, an initial breakwater was built and a deep-water harbor was constructed (*figure 15.8B*). The initial breakwater was not tied to the shore, but sand moved by longshore drift began to pour through the gap and fill the harbor, which was protected from wave refraction and longshore drift by the breakwater (diagram *B*). This necessitated connecting the breakwater to the shore. Sand then accumulated behind the breakwater and built a smooth, curving beach (diagram *C*) as longshore drift moved excess sand around the end of the breakwater and deposited it inside the harbor. This produced two

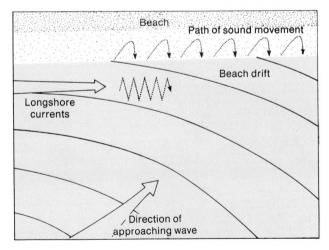

Figure 15.7 *The origin of longshore drift. When a wave moves toward the beach at an oblique angle, sediment is lifted by the surf and moved diagonally up the beach slope. The backwash then carries the particles back down the beach at right angle to the shoreline. This action transports the sediment along the coast in a zigzag pattern. Particles also are moved underwater in the breaker surf zone by this action.*

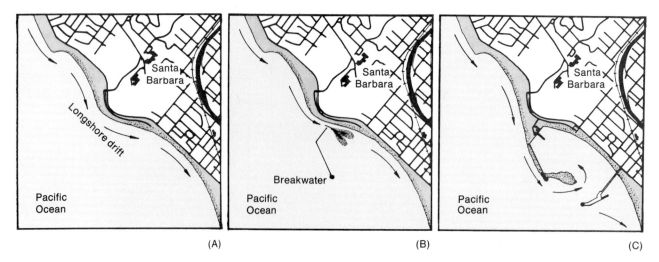

(A)   (B)   (C)

Figure 15.8 *The effect of a breakwater on longshore drift at Santa Barbara, California.*
(A) The Santa Barbara coast before the breakwater was built.
(B) After the initial breakwater was constructed, the harbor became filled with sand because longshore currents could not be generated in the protected area behind the breakwater.
(C) After the breakwater was connected to the shore, longshore currents moved sand around the breakwater and filled the mouth of the harbor. Sand is now dredged from the harbor and pumped down the coast.

disastrous effects: (1) the harbor became so choked with sand that it could accommodate only very shallow draft vessels, and (2) the beaches downcoast between Santa Barbara and Monterey were deprived of their source of sand and began to erode. Within 12 years, the property damage exceeded $2 million as the beach in some areas was cut back 75 m. To solve the problem, a dredge was installed in the Santa Barbara harbor to pump out the sand and return it into the system of longshore drift on the downcurrent side of the harbor. Most of the beaches have been partly replenished, but the cost of dredging exceeds $30,000 per year.

Erosion Along Coasts

Statement

Coast is a general term used to designate the broad region of the land next to the ocean. It is an area sculptured into many shapes and forms such as steep rocky cliffs, low beaches, quiet bays, tidal flats, and marshes. The topography of the coast results from the same forces that shape other land surfaces, that is, erosion, deposition, tectonic uplift, and subsidence.

Wave action is the major agent of erosion, and its power is especially awesome during storms. When a wave breaks against a **sea cliff,** the sheer impact of water alone is capable of exerting pressure of over 250 kg/m³. Thus, water is driven into every crack and crevice and tightly compresses the air within the fractures of the rocks. The compressed air then acts as a wedge that widens the cracks and loosens the blocks.

Solution activity also takes place along the coast and is especially effective in eroding limestone. Even noncalcareous rocks can be weathered rapidly by solution activity, because the chemical action of seawater is greater than that of fresh water.

The most effective process of wave erosion is probably the abrasive action of sand and gravel moved about by the waves. These tools of erosion operate in a manner similar to that of the bed load of a river; however, instead of cutting a vertical channel, the sand and gravel moved by waves cut horizontally to form **wave-cut cliffs** and **wave-cut platforms.**

Discussion

In order to understand the nature of wave erosion and the principal features it forms, first consider what happens along a profile at right angles to a shore (*figure 15.9*). In coastal areas where steeply sloping land descends beneath the water, wave action acts like a horizontal saw and cuts a notch into the bedrock at sea level (*figure 15.10*). This undercutting produces an overhanging cliff that ultimately collapses. The fallen debris soon is broken up and removed by wave action, and the process is repeated on the fresh surface of the new face on the cliff. As the sea cliff retreats, a wave-cut platform is produced at its base, the upper part of which commonly is visible near shore at low tide. Sediment derived from erosion of the cliff and transported by longshore drift can be deposited in deeper water to form a **wave-built terrace.** Stream valleys that formerly reached the coast at sea level are shortened and left as hanging valleys as the cliffs recede.

As the platform is enlarged, the waves break progressively farther from shore and lose much of their energy by friction as they travel across the shallow platform. Consequently, wave attack upon the cliff is reduced greatly. Beaches then can develop, and the cliff face is worn down by weathering and mass movement. The wave-cut platforms effectively dissipate wave energy, thus limiting the size to which they can grow. Some volcanic islands, however, appear to have been truncated completely by wave action and slope retreat, leaving a flat-topped platform near low tide.

Figure 15.9 *Profile of a wave-cut platform. Wave action acts like a horizontal saw cutting at the base of the cliff. The cliff is undermined and collapses. The debris soon is removed by wave action and undercutting continues. As erosion continues, the cliff recedes, leaving a gently sloping wave-cut terrace. Some sediment eroded from the shore may be deposited in deeper water to form a complementary wave-built terrace.*

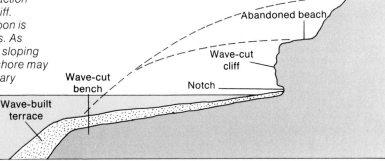

Figure 15.10 *Wave erosion along a coast of Mexico.*

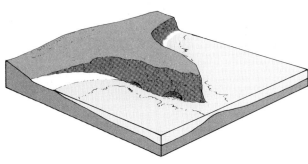

(A) Wave energy is concentrated on the headlands as a result of wave refraction. Zones of weakness such as joints, faults, and nonresistant beds erode faster and develop sea caves.

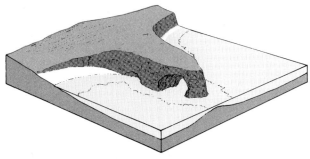

(B) Sea caves enlarge to form an arch.

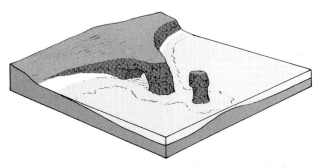

(C) Eventually, the arch collapses to form a sea stack. A new arch may develop from the remaining headland.

Figure 15.11 *Evolution of sea caves, arches, and stacks.*

Sea Caves, Arches, and Stacks. The rate at which erosion of a sea cliff proceeds depends on the durability of the rock and the degree of exposure of the coast to wave attack. Zones of weakness (such as outcrops with joint systems, fault planes, and beds of shale between harder sandstones) provide loci for accelerated erosion. If a joint extends across a headland, wave action may hollow out an alcove that later enlarges to a **sea cave.** Since the headland commonly is subjected to erosion from two sides, caves excavated along a zone of weakness may join to form a **sea arch** (*figures 15.11* and *15.12*). Eventually, the arch collapses, leaving an isolated pin-

nacle called a **sea stack** in front of the cliff. It should be emphasized that, in the erosion of a shoreline, marine and subaerial (on the exposed land surface) agents operate together to produce erosion above wave level. Seepage of ground water, frost action, wind, and mass movement all combine with the undercutting action of waves to erode back the coast.

Shoreline of Equilibrium. Erosion along shores tends to develop a profile of equilibrium in much the same way that a river system approaches a state of equilibrium with its discharge, gradient, and sediment load. In a shoreline at equilibrium, energy available in the waves is sufficient to transport available sediment but not great enough to erode back the shore. When more sediment is available than can be transported, the shoreline is built out into the ocean. When wave action is strong enough to erode back the sea cliffs, it does so until the wave-cut platform becomes large enough to limit further growth. Naturally, the rate at which a sea cliff retreats to leave a wave-cut platform depends upon the physical characteristics of the bedrock and the strength of the waves. In poorly consolidated material, the average rate of cliff retreat can be 1.5 to 2 m per year. Since Roman times, parts of the British coast have been worn back more than 5 km and many villages and ancient landmarks have been swept away. New volcanic islands such as Surtsey (near Iceland), composed almost entirely of volcanic ash, soon are planed off completely by wave action.

The development of a coast by wave erosion is shown in the series of diagrams in *figure 15.13.* In the initial stage (block *A*), sea level rises over a landscape eroded by stream action, forming an irregular shoreline. Wave action develops a small notch, and abrasion of the plat-

Figure 15.12 Sea stacks and an arch along the coast of Iceland.

form begins. Continued wave erosion enlarges the platform and develops a high wave-cut cliff (block *B*). Sea caves, sea arches, and sea stacks form by differential erosion in weak places in the bedrock. In the advanced stages of development (block *C*), the platform is enlarged to such a degree that it absorbs most of the wave energy. Weathering, mass movement, and stream erosion subdue the cliff, and a beach is developed as a result of the low level of energy available along the coast.

Deposition Along Coasts

Statement

Sediment transported by waves and longshore currents is deposited in areas of low energy to form a variety of landforms, including **beaches, spits, tombolos,** and **barrier islands.** The beach and associated features are parts of a dynamic shoreline system in which sediment (mostly sand and gravel) is in constant motion and the configuration of the coast is being modified continually by both erosion and deposition. Changes occur until the configuration of the coast arrives at equilibrium with the available wave energy. This is usually a smooth and straight or a gently curving coastline.

Discussion

Beaches. A beach can be defined as a shore built of unconsolidated sediment. Sand is the most common material, but some beaches are composed of cobbles and boulders and others of fine silt and clay. The physical

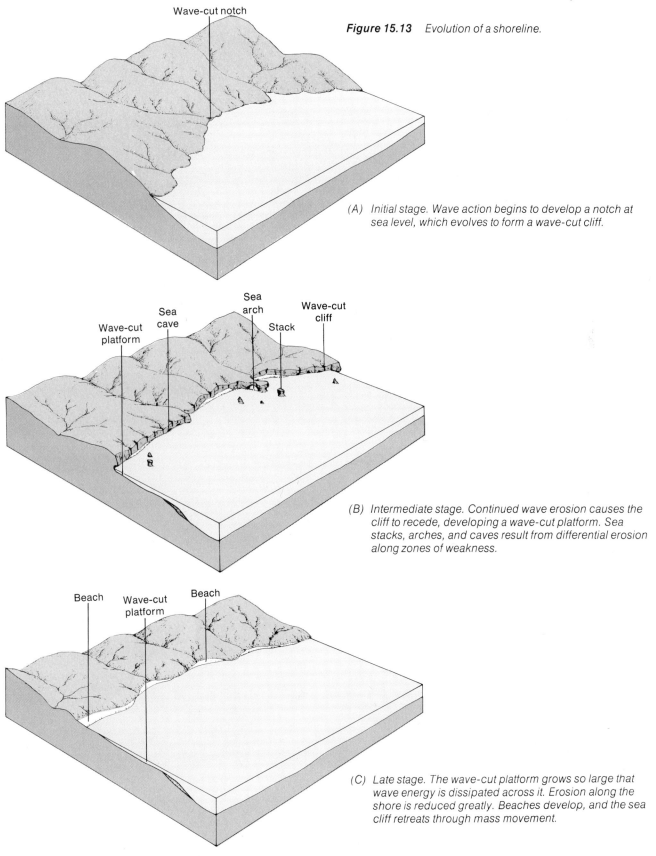

Figure 15.13 *Evolution of a shoreline.*

(A) *Initial stage. Wave action begins to develop a notch at sea level, which evolves to form a wave-cut cliff.*

(B) *Intermediate stage. Continued wave erosion causes the cliff to recede, developing a wave-cut platform. Sea stacks, arches, and caves result from differential erosion along zones of weakness.*

(C) *Late stage. The wave-cut platform grows so large that wave energy is dissipated across it. Erosion along the shore is reduced greatly. Beaches develop, and the sea cliff retreats through mass movement.*

characteristics of a beach (such as slope, composition, and shape) depend largely upon wave energy, but the supply of sediment and the size of particles available are also important. As a rule, beaches composed of fine-grained material are flatter than those composed of coarse sand and gravel.

The primary source of sediment for beaches and associated depositional features is the rivers that drain the continents (*figure 15.14*). The sediment input from the rivers is transported along the shore by longshore drift and is deposited in areas of low energy. Erosion of headlands and sea cliffs is also a source of sediment, and, in the tropical areas, the greatest source of sand commonly is shell debris derived from wave erosion of nearshore coral reefs.

The beach is far from being a static, permanent feature, and sediment moving along the beach can leave the system entirely. Wind commonly transports large volumes of beach sediment inland to form coastal dunes, and oceanic currents transport it offshore into deeper water. As was mentioned earlier, some of the sediment transported along the coast can be deposited permanently in areas of low wave energy and thus can modify configurations of the shoreline.

Spits. In areas where a straight shoreline is indented by bays or estuaries, longshore drift can extend the beach from the mainland to form a spit. A spit can continue to grow far out into the bay as material is deposited at its end and it migrates; it tends to become curved as a result of wave refraction (*figure 15.15*). With continued growth, the spit can extend completely across the front of the bay to form a baymouth bar.

Figure 15.14 *The shore as a system of moving sediment. Most of the sediment involved in a shoreline system is supplied from rivers bringing erosional debris from the continent and by erosion of the sea cliff by wave action. This material is transported by longshore drift and may be deposited in growing beaches, spits, and bars. Some of it, however, leaves the system as it is transported to deeper water. Some material also may leave the system by landward migration of coastal sand dunes.*

Figure 15.15 *A curved spit developed by longshore drift moving sand beyond the mainland.*

Tombolos. Spits also can grow outward and connect the shore with an offshore island to form a tombolo. This feature commonly is produced by the effect the island has on wave refraction and longshore drift (*figure 15.16*). An island near shore can cause wave refraction around it so that little or no wave energy strikes the shore behind it. Longshore drift is not generated in this wave shadow zone, so deposition of sediment carried along the coast occurs behind the island. Ultimately, the sediment carried by longshore drift can build up a deposit that be-

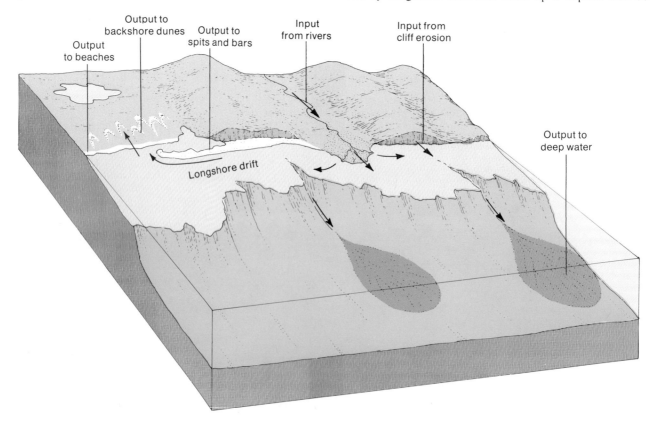

Output to beaches

Output to backshore dunes

Output to spits and bars

Input from rivers

Input from cliff erosion

Output to deep water

Longshore drift

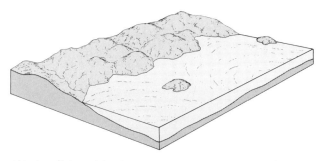

(A) An offshore island acts as a breakwater to incoming waves and creates a wave shadow along the coast behind it where longshore drift cannot occur.

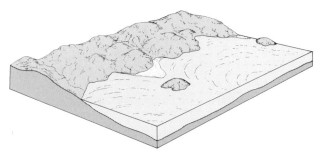

(B) Sediment moved by longshore drift is trapped in the shadow zone.

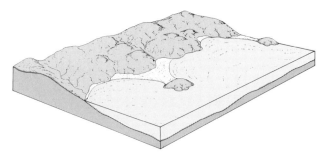

(C) The zone of deposit ultimately grows until it connects with the island. Longshore drift then is able to move sediment along the shore and around the tombolo.

Figure 15.16 *Development of a tombolo.*

Figure 15.17 *A barrier island along the Atlantic coast (see also plates 19 and 20).*

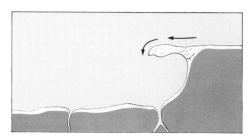

(A) Sediment moves along the shore and is deposited as a spit in the deeper water near a bay.

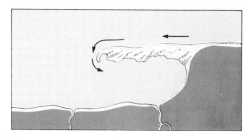

(B) The spit grows parallel to the shore by longshore drift.

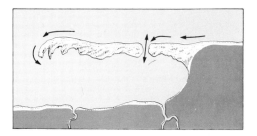

(C) Tidal inlets cut the spit, which is then long enough to be considered a barrier island.

Figure 15.18 *The origin of a barrier island from longshore migration of a spit.*

comes connected to the island to form the tombolo. Longshore drift then moves uninterrupted along the shore and around the tombolo.

Barrier Islands. Barrier islands are long, low islands of sediment trending parallel to the shore (*figure 15.17*). Almost invariably, they are present along shorelines adjacent to gently sloping coastal plains and typically are separated from the mainland by a lagoon. Most barrier islands are cut by one or more tidal inlets.

The origin of barrier islands has been a controversial subject for many years. One theory contends that they result from the growth of spits across irregularities in the shoreline, such as that illustrated in *figure 15.18*. Certainly, some barrier islands develop in this manner. Others may form by the shoreward migration of offshore bars. Another hypothesis contends that barrier islands are the result of the worldwide rise of sea level that

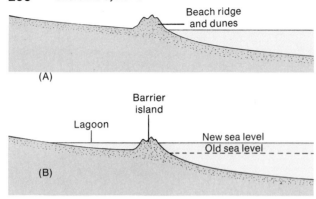

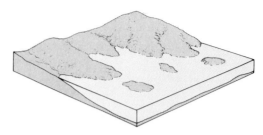

(A) A rise in sea level floods a topography eroded by a river system and forms bays, headlands, and islands.

Figure 15.19 *Development of a barrier island by sea level rising over former beach ridges.*
(A) *On a gently inclined coast, wave action builds a beach ridge at sea level. The ridge may grow over 32 m high as sand is piled up by wind action.*
(B) *A rise in sea level during the melting of glaciers drowns the former beach ridge and produces a barrier island that then is modified by wave action.*

(B) Wave erosion cuts cliffs on the islands and peninsulas.

drowned early-formed beaches and isolated them off-shore. The process is illustrated in *figure 15.19*. During a lower stand of sea level, such as was produced during the ice age, a beach would form along a low coastal plain. Sand from the beach would be piled up into a dune ridge parallel to the shore. As sea level rose (with the melting of the ice), the dune ridge would be drowned and become a barrier island, later to become reworked, enlarged, and modified by wave action and longshore drift.

(C) The wave-cut cliffs recede and grow higher, and headlands are eroded back to a sea cliff. Sediment begins to accumulate as beaches and spits.

Evolution of Shorelines

Statement

Processes of erosion and deposition along the shore combine to develop a long and straight or gently curving coastline. Headlands are eroded back and bays and estuaries are filled in with sediment so that the configuration of the shoreline is such that energy is distributed equally along the coast and neither large-scale erosion nor deposition occurs. A shoreline with such a balance of forces commonly is called a shoreline of equilibrium and, in a sense, is similar to a stream profile of equilibrium in that there is a delicate balance between the landform and the geologic processes operating upon it. On a shoreline of equilibrium, the energy in the waves and longshore drift is just sufficient to transport the sediment supplied.

Very few present-day shorelines, however, approach this state of equilibrium, because they are all very young, having been formed during the last few thousand years as a result of the worldwide rise in sea level associated with melting of the Pleistocene glaciers. Most coastlines presently are being modified by erosion and deposition and are changing or evolving in the direction of becoming shorelines of equilibrium.

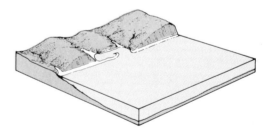

(D) The islands eventually are eroded away completely, beaches and spits enlarge, and lagoons form in the bays.

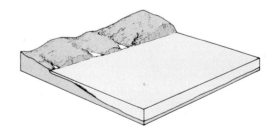

(E) A straight shoreline is produced by additional retreat of the cliffs and by sedimentation in bays and lagoons. The large wave-cut platform then limits further erosion by wave action.

Discussion

A simple conceptual model of the evolution of a shoreline to a state of equilibrium can be constructed to

Figure 15.20 *Evolution of an embayed coastline.*

show the types of changes that would be expected to occur as the processes of erosion and deposition operate. One such model is shown in *figure 15.20.* Diagram *A* shows an area originally shaped by stream erosion and subsequently partly flooded by rising sea level. River valleys are drowned to form irregular, branching bays and some hilltops form peninsulas and islands. In diagram *B,* marine erosion begins to attack the shore, and the islands and headlands are eroded to produce sea cliffs. As erosion proceeds (diagram *C*), the islands and headlands are worn back, and the cliffs increase in height. A wave-cut platform develops, and sediment begins to accumulate as spits and beaches. The wave-cut platform enlarges, reducing wave energy so that a beach forms at the base of the cliff. In a more advanced stage of development (diagram *D*), the islands are eroded away completely and bays become sealed off, partly by growth of spits, forming lagoons. The shoreline then becomes straight and simple. In the final stages of marine development (diagram *E*), the shoreline is cut back beyond the limits of the bay. Sediment moves along the coast by longshore drift, but the wave-cut platform is so wide that it effectively eliminates further erosion of the cliff by wave action. The shoreline is straight and approaches a state of equilibrium with the energy acting upon it. Further modification of the cliffs results from subaerial weathering and erosion.

Naturally, the development of the idealized shoreline would be modified by special conditions of structure and topography and would be adjusted with fluctuations of sea level, but the general process of erosion of the headlands by wave action and the straightening of the shoreline by both erosion and deposition basically would follow the generalized sequence. An area rarely proceeds through all of these stages, however, because fluctuations in sea level upset the previously established balance.

The development of a shoreline is interrupted in many areas by tectonic uplift, and sea cliffs and wave-cut platforms are abruptly elevated above the level of wave attack. When this occurs, the processes of wave erosion begin at a new level and the elevated marine terraces are stranded high above sea level and are eroded and obliterated eventually by weathering and stream erosion (*figure 15.21*).

Reefs

Statement

Reefs are a unique type of coastal feature, because they are built by organisms. Modern reefs are built by a complex community of corals, algae, sponges, and other marine invertebrates, and most grow and thrive in the warm, shallow waters of the semitropical and tropical regions. Only the upper part of a reef is organically active, because sunlight is required for vigorous growth. The living animals and plants of the reef community build new structures on the old, once-living shells and the accumulation of shell debris. Wave action continually breaks up the reef so that large bodies of calcium carbonate shell debris, originally secreted by the organisms of the reef community, accumulate along the reef flanks and add mass to the structure. Reefs can grow upward with rising sea level, provided that the rate of rise is not excessive,

Figure 15.21   *A series of elevated beach terraces that result from tectonic uplift along the southern coast of California and the offshore islands.*

and they also can grow seaward over the flanks of reef debris.

Discussion

Reef Ecology. The marine life that forms a reef can flourish only under strict conditions of temperature, salinity, and water depth. Most modern coral reefs occur in warm tropical waters between the limits of 25° South latitude and 30° North. Colonial corals need sunlight and cannot live in water deeper than about 76 m. They grow most luxuriantly just a few meters below sea level. Dirty water inhibits rapid, healthy growth of corals, because it cuts off sunlight. Therefore, corals are absent or are stunted near the mouths of large rivers that bring in muddy water. Salinity levels must range from 27 to 40 parts per thousand. As a result, a reef can be killed by a flood of fresh water from the land that appreciably alters the degree of salinity. Coral reefs are remarkably flat on top, the upper surface being positioned at the level of the upper third of the tidal range. They usually are exposed at low tide but must be covered at high tide. In summary, corals thrive only in clear, warm, shallow oceans where wave action brings sufficient oxygen and food.

The fact that reefs form in such restricted environments makes them especially important as indicators of past climate and geographic and tectonic conditions.

Reef Types. The most common types of reefs in present oceans are fringing reefs, barrier reefs, and atolls (*figure 15.22*).

Fringing reefs, generally ranging from 0.5 to 1 km wide, are attached to such landmasses as the shores of volcanic islands. The corals grow seaward toward their food supply and are usually absent near the deltas and the mouths of rivers, where the waters are muddy. Heavy sedimentation and high runoff make some tropical coasts of continents less attractive than the small oceanic islands for fringing reefs.

Barrier reefs are separated from the mainland by a lagoon that can be more than 20 km wide. As seen from the air, the barrier reefs of the South Pacific islands are marked by a zone of white breakers (see *plate 9*). At intervals, narrow gaps occur through which excess shore and tidal water can exit. The finest example of this type of reef is the Great Barrier Reef of Australia, which stretches 800 km along the northern shore of Australia at a distance of 30 to 160 km off the Queensland coast.

Atolls are roughly circular reefs that rise from deep water and enclose a shallow lagoon in which there is no exposed central landmass. The outer margin of an atoll is naturally the site of most vigorous coral growth and commonly forms an overhanging rim from which pieces of coral rock break off and accumulate as talus on the slopes below. A cross-sectional view of a typical atoll shows that the lagoon floor is shallow and is composed of calcareous sand and silt with rubble derived from erosion of the outer side. (See front of *figure 15.22C.*)

Atolls are by far the most common type of coral reef. Over 330 are known, of which all but 10 lie within the Indo-Pacific tropical area. Drilling indicates that coral reef material in some atolls extends down as much as 1400 m below sea level, where it rests on a basalt plat-

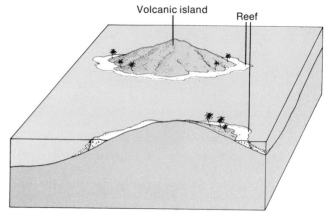

(A) The reef begins to grow along the coast of a newly formed volcanic island.

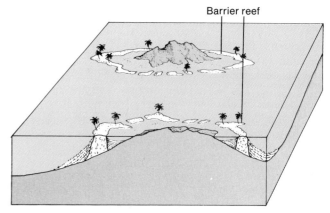

(B) As the island subsides, the reef grows upward and develops a barrier that separates the lagoon from open water.

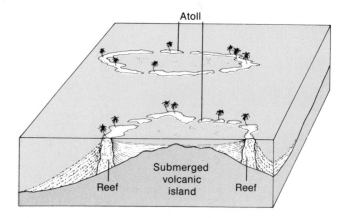

(C) Further subsidence completely submerges the island, but the reef continues to grow upward to form an atoll if subsidence is not too rapid.

Figure 15.22 *Evolution of an atoll from a fringing reef.*

form carved on an ancient volcanic island. Upward growth of the reef to form such an atoll requires a gradual and continuous relative subsidence. Reefs this thick presumably represent accumulations over a period of 40 to 50 million years.

Platform reefs are reefs that grow in isolated oval

patches in warm, shallow water on the continental shelf and were apparently more abundant during various past geologic periods. Modern platform reefs appear to be somewhat randomly distributed, although some appear to be oriented in belts—a feature suggesting control from submarine topographic highs, such as drowned shorelines.

Origin of Atolls. In 1842, Charles Darwin first proposed a theory to explain the origin of atolls. As indicated in *figure 15.22*, the theory is based on continued relative subsidence of a volcanic island. The theory suggests that coral reefs are established originally as fringing reefs along the shores of new volcanic islands. As the island gradually subsides, the coral reef grows upward along the outer margins; the rate of upward growth essentially keeps pace with subsidence. With continued subsidence, the area of the island becomes smaller, and the reef becomes a barrier reef. Ultimately, the island is submerged completely and the upward growth of the reef forms an atoll. Erosional debris from the reef fills the enclosed area of the atoll to form a shallow lagoon. The depth of the reef material in atolls tends to confirm Darwin's theory of the evolution of atolls.

Types of Coasts

Statement

In previous sections of this chapter, it has been shown that erosion and deposition by waves and currents operate to varying degrees along all coasts; therefore, it might seem logical to classify a coast according to the stage to which it has evolved. However, nearly all coastlines have been affected profoundly by the worldwide rise of sea level associated with the melting of the Pleistocene glaciers. As a result, the *configuration of many coasts is not the result of marine processes at all* but is the result of geologic processes that operated in the area before the rise of sea level. Therefore, a meaningful classification of coasts can be made on the basis of the geologic processes responsible for the coastal topography. Two principle subdivisions are recognized.
1. **Primary coasts,** which have been shaped mainly by terrestrial processes.
2. **Secondary coasts,** which have been formed by marine processes.
Further subdivisions of each group can be made according to the specific geologic agent that has had the greatest influence. One of the important values of this classification is that it can be used easily; and, with the aid of aerial photographs or maps, most of the proper classification of most coasts can be recognized, even by the beginning student.

Discussion

Primary Coasts. The configuration of primary coasts is largely the result of subaerial geologic agents, such as streams, glaciers, volcanism, and earth movements. These agents produce a highly irregular coastline that is characterized by bays, estuaries, fiords, headlands, peninsulas, and offshore islands. The landforms can be either erosional or depositional, but very little

(A) Stream erosion.

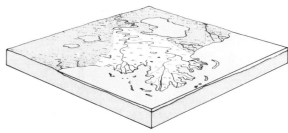

(B) Stream deposition (see plate 16).

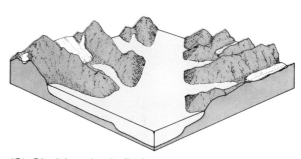

(C) Glacial erosion (valley).

Figure 15.23 *Primary coasts—shape of coasts produced by nonmarine processes.*

modification has resulted from marine processes. Some of the more common types are illustrated in *figure 15.23*.
1. *Stream Erosion.* When an area is eroded by running water and is flooded subsequently by the rise of sea level, the landscape becomes partly drowned. Stream valleys become bays or estuaries, and former hills become islands. The bays following the tributary valley system inland result in a coastline having a dendritic pattern. Chesapeake Bay is a well-known example.
2. *Stream Deposition—Deltaic Coasts.* At the mouths of major rivers, fluvial deposition results in the buildup of deltas out into the ocean and dominates the configuration of the coast. Deltas can assume a variety of shapes and are modified locally by marine erosion and deposition (see *plate 16*).
3. *Glacial Erosion.* Drowned glacial valleys usually are called **fiords** and form some of the most rugged and scenic shorelines in the world. Fiords are characterized by long, troughlike bays that cut into mountainous coasts and can extend inland as much as a hundred kilometers. In polar areas, glaciers still remain at the heads of many fiords. Walls of fiords are steep

and straight. Hanging valleys with spectacular waterfalls are common.

4. *Glacial Deposition.* Glacial deposition dominates coasts in the northern latitudes, where continental glaciation once extended out beyond the present shoreline onto the continental shelf. The ice sheet left drumlins and moraines to be drowned by the subsequent rise in sea level. One example, Long Island, is partly submerged moraine, and, in Boston Harbor, partly submerged drumlins form elliptical islands.

As erosion and deposition continue, marine processes ultimately control the configuration of the coast and primary coasts become secondary coasts.

Secondary Coasts. Secondary coasts are shaped by marine erosion and deposition and are characterized by wave-cut cliffs, beaches, barrier bars, spits, and (in some cases) sediment deposited through the action of biologic agents such as marsh grass, mangroves, and coral reefs. Marine erosion and deposition smooth out and straighten the shoreline and establish a balance between the energy of the waves and the configuration of the shore (see *plate 20*). The most common types are illustrated in *figure 15.24*.

(A) Marine erosion.

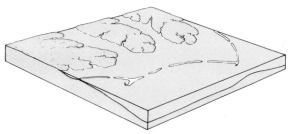

(B) Marine deposition (see plates 19 and 20).

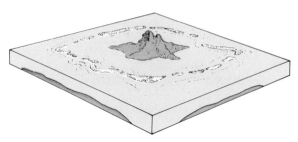

(C) Organic coast (reef) (see plate 9).

Figure 15.24 *Secondary coasts—shape of coasts produced by marine processes.*

1. *Wave Erosion Coasts.* Wave erosion begins to modify primary coasts as soon as the landscape produced by other agents is submerged. The energy of waves is concentrated on the headlands and develops a wave-cut platform slightly below sea level. Ultimately, a straight cliff is formed with hanging stream valleys and a large wave-cut platform. The cliffs of Dover, England, are an excellent example.

2. *Marine Deposition Coasts.* In areas where abundant sediment is supplied by streams or marine currents, various marine deposits determine the characteristics of the coast. Barrier islands and beaches are the dominant features (see *plate 20*). The shoreline is modified as waves break across the barriers and drive them inland. The barrier also increases in both length and width as sand is added. The lagoons behind the barrier receive sediment and fresh water from streams and, thus, often are capable of supporting dense marsh vegetation. Gradually, the lagoon fills through deposition of stream sediment or from the invasion of sand from the barrier bar through tidal deltas, as well as from plant debris swamps. The barrier coasts of the south Atlantic and Gulf states are excellent examples.

3. *Coasts Built by Organisms.* Coral reefs develop another type of coast that is very prominent in the islands of the southwestern Pacific (see *plate 9*). As was shown earlier, the reefs are built up to the surface by corals and algae and ultimately can evolve into an atoll. Another type of organic coast prevalent in the tropics results from the growth of mangrove trees out into the water, particularly in shallow bays. Mangroves are plants whose roots must extend into water of high salinity. The root systems form an interlocking mesh, an impressive barrier that acts as a breakwater. Sediment is trapped in the spaces between the roots, and the mangrove barrier grows seaward.

Tides

Statement

The gravitational attraction of the moon and the sun pull the water away from the earth to form a tidal bulge. Many variables affect the height of tides, including the relative position of the moon and the sun, the latitude, the Coriolis effect, and the configuration of the coast. The gravitational attraction of the moon also produces tides in the atmosphere and in the solid body of the earth, but the main concern here is the effect of ocean tides upon the coasts. Tides affect beach processes in two ways.
1. By initiating a rise and fall of the water level.
2. By generating currents.

Discussion

Since the first century A.D., it has been known that the moon controls the tides in some manner, but it was not until Sir Isaac Newton stated the law of gravitation in 1687 that the true explanation of tide became known. The law of gravitation states that two bodies attract each other with a force directly proportional to the product of their masses and inversely proportional to the square of the distance between them.

The forces involved in tides include (1) the gravitational attraction of the moon on the earth and (2) the centrifugal force on the earth resulting from the earth-moon system revolving about a common center of gravity. The centrifugal force caused by the earth's revolving about the earth-moon system is constant and in opposition to the gravitational force exerted by the moon. The water in the oceans is free to move and so is deformed by these forces. As illustrated in *figure 15.25,* on the side of the earth facing the moon, water is pulled toward the moon, causing a high tide. As the moon revolves around the earth, the tidal bulge in the ocean follows the moon. The net effect of the bulge as it moves around the earth is to raise the sea level. Thus, a high tide results as the bulge passes at a given point, and thus there is always somewhere on the earth where there is high tide. There is also a tidal bulge on the side of the earth opposite the moon that is caused by centrifugal force. The balance between gravitational attraction and centrifugal force is exact at the center of the two bodies; but, at the earth's surface, the two forces are not equal. On the moon side of the earth, the gravitational force is greater; while, on the opposite side of the earth, the centrifugal force is greater and causes a bulge in the ocean surface. Thus, there are two high tides produced as the moon orbits the earth.

The movement of tides produces a significant effect on many coasts. Although the difference between high and low tide in the open ocean is small, exceptionally high tides are produced in shallow seas where the rising water is funneled into bays and estuaries. In the Bay of Fundy, between New Brunswick and Nova Scotia, the range between high and low tide is as much as 21 m. Off the coast of Brittany, tides of 12 m, with current velocities as high as 20 km per hour, are produced. In some rivers, high tide can reverse the flow of water as it rushes upstream in a breaking wave called a **tidal bore.** Thus, tidal currents are capable of transporting sediment and eroding the seafloor near shore.

Tsunamis

Statement

Waves produced by the wind are the most significant agents in the development of coastal landforms, but earthquakes, volcanic eruptions, and submarine landslides frequently produce an unusual wave called a seismic sea wave, or a **tsunami.** Tsunamis have long wave lengths and travel across the open ocean at high speeds. As they approach shore, the wave length decreases and the wave height increases so that they can be a formidable agent of destruction along the shoreline.

Discussion

A tsunami wave differs from a wind wave in that the energy is transferred to the water from the seafloor so that the entire depth of water is involved in the wave motion. The wave front travels out from its point of origin at very high speeds, ranging from 500 to 800 km per hour, and may traverse an entire ocean. The wave height is only 30 to 60 cm, and the wave length ranges from 55 to 200 km so that, in an open ocean, a tsunami may pass unnoticed. As the wave approaches the shore, however,

Figure 15.25 *The origin of tides is related to the gravitational attraction of the moon and the centrifugal force of the earth-moon system. On the side of the earth facing the moon, gravitational attraction is greater, forming a tidal bulge in the ocean's water. On the other side of the earth, centrifugal force is greater, causing another tidal bulge. The two forces balance at the center of the earth. Rotation of tidal bulges follows the motion of the moon around the earth.*

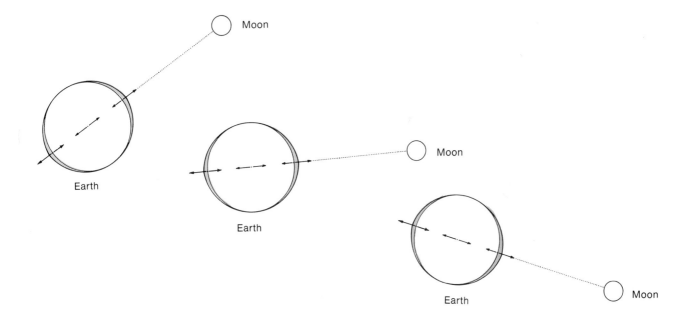

some very important changes take place. The energy distributed in the deep column of water becomes concentrated in an increasingly shorter column, which results in a rapid increase in wave height. The waves that were less than 60 cm high in the deep ocean build rapidly to heights of more than 15 m in many cases and well over 30 m in rare instances. Thus, they exert an enormous force against the shore and can inflict serious damage and great loss of life.

A number of tsunamis are well documented by records from seismic stations and coastal observations. For example, the tsunami that hit Hawaii on April 1, 1946, originated in the Aleutian trench off the island of Unimak. The waves moved across the open ocean, imperceptible to ships that lay in their path because the wave height was only 30 cm. Moving at an average speed of 760 km per hour, they reached the Hawaiian Islands, a distance of 3200 km away, in less than five hours. Since the wave length was 150 km, the wave crests arrived about 12 minutes apart. As the waves approached the island, their height increased to a minimum of 17 m, thus producing an extremely destructive surf that swept inland, completely demolishing houses, trees, and almost everything in its path.

Summary

The shoreline is a dynamic system that is modified continually by the forces of waves and currents. Wind-generated currents provide most of the energy for erosion, transportation, and deposition of the sediment, but tides and tsunamis can be important locally.

Waves approaching a shore are bent (or refracted) so that energy is concentrated on headlands and dispersed in bays.

Longshore drift is one of the most important shoreline processes. It is generated as waves strike the shore at an angle. Water and sediment move obliquely up the beach face but return with the backwash directly down the beach perpendicular to the shoreline. This results in a net transport parallel to the shore.

Erosion along coasts results from abrasive action of sand and gravel moved by the waves and currents, and to a lesser extent, from solution and hydraulic action. Erosion along coasts typically develops a sea cliff by the undercutting action of waves and currents. As the sea cliff recedes, a wave-cut platform develops. Minor erosional forms of sea cliffs include sea caves, arches, and stacks.

Sediment transported by waves and longshore drift is deposited in areas of low energy to form a variety of landforms, including beaches, spits, tombolos, and barrier islands.

Processes of erosion and deposition tend to develop a long and straight or a gently curving coastline so that neither large-scale erosion nor large-scale deposition occurs. A shoreline with such a balance of forces is called a shoreline of equilibrium.

Reefs form a unique type of coastal feature, because they are organic. Commonly, a reef evolves through a series of types from fringing reefs, to barrier reefs, to atolls.

The worldwide rise in sea level associated with melting of the Pleistocene glaciers has drowned many coasts, but some are emerging as a result of tectonic uplift. Therefore, coasts are classified on the basis of the process that has been most significant in developing their configurations.

The wind has been emphasized as the major force in developing waves that in turn modify the coasts. Wind is also an effective agent in transporting and depositing loose sand and dust on the continents and is responsible for the major landforms in the great desert regions of the world. In the next chapter, the geologic work of wind that is part of the moving surface fluids (gas and liquid) on the earth will be considered.

Additional Readings

Bascom, W. 1959. "Ocean Waves." Sci. Amer. 201(2): 74-84. (Reprint no. 828.) San Francisco: W. H. Freeman and Company.

Bascom, W. 1964. Waves and Beaches. New York: Anchor Books (Doubleday and Company).

Oceanography. 1971. A Scientific American Book. San Francisco: W. H. Freeman and Company.

Weyl, P. K. 1970. Oceanography: An Introduction to the Marine Environment. New York: John Wiley and Sons.

Eolian Systems

Where loose sediment is abundant and climatic conditions are favorable, wind transports sand and dust, producing depositional features that locally dominate the landscape. Foremost among these are the great dune fields in the large deserts, sand dunes along many of the world's coasts, and blankets of windblown dust called loess that cover millions of square kilometers in parts of the middle-latitude continents.

In spite of this, wind is probably the least effective agent of erosion, although many curious erosional landforms mistakenly are attributed to it. Even in the desert, most erosional landforms are the products of running water and the greatest effects produced by wind are expressed in shifting sand dunes. The most significant process involving the wind is the transportation and deposition of loose, fine-grained sediment eroded by other geologic agents.

This chapter is concerned primarily with how wind transports and deposits sediments to form distinctive landforms.

Major Concepts

1. Wind is not an effective agent of erosion but is capable of transporting loose, unconsolidated fragments of sand and dust.
2. Sand is transported by saltation and surface creep. Dust is transported in suspension and can remain high in the atmosphere for long periods of time.
3. Sand dunes migrate as sand grains are blown up and over the windward side of the dune and accumulate on the lee slope. The internal structure of a dune consists of strata inclined in a downwind direction.
4. Various types of dunes form in response to wind velocity, sand supply, constancy of wind direction, and characteristics of the surface.
5. Windblown dust (loess) forms blanket deposits that can mask the former landscape. It originates from deserts or fine rock debris deposited by glaciers.

Wind as a Geologic Agent

Statement

Wind is an effective local geologic agent because it is capable of lifting and transporting loose sand and dust, but its ability to erode solid rock is very limited. The main action of wind as a geologic agent is in transportation and deposition in arid regions.

Discussion

Formerly, it was thought that the wind, like running water and glaciers, had great erosional power—power to abrade and wear down the earth's surface—but it has become increasingly apparent that few major erosional topographical features are formed by wind erosion. Even in the desert, where water is not an obvious geologic agent, wind action is not the major erosional process; its major function is to transport loose sediment. Most erosional landforms in deserts were produced by weathering and running water in times of wetter climate and are, in a sense, "fossil" landscapes formed by processes no longer active. Sand can be transported by the wind with considerable velocity, and, locally, wind abrasion can aid in eroding and shaping some details of landforms. But wind erosion is hardly capable of producing erosional features of great areal extent. Even minor alcoves and niches (or wind caves), as well as certain topographic features called pedestal rocks (often thought to be caused by wind erosion), are produced by differential weathering, not by wind abrasion.

Although relatively insignificant as an erosional agent, the wind is effective in transporting loose, unconsolidated sand, silt, and dust and is responsible for the formation of great "seas of sand" in the Sahara, Arabian, and other deserts, as well as the blankets of windblown dust covering millions of square kilometers in China, the central United States, and parts of Europe.

Prevailing wind patterns are the result of the earth's rotation and the resulting **Coriolis effect** (deflection due to rotation), plus the effect of the configuration of continents and oceans, the location of mountain ranges, and variations in solar radiation with latitude (*figure 16.1*). The great deserts of the world, such as the Sahara and other deserts in northern Africa and the deserts of Asia, are located largely in belts 15° to 35° north and south of the equator (*figure 16.2*). Air rises at the equator and descends about 30° to 35° north and south of the equator. As it descends, it is heated by compression. The air is heated further as it moves along the ground toward the equator. The increased temperature of the air increases its ability to hold water. As a result, evaporation of the surface moisture, rather than precipitation, occurs. When the air reaches the equator, it again rises and cools and thus releases this moisture as rain.

Other deserts such as the Kalahari in South Africa and the Basin and Range in the western United States lie in the interior of continents or where high mountain ranges intercept the moisture-laden air, forcing it to rise and the moisture to condense and fall as rain before reaching the desert region.

Figure 16.1 *Atmospheric circulation and prevailing wind patterns. Heated air from the equatorial regions rises and moves toward the poles, where it is cooled, compressed, and forced to descend, forming the subtropical high-pressure belts. The air then moves toward the equator as trade winds. In the Northern Hemisphere, this air is deflected by the earth's rotation to flow southwest. In the Southern Hemisphere, it is deflected to flow northwest. The cold polar air tends to wedge itself toward the lower latitudes and form the polar fronts.*

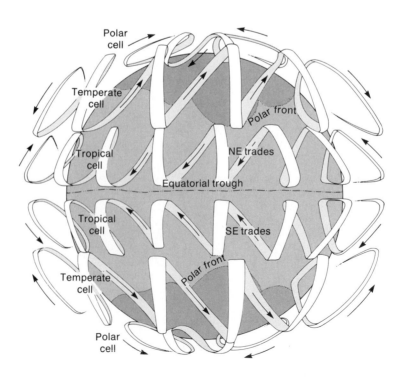

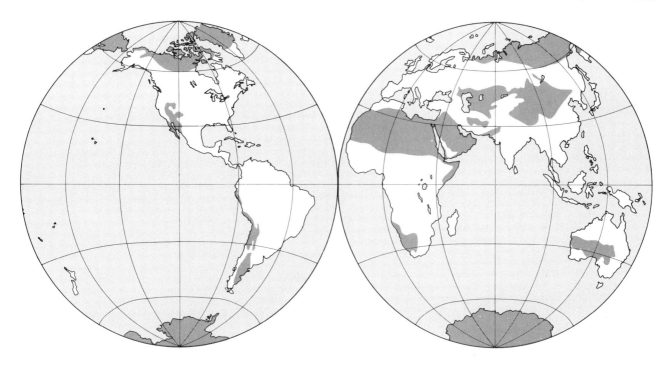

Figure 16.2 *Major desert areas of the world. The great deserts such as the Sahara, Arabian, Kalahari, and Australian all lie between 10° and 30° north and south of the equator. These are areas under almost constant high atmospheric pressure, which is characterized by subsiding air that warms as it descends and produces low humidity. Desert, or near-desert, areas cover nearly one-third of the land surface.*

Wind action is most significant in these desert areas but is not confined to them. Most coasts are modified by winds that pick up loose sand on the beach and transport it inland.

It is interesting to note the significance of wind action on the planet Mars, which was shown so dramatically by the satellite studies of 1971, 1972, and 1976. Without water and vegetation, wind action is probably the dominant process on the surface of Mars. In former geologic periods of the earth when climatic conditions were much different than now, the work of the wind may have been more significant in various regions of the earth.

Wind Erosion

Statement

Wind is an effective agent in transporting loose sand and dust and is responsible for the great dune fields of the deserts, but its ability to *erode* (abrade and wear away) solid bedrock is limited and only of local importance.

Wind erosion is manifested in two ways.
1. By **deflation,** the lifting and removing of loose sand and dust particles from the earth's surface.
2. By **abrasion,** the sandblast action of windblown sand.

Discussion

Deflation.   Deflation is responsible for many depres-

sions called **blowouts** in areas where weak, unconsolidated sediment is exposed at the surface. In the Great Plains area of the United States, tens of thousands of small deflation basins dot the landscape (*figure 16.3*). They may be shallow depressions several meters in

Figure 16.3 *Deflation basins in the Great Plains are produced where solution activity has dissolved the cement that binds the sand grains together in the horizontal rocks. The loose sand is removed by the wind to form a basin. Water trapped in the basin further dissolves cement, and the basin is enlarged.*

diameter or large ones more than 50 m deep and several kilometers across. The basins develop where calcium carbonate cement in the bedrock is dissolved by ground water, leaving loose material to be picked up and transported by the wind.

Generally, wind is capable of moving only sand- and dust-size particles so that deflation leaves concentrations of coarser material called **lag deposits** or **desert pavements** (*figure 16.4*). Locally, these are very striking desert features and stand out in contrast to dune fields and playa lake deposits.

Deflation occurs only where unconsolidated material is exposed at the surface; it does not occur where there are thick covers of vegetation or layers of gravel. Therefore, the process is limited to areas such as deserts, beaches, and barren fields.

Abrasion.   The effects of wind abrasion can be seen on the surface of bedrock in most desert regions, and, in some areas where soft, poorly consolidated rock is exposed, wind erosion can be both spectacular and distinctive. The process of wind abrasion is essentially the same as that used in artificial sandblasting to clean building stone. Some pebbles called **ventifacts** are shaped and polished by the wind (*figure 16.5*). Such pebbles are distinctive in that they have two or more flat faces that meet at sharp ridges. Generally, they are well polished and can have surface irregularities and grooving aligned with the wind direction.

Some of the most notable, but somewhat rare, features of wind erosion are elongate ridges and grooves parallel to the strongest prevailing wind directions. These features, called **yardangs,** commonly occur in groups where relatively unconsolidated rock is exposed at the surface.

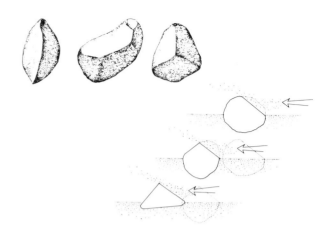

Figure 16.5   *Ventifacts are pebbles shaped and polished by wind action.*

Transportation of Sediment by Wind

Statement

Field observations and wind-tunnel experiments indicate that windblown sand moves by skipping or bounding into the air (a process called **saltation**) and by rolling or sliding along the surface (called surface **creep**). Fine silt and dust are carried in suspension over great distances and settle back to the ground only after the turbulent wind ceases.

Discussion

Movement of Sand.   Although both wind and water transport sand by saltation, the mechanics of motion involved are quite different because the viscosity of water is so much greater than that of air. In water, saltation involves a hydraulic lift; in air, saltation results in a series of elastic bounces (*figure 16.6*). The initial energy that lifts the grains into the air comes from impact with other grains. When the wind reaches a critical velocity, the sand grains begin to move by rolling or sliding. As a grain moves along the surface, it strikes the other grains. The impact can cause one or both of the grains to bounce into the air, where they are forced forward by the stronger wind above the ground surface. The force of gravity soon pulls the grain back to earth at an angle of impact generally ranging from 10° to 16°. If the surface over which the sand is moving is solid rock, the grain bounces back into the air. If the surface consists of loose sand, the impact of the falling grain can eject other grains into the air; these then are moved forward by the wind in a parabolic path. In this way, a sandy surface is set in motion by individual saltating grains.

Some sand grains, too large to be ejected into the air by impact from other grains, move by creep (rolling and sliding). Their momentum comes by impact from saltating grains, not directly from the wind. Approximately one-fifth to one-fourth of the sand moved by a sandstorm travels by rolling and sliding. Particles over 1 cm in diameter rarely are moved by the wind.

Movement of Dust.   Dust-size particles are light enough to be lifted by an eddy of turbulent air, but it is

Figure 16.4   *Wind action selectively transports the sand and fine sediment but leaves the coarser gravels to form a desert pavement.*

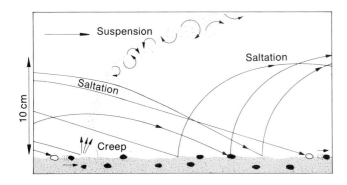

Figure 16.6 *Transportation of sediment by wind. Coarse grains move by impact from other grains and slide or roll (creep). Medium grains move by skipping or bouncing (saltation). Fine silt or dust moves in suspension.*

difficult for turbulent air alone to lift dust particles from the surface initially. This is because dust lies within a thin zero velocity layer close to the ground. Laboratory experiments show that particles less than 0.03 mm in diameter cannot be lifted from the surface by the wind once they have settled to the ground, regardless of wind velocity. Initially, dust is ejected into the air by impact from saltating sand grains or other disturbances. Once in the air, dust can be lifted high into the atmosphere and transported many kilometers.

Estimates on the quantity of material transported by the wind are based on numerous samples throughout many parts of the world. For example, some dustfalls in the eastern United States originated 3200 km to the west, and dust from the Sahara is known to have fallen on Great Britain. In a single duststorm originating in the Sahara, as much as 1.8 metric tons of dust fell on Europe and much more on northern Africa. During the dust bowl years of the 1930s in the United States, a dust cloud extending 3650 m above the ground near Wichita, Kansas, was estimated to contain 50,000 metric tons of dust per cubic kilometer. Thus, locally, the atmosphere can be the dominant factor in developing the surface features of an area, surface features that are largely the result of transportation and deposition of sediment.

Migration of Sand Dunes

Statement

Wind commonly deposits sand in the form of dunes (mounds or ridges), which generally migrate downwind. In many respects, dunes are similar to ripple marks (formed either in air or in water) and large sand waves or sandbars common in many streams and in shallow marine water.

Many dunes originate where an obstacle creates a protected area, or **wind shadow,** which reduces the wind velocity to a point where deposition occurs. Once formed, the dune itself acts as a wind shadow and migrates downwind as sand is carried up the windward side and accumulates on the leeward side. Thus, the internal structure of a dune is characterized by cross-stratification inclined in the direction the wind was blowing.

Discussion

Sand dunes are probably the most familiar type of wind deposit, but the factors that control their development are far from simple. Commonly, a sand dune originates when an obstacle (such as a bush or large rock) deflects the wind and causes a pocket of quieter air to form behind it (*figure 16.7*). As sand is blown over or around the obstruction into the wind shadow, it is deposited and a mound of sand accumulates. Once a small dune is formed, it acts as a barrier itself and disrupts the flow of air, causing continued deposition downwind (*figure 16.8*). Dunes range in size from a few meters high to huge deposits as much as 200 m high and 1 km wide.

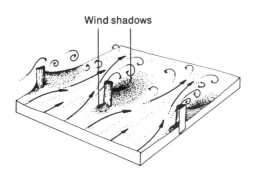

Figure 16.7 *Wind shadows form behind an obstacle that diverts the wind. Sand accumulates in the protected areas, and the accumulation may grow large enough to form a dune.*

Figure 16.8 *A wind shadow from a fence post and the resulting sand accumulation on the Great Salt Lake Desert.*

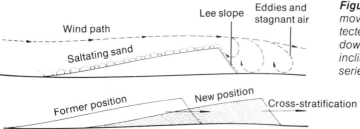

Figure 16.9 *A sand dune migrates as sand grains are moved up the slope of the dune and accumulate in a protected area on the downwind face. The dune slowly moves downwind, grain by grain. As the grains accumulate in inclined layers on the downwind slope, they produce a series of layers (crossbeds) inclined in a downwind direction.*

Figure 16.10 *Crossbedding in the windblown sandstone of the Navajo Formation in Zion National Park, Utah. The inclination of the strata shows that the wind blew from the left to right throughout most of the time this formation was deposited.*

A profile showing movement of sand in a typical dune is diagrammed in *figure 16.9*. Dunes are asymmetrical, with a gently inclined windward slope and a steeper downwind slope called the **lee slope** or slip face. The steep slip face of the dune indicates the direction of the prevailing wind. As the wind blows over the dune, it transports the sand by saltation and by surface creep up the windward slope to the crest of the dune. The wind continues upward past the crest, creating divergent air flow and eddies just over the lee slope. Beyond the crest of the dune, the sand drops out of the wind stream and accumulates on the slip face. The maximum slope of the slip face is 34°, the angle of repose for dry, well-sorted sand. Therefore, slumping is common on this unstable slope. As more sand is transported from the windward slope and accumulates on the lee slope, the dune migrates downwind. The internal structure of a migrating dune consists of cross-stratification, which forms on the lee slope and is, therefore, inclined in a downwind direction. This structure enables geologists to map the direction of ancient wind systems in windblown sandstone simply by measuring the direction in which the cross-strata are inclined (*figure 16.10*).

Types of Sand Dunes

Statement

Sand dunes can assume a variety of fascinating shapes and patterns depending upon a number of factors such as:
1. Sand supply.
2. Wind velocity.
3. Variability of wind direction.
4. The characteristics of the surface over which the sand moves.

The most common dune varieties are **transverse dunes, barchan dunes, longitudinal dunes, star dunes,** and **parabolic dunes** (*figures 16.11* and *16.12*).

Discussion

Transverse Dunes (*figure 16.11A*).   Transverse dunes are typical of areas where there is a large supply of sand and a constant wind direction. These dunes completely cover large areas and develop a wavelike form, with sinuous ridges and troughs perpendicular to the prevailing wind. Transverse dunes commonly form in desert regions that have exposed ancient sandstone formations that provide an ample supply of sand or along beaches where sand is transported landward by strong onshore winds. These dunes commonly cover large areas known as sand seas, so called because the wavelike form of the dune produces a surface resembling a stormy sea. Typically, the dune develops a gentle windward slope and a steep leeward slope characteristic of other dunes.

Barchan Dunes (*figure 16.11B*).   Where the supply of sand is limited and winds of moderate velocity blow in a constant direction, crescent-shaped dunes (called barchans) tend to develop. Typically, the barchan is a small, isolated dune from 1 to 50 m high. The tips (or horns) of the barchan point downwind, and sand grains are swept around the barchan as well as up and over the crest.

(A) Transverse dunes develop when the wind direction is constant and the sand supply is large.

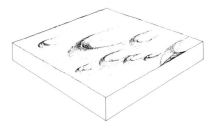

(B) Barchan dunes develop when the wind direction is constant but the sand supply is limited.

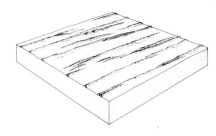

(C) Longitudinal dunes are formed by converging wind directions in an area where there is a limited sand supply.

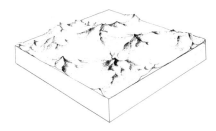

(D) Star dunes develop where the wind direction is variable.

(E) Parabolic (or blowout) dunes are formed by strong onshore winds.

Figure 16.11 *Schematic diagrams showing some of the major dune types.*

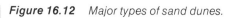

Figure 16.12 *Major types of sand dunes.*

(A) Transverse.

(B) Barchan.

(D) Star.

(C) Longitudinal.

With a constant wind direction, a beautifully symmetrical crescent can be formed; but, with shifts in wind direction, one horn can become much larger than the other. Although barchans typically are isolated dunes, they often are arranged in a chainlike fashion extending downwind from the source of sand. They can migrate 7 to 15 m a year.

Longitudinal Dunes (figure 16.11C; see also plate 22). Longitudinal dunes, also called **seif** (Arabic word for sword) **dunes,** are long, parallel ridges of sand, elongate in a direction parallel to the vector resulting from two converging wind directions. They develop where strong prevailing winds converge and blow in a constant direction over an area having a limited supply of sand. Many longitudinal dunes are less than 4 m high, but they can extend downwind for several kilometers. In larger desert areas, they can become 100 m high and 120 km long and are usually spaced 0.5 to 3 km apart. Longitudinal dunes occupy the vast area of central Australia called the sand ridge desert and are especially well developed in some desert regions of North Africa and the Arabian peninsula.

Star Dunes (figure 16.11D). A star dune is a mound of sand having a high central point from which three or four arms or ridges radiate in various directions. This dune type is typical of parts of North Africa and Saudi Arabia. The internal structure of these dunes suggests that they were formed by winds blowing in three or more directions.

Parabolic (or Blowout) Dunes (figure 16.11E). Parabolic dunes develop along coastlines where vegetation partly covers the sand. In spots where vegetation is absent, small deflation basins are produced by strong onshore winds. The blowout depression grows larger as more sand is exposed and removed. Usually, the sand piles up on the lee slope of the shallow deflation hollow and forms a crescent-shaped ridge. In map view, the parabolic dune is similar to a barchan but the tips of the parabolic dune point upwind. Because the form resembles a hairpin, it also is called a hairpin dune.

Loess

Statement

Loess is a deposit of windblown dust that slowly accumulates over large areas as a blanket deposit, often masking former landforms. It covers possibly as much as one-tenth of the land surface of the world and is particularly widespread in the semiarid regions marginal to the great deserts of the world (*figure 16.13*). Areas free from loess are the equatorial tropics and the areas formerly covered by continental glaciers.

Loess is a distinctive sedimentary deposit. It is composed of particles very similar, if not identical, to the dust in the air at the present time; that is, angular fragments of quartz with lesser amounts of feldspar, mica, and calcite. Normally, loess lacks stratification and erodes into vertical cliffs. Large loess deposits are derived from:
1. Desert regions.
2. Glacial deposits.

Discussion

Desert Loess. In northern China, extensive loess deposits consist primarily of disintegrated rock material

Figure 16.13 *Major loess deposits of the world. Loess is particularly widespread in semiarid regions marginal to the great deserts (such as those in China and South America). The loess deposits of Europe and the United States are associated with sediment deposited along the margins of continental glaciers. Note that the tropics and the areas formerly covered by glaciers are free from loess.*

brought by the prevailing westerly winds from the Gobi Desert in central Mongolia. The yellow-colored loess reaches a thickness of more than 60 m and blankets a large area. It is responsible for the characteristic color of the Yellow River and the Yellow Sea, since it is eroded easily and transported in suspension by running water.

Loess in the eastern Sudan of North Africa also is considered to have originated from a desert, probably the Sahara to the west. Similarly, the loess deposits of Argentina have been derived from the arid regions to the west, rather than from glacial deposits to the south.

Glacial Loess. Much of the loess of North America and Europe appears to have originated from rock debris pulverized by glaciers and deposited as outwash. This sediment, commonly called **rock flour,** is transported readily by wind. In the central United States, the great bulk of loess occurs in the bluffs and uplands of the Mississippi valley, the major drainage system through which the meltwaters of the glaciers drained. Near the rivers, the loess is 30 m or more thick, but it thins out away from the river channels. The greatest accumulations are east of the flood plains. In many areas, loess lies directly upon glacial or glaciofluvial deposits. Apparently, as the glaciers melted, the fine rock flour ground up by the glaciers was deposited so rapidly as glacial outwash that a protective plant cover could not be established—a situation typical of many outwash plains. Winds then picked up the dust-size particles and deposited them east (downwind). In Europe, the loess deposits also appear to have been derived from the adjacent glaciated areas.

Summary

Wind is capable of transporting huge quantities of sand and dust that, when deposited, form distinctive landforms (sand dunes) and thick layers of loess. It is an important geologic agent of sediment transport and deposition in the arid regions of the world, but it is not a major agent in abrading solid bedrock.

Sand is transported by the wind through saltation and surface creep; dust is transported in suspension. Wind-blown sand typically accumulates in dunes that migrate downwind as sand is transported particle by particle up the windward slope and accumulates in the relatively quiet area of the lee slope of the dune. This produces crossbedding that is inclined downwind in the dune.

A variety of dune types result from variations in sand supply, wind direction and velocity, and characteristics of the desert surface. The most significant include (1) transverse dunes, (2) barchan dunes, (3) longitudinal dunes, (4) star dunes, and (5) parabolic dunes.

Loess accumulates as a blanket deposit that can cover completely the preexisting surface. The dust is derived from rock flour near glacial margins or from desert regions.

Additional Reading

Bagnold, R. A. 1941. The Physics of Blown Sand and Desert Dunes. London: Methuen Publishing Company.

Plate Tectonics

The theory of plate tectonics is accepted by most geologists today; yet, only a few years ago, the continents and the ocean basins were considered to be permanent features that had existed since the beginning of the earth. We now believe that the lithosphere is in constant motion and that the continents have drifted thousands of kilometers across the earth's surface. Movement of lithospheric plates causes earthquakes, mountain-building, metamorphism, and igneous activity. What is the nature of the mobile plates? What causes them to move? If the lithosphere is in motion, is it possible to measure rates and direction of movement? Have the plates always been in motion? Will they continue to move in the future? Why have we discarded the older concepts of permanent continents and ocean basins? These are some of the questions this and subsequent chapters will try to answer.

Although continental drift was proposed more than 50 years ago, the theory of plate tectonics was not developed until the early 1960s, when oceanographic surveys had provided enough data to make accurate regional topographic maps of the ocean floors. This data shows that the ocean floors are not flat, featureless areas covered with sediment; nor are they like the continents. Instead, there is a worldwide rift system along the crest of the oceanic ridge and a system of deep trenches along some of the oceanic margins. These two systems are the most seismically active areas in the world.

In this chapter, the development of the theory of plate tectonics, and the evidence upon which it is based, will be reviewed briefly. Then, the nature of the lithospheric plates and their boundaries and the way they move relative to each other to produce the tectonic features of the earth will be considered.

Major Concepts

1. The theory of continental drift was proposed in the early 1900s and was supported by a variety of geologic evidence. However, without a knowledge of the nature of the oceanic crust, a complete theory of earth dynamics could not be developed.
2. A major breakthrough in the development of the plate tectonic theory occurred in the early 1960s, when the topography of the ocean floors was mapped and magnetic and seismic characteristics were determined.
3. The lithosphere can be divided into a series of plates bounded by the oceanic ridge, trenches, mountain ranges, and transform faults.
4. The plates move apart where the convecting mantle rises and spread laterally beneath the oceanic ridge. The plates descend into the mantle beneath the trenches and are consumed.
5. The energy for plate tectonics is internal heat, probably generated by radioactivity in the asthenosphere.

Continental Drift

Statement

The theory of plate tectonics has brought about a sweeping change in our understanding of the earth and the forces that shape it. Some scientists have considered this change in thought to be as profound as those that occurred when Darwin reorganized biology in the nineteenth century and when Copernicus determined that the earth is not the center of the universe. Yet, continental drift is an old idea that was formulated in the early 1900s and was supported by some very convincing evidence. Soon after the first reliable world maps were made, scientists noted that the continents, particularly Africa and South America, would fit together like a jigsaw puzzle if they could be moved. One of the first men to give the idea serious study was a Frenchman named Antonio Snider-Pellegrini, who, in his book *Creation and Its Mysteries Revealed* (1858), showed how the continents looked before separation. He cited fossil evidence in North America and Europe but based his reasoning on the catastrophies of Noah's flood. The concept of continental drift was not considered seriously, however, until 1908, when American geologist Frank B. Taylor pointed out a number of geologic facts that could be explained by continental drift.

Perhaps the ideas were best explained by Alfred Wegener, a German meteorologist, in a book published in 1915—*The Origins of Continents and Oceans*. Wegener based his theory not only on the shape of the continents but on geologic evidence such as similarities in the fossils found in Brazil and Africa. He drew a series of maps showing three stages in the drift process and called the original large landmass Pangaea (meaning all lands). Wegener believed that the continents, composed of light granitic rock, somehow plowed through the denser basalts of the ocean floor, driven by forces related to the rotation of the earth (*figure 17.1*).

Most geologists and geophysicists rejected Wegener's theories, although many scientific observations supporting continental drift were known in Wegener's time. A few noted scholars, however, seriously considered continental drift, especially Arthur Holmes of England, who developed the theory in his textbook *Principles of Physical Geology;* and a South African, Alex L. Du Toit, who compared the landforms and fossils of Africa and South America and further expounded the theory in his book *Our Wandering Continents*.

Discussion

The early arguments for continental drift were supported by some important and imposing lines of evidence, most of which resulted from regional geologic studies.

Paleontologic Evidence. The striking similarity of certain fossils found on the continents on both sides of the Atlantic is difficult to explain unless the continents were once connected. The record of life indicates that new species appear at one point and disperse outward from that point. Floating and swimming organisms could migrate in the ocean from the shore of one continent to another, but the Atlantic Ocean would present an in-

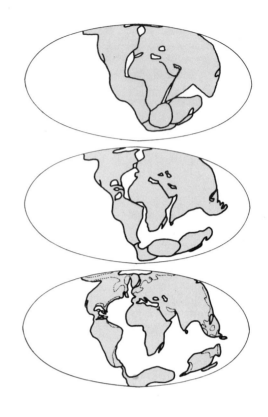

Figure 17.1 *Continental drift as visualized by Wegener in 1915.*

surmountable obstacle for migration of land-dwelling animals such as reptiles and insects and certain land plants. Consider two examples.

Fossils of the seed fern *Glossopteris* have been found in rocks from South America, South Africa, Australia, and India and within 480 km of the south pole in Antarctica. Mature seeds of this plant were several millimeters in diameter, too large to have been dispersed across the ocean by winds. The presence of *Glossopteris* in rocks of the same age in the southern continents, therefore, is considered strong supporting evidence for continental drift.

The distribution of Paleozoic and Mesozoic reptiles provides similar arguments in favor of continental drift, because fossils of several species have been found in the now separated southern continents. An example is a mammallike reptile belonging to the genus *Lystrosaurus*. This creature was strictly a land dweller, and its fossils are found in abundance in South Africa and South America, as well as Asia. In 1969, a United States expedition discovered fossils of this animal in Antarctica, so members of the family are found on all of the southern continents. It is clear that this reptile could not have swum thousands of kilometers across the Atlantic and Antarctic Oceans, so some previous connection must be postulated. Continental drift supplies one possible explanation. The existence of a land bridge, similar to Central America but now submerged, is another possi-

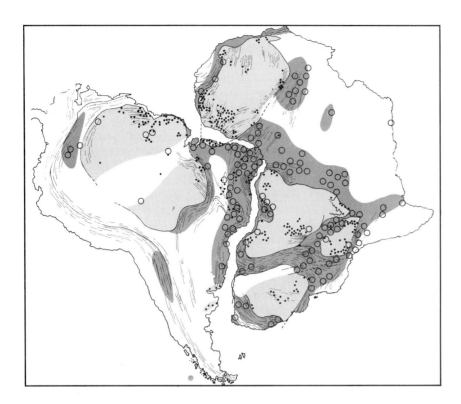

Figure 17.2 *South America and Africa fit together, not only in outline, but in rock types and geologic structure. The light gray areas represent the shields of metamorphic and igneous rocks formed at least 2 billion years ago. The dark gray areas represent younger rock, much of which has been deformed by mountain-building. Structural trends such as fold axes are shown by dashed lines. Most of the deformation occurred 450 to 650 million years ago. Dots show sites of rocks dated by radiometric means. Black dots represent rocks older than 2 billion years. Open dots denote younger rocks. Note that several fragments of the African Shield are left stranded along the coast of Brazil.*

bility; however, surveys of the ocean floor show no evidence of such a submerged land bridge.

Evidence from Structure and Rock Type. There are a number of geologic features that abruptly terminate at the coast of one continent and appear on the facing continent across the Atlantic (*figure 17.2*). Folded mountain ranges at the Cape of Good Hope at the southern end of Africa trend east-west and terminate abruptly at the coast; but an equivalent of this structure, recognized by age and style of deformation, is found near Buenos Aires, Argentina. Other structures and rock types—including igneous, sedimentary, and metamorphic rocks—on the two continents match.

Another example is the folded Appalachian Mountains, which extend northeastward across the eastern United States and up through Newfoundland and abruptly terminate at the ocean but reappear at the coast of Ireland and Brittany. There are other examples that could be cited, but the important point is that the continents on both sides of the Atlantic fit together not only in outline but in rock type and structure. This fit is similar to matching pieces of a torn newspaper (*figure 17.3*).

Figure 17.3 *The continents fit together like a jigsaw puzzle or pieces of a torn newspaper. Not only do the outlines of the pieces fit together, but the printing on the pieces (analogous to the ages and structural features of the continents) also matches and fits together across the edges of the separate pieces.*

Not only do the jagged edges fit, the printed lines come together to form a single sheet. In this analogy, the structure and the rock types of the continents correspond to the printed lines. One important point should be emphasized. The geologic similarities on opposite sides of the Atlantic are found only in rocks older than the Cretaceous period. (The continents are believed to have split and begun to drift apart in Jurassic time.)

Evidence from Glaciation.   During the latter part of the Paleozoic era (about 300 million years ago), glaciation occurred throughout a large portion of the continents in the Southern Hemisphere. The deposits left by these ancient glaciers can be recognized readily, and striations and grooves on the underlying rocks show the direction in which the ice moved (*figure 17.4*). All the continents in this hemisphere except Antarctica now lie close to the equator. On the other hand, continents in the Northern Hemisphere show no trace of glaciation during this time, but fossil plants indicate a tropical climate in that area. These facts are difficult to explain in the context of fixed continents and the climatic belts that are determined by latitude. Even more difficult to explain is the direction in which the glaciers moved, as indicated from regional mapping of striations and grooves. In South America, India, and Australia, the ice moved inland from the oceans, a situation quite impossible unless there were a landmass where the ocean now exists.

If the continents were grouped together as Wegener proposed, the glaciated areas would be grouped in a neat package near the south pole (*figure 17.5*) and would explain Paleozoic glaciation very nicely. The pattern of glaciation was considered to be strong evidence for continental drift, and many geologists who worked in the Southern Hemisphere became ardent supporters of the theory because they could see the evidence with their own eyes.

Evidence from Paleoclimates.   Other evidence of striking climatic changes tends to support the drift theory. Great coal deposits in Antarctica show that the area now mostly covered with ice once flourished with abundant plant life.

Rock salt deposits and formations of windblown sandstone on other continents provide additional paleoclimatic indications that permit a reconstruction of the climatic zones of the past. When the continents are grouped together (*figure 17.6*), it is clear that the pattern is easily explained. With the continents in their present positions, the patterns are quite baffling.

This evidence was considered and debated for years. While it was convincing to many, most geophysicists objected to a theory of continental drift because it simply was impossible for continents to drift through the solid rocks of the oceanic crust. In the absence of a reasonable mechanism for the drift process, there was little further development in the theory until after World War II. Then, an explosion of knowledge took place that not only provided renewed support for the drift hypothesis but also led to the discovery of a possible mechanism.

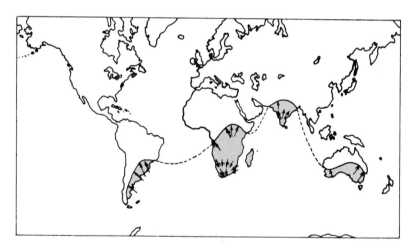

Figure 17.4   *The distribution of late Paleozoic glacial deposits is restricted to the Southern Hemisphere (except for India). Arrows show the direction of ice movement. These areas are now close to the tropics. The present-day cold latitudes in the Northern Hemisphere show no evidence of glaciation during this period.*

Figure 17.5   *If the continents were restored to their former positions according to Wegener's theory of continental drift and if the former south pole were located approximately where South Africa and Antarctica meet, the location of late Paleozoic glacial deposits and the flow directions of the ice would be explained very nicely.*

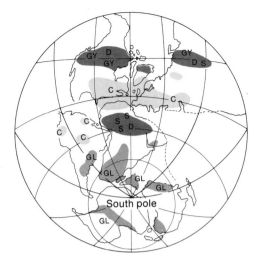

Figure 17.6 *Paleoclimatic evidence for continental drift. Deposits of coal (C), desert sandstone (D), rock salt (S), windblown sand (dotted area), gypsum (GY), and glacial deposits (GL) each indicate a specific climatic condition at the time of their formation. The distribution of these deposits is best explained if we assume that the continents were once grouped together as shown in this diagram.*

Development of the Theory of Plate Tectonics

Statement

Although the theory of continental drift was supported by some very convincing evidence, the data upon which it was based came only from the continents because, prior to the 1950s, there was no effective means of studying the ocean floor. Before 1950, therefore, geologists were faced with an almost total absence of data about the geology of 70% of the earth's surface. Then, in the 1950s and 1960s, an outburst of new data and new ideas resulted from research efforts in two areas.

1. The topography and geology of the ocean floor.
2. Rock magnetism.

Discussion

Geology of the Ocean Floor. With newly developed echo-sounding devices, marine geologists and geophysicists were able to map the topography of the ocean floor in considerable detail (see chapter 21). When the results of these studies were compiled, it was found that the ocean basin is divided by a great ridge approximately 65,000 km long. Moreover, at the crest of the ridge is a central valley 1 to 3 km deep. This feature appears to be a rift valley, under a state of tension, that is splitting apart. Other evidence shows that ocean basins are relatively young. Seismic work established that the oceanic crust (composed largely of basalt) had a completely different composition (and was much thinner) than the continental crust. Furthermore, the oceanic crust was not deformed into folded mountain structures and apparently had not been subjected to strong compressional forces.

In 1960, H. H. Hess, a noted geologist from Princeton University, considered this new data and proposed that the ocean floors were spreading apart and moving symmetrically away from the oceanic ridge, propelled by convection cells in the mantle. Hess believed that fractures in the rift valley, produced by the spreading, were being filled continually with magma as the seafloor

Figure 17.7 *The earth's magnetic field.*
(A) The lines of force in the earth's magnetic field. If a magnetic needle were free to move in space, it would be deflected by the earth's magnetic field (as shown by the arrows). Close to the equator, the needle would be horizontal and would point toward the poles. At the magnetic poles, the needle would be vertical.
(B) Schematic diagram showing how electrical currents in the earth's core could produce the magnetic field. Theoretically, convection in the core can generate an electrical current (in a manner similar to a dynamo) that produces a magnetic field.

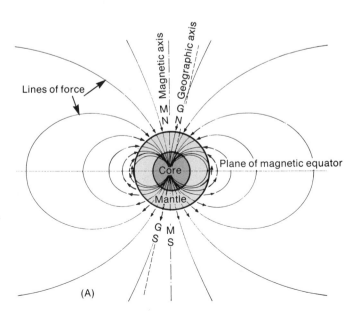

(A)

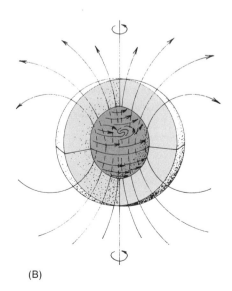

(B)

spread apart. In this way, new oceanic crust was being generated continually. The continents and the seafloor were considered to be carried passively by convection currents away from the oceanic ridge and toward the deep-sea trenches, where the oceanic crust descended into the mantle with the descending convection. In this method, the entire ocean floor could be regenerated completely within 200 to 300 million years. In a sense, what Hess did was to elaborate on the theory of continental drift in the light of fresh new knowledge and redefine it in the scheme of seafloor spreading. A test for his ideas, utilizing new studies in paleomagnetism, was soon to follow.

Rock Magnetism. The study of rock magnetism developed during the 1950s with the perfection of new, highly sensitive magnetometers. Certain rocks, such as basalt, are fairly rich in iron and become weakly magnetized by the earth's magnetic field when they cool. These rocks retain an imprint of the earth's magnetic field at the time they were formed, and the mineral grains, in a sense, become a fossil magnet preserving a record of the earth's magnetic field. Red sandstone also contains enough iron to become oriented in the earth's magnetic field as the sediment is deposited, and it can be used in a similar way to indicate the orientation of the paleomagnetic fields.

The magnetic field of the earth is like that which would be produced by a simple bar magnet whose axis is inclined 11° from the earth's geographic axis (*figure 17.7A*). The temperature of the mantle and the core of the earth, however, is far too high for the earth's magnetic field to be produced by a permanent magnet. The earth's magnetism, therefore, must be generated electromagnetically. The electromagnetic, or dynamo, theory postulates that the outer core of liquid iron is slowly rotating with respect to the surrounding mantle. Such motion would generate strong electrical currents, which would establish a magnetic field (*figure 17.7B*).

Studies of paleomagnetism in rocks of widely different ages from Europe showed that the earth's north magnetic pole apparently has steadily changed its position with time. In the Precambrian era, the pole appears to have been located near Hawaii. It slowly migrated northward

and westward to its present position (*figure 17.8*). The difference was systematic, not random. A similar migration of the magnetic pole was found from paleomagnetic work in North America, but the path of migration was systematically different although it paralleled that of the European data.

These observations could be explained very nicely by drifting continents, and students of paleomagnetism became leading proponents of the theory of continental drift. Soon, results collected from the southern continents were reported, and again a systematic change in the magnetic pole through time was reported—but with different paths for different continents. These differences could be resolved only if the continents were conceived of in the arrangement in which they were believed to have been before drifting. Since it is impossible that there were numerous magnetic poles that migrated systematically and eventually merged, the most logical explanation was that there was one pole throughout time and that the continents were moving with respect to the pole. This discovery led to renewed interest in the theory of continental drift and the conclusion that the Atlantic Ocean opened relatively recently.

Magnetic Reversals. Recent studies of the magnetic properties of numerous samples of basalt from many parts of the world demonstrate that the earth's magnetic field has been reversed many times throughout the last

Figure 17.8

(A) The magnetic properties in the rocks of North America suggest that the magnetic pole has migrated in a sinuous path over the last several hundred million years. Evidence from other continents shows similar migration but along different paths. If the continents had remained fixed, how could different continents have different magnetic poles at the same time?

(B) The question can be answered if the pole has remained fixed while the continents have drifted. If, for example, Europe and North America were joined, the paleomagnetic field preserved in their rocks would indicate a single pole location until they drifted apart. The sequence of rocks on each continent would show different paths of migration to the present position of the poles.

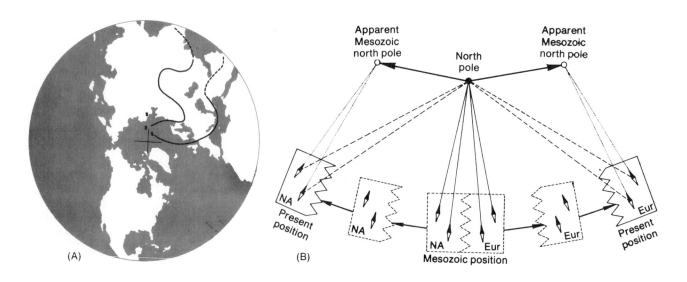

(A)

(B)

70 to 80 million years; that is, there have been periods of 1 to 3 million years during which magnetic poles have been close to their present location, followed by similar periods during which the north and south poles have been reversed. At least nine reversals have occurred in the last 4.5 million years. The present period of "normal" polarity began about 700,000 years ago and was preceded by a period of reversed polarity that began 2.5 million

Figure 17.9 *Reversals in lines of force in the earth's magnetic field. (A) With normal polarity. (B) With reverse polarity. (C) Changes in the earth's magnetic field with time. Many rocks preserve a fossil record of the earth's magnetic field at the time they were formed. Some show normal polarity, whereas others show reversed polarity. By dating the rocks radiometrically, it is possible to construct a geomagnetic time scale for polarity reversals. In this diagram, the patterns of alternating polarities are shown.*

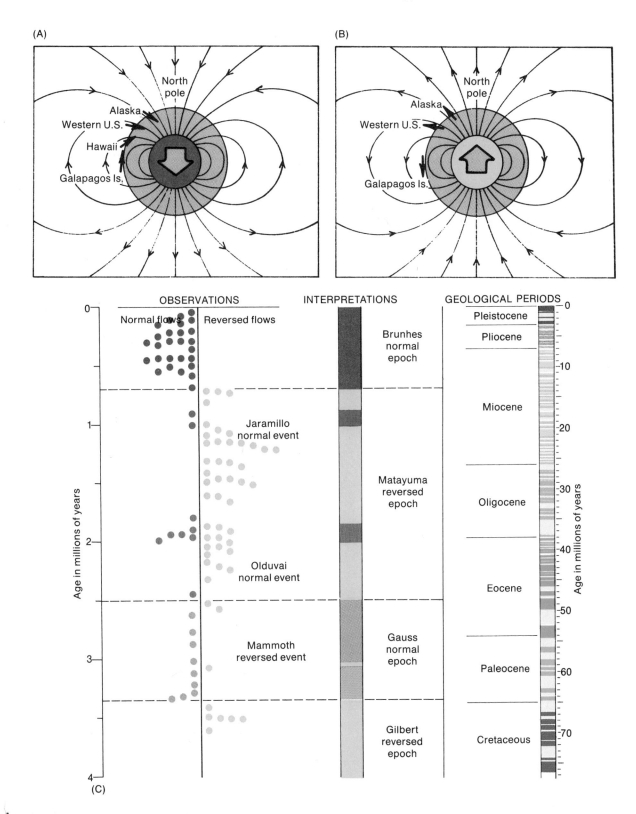

years ago and contained two short periods of normal polarity. The major intervals of alternating polarity (about a million years apart) are termed **polarity epochs,** and intervals of shorter duration are termed **polarity events.** The pattern of alternating polarities has been clearly defined, and evidence of the occurrence of polarity epochs has been found in widely spaced parts of the earth. The sequence of magnetic reversals has been well documented, and their radiometric ages have been determined so that a reliable chronology of magnetic reversals for the most recent 4 million years has been established (*figure 17.9*). In addition, extrapolation as far back as 76 million years reveals a sequence of at least 171 reversals.

In 1963, Fred Vine and D. H. Matthews saw a way to test the idea of seafloor spreading put forth by Hess and suggested that, if seafloor spreading had occurred, it might be recorded in the magnetism of the basalts on the ocean's crust. (This same idea was developed independently by L. W. Morley.) They postulated that, if the earth's magnetic field had been reversed intermittently, new basalt forming at the crest of the oceanic ridge would have been magnetized according to the polarity at the time. As spreading of the ocean floor continued, a series of normal and reversed magnetic stripes of oceanic crust should have been produced symmetrically from the center of the ridge. Each band of new material moving out across the ocean floor should have retained its original magnetic orientation, an orientation shared by a corresponding band on the other side of the ridge. The result should have been a matching set of parallel bands of normally and reversely magnetized rocks on

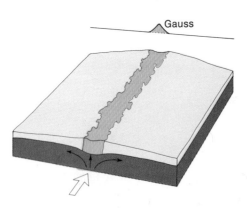

(A) The oceanic ridge during the Gauss normal polarity epoch, 2.75 million years ago. As magma cooled and solidified along the ridge in dikes and flows, it became magnetized in the direction of the magnetic field existing at the time (normal).

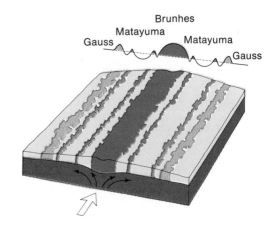

(C) The oceanic ridge at the present time, with polarity reversals of the past 3 million years. The alternating directions of magnetism in the seafloor produced a symmetrical sequence of positive and negative anomalies on each side of the ridge. Note that the pattern of magnetic reversals away from the ridge is the same as the pattern produced in a sequence of basalt flows on the continent. This permits one to correlate rocks on the continent with rocks of equivalent age on the seafloor.

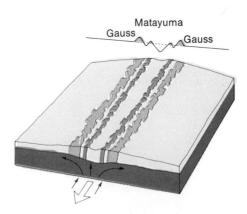

(B) The oceanic ridge during the Matuyama reversed polarity epoch, 2.25 million years ago. As seafloor spreading occurred, the magnetized crust formed during the Gauss epoch separated into two blocks. Each was transported laterally away from the ridge as though on a conveyor belt. New crust being formed at the ridge became magnetized in the opposite direction.

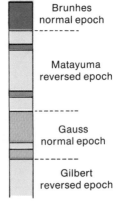

(D) Patterns of magnetic reversals in a vertical sequence of basalts on the continents.

Figure 17.10   Schematic representation of the magnetisms preserved in new crust generated at the oceanic ridge as the lithosphere is transported laterally away from the ridge.

either side of the oceanic ridge. Subsequent investigations have proven conclusively the theory proposed by Vine and Matthews and Morley.

The origin of magnetic patterns on the seafloor perhaps is best understood by considering how the seafloor would have evolved during the last few million years according to the theory of plate tectonics and seafloor spreading. In *figure 17.10A,* the seafloor is shown as it is considered to have been during the Gauss normal polarity epoch (named for German mathematician K. F. Gauss). As basalt was injected into the fractures of the oceanic ridge to form dikes or was extruded over the seafloor as submarine flows, it solidified and became magnetized in the direction of the existing (normal) magnetic field. This basalt produced a zone of new crust with normal magnetic polarity along the oceanic ridge. As seafloor spreading continued, this zone of new crust migrated away from the rift zone but remained parallel to it. About 2.5 million years ago, the magnetic polarity was reversed so that new crust generated at the oceanic ridge became polarized in the opposite direction (*figure 17.10B*). This produced a zone of crust with a reversed polarity. When polarity was changed to normal again, the new crust being created at the ridge was magnetized in a normal direction. In this way, the sequence of polarity reversals for the whole earth became imprinted as magnetic stripes on the ocean crust.

It should be noted that the pattern of magnetic stripes on the ocean floor on either side of the ridge is the same as that found in a sequence of recent basalts on the continents (*figure 17.10C;* see also *figure 17.9C*). That is, the crest of the ridge shows normal polarity and is flanked on either side by a reversed epoch containing two short, normal events. This is followed by a normal epoch with one brief reversal event, and so on. In brief, the pattern of magnetic reversals away from the crest of the ridge is the same as that found in a vertical sequence of rocks on the continents, going from youngest to oldest. These data provide compelling evidence that the seafloor is spreading.

An important aspect of these patterns is that it is possible to determine the interval of time represented by the anomalies and the rates of plate movement. Studies of magnetic reversals of rock sequences on the continents where radiometric dates have been determined show that the present normal polarity has existed for the last 0.7 million years. This normal polarity was preceded by the pattern shown in *figure 17.9C.* Since the same pattern exists on the ocean floor, provisional ages can be assigned to the magnetic anomalies on the ocean floor.

Many magnetic surveys of the ocean floor have now been made, and the patterns of magnetic reversals have been determined over much of the ocean floor (*figure 17.11*). These studies show that almost the total area of the deep seafloor was formed during Cenozoic time (during the last 65 million years), and it now seems probable that very little of the ocean basin is older than Jurassic. From the patterns of magnetic reversals, the rate of seafloor spreading appears to range from 1 to 16 cm per year.

Iceland provides an especially interesting example of seafloor spreading, because it is, in essence, a large exposure of the mid-Atlantic ridge and offers a unique opportunity to study the physical mechanism of seafloor spreading. Geologic work in Iceland shows that the island is being pulled apart by the spreading seafloor beneath it. The tension causes normal faults and fissures parallel to the ridge axis. Through these fissures, volcanic eruptions occur. Swarms of parallel dikes are injected into the fissures with each increment of crustal extension. The aggregate width of these dikes is about 400 km and corresponds to the total amount of crustal extension since the beginning of Tertiary time, about 65 million years ago. A geologic map of Iceland (*figure 17.12*) shows that the rocks are oldest in the extreme east and west ends of the island and become progressively younger toward the center, where present-day volcanism is almost entirely confined.

Evidence from Sediment on the Ocean Floor. To many, some of the most convincing evidence of all comes from recent drilling into the sediment on the ocean floor. The deep-sea drilling project is a truly remarkable expedition in scientific exploration. It began in 1968 with a special ship named the *Glomar Challenger,* designed by a California offshore drilling company. The *Challenger* can lower more than 6100 m of drilling pipe into the open ocean, bore a hole in the seafloor, and bring up bottom cores and samples. The project was funded by the National Science Foundation and was planned by JOIDES—Joint Oceanographic Institutions for Deep Earth Sampling—under the direction of the Scripps Institute of Oceanography. Since 1968, the *Challenger* has drilled more than 400 holes into the seafloor and has provided many data in support of plate tectonics. As predicted by the plate tectonics theory, the youngest sediments were found near the oceanic ridges, where new crust is being created. Away from the ridge, the sediment that lies directly above the basalt becomes progressively older, with the oldest sediment found near the continental borders. The deep-sea drilling projects confirmed the conclusions drawn from paleomagnetic studies by checking the age of the fossils that first accumulated on each portion of the ocean floor after its formation. According to measurements of the rates of sedimentation in the open ocean, red clay and organic ooze accumulate at the rate of 0.9 to 1.2 cm per 1000 years. If the ocean basins of today had existed since Cambrian time, for example, the sediments would be at least 5 km thick (*figure 17.13A*). The greatest thickness of deep-ocean sediments measured to date is only 300 m, a fact that would suggest that the ocean basins are young geologic features indeed (*figure 17.13B*). The oldest sediments yet found on any ocean floor are only 160 million years old. In contrast, the metamorphic rocks of the continental shields are as much as 3.9 billion years old.

Not only do the thickness and age of sediment increase away from the crest of the oceanic ridge, but certain types of sediment also indicate seafloor spreading. For example, upwelling, warm, nutrient-rich water from the Pacific equatorial zone permits planktonic life to thrive there. As the creatures die, their tiny skeletons

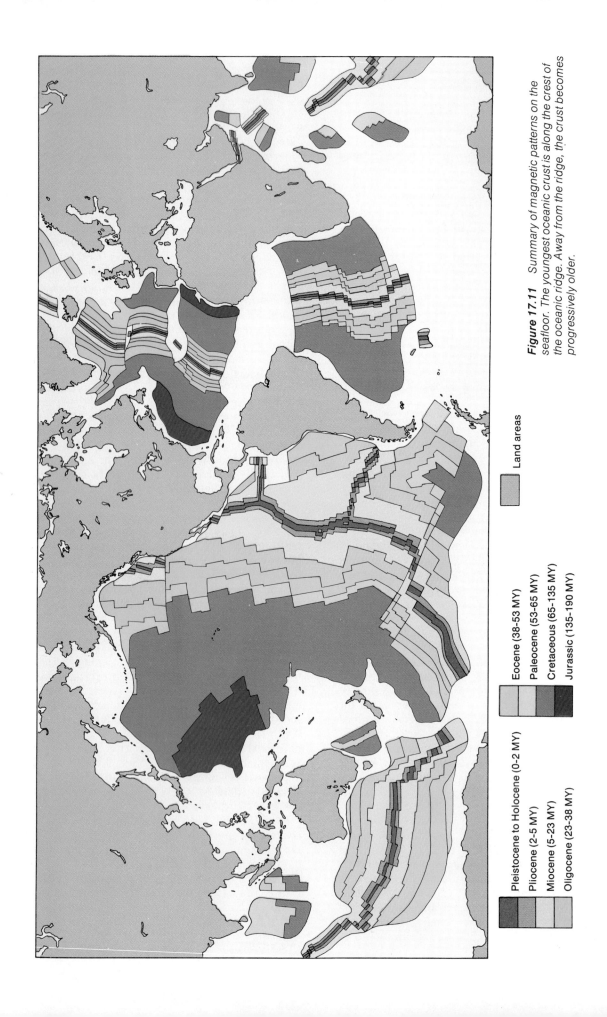

Figure 17.11 Summary of magnetic patterns on the seafloor. The youngest oceanic crust is along the crest of the oceanic ridge. Away from the ridge, the crust becomes progressively older.

Pleistocene to Holocene (0–2 MY)

Pliocene (2–5 MY)

Miocene (5–23 MY)

Oligocene (23–38 MY)

Eocene (38–53 MY)

Paleocene (53–65 MY)

Cretaceous (65–135 MY)

Jurassic (135–190 MY)

Land areas

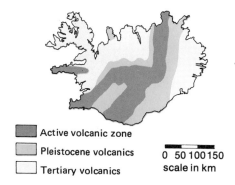

Active volcanic zone

Pleistocene volcanics

Tertiary volcanics

0 50 100 150
scale in km

Figure 17.12 *A geologic map of Iceland shows that the oldest rocks are along the eastern and western margins and the youngest rocks are near the center of the island.*

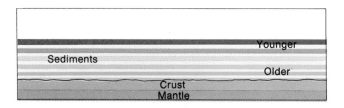

Younger

Sediments

Older

Crust
Mantle

(A) *With no seafloor spreading, the entire ocean floor would be covered with a thick sequence of oceanic sediment, with alternating polarity extending back into the Precambrian.*

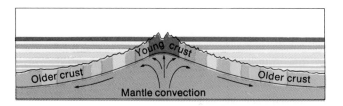

Young crust

Older crust

Older crust

Mantle convection

(B) *With seafloor spreading, the blanket of oceanic sediment thins progressively toward the crest of the oceanic ridge and essentially is nonexistent in the rift valley. Part of each layer of magnetized sediment lies upon a part of the basaltic crust generated at the spreading center at the same time interval the sediment was deposited and subsequently migrated away from the ridge.*

Figure 17.13 *Thickness of sediment and magnetic reversals on the seafloor.*

rain down unceasingly to build a layer of soft, white chalk on the seafloor. The chalk marks an environmental zone restricted to the equatorial belt, one that cannot form in the colder waters of higher latitudes. Yet, drilling by the *Challenger* has shown that the chalk line in the Pacific occurs at varying distances north of today's equator. The only logical conclusion is that the seafloor moved and that the Pacific floor has been migrating northward for at least 100 million years.

The theory of plate tectonics now is firmly established and is used as the fundamental theory of the earth's dynamics. Its first achievement was to explain the meaning of features on the ocean floor, but now the emphasis has switched to the continents, where most previous geologic observations are being interpreted in light of plate tectonics.

Plate Geography

Statement

The shorelines of the continents are major geographic features but have little significance from the standpoint of the earth's tectonics. Plate boundaries are the most significant structural elements of the planet, and, in order to understand plate tectonics, a new geography must be learned—the geography of tectonic plates. This should not be difficult, because plate boundaries generally are marked by major topographic features. Understanding requires only that you focus your attention on the structural features of the earth, rather than on the boundaries between land and ocean.

Discussion

The new geography of tectonic plates is illustrated in *figure 17.14*. The outer, rigid layer of the earth—the lithosphere—is divided into a mosaic of seven major plates and a number of smaller subplates. The major plates are outlined by oceanic ridges, trenches, and young mountain systems. These include the European, North American, South American, Pacific, African, Australian, and Antarctic plates. The largest is the Pacific plate, which is composed entirely of oceanic crust and covers about one-fifth of the earth's surface. The other plates contain both continental and oceanic crust, but there are no major plates composed entirely of continental crust. Smaller plates include the China, Philippine, Arabian, Iran, Nazca, Cocos, Caribbean, and Scotia plates, plus a number of smaller ones that have not yet been defined precisely. The smaller plates appear to form near convergent boundaries of the major plates that involve colliding continents or a continent and an island arc. Thus, smaller plates are characterized by rapid and complex movement.

Plates range in thickness from approximately 70 km beneath oceanic areas to 150 km beneath continents.

Individual plates are not permanent features but are in constant motion and continually change in size and shape. Those without continents can be consumed completely in the subduction zone. The plate changes its shape by splitting along new lines or by welding itself to another plate. In addition, plate margins are not fixed. They move and modify the size and shape of the plates themselves.

Plate Boundaries

Statement

Each tectonic plate is rigid and moves as a single mechanical unit; that is, when one part moves, the entire plate moves. The plates can be warped or flexed slightly

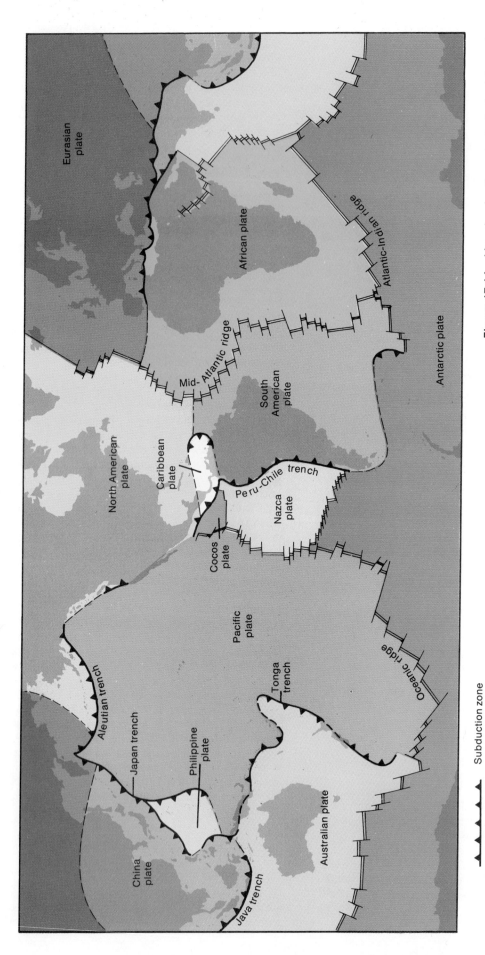

Figure 17.14 Map showing the major tectonic plates and the nature of their boundaries. The plates are delineated by the major tectonic features of the globe, the oceanic ridges, deep-sea trenches, and young mountain belts. Plate boundaries are outlined by earthquake belts and volcanic activity. Most plates (such as the North American, African, Australian, et cetera) contain both continental and oceanic crust, but the Pacific, Cocos, and Nazca plates contain only oceanic crust.

Eurasian plate

African plate

Atlantic-Indian ridge

Antarctic plate

Mid-Atlantic ridge

South American plate

North American plate

Caribbean plate

Peru-Chile trench

Nazca plate

Cocos plate

Pacific plate

Oceanic ridge

Aleutian trench

Japan trench

Tonga trench

Philippine plate

China plate

Australian plate

Java trench

Subduction zone

Ridge axis

Transform fault

Uncertain plate boundary

as they move, but relatively little change occurs in the middle of the plates. Nearly all of the major tectonic activity is accomplished along plate margins, so it is on the plate margins that geologists and students of geology should focus their attention.

Three kinds of plate boundaries are recognized and define three fundamental kinds of deformation and geologic activity (*figure 17.15*).

1. **Divergent plate boundaries** (also called **spreading edges**) are zones of tension where plates split and spread apart.
2. **Convergent plate boundaries** (also called **subduction zones** or **edges of consumption**) are zones of compression where plates collide and one plate moves down into the mantle.
3. **Transform fault boundaries** (also called **passive plate margins**) are zones of shearing where plates slide past each other without diverging or converging.

Discussion

Processes at Divergent Plate Boundaries. Divergent plate boundaries form where a previous plate has split and pulled apart. Where the zone of spreading intersects a continent, rifting occurs and the continent splits (*figure 17.16*). The separate continental fragments drift apart with the separating plates, creating a new, and continually enlarging, ocean basin in the site of the initial rift zone. Thus, divergent plate boundaries are characterized by tensional stresses that continually produce block faulting, fractures, and open fissures along the margins of the separating plates. Basaltic magma derived from the partial melting of the mantle is injected into the fissures, where it cools and becomes part of the moving plates. These are some of the most active volcanic areas on the earth.

Several examples showing various stages of continental rifting and the development of new ocean basins can be cited (*figure 17.17*). The initial stage is represented by the system of great rift valleys in East Africa. The

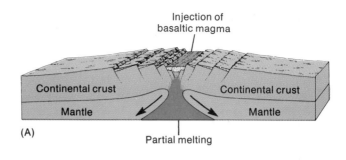

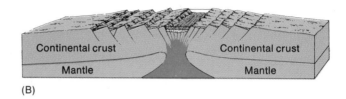

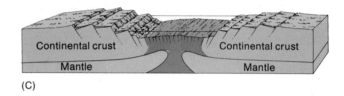

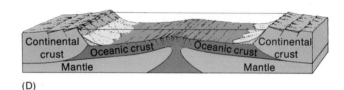

Figure 17.16 *Stages in the splitting and separation of continents and the development of a new oceanic basin. (A) Initial stage of rifting. The continental crust is uparched and stretched, producing block faulting. Continental sediment accumulates in the depressions of the downfaulted blocks, and basaltic magma is injected into the rift system. Flood basalt can be extruded over large areas of the rift zone during this phase. (B) Rifting continues, and the continents separate enough for a narrow arm of the ocean to invade the rift zone. Injection of basaltic magma continues and begins to develop new oceanic crust. (C) As the continents separate, new oceanic crust and new lithosphere are formed in the rift zone and the oceanic basin becomes wider. Remnants of continental sediment may be preserved in the downdropped blocks of the new continental margins. (D) As spreading continues, the oceanic basin grows larger and the continents move off from the uparched spreading zone and parts of the continental crust may be covered by the ocean.*

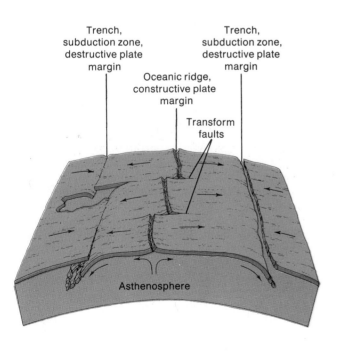

Figure 17.15 *Types of plate margins. Constructive margins (divergent plate boundaries) occur along the oceanic ridges where plates move apart. Destructive margins (convergent plate boundaries) occur along the deep trenches. Passive plate margins are along the fracture zones.*

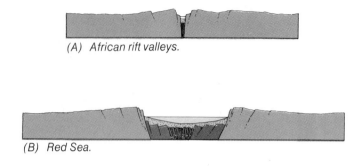

(A) African rift valleys.

(B) Red Sea.

(C) Atlantic Ocean.

Figure 17.17 *Examples of the stages of continental rifting.*

long, linear valleys (which are occupied partly by lakes) are huge, downdropped fault blocks that result from the initial tensional stress. Volcanism along the rift zone, including the great volcanoes of Mount Kenya and Mount Kilimanjaro, occurs as magma is injected into the rift zone. A more advanced stage of rifting is exemplified by the Red Sea, where the Arabian peninsula has been separated completely from Africa and a new linear ocean basin is just beginning to develop. A still more advanced stage of continental drifting and seafloor spreading is exemplified by the Atlantic Ocean, where the American continents have been separated from Africa and Europe by thousands of kilometers. The mid-Atlantic ridge is the boundary between the diverging plates, with the American plates moving westward relative to Africa and Europe. The eastern margins of both North and South America as well as the western margins of Africa and Europe are *not* plate boundaries; they are passive continental margins. The western margins of North and South America, in contrast, are located along converging plate boundaries, and they are geologically active.

Processes at Converging Plate Boundaries. The boundary between converging plates is a zone of complicated geologic processes, which include igneous activity, crustal deformation, and mountain-building. The specific geologic processes acting in this area depend upon the nature of the converging plates and how they interact with each other.

When both plates at the convergent boundary are oceanic, one is thrust under the margin of the other and descends into the asthenosphere, where it is heated and, ultimately, absorbed into the mantle. When one plate contains a continent, the lighter continental crust resists subduction and always overrides the oceanic plate. When continental crust exists on both converging plates, neither can subside into the mantle. Both plates are subjected to compression, although one may override the other for

a short distance. The continents are ultimately "fused" or "welded" together into a single continental block with a mountain range marking the line of suture.

The major processes and geologic phenomena characteristic of converging plate margins are shown in *figure 17.18.* The subduction zone (or zone of underthrusting) usually is marked by a deep-sea trench, and the movement of the descending plate generates an inclined zone of seismic activity. As the plate moves down into the hot asthenosphere, partial melting of the oceanic crust generates silica-rich magma, which (being less dense than the surrounding material) moves upward. Some magma is extruded at the surface as lava to form an island arc or a chain of volcanoes in the mountain belt of the overriding plate, but most usually is intruded in the deformed mountain belt to form batholiths. In both cases new material is added to the continental plate so that continents grow by accretion. This is an important mechanism in the differentiation of the earth, whereby lighter material is concentrated in the upper layers of the planet.

The zone of collision between two plates is a zone of deformation, mountain-building, and metamorphism. When continental crust exists on the overriding plate, compression deforms the margins into a folded mountain belt and the deeper roots of the mountains are metamorphosed.

It has been emphasized in other chapters that continents are composed of light granitic material and ride as passive passengers on the moving plates. One of their most important characteristics is that they cannot sink, because they are not heavy enough to be submerged in the denser material of the earth's mantle. This is why the continents are much older than the ocean basins. Because continents are never consumed into the mantle, they preserve records of plate movements in the early history of the earth—records in the forms of ancient faults, old mountain belts, granitic batholiths, and sediments deposited along ancient continental margins.

Processes at Transform Fault Boundaries. The third type of plate boundary occurs where the plates slide horizontally past each other along a special type of fault called a **transform fault.** The term *transform* is used be-

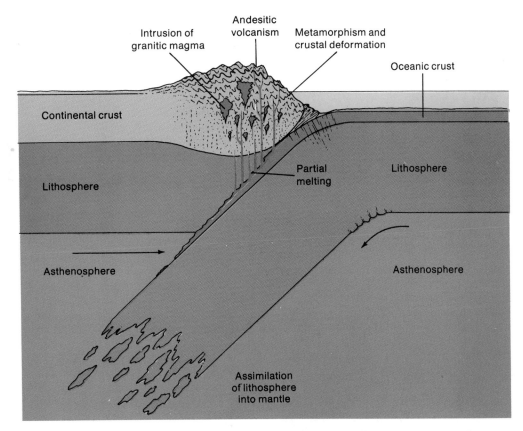

Figure 17.18 *Geologic processes along converging plate margins.*

cause one type of relative motion between plates can be changed—transformed—to a different type of motion from one end of the fault to another. For example, the diverging motion between plates at an oceanic ridge can be transformed along the fault to the converging motion between plates at a subduction zone. Transform faults are simply strike-slip faults between two plates, and they can join converging and diverging plate boundaries together in various combinations.

Where transform faults connect two diverging plate margins, they create a major topographic feature called a **fracture zone.** Fracture zones, however, are not what they might seem to be at first sight, and one must keep in mind the relative motion between plates produced at the spreading center. The *apparent* offset of the oceanic ridge would suggest a simple strike-slip fault with displacements of thousands of kilometers. Careful study of *figure 17.19,* however, shows that the actual relative motion is produced by the plates moving out from the spreading center. The only relative motion between plates occurs in the area between spreading centers, the only place where the fault forms a boundary between plates. Beyond this zone, the plates on either side of the fracture are moving in the same direction and at the same rate and can be considered to be linked together. The spreading center was not offset by motion along the transform fault; it was offset from the beginning and probably represents an old line of weakness. The only movement and seismic activity occur in the part of the fracture between the two oceanic ridges. Beyond the ends of the ridges,

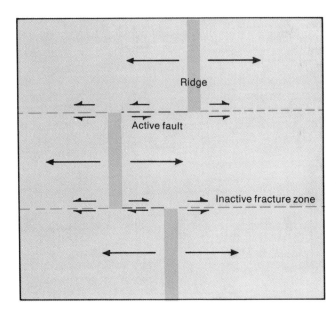

Figure 17.19 *Relative movement of plates associated with a ridge-to-ridge transform fault. The plates are moving away from the ridge, but the relative movement between the plates along the transform faults depends upon the position of the spreading centers. Between the spreading centers along A-A', movement on the two sides of the faults is in the opposite direction. Beyond the spreading centers, however, along lines A-B and A'-B', movement on both sides of the fault is in the same direction, with no relative motion along the fault plane.*

the fracture is inactive. Moreover, there is no volcanic activity associated with transform faults; but, as the plates slide past each other, their boundaries are fractured and broken. This produces parallel ridges and troughs along the fault zone.

Transform faults also can join ridges to trenches and trenches to trenches. In all cases, transform faults are parallel to the direction of relative plate motion, so there is neither divergence nor convergence along this type of boundary. As a result, rocks are not generated, volcanism does not occur, and plates are neither enlarged nor destroyed. The plates slide passively along the fracture system, producing only fracturing and seismic activity.

Plate Motion

Statement

The motion of a series of rigid plates on a sphere can become very complex, because the plates move as independent units, in different directions and at different velocities. Each plate can be thought of as a piece of shell moving on a sphere. The pieces are not flat but are curved; so, as they move, different parts of a plate move at different velocities. The motion of each plate can be described in terms of a pole of rotation.

Discussion

The spherical geometry of a curved plate over a sphere was worked out nearly 200 years ago by a Swiss mathematician named Leonhard Euler (1707-1783). As can be seen in *figure 17.20,* the motion of plate A relative to plate B is a rotation around pole P. This is an imaginary pole that is the axis around which plate A rotates. It should be emphasized that the pole of the plate rotation is completely independent of the axis around which the earth rotates and, of course, has nothing to do with the magnetic poles.

Several important relationships concerning plate motion are immediately apparent in *figure 17.20.* First, different parts of a plate move with different velocities. Maximum velocity occurs at the equator of rotation, and minimum movement occurs at the poles. This perhaps is best illustrated by considering a plate so large that it covers an entire hemisphere. All motion occurs around the axis of rotation. The pole of rotation has no velocity because it is a fixed point around which the hemispheric shell moves. Points 2, 3, and 4 have progressively higher velocities, with a maximum velocity at point 5, which is on the equator of rotation.

Important points to note are that transform faults should lie on lines of latitude about the pole of rotation and that the length of a fracture zone should be at a maximum at the equator and at a minimum toward the pole. This characteristic is closely approached for most transform faults in nature, as can be seen on the physiographic map of the Atlantic (*figure 21.11*). As a result, the relationship between transform faults and spreading poles can be used to establish the position of the pole of rotation for each plate.

The direction of movement of the major plates, relative to their neighbors, can be determined from the trends of the oceanic ridge and the associated fracture zones, as well as from seismic data (page 300) and from the age relationship of the seafloor and chains of volcanic islands and seamounts (page 316). Utilizing this data, geologists have determined the motion of the present tectonic plates. The motion is summarized in *figure 17.21.*

The Pacific plate is moving generally in a northwestern direction from the eastern Pacific rise toward the system of trenches in the western Pacific. It is bordered by several different plates along the subduction zone, so the relative motion at each of the trenches differs from this general trend. The American plates are moving westward from the mid-Atlantic ridge, converging with the Pacific, Cocos, and Nazca plates. The Australian plate is moving northward, following the same basic model.

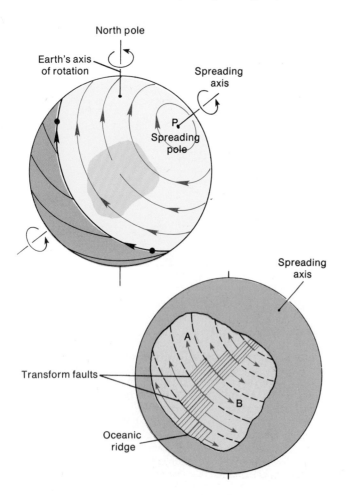

Figure 17.20 *Plate motion on a sphere. The idea of plate motion can be easily understood if the plate covers an entire hemisphere (yellow). Each point on the plate would move along lines of latitude about a pole of rotation (P). The motion of plate A relative to plate B also can be described as rotation around some imaginary pole extending through the earth. Each part of the oceanic ridge lies along a line of longitude that passes through that pole, and each transform fault lies on a line of latitude with respect to that pole. The amount and rate of spreading are at a maximum at the equatorial line and are zero at the pole.*

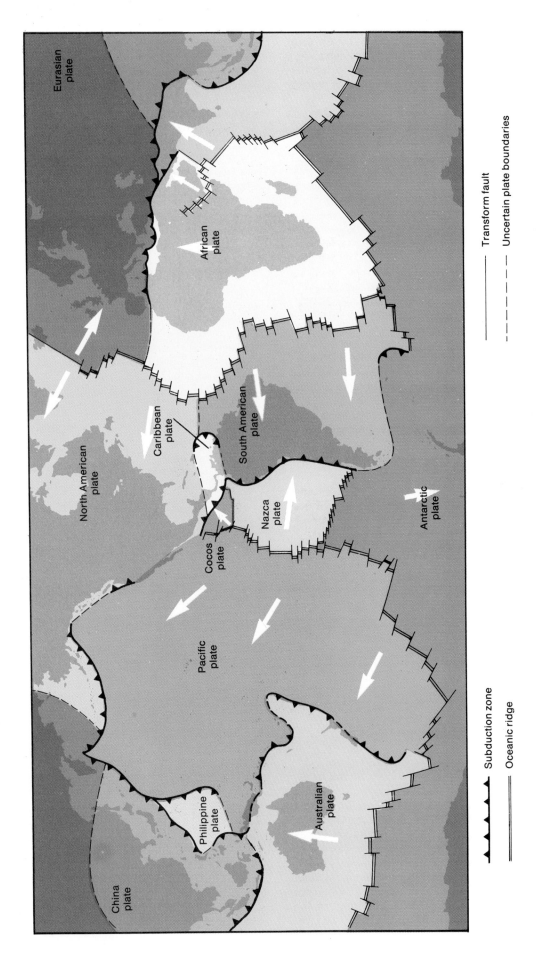

Figure 17.21 *Direction of present motion of plates. The lengths of arrows are proportional to the rate of movement.*

Eurasian plate

African plate

North American plate

Caribbean plate

South American plate

Cocos plate

Nazca plate

Antarctic plate

China plate

Pacific plate

Philippine plate

Australian plate

—— Subduction zone

—— Oceanic ridge

—— Transform fault

- - - Uncertain plate boundaries

Africa and Antarctica, however, present a different situation. Both are nearly surrounded by spreading centers with no associated subduction zones to accommodate the new lithosphere generated along the spreading centers. Some northward movement of Africa toward the convergent boundary in the Mediterranean area can be demonstrated, but this does not accommodate the east-west spreading from the Atlantic and Indian ridges. Apparently, the African and Antarctic plates are being enlarged as new lithosphere is generated around their surrounding divergent margins. Without subduction zones, the spreading centers surrounding these plates must be moving out relative to Africa and Antarctica. This brings up a very important point! *Plate margins are not fixed but can move as well as the plate itself.* As is illustrated in *figure 17.22,* when two diverging plate margins are *not* separated by a subduction zone, new lithosphere is formed at each spreading center but none is destroyed between them. The plate between the spreading centers is continually enlarged, so the spreading centers themselves must move apart.

As the spreading centers move toward the subduction zone with the migrating plate, both the spreading center and the subduction zone ultimately will be destroyed with the formation of a single plate where once there were three (*figure 17.22*).

Subduction zones also are temporary features, as an active zone can be abandoned and a new one created in a different position (*figure 17.23*). This has apparently occurred several times in the western Pacific, where zones of double trenches are common.

Another important change is in the lengths of plate margins. An oceanic ridge is, in essence, a fracture in the lithosphere. It can not only grow larger with time, it can extend its length as well. A good example is the spreading center in the Atlantic Ocean. It is much longer now than it was when spreading began to separate South America from Africa.

These examples of the changes in the shapes and patterns of plates underline the important fact that the margins of plates are not permanent but can be displaced and migrate to different positions. As the plate margins move, they change the shape and configuration of the plates, and, in many cases, the plate itself ultimately is destroyed.

Convection Currents in the Mantle

Statement

Various sources of energy have been proposed to explain convection in the mantle. Some geologists believe that the energy is associated with either the earth's rotation, the earth's tides, or the heat from the core. Most advocates of the plate tectonic theory, however, favor radiogenic heat produced in the upper mantle as the source of the energy. In chapter 1 it was shown that the heavy elements of the earth are concentrated in the core. Less dense elements, which form ferromagnesian silicate minerals, are concentrated in the mantle; and the light silica-rich minerals are concentrated in the continental crust. Radioactive elements probably are concentrated in the asthenosphere. As these elements generate heat, the rocks within the asthenosphere yield to plastic flow and rise in a convection system. Thus, the internal heat is carried to the surface and escapes into space.

The energy for this system was inherited from the early phases of the formation of the earth and is steadily diminishing. The system is irreversible. Energy within the earth, like that in the sun, ultimately will be spent, and the lithospheric plates will cease to move. When this happens in the far-distant future, mountains will no longer form and all types of volcanic activity and earthquakes will end. The earth then will be a dead planet changed only by energy from external sources.

Discussion

Little is known about the actual movement of material in the mantle, but several models of convection systems have been proposed to try to explain the observed facts. Some of these are shown in *figure 17.24.* In diagram *A,* the entire thickness of the mantle is considered to be involved in a series of convection cells. Where the currents rise and move apart horizontally, they initiate the horizontal motion of the lithospheric plates, which move away from the oceanic ridge. Where convection cells descend, they drag the lithosphere down into the mantle. Diagram *B* shows a convection system limited to the asthenosphere. There, heat is considered to originate from the concentration of radioactive elements near the outer part of the mantle.

Summary

The idea of drifting continents was proposed in the early 1900s and was best developed by Alfred Wegener in his book *The Origins of Continents and Oceans.* It was supported by various types of geologic evidence including (1) parallel geologic features in the continents, (2) paleontology, (3) Paleozoic glaciation, and (4) paleoclimates. Without a knowledge of the oceanic crust, however, a complete theory of the earth's dynamics could not be developed.

During the 1960s, new information on paleomagnetics and on the topography of the ocean floor led to the development of the plate tectonic theory. Since then, the theory has been supported by a wide variety of other geologic and geophysical data from both the continents and the oceans. Paleomagnetic studies show that the continents have changed positions relative to the magnetic poles throughout geologic time and that the continents followed different paths relative to the poles. This indicates that the continents have moved relative to each other. Paleomagnetic reversals on the seafloor show symmetrically matching sets on both sides of the oceanic ridge, a fact explained by seafloor spreading.

The theory of plate tectonics explains the earth's crustal dynamics in terms of a series of rigid lithospheric plates that move because of convection in the astheno-

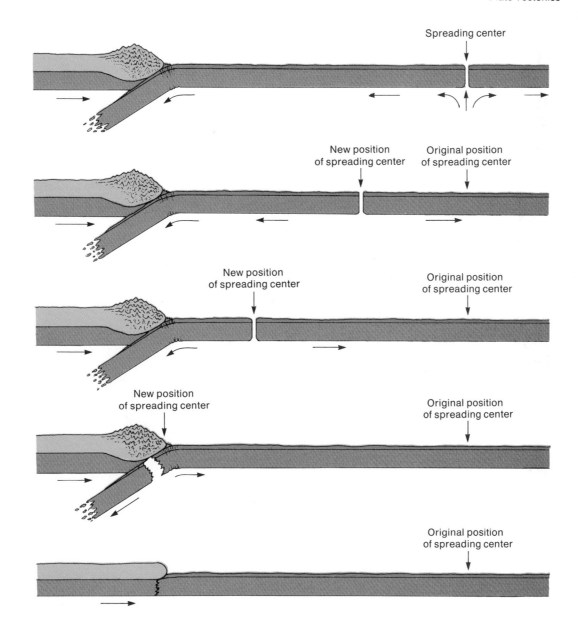

Figure 17.22 *Migration of spreading centers can result in the ultimate destruction of both the spreading center and the subduction zone and the formation of a single plate where there originally were three.*

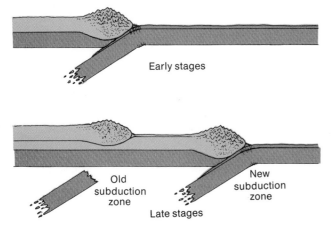

Figure 17.23 *Migration of subduction zones. A subduction zone can be abandoned and a new one created in its place to leave an inactive trench, a dead mountain belt, and isolated fragments of the descending lithosphere.*

sphere. New crust is created at the oceanic ridge where plates split and spread apart and basaltic magma wells up in the rift. The oceanic crust is consumed into the mantle, where the plate descends at the subduction zone. The major structural features of the earth are formed along plate boundaries. At divergent plate boundaries, the lithosphere is under tension, and the major geologic processes are (1) rifting and block faulting, (2) generation of basaltic magma and formation of new oceanic crust, and (3) rifting of continents. At converging plate boundaries, the important geologic processes include (1) subduction, (2) partial melting of the descending plate and the formation of granitic magma, and (3) compression and mountain-building. Transform fault boundaries are zones where plates slide passively past each other.

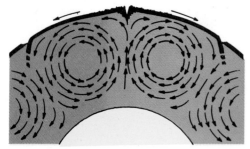

(A) *Large, deep convection cells that involve most of the mantle.*

(B) *Shallow convection cells that are restricted largely to the asthenosphere.*

Figure 17.24 *Possible convection systems in the mantle.*

Each plate is part of a curved shell on a sphere, so different parts of the plate move at different velocities. The direction of plate motion can be determined from (1) trends of spreading centers and fracture zones, (2) seismic data, and (3) ages of the seafloor on either side of the ridge and ages of island chains. Plates move at a rate of 1 to 16 cm per year.

Slow convection in the asthenosphere is believed to be the driving force for plate motion. As plates collide, diverge, and slip past each other they generate earthquakes, and it is the study of earthquakes that tells us most about present-day plate motion. This subject will be discussed in the following chapter.

Additional Readings

Anderson, D. L. 1971. "The San Andreas Fault." Sci. Amer. 225(5):52-68. (Reprint no. 896.) San Francisco: W. H. Freeman and Company.

Barazangi, M., and J. Dorman. 1969. "World Seismicity Maps Compiled from ESSA, Coast and Geodetic Survey, Epicenter Data 1961-1967." Bull. Seismol. Soc. Amer. 59(1): 369-80 (precise plots).

Bullard, E., J. E. Everett, and A. G. Smith. 1965. "The Fit of the Continents around the Atlantic." Phil. Trans. Roy. Soc. London Ser. A Math. Phy. Sci. 258:27-41.

Calder, N. 1973. The Restless Earth. New York: Viking Press.

Condie, K. C. 1976. Plate Tectonics and Crustal Evolution. New York: Pergamon Press.

Cox, A., ed. 1973. Plate Tectonics and Geomagnetic Reversals. San Francisco: W. H. Freeman and Company.

Cox, A., G. B. Dalrymple, and R. R. Doell. 1967. "Reversals of the Earth's Magnetic Field." Sci. Amer. 216(2):44-54.

Dewey, J. F. 1972. "Plate Tectonics." Sci. Amer. 226(5):56-68. (Reprint no. 900.) San Francisco: W. H. Freeman and Company.

Dewey, J. F., and J. M. Bird. 1970. "Mountain Belts and the New Global Tectonics." J. Geophys. Res. 75(14):2625-47.

Dickinson, W. R. 1971. "Plate Tectonics in Geologic History." Science 174:107-13.

Dietz, R. S., and J. C. Holden. 1970. "The Breakup of Pangaea." Sci. Amer. 223(4):30-41.

Hallam, A. 1973. A Revolution in the Earth Sciences. Oxford: Clarendon Press.

Heirtzler, J. R. 1968. "Sea-Floor Spreading." Sci. Amer. 219(6):60-70. (Reprint no. 875.) San Francisco: W. H. Freeman and Company.

Hess, H. H. 1962. History of Ocean Basins. In A. E. J. Engel et al., eds. Petrologic Studies: A Volume in Honor of A. F. Buddington, pp. 599-620. Boulder, Colo.: Geological Society of America.

Hurley, P. M. 1968. "The Confirmation of Continental Drift." Sci. Amer. 218(4):52-64. (Reprint no. 874.) San Francisco: W. H. Freeman and Company.

Isacks, B., J. Oliver, and L. R. Sykes. 1968. "Seismology and the New Global Tectonics." J. Geophys. Res. 73(18):5855-99.

Le Pichon, X. 1968. "Sea-Floor Spreading and Continental Drift." J. Geophys. Res. 73(12):3661-97.

Marvin, U. B. 1973. Continental Drift. Washington, D.C.: The Smithsonian Institution.

McKenzie, D. P. 1972. "Plate Tectonics and Sea Floor Spreading." Amer. Scientist 60:425-35.

Sullivan, W. 1974. Continents in Motion: The New Earth Debate. New York: McGraw-Hill.

Takeuchi, H., S. Uyeda, and H. Kanamori. 1970. Debate about the Earth. San Francisco: Freeman, Cooper and Company.

Vine, F. J. 1966. "Spreading of the Ocean Floor: New Evidence." Science 154(3744):1405-15.

Vine, F. J., and D. H. Matthews. 1963. "Magnetic Anomalies over Oceanic Ridges." Nature 199:947-49.

Wilson, J. Tuzo, compiler. 1970. Continents Adrift: Readings from Scientific American. San Francisco: W. H. Freeman and Company.

Wyllie, P. J. 1976. The Way the Earth Works. New York: John Wiley and Sons.

York, D. 1975. Planet Earth. New York: McGraw-Hill.

The Earth's Seismicity 18

Each year, more than 150,000 earthquakes are recorded by the worldwide network of seismic stations and are analyzed, with the aid of computers, at the earthquake data center in Boulder, Colorado. With this system, the exact location, depth, and magnitude of all detectable earthquakes are plotted on regional maps, and information about the direction of fault movement associated with the shock is determined. As a result, we are able, literally, to monitor the details of present plate motion. Indeed, seismology provides some of the most convincing evidence in support of the theory of plate tectonics. But that's not all. In addition, seismic waves provide the most effective probe to the earth's interior and the basic data upon which our present concepts of the internal structure of the earth are based.

Major Concepts

1. Seismic waves are vibrations in the earth that are caused by the rupture and sudden movement of rocks that have been strained beyond their elastic limits.
2. Three types of seismic waves are produced by an earthquake shock: (a) P waves, (b) S waves, and (c) surface waves.
3. The primary effect of an earthquake is ground motion; secondary effects include (a) landslides, (b) tsunamis, and (c) regional or local uplift and subsidence.
4. Most earthquakes occur along plate boundaries. Divergent plate boundaries produce shallow earthquakes and convergent plate boundaries produce a zone of shallow, intermediate, and deep-focus earthquakes.
5. Infrequent shallow earthquakes occur within plates rather than along plate boundaries.
6. The velocities at which P and S waves travel through the earth indicate that the earth has a solid inner core, a liquid outer core, a thick mantle, a soft asthenosphere, and a rigid lithosphere. The most striking discontinuity is between the core and the mantle.

Characteristics of Earthquakes

Statement

Earthquakes are vibrations of the earth caused by the rupture and sudden movement of rocks that have been strained beyond their elastic limits. The vibration of the earth results when the rock breaks and then snaps into a new position. In the process of rebounding, vibrations called seismic waves are generated. These vibrations are somewhat analogous to those produced by a pebble dropped into a pool of water. The stone introduces energy into the system, and concentric waves spread out in all directions from the point of disturbance. The land waves produced by an earthquake have been reported with heights of more than 0.5 m and with wavelengths of 8 m. The period between passage of successive waves can be as much as 10 seconds, and all of the vibrations can continue for as long as an hour before the wave dies out.

Three types of seismic waves are generated by an earthquake shock.

1. **P** (primary) **waves,** in which particles move back and forth in the direction the wave travels.
2. **S** (secondary) **waves,** in which particles move back and forth at right angles to the direction the wave travels.
3. **Surface waves,** which travel only in the outer layers of the earth and are similar to waves on water.

Surface waves cause the most damage during an earthquake and (together with secondary effects from associated landslides, **tsunamis** [tidal waves], and fire) result in the loss of approximately 10,000 lives and $100 million each year.

Discussion

Elastic Rebound Theory. Earthquakes are caused by ruptures that occur where rocks are strained beyond their elastic limits. The concept can be illustrated by a simple experiment with a stick that is bent until it snaps. Energy is stored in elastic bending and is released when rupture occurs, causing the fractured ends to vibrate and send out sound waves. Detailed studies of active faults show that this model, known as the **elastic rebound theory,** appears to hold for all major earthquakes (*figure 18.1*). Precision surveys across the San Andreas fault in California show that railroads, fence lines, and streets first are slowly deformed and then are offset when movement along the fault occurs and the elastic strain is released. The San Andreas fault is the boundary between the Pacific and North American plates. Movement is horizontal, with the Pacific plate moving toward the northwest. On a regional basis, the plates move quite steadily at a rate of roughly 5 cm per year; but, along the fault, movement can be smooth or it can occur in a series of jerks because sections along the fault can be "locked" together until enough strain accumulates to cause displacement. The elastic rebound theory explains the origin of earthquakes as a result of rupture and sudden movement along a fault or of recurrent movement along existing fractures.

The point within the earth where the initial slippage

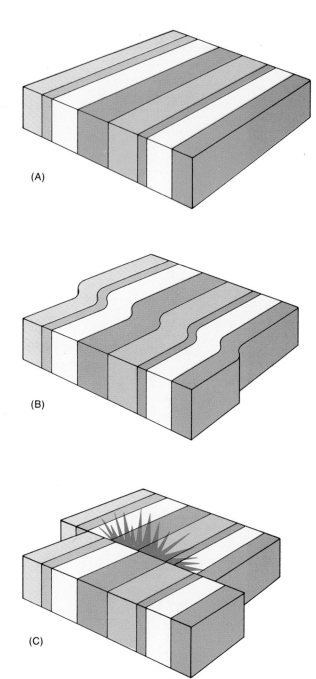

Figure 18.1 *The origin of earthquakes. (A) Undeformed rocks. (B) Strain is built up in the rocks until they rupture or move along preexisting fractures. (C) Energy then is released, and seismic waves move out from the point of rupture.*

occurs to generate earthquake energy is called the **focus,** and the point on the earth's surface directly above the focus is called the **epicenter** (*figure 18.2*).

Types of Seismic Waves. Three major types of seismic waves are generated by an earthquake shock (*figure 18.3*). Each type travels at a different speed through the earth so that each arrives at a seismograph hundreds of kilometers away from the others, at a different time. The first wave to arrive is called the primary (or P) wave. P waves are identical in character to sound

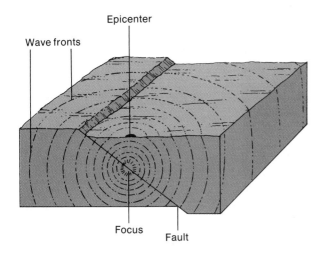

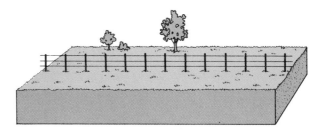

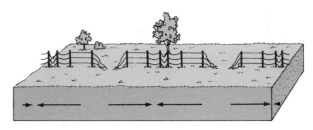

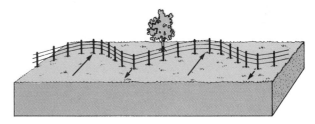

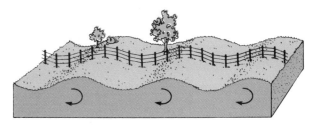

Figure 18.2 *Diagram showing the relationship between the focus and the epicenter of an earthquake. The focus is the point of initial movement on the fault. Seismic waves radiate out from this point. The epicenter is a point on the earth's surface directly above the focus.*

waves passing through a liquid or a gas. The particles involved in these waves move forward and backward in the direction the waves travel and cause relatively small movements. The second waves to arrive are called secondary (or S) waves. In these waves, particles move back and forth at right angles to the direction the wave travels. These cause strong movements to be recorded on the seismograph. The last waves to arrive are surface waves, which travel relatively slowly over the earth's surface. The motion of surface waves is in an orbit similar to waves on water. Surface waves cause the most damage.

Occurrence. The focus of an earthquake is important in the study of plate tectonics, because it indicates the depth within the earth at which rupture and movement occur. Although movement of material occurs throughout the entire earth (through its depth of 6371 km), earthquakes are concentrated in the upper 700 km, where the brittle lithosphere fractures and undergoes sudden slippage.

Within the 700-km range, earthquakes can be grouped according to depth of focus. Shallow-focus earthquakes occur from the surface to a depth of 70 km. They occur in all seismic belts and produce the largest percentage of earthquakes. Intermediate-focus earthquakes occur between 70 and 300 km. And deep-focus earthquakes between 300 and 700 km. Both intermediate and deep earthquakes are limited in number and distribution and generally are confined to converging plate margins. The maximum energy released by an earthquake tends to become progressively smaller as the depth of focus increases, and seismic energy from a source deeper than 70 km is largely dissipated by the time it reaches the surface. Most large earthquakes are shallow focus and originate in the crust. The focus of an earthquake is determined from the time elapsing between the arrivals of the three major types of seismic waves.

The method of locating the epicenter of an earthquake is relatively simple and can be understood easily

by reference to *figure 18.4*. Since the P wave travels faster than the S wave, the P wave is the first to be recorded at a seismic station. The time interval between the arrival of the P wave and the S wave is a function of the distance from the epicenter. By tabulating the travel times of P and S waves from earthquakes of known sources, seismologists have constructed time-distance

(A) Fence line prior to seismic disturbance.

(B) Motion produced by a P wave. Particles are compressed and then expanded in the line of wave progression. P waves can travel through any earth material.

(C) Motion produced by an S wave. Particles move back and forth at right angles to the line of wave progress. S waves travel only through solids.

(D) Motion produced by a surface wave. Particles move in a circular path at the surface and diminish with depth.

Figure 18.3 *Motion produced by the various types of seismic waves. Each type of seismic wave produces a characteristic motion that can be illustrated by the distortions they produce in a straight fence line.*

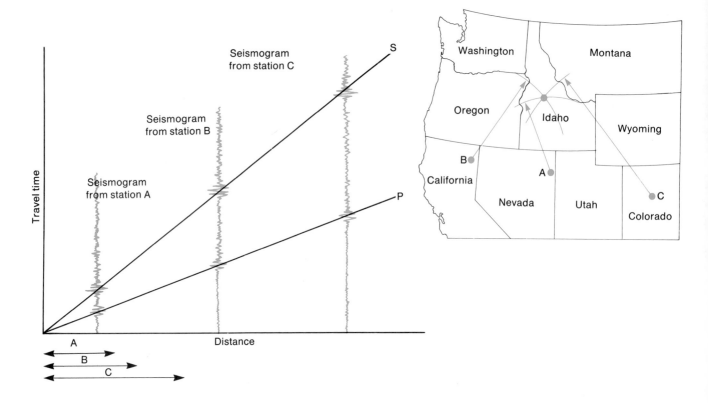

Figure 18.4 *Locating the epicenter of an earthquake. (A) Graph showing the time intervals between arrivals of P and S waves. By tabulating the arrival times of P and S waves, seismologists have constructed a time-distance graph with which the distance to an earthquake's epicenter can be determined. (B) When the distance to the epicenter is determined from three seismic stations, the location of the epicenter can be determined.*

graphs that can be used to determine the distance to the epicenter of a new quake. The seismic records indicate the distance, but not the direction, to the epicenter; therefore, records from at least three stations are necessary to determine the precise location of the epicenter.

Intensity. The intensity, or destructive power, of an earthquake is an evaluation of the severity of ground motion at a given location. It is measured in relation to the effects the earthquake had on human life and generally is described in terms of the destruction—buildings, dams, bridges, et cetera—caused, based upon reports by witnesses. An intensity scale commonly used in the United States and a description of some of its criteria are listed in *table 18.1*. The intensity of an earthquake is, of course, dependent upon a number of factors, foremost among which are (1) the total amount of energy released, (2) the distance from the epicenter, and (3) the type of rock and degree of consolidation. In general, larger wave amplitude and greater destruction occur in soft, unconsolidated material than in dense, crystalline rock. The intensity is greatest close to the epicenter.

Magnitude. The magnitude of an earthquake is a measure of the amount of energy released and is a much more precise measure of earthquakes than intensity. Based on direct measurements of the size (amplitude)

Table 18.1 Scale of Earthquake Intensity

I	Not felt except by very few people under special conditions. Detected mostly by instruments.
II	Felt by a few people, especially those on upper floors of buildings. Suspended objects may swing.
III	Felt quite noticeably indoors. Standing motorcars may rock slightly.
IV	Felt by many who are indoors, felt by a few outdoors. At night, some awakened. Dishes, windows, and doors rattle.
V	Felt by nearly everyone, many awakened. Some dishes and windows broken. Unstable objects overturned.
VI	Felt by everyone; many people become frightened and run outdoors. Some heavy furniture moved. Some fallen plaster.
VII	Most people in alarm and run outside. Damage negligible in buildings of good construction; considerable in buildings of poor construction.
VIII	Damage slight in specially designed structures; considerable in ordinary buildings; great in poorly built structures. Heavy furniture overturned.
IX	Damage considerable in specially designed structures. Buildings shift from foundation and partly collapse. Underground pipes broken.
X	Some well-built wooden structures destroyed. Most masonry structures destroyed. Ground badly cracked. Considerable landslides on steep slopes.
XI	Few, if any, masonry structures remain standing. Rails bent, broad fissures in ground.
XII	Virtually total destruction; waves seen on ground surface. Objects thrown into air.

of seismic waves, it is made with recording instruments and is not an expression of destruction based on subjective observations. The total energy released by an earthquake can be measured by recording the amplitude of the wave and the distance from the epicenter. Seismologists use the Richter scale as an expression of magnitude, with an arbitrary zero for the lowest limits of detection. Each step on the scale represents an increase in amplitude by a factor of 10 (*table 18.2*). Thus, an earthquake with a magnitude of 2 produces vibrations with 10 times the amplitude of one with a magnitude of 1, and an earthquake with a magnitude of 8 has an amplitude 1 million times greater than one with a magnitude of 2.

The largest earthquake yet recorded had a magnitude of approximately 8.5 on the Richter scale. Significantly larger earthquakes are not likely to occur, because there is insufficient strength in the rocks of the earth to allow enough energy to accumulate.

Effects of Earthquakes. The primary effect of earthquakes is the violent ground motion accompanying movement along a fracture. This motion can shear and collapse buildings, dams, tunnels, and other rigid man-made structures. Secondary effects include landslides, tsunamis, and regional or local submergence of the land. The following are a few examples of well-documented earthquakes in historical times.

The earthquake that shook Guatemala on February 4, 1976, had a magnitude of 7.5. It claimed the lives of 22,700 people and caused damage estimated at $1.1 billion. The earthquake occurred along the Motagua fault, which bisects Guatemala from east to west and is similar to the well-known San Andreas fault in California. Most of the damage and loss of life was caused by the collapse of buildings, many of which were constructed of adobe and so were easily destroyed by ground motion.

Another devastating earthquake was the one that hit Peru on May 31, 1970, and killed an estimated 50,000 people in 5 minutes. This is considered to have been the deadliest earthquake in the history of Latin America. The magnitude was 7.8, and, like the Guatemala earthquake, it caused extensive damage to towns and villages. Eighty percent of the adobe houses in an area of 65,000 km² were destroyed. Vibrations caused most of the destruction to buildings, but the massive landslides triggered by the quake in the Andes were a second major

cause of fatalities. Ninety percent of the resort town of Yungay, with a population of 20,000, was buried by a huge debris flow in a matter of a few minutes. Similar debris flows triggered by earthquakes killed 41,000 in Ecuador and Peru in 1797 and, in 1939, killed 40,000 in Chile.

An earthquake in Chile in 1960 produced extensive damage from ground motion, landslides, and flooding, but it also produced a spectacular tsunami, which devastated seaports with a series of waves 7 m high. This tsunami traveled across the earth's surface at speeds of approximately 1000 km per hour. It was 11 m high at Hilo, Hawaii, and, 22 hours after the quake, it reached Japan and caused approximately $70 million worth of damage.

One of the greatest tectonic events of modern times was the earthquake that occurred in southern Alaska late in the afternoon of March 27, 1964. The magnitude of the quake was between 8.30 and 8.75, and its duration ranged from 3 to 4 minutes at the epicenter (by comparison, the San Francisco earthquake of 1906 lasted about 1 minute). This was the most violent earthquake to occur in North America in the twentieth century. It is important because it had such marked effects on the earth's surface and because perhaps no other earthquake was better documented. Despite the earthquake's magnitude and severe tectonic effects, property damage and loss of life ranked far below other national disasters (114 lives lost and $311 million property damage), because much of the area affected was uninhabited. The crustal deformation associated with the Alaskan earthquake was more extensive than any known deformation related to previous earthquakes. The level of the land was changed in a zone 1000 km long and 500 km wide (an area of 500,000 km² was elevated or depressed). Submarine and subaerial landslides triggered by the earthquake caused spectacular damage to communities, and the shaking spontaneously liquified deltaic materials along the coast, causing slumping of the waterfronts in Valdez and Seward.

One of the most famous earthquakes in historical times was the San Francisco quake of 1906. The quake lasted only a minute, with a magnitude of 8.3. The fire that followed caused most of the destruction (an estimated $400 million in damage and a reported loss of 700 lives). From a geologic point of view, the earthquake was important because of the visible effects it produced along the San Andreas fault. Horizontal displacement occurred over a distance of about 400 km and offset roads, fences, and buildings by as much as 7 m.

Earthquakes and Plate Tectonics

Statement

The establishment in 1966 of a worldwide network of sensitive seismic stations that monitor nuclear testing has enabled seismologists to amass an amazing amount of information concerning earthquakes and plate tectonics. The data is analyzed with the aid of computers at the seismic data center in Boulder, Colorado. There, the location and magnitude of the thousands of earthquakes that occur each year are plotted on regional maps, and

Table 18.2 Richter Scale of Earthquake Magnitude

Magnitude	Approximate Number Per Year
1	700,000
2	300,000
3	300,000
4	50,000
5	6,000
6	800
7	120
8	20
over 8	1 every few years

other information about the direction of displacement on the faults where the earthquakes occurred is synthesized. The result is a new and important insight into the details of current plate motion. This is what has been found.

1. The distribution of earthquakes delineates plate boundaries.
2. Shallow earthquakes coincide with the crest of the oceanic ridge and with transform faults between ridge segments.
3. The earthquakes along the ridge crest originate in normal faults trending parallel to the ridge crest, indicating tensional stresses perpendicular to the trend of the faults.
4. Earthquakes in transform faults originate from lateral slips.
5. Beyond the ridge, the transform faults do not produce earthquakes and are not active.
6. Earthquakes at converging plate margins occur in a zone inclined downward beneath the adjacent continent or island arc.

Discussion

Global Patterns of the Earth's Seismicity.   The location and depth of tens of thousands of earthquakes that have occurred since the establishment of a worldwide

network of seismic observation stations are summarized in the seismicity map in *figure 18.5*. From the standpoint of the earth's dynamics, this map is an extremely significant compilation because it shows where and how the crust of the earth is moving at the present time. The most obvious and significant facts about the earth's seismicity are that earthquakes occur in narrow belts that coincide precisely with tectonic plate boundaries and that the type of seismicity is governed by the type of plate boundaries. In addition, with a sufficient number of seismic recording stations, it is possible to determine the direction of movement on the fault at the time the earthquake occurred. This greatly enhances our knowledge of plate motion.

Seismicity at Divergent Plate Boundaries.   The global pattern of the earth's seismicity shows a narrow belt of shallow earthquakes that coincides almost precisely with the crest of the oceanic ridge and marks the boundaries between diverging plates. This zone is remarkably narrow when compared to the zone of seismicity that follows the trends of young mountain belts and island arcs. The shallow earthquakes in this zone are less than 70 km deep and typically are small in magnitude. Although the zone appears as a nearly continuous line on regional maps, there are two types of seismic boundaries that are distinguishable on the basis of fault motion. These are (1) spreading centers and (2) transform faults (*figure 18.6*). The earthquakes associated with the crest of the oceanic ridge occur within, or near, the rift valley and appear to be associated with normal faulting and intrusions of basaltic magmas. Locally, earthquakes occur in swarms. Detailed studies indicate that the earthquakes associated with the ridge crest are produced by vertical faulting (a process that appears to be responsible for the ridge topography [see p. 337]).

Figure 18.5   *Seismicity of the earth. The locations of tens of thousands of earthquakes that occurred between 1961 and 1967 show that earthquakes occur most frequently along plate margins. Shallow earthquakes occur at both diverging and converging margins, whereas intermediate- and deep-focus earthquakes are restricted to the subduction zones of converging plates.*

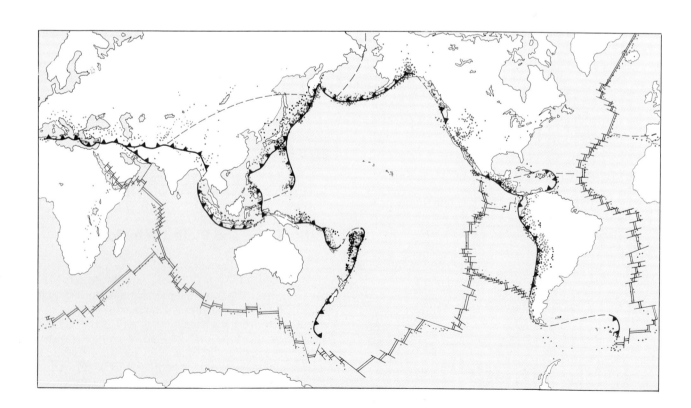

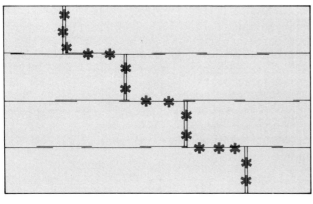

—·—·— Fracture zone

════ Ridge crest

✳ Earthquake epicenters

Figure 18.6 *Map showing the distribution of earthquakes associated with divergent plate boundaries. Two seismic boundaries are recognized: (1) seismicity along the spreading centers resulting from normal faulting and (2) seismicity along transform faults resulting from strike-slip movement.*

Figure 18.7 *(A) Map showing the depth of earthquakes in the Tonga region of the South Pacific. (B) A vertical cross section across this area shows the inclination of the zone of earthquakes beneath the Tonga trench.*

Shallow-focus earthquakes also follow the transform faults that separate the ridge crest but generally are not associated with volcanic activity. Studies of fault motion indicate horizontal displacement in a direction away from the ridge crest. Moreover, as would be predicted from the plate tectonic theory, earthquakes are restricted to the active transform-fault zone—the area between ridge axes—and do not occur in the inactive fracture zones.

Seismicity at Converging Plate Boundaries. The most widespread and intense zone of earthquake activity on the earth occurs along the subduction zones at converging plate boundaries. This is immediately apparent from the world seismicity map shown in *figure 18.5*, where there is a strong concentration of shallow, intermediate, and deep earthquakes coinciding with the subduction zones of the Pacific Ocean. The three-dimensional distribution of earthquakes in this belt defines a seismic zone that is inclined at moderate to steep angles from the trench down under the adjacent island arcs or continental borders. This is well illustrated in the Tonga trench in the South Pacific, where the zone of seismicity forms a nearly planar surface that plunges down into the mantle to a depth of more than 600 km (see *figure 18.7*).

Studies of fault motion of seismic waves generated in this zone indicate variations in faulting as a function of depth. Near the walls of the trench, normal faulting is typical as a result of tensional stresses generated by the initial bending of the plate. In the zone of the shallow earthquakes, thrust faulting dominates as the descending lithosphere slides beneath the upper plate. At inter-

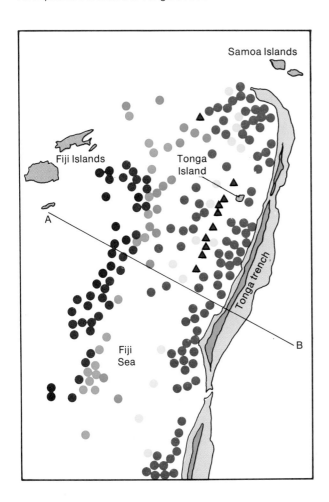

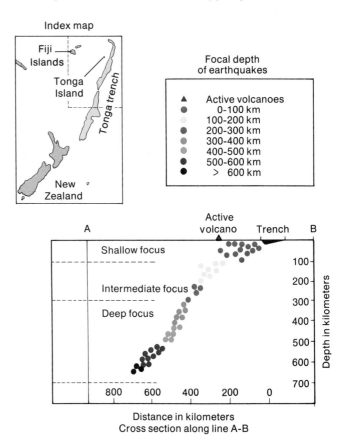

mediate depths, extension or compression can occur, depending on the specific characteristics of the subduction zone. Extension and normal faulting can occur when the descending slab is denser than the surrounding mantle and sinks under its own weight. Compression can result if the mantle resists the downward motion of the descending plate. The deep zone of earthquakes shows compression within the descending slab of lithosphere, indicating that the material of the mantle at that depth resists movement of the descending lithosphere.

Intraplate Seismicity.   Although most of the world's seismicity occurs along plate boundaries, the continental platforms do experience infrequent and scattered shallow-focus earthquakes. The zones of seismicity of East Africa and the western United States are most striking and are, in all probability, associated with the spreading centers, which can be projected into those regions. The minor shallow earthquakes in the eastern United States and Australia, however, are more difficult to explain. Apparently, lateral motion of a plate across the asthenosphere involves minor vertical movement. Stress could build up and exceed the strength of the rocks within the lithospheric plate and cause infrequent faulting and seismicity. In contrast, the ocean floors beyond the spreading centers are seismically inactive, except for isolated earthquakes associated with oceanic volcanoes.

Plate Motion as Determined from Seismicity.   From the data shown in *figure 18.5,* the outline of the six major lithospheric plates and the direction of their present movement can be seen. The Pacific plate, consisting entirely of oceanic crust, is moving northeastward away from the spreading center along the east Pacific rise. Local variations in the direction of movement occur along the converging plate margins, because the Pacific plate is bordered by several different plates. The width of the zone of shallow-, intermediate-, and deep-focus earthquakes along the subduction zone indicates that the plate is descending down into the mantle at an angle of roughly 45°. The existence of the slab of lithosphere down to depths of 700 km at convergent boundaries indicates that the present movement has continued long enough for 700 km of new crust to have been generated at the oceanic ridges.

The American plates are moving eastward from the mid-Atlantic ridge and encounter the Pacific and adjacent plates along the trench on the east coast of South and Central America.

The African plate is moving northward toward the convergent boundary in the Mediterranean region by spreading from the ridge that essentially surrounds it in the Atlantic and Indian Oceans. There are no subduction zones between the Atlantic and Indian Oceans' spreading centers, so the plate boundaries surrounding Africa themselves apparently are moving relative to the African plate and to each other. The same is probably true for the Antarctic plate, which is surrounded completely by the spreading center of the oceanic ridge.

The Himalayas and the Tibetan Plateau define a wide belt of shallow earthquakes. This is an area where the converging plates produce a continent-to-continent relationship. India moved northward from the south until it collided with Asia, and the convergence of the plates

caused Asia to ride up and over the Indian plate, resulting in a double thickness of continental rocks in this area. This produced the wide zone of exceptionally high topography in the Himalayas and the Tibetan Plateau.

Seismicity, therefore, defines the present movement of the lithosphere and provides some of the most convincing support of the theory of plate tectonics.

Seismic Waves as Probes of the Earth's Interior

Statement

Speculations regarding the interior of the earth have stimulated man's imagination for centuries, but it was not until we learned how to use seismic waves to obtain an X-ray-type picture of the earth's interior that we have been able to probe the deep interior of the earth and formulate models of its structure and composition. Seismic waves can be used to study the earth's interior, because both P and S waves travel faster through rigid material than through that which is soft or plastic. Therefore, the velocity at which the P and S waves travel through a specific part of the earth gives some indication of the types of rock involved, and abrupt changes in velocities indicate significant changes in the earth's interior. One difference between P and S waves is particularly significant. P waves pass through any substance —solid, liquid, or gas—but S waves are transmitted only through solids that have enough strength to return to their former shape after being distorted by the wave motion. S waves, therefore, cannot be transmitted through a liquid.

Seismic waves passing through the earth are refracted and show distinct discontinuities that provide the basis for believing the earth has a solid inner core, a liquid outer core, a thick mantle, a soft asthenosphere, and a rigid lithosphere.

Discussion

Seismic waves are similar in many respects to light waves, and the paths they follow are governed by laws similar to those of optics. Put simply, both seismic and light vibrations move in a straight line through a homogeneous body; but, when they encounter a boundary between different substances, they are reflected or refracted. Familiar examples are light waves reflected from a mirror or refracted (bent) as they pass from air to water.

If the earth were a homogeneous solid, then seismic waves would travel through it at a constant speed and the **seismic ray** (a line drawn perpendicular to the wave front) would be a straight line like the one shown in *figure 18.8.* Early investigations, however, found that seismic waves arrived progressively earlier than they were expected to at stations successively farther from the earthquake source. As can be seen in *figure 18.8,* the rays arriving at distant stations travel deeper through the earth than those reaching stations close to the earthquake source. Obviously, then, if the long-distance waves arrive progressively earlier as they go down deeper into the earth,

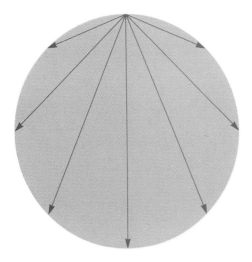

Figure 18.8 *The paths followed by seismic waves in a homogeneous planet. If a planet had uniform properties throughout, seismic waves would follow a linear path.*

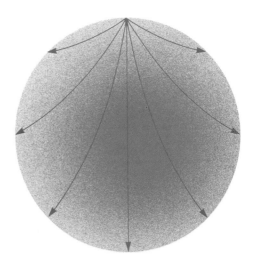

Figure 18.9 *The paths followed by seismic waves in a planet in which wave velocities increase steadily with depth. The change in velocity would cause the waves to be refracted, so the waves would follow a curved path.*

they must travel more rapidly at depth than they do nearer the surface. The significant conclusion from these studies is that the earth is not a homogeneous, uniform mass but that its physical properties change with depth. As a result, the seismic rays passing through the earth follow a curved path (*figure 18.9*).

Another important discovery was made in 1906, when it was recognized that whenever an earthquake occurs there is a large region on the opposite side of the planet where the seismic waves are not detectable. The nature and significance of this shadow zone perhaps can be best understood by referring to *figure 18.10*. If an earthquake occurs at a particular spot (labeled 0°), a shadow zone

Figure 18.10 *Interpretation of the shadow zone. The best way to explain the P-wave shadow zone is to postulate that the earth has a central core through which the P waves travel relatively slowly. Ray 1 would just miss the core and would be received at a station located at 103° from the earthquake's focus. Steeper rays, such as ray 2, would encounter the boundary of the core and would be refracted. Ray 2 would travel through the core, would be refracted again at the core's boundary, and would be received at a station less than 180° from the focus. Steeper rays would do the same until ray 3, which emerges at the surface 145° from the earthquake focus. Rays steeper than ray 3 would be bent severely by the core and would not be received in the shadow zone.*

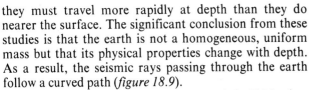

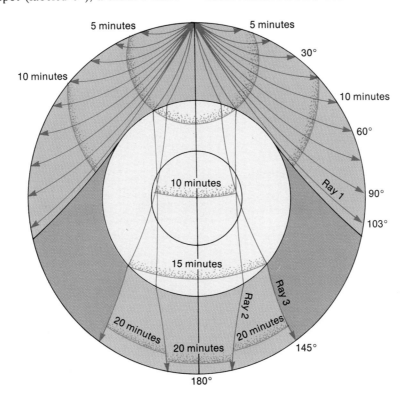

for P waves invariably exists between 103° and 143° distant from the earthquake focus. Evidently, something deflects the waves away from their expected paths. The best way to explain this shadow zone is to postulate that the earth has a central core through which the P waves travel relatively slowly. Seismic rays would follow a curved path from the earthquake's focus to a distance of 103° (slightly more than a quarter of the distance around the earth). The ray labeled 1 would just miss the core and would be received by a station located at 103°. Ray 2, however, would encounter the boundary of the core and would be refracted. It would travel through the core, be refracted again at the core's boundary, and would be received at a station on the opposite side of the earth. Steeper rays (3, 4, et cetera) also would be refracted through the core but would emerge on the opposite side of the earth beyond the shadow zone. *Figure 18.10* is a cross section through the earth; the true nature of the shadow zone is a band across the planet (as shown in *figure 18.11*).

In 1914, Gutenberg, a German seismologist, calculated the depth to the surface of the core to be 2900 km. Later analysis of much greater and more reliable seismic data showed that Gutenberg's original estimate was remarkably accurate, with a possible error of less than two-thirds of 1%. More recent studies of the P-wave shadow zone show that some weak P waves of low amplitude are received in this zone. This suggests an inner core that deflects the deep, penetrating P waves in a manner shown in *figure 18.12*.

The surface of the core has even a more pronounced effect on S waves, but this effect cannot be explained by reflection or refraction. S waves simply do not pass through the core, and they produce a huge shadow zone extending almost halfway around the earth opposite the earthquake's center (*figure 18.13*). Since S waves cannot pass through a liquid, the shadow zone generally is assumed to indicate that the core is liquid.

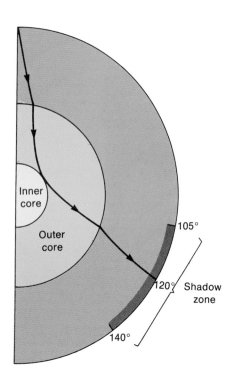

Figure 18.12   *Deflection of P waves by the inner core. Weak P waves received in the shadow zone suggest a solid inner core.*

With the present worldwide network of seismic stations, the variations in velocity of seismic waves as they travel through the earth can be determined with considerable accuracy. This data can be summarized graphically in velocity-depth curves like the one in *figure 18.14*, and it provides important additional information about the earth's interior. The most striking variation, of course, is that seen at the core's boundary at a depth of 2900 km. There, the S wave stops and the velocity of the P wave is drastically reduced. Other variations are apparent but are less striking. The first discontinuity occurs at a depth of 5 to 70 km below the surface. This is called the **Mohorovičić discontinuity** (or simply Moho), after the Yugoslavian seismologist who first recognized it, and is considered to be the base of the crust.

Perhaps the most significant discontinuity, however, is the low-velocity zone 100 to 200 km below the surface (*figure 18.15*). This zone was recognized in the 1920s by Gutenberg, but it was viewed with skepticism by most other seismologists at that time. The normal trend is for seismic velocities to increase with depth. At these depths, however, the trend is reversed and the seismic waves travel about 6% slower than they do in adjacent regions. More recent seismic data again confirms Gutenberg's observations and has convinced most seismologists that this zone in the earth is more plastic than areas above and below it.

The generally accepted explanation for the low-velocity zone is that the temperature and the pressure are such that part of the material, perhaps 1% to 10% is melted, producing a crystal-liquid mixture. A small amount of liquid film around the mineral grains would serve as a lubricant and increase the plastic nature of the material.

Figure 18.11   *The shadow zone produced by an earthquake in Japan. The shadow zone in which no direct P waves are received is a band on the earth's surface. From it, seismologists calculate that the depth to the core must be 2900 km.*

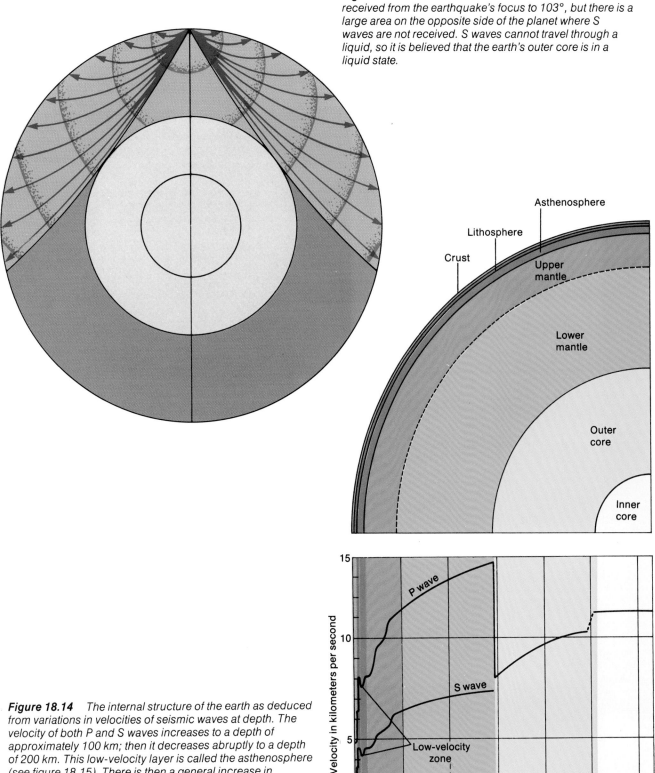

Figure 18.13 Shadow zone for S waves. S waves are received from the earthquake's focus to 103°, but there is a large area on the opposite side of the planet where S waves are not received. S waves cannot travel through a liquid, so it is believed that the earth's outer core is in a liquid state.

Asthenosphere

Lithosphere

Crust

Upper mantle

Lower mantle

Outer core

Inner core

Figure 18.14 The internal structure of the earth as deduced from variations in velocities of seismic waves at depth. The velocity of both P and S waves increases to a depth of approximately 100 km; then it decreases abruptly to a depth of 200 km. This low-velocity layer is called the asthenosphere (see figure 18.15). There is then a general increase in velocities of P and S waves to a depth of about 3000 km, where both change abruptly. The S wave does not travel through the central part of the earth, and the velocity of the P wave decreases drastically. This is the most striking discontinuity in the earth and is considered to be the boundary between the core and the mantle. Another discontinuity in P waves occurs at a depth of 5000 km, indicating an inner core.

P wave

S wave

Low-velocity zone

Velocity in kilometers per second

15

10

5

0

1000   2000   3000   4000   5000   6000

Depth in kilometers

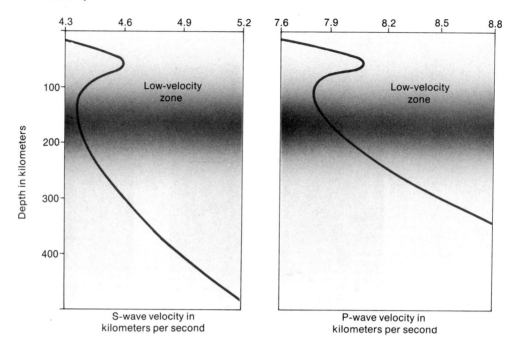

Figure 18.15 *Changes in the velocity of seismic waves in the asthenosphere. The rapid decrease in velocity of both P and S waves suggests a zone of low strength in the upper mantle at depths between 100 and 200 km below the surface.*

This low-velocity zone, as will be seen in later chapters, plays a key role in the theories of motion of material at the earth's surface. Other rapid changes in seismic-wave velocities occur at depths of about 350 km and 700 km and are interpreted to be the results of phase changes in the minerals, that is, rearrangements of the atomic packing into denser and more compact units.

Summary

Earthquakes are vibrations in the earth caused by the sudden movement of rocks that have been strained beyond their elastic limits. Three types of seismic waves are generated by an earthquake shock: P waves, S waves, and surface waves.

The intensity of an earthquake is a measure of its destructive power and depends on the total energy released, the distance from the quake's epicenter, and the nature of rocks in the crust. The magnitude of an earthquake is a measure of the total energy released. The primary effect of earthquakes is ground motion; secondary effects include (1) landslides, (2) tsunamis, and (3) regional uplift and subsidence.

The establishment in 1966 of a worldwide network of sensitive seismic stations has enabled seismologists to monitor the thousands of earthquakes that occur each year. This new insight about the earth's seismicity verifies, in a most remarkable way, the theory of plate tectonics, because it shows the present-day movement of plates. The distribution of earthquakes outlines with dramatic clarity the plate margins. Shallow earthquakes develop in a narrow zone along diverging plate margins, where the mantle rises and pulls the plates apart. Where plates collide and one is thrust under the other, a zone of shallow, intermediate, and deep earthquakes is produced as the oceanic plate moves down into the mantle. In the central part of the plates, there is little differential movement, and few earthquakes occur in these stable areas.

The velocities at which the P and S waves travel through the earth indicate that the earth has a solid inner core, a liquid outer core, a thick mantle, a soft asthenosphere, and a rigid lithosphere.

Additional Readings

Eckel, E. B. 1970. The Alaskan Earthquake, March 27, 1964: Lessons and Conclusions. Washington, D.C.: U.S. Geological Survey. Professional Paper 546.

Iacopi, Robert. 1964. Earthquake Country. Menlo Park, Calif.: Lane Book Company.

Nicholas, T. C. 1974. "Global Summary of Human Response to Natural Hazards: Earthquakes." *In* Natural Hazards, ed. G. F. White, pp. 274-84. New York: Oxford University Press.

Press, F. 1975. "Earthquake Prediction." Sci. Amer. 232(5): 14-23.

Wyllie, P. J. 1976. The Way the Earth Works. New York: John Wiley and Sons.

Volcanism

Volcanic eruptions are one of the most spectacular of all geologic phenomena, and for centuries they have dismayed and terrified mankind. Although they always have attracted attention because of their sometimes violent and catastrophic destruction, the significance of volcanic eruptions was only partly understood. It is now clear that volcanic activity is one of the key processes in the earth's dynamics. It is not a rare or abnormal event. It has occurred on the earth throughout most of its history and undoubtedly will continue far into the future. The importance of volcanic activity in the dynamics of the earth is readily appreciated by considering the volume of rock produced by it. Over two-thirds of the face of the earth—the ocean floors—consists entirely of rocks derived from lava, and volcanic activity is the most significant process in building island arcs. In addition, volcanism is important in many mountain chains and has constructed large lava fields along the margins of many continents.

The significance of volcanic activity, however, is not only in the quantity of lava that has been produced but in the close association it has with the movements of the tectonic plates, movements that also produce earthquakes, ocean basins, continents, and mountain belts. Moreover, volcanic activity is a window into the earth's interior that provides us with tangible products of the processes operating far below the planet's surface.

Major Concepts

1. Most volcanic activity corresponds to active seismic zones and is clearly associated with plate boundaries.
2. The type of volcanic activity depends upon the type of plate boundary.
3. Basaltic magma is generated at diverging plate boundaries by the partial melting of the asthenosphere. It is extruded mostly by quiet fissure eruptions.
4. Granitic magma is generated at converging plate margins by the partial melting of the oceanic crust.
5. Intraplate volcanic activity probably represents local hot spots in the upper mantle.

Global Patterns of Volcanism

Statement

Volcanoes occur in many different areas of the world, including the islands of the ocean and the young mountain ranges as well as the plains and the plateaus of the continents. The distribution of volcanic activity, however, is neither haphazard nor random. When the locations of active or recently active volcanoes are plotted on a tectonic map, two important facts stand out.

1. Most volcanic activity is coincident with the active seismic regions of the world and is clearly associated with the plate boundaries.
2. The type of volcanic activity depends upon the type of plate boundary.

These facts indicate that volcanic activity is a natural product of the tectonic system.

Discussion

The distribution of recently active volcanoes* and their relationships to the major plate boundaries are shown in *figure 19.1*. A worldwide belt of volcanic activity occurs along diverging plate margins but is largely concealed beneath the ocean. Locally, however, enough lava is extruded along this zone to build a volcanic pile that rises above sea level and can be observed directly. Iceland, the best example, is built entirely of volcanic

*When geologists speak of recently active volcanoes, they are referring to those that have erupted during the last 10,000 years or so (which in a geologic time frame is recent).

rock. Surtsey, a new island off Iceland's southern coast, was built by eruptions that started in 1963; a volcano erupted suddenly on the tiny island of Heimaey and buried much of Iceland's largest fishing port. Only a few submarine volcanoes rise above sea level south of Iceland, but submarine eruptions along the mid-Atlantic ridge have been reported when gas and ash were seen erupting above sea level by officers of ships.

The volcanoes located along divergent plate margins also would include those associated with the East African rift valleys and perhaps those along the margins of the Basin and Range Province in the western United States.

The most notable belt of volcanic activity occurs along converging plate margins and is intimately related to a belt of high seismicity that occurs along subduction zones. The most spectacular belt of volcanic activity practically surrounds the Pacific basin and has long been referred to as the "ring of fire." The volcanoes in this belt are unquestionably among the earth's great physical

Figure 19.1 *Active volcanoes and their relationship to plate boundaries. Basaltic volcanism and shallow intrusions occur all along the oceanic ridge where plates are moving apart. However, except for Iceland and a few small islands, the activity at spreading centers is concealed beneath the ocean. The most conspicuous volcanic activity occurs as chains of andesitic volcanoes at converging plate margins over subduction zones. These include the "ring of fire" around the Pacific Ocean and the volcanism of the Mediterranean and Near East. Intraplate volcanism is most active in the Pacific where plates move over hot spots in the mantle to produce volcanic islands and seamounts.*

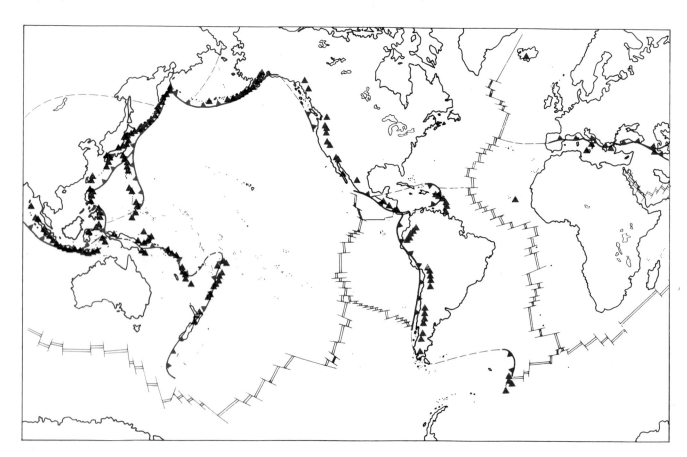

features. Their distribution is controlled by the subduction zones of the three major plates that make up the Pacific basin and the associated minor plates (such as the Caribbean and the Philippines).

Another similar belt of volcanic activity follows the converging margins of the African and Australian plates and extends through southern Europe to the Middle East and the island arcs of Indonesia (see *plate 31*).

Most of the world's famous volcanoes are located along the plate margins and correspond to the belt of high seismic activity, but some volcanoes occur within the central parts of tectonic plates. Most of these are in the Pacific. The islands of Hawaii are the most notable example.

Volcanism at Diverging Plate Margins

Statement

The diverging plate margins mark the site of the most voluminous extrusion of volcanic material on the earth. Yet, until recent oceanographic surveys were made, little was known about this important zone of volcanic activity and the little information that was gleaned from it generally was considered an abnormal curiosity. This is not surprising, because diverging plate margins are largely obscured beneath the oceans. However, we have obtained factual data on the processes operating in this zone from three different lines of study.

1. By studies of the seafloor, such as submarine profiles, drilling, dredging, and submarine photography.
2. By geologic studies in areas where parts of the diverging plates are elevated above sea level.
3. By studies of diverging plates along continental rift systems.

These studies indicate that volcanic activity along diverging plate margins is mostly quiet eruptions of basaltic magma and that volcanic activity has been occurring along divergent plate margins throughout the immensity of geologic time. This activity now is considered to be a fundamental process in the differentiation of the planet.

Discussion

Direct Observation of the Seafloor. Extensive studies of dredged samples and cores obtained from deep-sea drilling have been made in recent years in an attempt to understand the processes operating along the oceanic ridges. One of the most successful projects was a combined effort of French and American scientists who studied and photographed parts of the mid-Atlantic ridge in an area south of Iceland. The study, known as Project FAMOUS (French-American Mid-Ocean Undersea Study), began in 1971 and culminated in 1974 with a series of 42 dives in deep-diving vessels capable of reaching the seafloor for sampling and photography. Thousands of photographs show that the oceanic ridge, as expected, is composed of innumerable pillow structures (*figure 19.2*). Instead of forming a single flow unit like basalt flows on land, submarine flows characteristically form multitudes of bubblelike structures that resemble a jumbled mass of pillows. Pillow basalt forms because underwater eruptions are rapidly chilled or quenched,

Figure 19.2 *Pillow basalt along the mid-Atlantic ridge photographed at close range by scientists in the deep-diving submersible* Alvin.

producing a rounded frozen skin on the flow as soon as it comes in contact with the water. Breaks through the crust (or skin) of the flow produce a succession of swelling, pillow-shaped bodies that can become detached from the parent flow and come to rest on the seafloor while they are still hot and plastic. In this manner, a succession of pillow basalts accumulates, which shows an unmistakable appearance of having been formed underwater.

Another feature of great significance observed during project FAMOUS was the presence of numerous open fissures in the crust along the oceanic ridge. More than 400 fractures were observed in an area of 6 km², some of which were as wide as 3 m (*figure 19.3*). These are con-

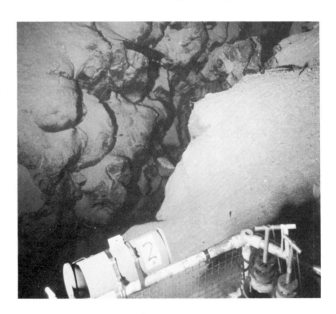

Figure 19.3 *Open fissure in basalt along the mid-Atlantic ridge clearly indicating that the rift zone in the oceanic ridge is pulling apart.*

sidered as conclusive evidence that the oceanic crust is being pulled apart.

Geologic Studies of Iceland. One of the best modern examples of the oceanic ridge rising above sea level is Iceland. There, we can examine in detail the surface expression of the spreading center. The island consists of a plateau of basalt with a well-marked, troughlike rift extending through the center (see *figure 17.12*). Although some cinder cones develop, the great floods of basalt have been extruded quietly along the fissures. The youngest rocks are located along the rift, with progressively older basalts occurring toward the east and west coasts. Of special importance is the presence of innumerable vertical basalt dikes called **sheet dikes.** The aggregate thickness of these dikes is about 400 km, which represents the total thickness of new crust created in Iceland during the last 65 million years. This information—together with data obtained from seismic studies, deep-sea drilling, and observations of the seafloor made by Project FAMOUS —confirms that the style of volcanism along divergent plate margins is largely fissure eruptions of fluid basaltic magma in which great floods of lava are extruded to form new crust.

Studies of Plateau Basalts. Where the spreading center passes beneath a continent, the continental crust is split and great volumes of basalt commonly are extruded and spread out over large areas near the rift system. These great floods of lava fill lowlands and depressions in the existing topography and, with time, are eroded into basalt plateaus (*figure 19.4*). The deposits, therefore, are referred to as **flood basalts** or **plateau basalts.** For example, in southern Brazil, over a million cubic kilometers of basalt were extruded in a relatively short period of geologic time (10 million years). Similar

Figure 19.4 *Flood basalts of the Columbia River Plateau.*

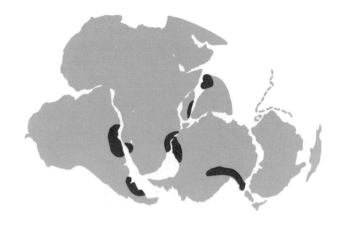

Figure 19.5 *The major plateau basalts are located near the continental margins and are believed to represent extrusions associated with the initial phases of continental rifting.*

floods have occurred in the Deccan Plateau of India, the Ethiopian Plateau of Africa, the Columbia River Plateau of the western United States (see also *plate 29*), and large areas of Siberia, Greenland, Antarctica, and northern Ireland. Much older flood basalts are found in northern Michigan and the Piedmont region of the eastern United States.

Plateau basalts are believed to represent the initial phases of continental drift and provide direct evidence of the nature of volcanic activity along divergent plate margins. Evidence for this can be seen in *figure 19.5,* a map of the southern continents prior to continental drift, which began in late Mesozoic time. The plateau basalts lie along the present-day continental margins, but they were extruded originally along rift systems that later developed into the oceanic ridge.

Excellent examples of plateau basalts are those of the Columbia Plateau in eastern Washington and Oregon

and western Idaho. These basalts lie along the northward extension of the east Pacific ridge and well may represent the initial stages of the break-up of the western part of North America. The Columbia River basalt covers an area of nearly 5 million km², with a total thickness of between 1 and 2 km (*figure 19.4*). This great accumulation of lava was not fed from central eruptions associated with a volcano, but instead the lava flowed to the surface through numerous fissures. Dikes grouped in vast swarms now mark some of the fissures through which this great volume of lava flowed.

The lavas of the Columbia River Plateau were extruded during the past 30 million years, with considerable activity taking place during the last million years. The most recent flows in the Crater of the Moon National Monument in southern Idaho occurred as fissure eruptions during the last few hundred years (*figure 19.6*). There, the lava forms extensive flows with pahoehoe or aa surfaces, and, locally along the fracture system, spatter cones or piles of cinder accumulate instead of the pillow lava that forms beneath the ocean.

An important point that these and other observations support is that the basic style of volcanic activity along divergent plate margins is through fissure eruptions of basaltic lava. The fluid lava flows readily from the cracks and fissures in the rift zone and tends to spread out laterally instead of building high volcanoes. Where extrusions occur beneath the sea, pillow lava forms, whereas eruptions of basalt on land produce floods of aa or pahoehoe flows. Regardless of where the lava is extruded, the basic type of fissure eruption is the dominant volcanic process along divergent plate margins.

Volume of Lava Extruded Along Divergent Plate Margins. The great volumes of lava extruded along diverging plate margins is difficult to comprehend, because most of the volcanic activity occurs beneath the

Figure 19.6 *Fissure eruptions in the Snake River plain, Idaho.*

Figure 19.7 *Map showing new crust created from volcanic activity along diverging plate margins during the last 10 million years.*

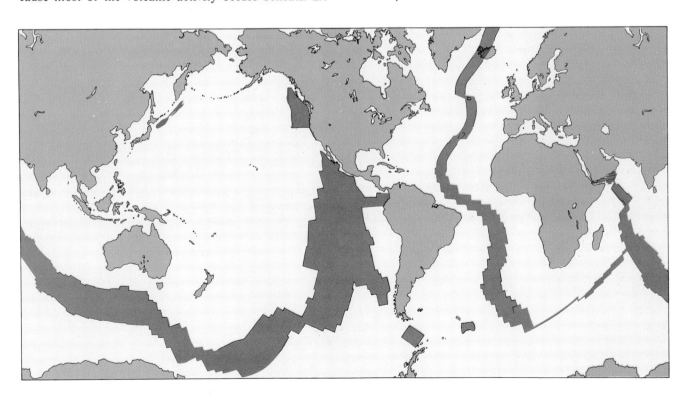

ocean and is never seen. The spreading centers, however, are the sites of the most extensive volcanic activity on the earth. To appreciate this fact, consider the amount of new oceanic crust created during the last 10 million years (*figure 19.7*). It has been estimated that approximately 20 km³ of basalt is extruded each year along this zone.

Generation of Magma. One of the most fundamental questions concerning volcanic activity along the spreading centers of the ocean floor is why and how magma is generated in this area rather than in some other place. Although many questions concerning this subject remain unanswered, the key to this problem lies in the special characteristics of temperature and pressure in the asthenosphere and their relationship to the melting of minerals in the mantle. The generation of large volumes of magma along diverging plate margins is not the result of high temperature alone but is related to the effects of pressure on the temperature at which melting occurs. Under high pressure, most materials require a higher temperature to induce melting. The asthenosphere is believed to be partly melted, because of the low velocity at which it transmits seismic waves and the degree to which seismic energy is attenuated. If it were rigid like the overlying lithosphere, it would transmit seismic waves at a much greater velocity. The physical characteristics of the asthenosphere might be compared to those of slushy snow; that is, a mixture of solid crystals and liquid

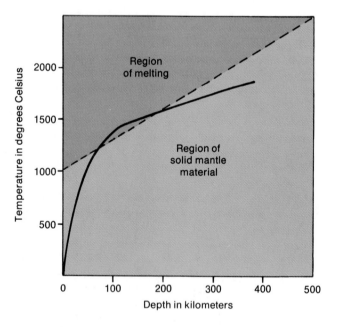

Figure 19.8 *Graph showing the relationship among temperature, pressure, and melting in the asthenosphere. In the upper part of the graph colored color, temperature and pressure conditions are such that melting would occur. The solid line shows theoretical temperature variations with depth. From the surface to a depth of about 100 km (the lithosphere), temperature is too low for melting. From a depth of 100 to 200 km (the asthenosphere), temperature passes into the zone where melting would occur. Although the temperature continues to rise below this depth, pressures are too great for melting to occur.*

or melt. This is probably the result of a critical balance in temperature, pressure, and composition at that depth (*figure 19.8*). In the overlying lithosphere, temperature is too low for melting to occur, while, in the underlying mantle, the confining pressure is so great that the rocks are well under their melting points.

An important factor influencing the generation of magma along the spreading centers is that, as the magma moves upward, the pressure is reduced and the decrease in pressure, in turn, lowers the temperature at which melting occurs. Thus, large bodies of magma are generated along the spreading center, in preference to other zones.

At this point, it should be reemphasized that each mineral in a rock has its own melting point and that, when the mantle—which is composed mostly of peridotite—is heated, pyroxene melts first. The magma produced by the partial melting of the asthenosphere would be basaltic in composition. The basaltic magma, being less dense than the solid mantle material, would move upward into the fractures formed between the spreading plates and would be extruded onto the ocean floor.

Volcanism at Converging Plate Margins

Statement

In previous sections, it has been shown that most of the world's volcanoes that erupt above sea level are clearly associated with subduction zones at converging plate margins (*figure 19.1*) The geographic setting for volcanic activity along this zone depends upon the type of plate interactions. Where two oceanic plates converge, an arcuate chain of islands builds up on the edge of the overriding plate parallel to the trench. Typical examples of this volcanic setting are Japan, the Aleutian Islands, and the Philippine Islands. Where a continent occurs on the leading edge of the overriding plate, similar volcanic activity develops in the mountain belt. The Andes Mountains of South America are an example of this setting. In both cases, the close association of volcanism and the converging plate margins is clear.

The type of volcanic activity produced along subduction zones, however, is quite different from the basaltic fissure eruptions that characteristically occur at spreading centers. The magma generated at the subduction zone is largely andesitic. It is somewhat richer in silica than basalt and thus is more viscous, so that entrapped gas cannot escape easily. This results in violent, explosive eruptions from central vents, which commonly produce ash flows, stratovolcanoes, and collapse calderas (see *plates 30* and *31*).

Discussion

An extremely large number of volcanic eruptions have occurred in historical times, but most were not recorded and many were not even observed. However, a few have had a major impact upon human affairs and have been described in minute detail. From these accounts and from studies of recent volcanic fields, we can gain some idea of the nature of volcanic activity associated with converging plates. Three classic examples will be considered: the eruption of Mount Vesuvius, Krakatoa, and Mount Pelée.

Vesuvius, A.D. 79. An extraordinarily vivid and accurate eyewitness account of the eruption of Mount Vesuvius in A.D. 79 comes from Pliny the Younger, a 17-year-old boy who related details of how his famous uncle, Pliny the Elder, died in the destruction of Pompeii and Herculaneum.

Beginning in A.D. 63 and continuing for 16 years, earthquakes shook the area around what is now Naples on the west coast of Italy. Then, on the morning of August 24, A.D. 79, Mount Vesuvius exploded with a devastating eruption of white-hot ash and gas. Within a few hours, it had asphyxiated and overwhelmed the population of Pompeii. Many people, suffocated by sulfurous fumes of the ash cloud, died in their homes or on the streets. The entire town and most of its 20,000 inhabitants were buried by ash and forgotton for more than a thousand years until Pompeii was excavated in 1748 (*figure 19.9*).

The great thickness of ash piled up on the slopes of Mount Vesuvius was saturated by rainstorms and soon gave rise to huge mudflows, which rushed down the mountainside. One of these overwhelmed the city of Herculaneum.

The ashfall that buried Pompeii is a first-class example of the type of violent eruption common in volcanoes along converging plates. The once-smooth and symmetrical cone of Mount Vesuvius was shattered by the explosion that created a large caldera where a peak once existed.

Krakatoa, 1883. Krakatoa is a relatively small volcanic island west of Java and is part of an island arc along the subduction zone associated with the Java trench. After remaining dormant for two centuries, it

Figure 19.9 *Excavation of Pompeii, Italy.*

began to erupt on May 20, 1883, and the eruption culminated in a series of four great explosions on August 26 and 27. One of the explosions was heard in Australia, 4800 km away. The explosions are considered to have been the greatest in recorded history. The whole northern part of the island, which stood about 600 m high, disappeared, and a huge caldera 300 m below sea level took its place (*figure 19.10*). Great quantities of ash were blown high into the atmosphere, and some of the ash circled the globe and took 2 years to fall. The island was uninhabited, but over 36,000 lives were lost in Java and Sumatra because of the huge tsunamis created by the explosion.

Mount Pelée, 1902. Ash flows are an important phenomenon associated with volcanism along converging plates, and the great eruption of Mount Pelée on the island of Martinique in the West Indies played an important part in initiating an interest in, and an understanding of, this type of eruption. The eruption of Pelée was preceded by nearly a month of extrusions of steam and fine ash from the volcanic vent. Then, on May 8, 1902, a gigantic explosion occurred in which ash and steam were blown thousands of meters into the air. The denser hot ash moved as a body and swept down the slopes like an avalanche. It took less than 2 minutes for the hot, incandescent ash flow to move 10 km from the side vent on Pelée and sweep past the city of St. Pierre, but it completely wiped out the population of more than 30,000 people, except for one prisoner deep underground in the city jail. Every flammable object was set aflame instantly, and as the ash moved over the waterfront all the ships were capsized.

The ash flow proper consisted of a mixture of hot pumice and ash that flowed at the base of a billowing cloud of gas. The fundamental force that caused the ash to flow and move so rapidly is simply the pull of gravity. Still, the mixture of hot gas and fragments of

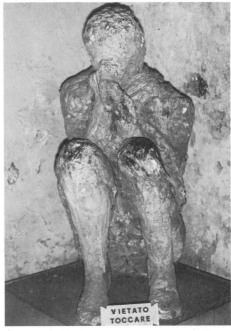

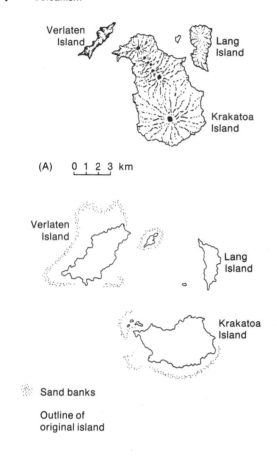

(A)   0 1 2 3 km

Verlaten Island

Lang Island

Krakatoa Island

Verlaten Island

Lang Island

Krakatoa Island

Sand banks

Outline of original island

(B)

Figure 19.10   *Maps showing Krakatoa before (A) and after (B) the eruption of 1883 (see also plate 31).*

ash and lava is highly mobile and practically friction-less, because each particle is separated from its neighbor and the surface over which it moves by a cushion of expanding gas.

Intermittent ash-flow eruptions continued for several months, and by October, a bulbous dome of lava too thick to flow formed in the crater. A spire of solidified lava then was slowly pushed out of a crack in the dome like toothpaste from a tube. This Spire of Pelée had a diameter of between 130 and 230 m and rose as much as 26 m per day to a maximum height of 340 m above the crater floor. The spire repeatedly crumbled and grew again from the lava dome, which glowed red in the night.

Generation of Magma.   The violent eruptions that characterize volcanoes at converging plate margins typically produce huge quantities of ash and thick viscous lava. Mobile ash flows are largely the result of the high silica content of magma generated in the subduction zone. Silicic magma is viscous, so dissolved gases cannot escape easily. As a result, tremendous pressure builds up in the magma, and, when eruptions occur, they are highly explosive. Some of the factors involved in the generation of magma in the subduction zone should be considered. *Figure 19.11* serves as a quick visual summary of the major factors involved. The basic idea is that magma originates by the partial melting of the basalts and sediments of the oceanic crust as the oceanic crust

Figure 19.11   *Diagram showing the major factors involved in generating magma in the subduction zone. At a critical depth, partial melting of the descending slab of oceanic lithosphere occurs. The melting generates andesitic magma, which rises buoyantly to form volcanic rocks of an island arc or granitic batholiths.*

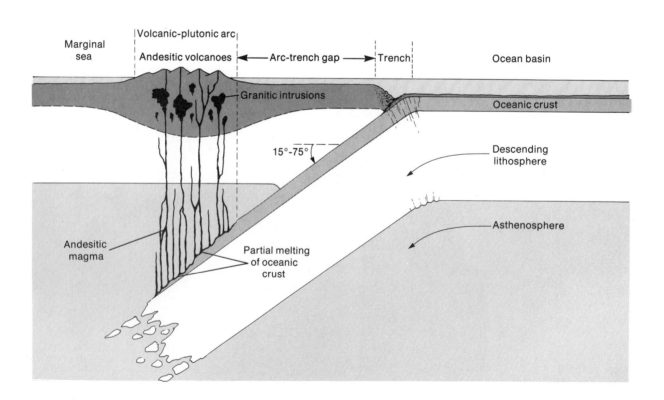

plunges diagonally into the hot asthenosphere. The key to understanding volcanic activity within this zone is the composition of magma produced by the partial melting of the basaltic oceanic crust. As was shown in the preceding section, basaltic magma is produced by the partial melting of the peridotite in the asthenosphere. As the oceanic lithosphere descends down the subduction zone, it is subjected to progressively higher temperatures and it begins to melt. The first material to melt is the layer of silica-rich sediment saturated with seawater, followed by Na-plagioclase, amphibole, and, finally, pyroxene. The residue containing minerals rich in magnesium and iron (olivine and some pyroxene) does not melt but continues to sink and becomes assimilated into the mantle. The magma produced by the partial melting of the oceanic crust thus is enriched in silica, sodium, and potassium (compared to basaltic magma generated from the partial melting of the upper mantle) and typically produces andesites and associated rocks.

As can be seen in *figure 19.11,* the depth of the descending plate increases away from the trench. This causes several systematic variations in volcanic activity across the arc. (1) Volcanic activity typically begins abruptly along a line 200 to 300 km inland from the trench axis. This line is called the **volcanic front.** It occurs because a critical depth of 100 km is necessary for the partial melting to produce enough magma to migrate to the surface. Thus, between the trench and the island arc there is a gap where volcanic activity does not occur. This area is called the **arc-trench gap.** The width of the gap is, of course, an expression of the angle of the descending plate. (2) The volume of magma that migrates to the surface and is erupted decreases rapidly behind the volcanic front, possibly as a result of a depletion of water and minerals with a low melting point. (3) The potassium content of the magma increases away from the trench toward the back of the volcanic arc. This may be related to differences in the depth at which the magma originates.

At this point, it may be well to emphasize the fact that, in the process of generating magma by partial melting at both the spreading centers and the subduction zones, the materials of the earth have been differentiated and segregated. The low-density material enriched in silicas, aluminum, sodium, potassium, and calcium is separated from the material of the mantle and concentrated first in the oceanic crust. Then, it is enriched further by a second episode of partial melting and concentrated in the island arcs or mountain belts of the continents. Because of its low density, this material cannot sink back into the mantle but is concentrated in the continental crust and is the principal method by which continents grow. More will be said about this in chapter 22.

Intraplate Volcanic Activity

Statement

Volcanic eruptions in the central parts of the plates beyond the active margins are trivial when compared to the vast volumes of magma extruded along the spreading centers and subduction zones, but they may be important as surface expressions of local thermal variations, or hot spots, in the mantle's material. The area of the most active intraplate volcanic activity is the floor of the South Pacific, where numerous submarine volcanoes and volcanic islands occur both as isolated features and in linear chains.

Igneous activity within continental platforms, in areas not associated with plate margins, is rare, and, where it does occur, it commonly is limited to scattered extrusions and small dikes and sills. These also are thought to be the result of **mantle plumes** (rising masses of hot mantle material that may or may not be parts of a large convection cell).

Discussion

A map showing the distribution of the major centers of intraplate volcanic activity is shown in *figure 19.12.* At first glance, the distribution of intraplate volcanoes may appear to be random, but obvious linear trends, or chains, are soon apparent. Excellent examples include the Hawaiian-Emperor chain, the Tuamotu-Line chain, and the Austral-Marshall-Gilbert chain. The best data available suggests that volcanic chains result when a lithospheric plate moves over a mantle plume, or hot spot (*figure 19.13*). Volcanism would occur over the hot spot and would produce a submarine volcano that could grow to become an island. If the position of the hot spot remained fixed in the mantle for long periods of time, the moving lithosphere would carry the volcano beyond the magma source, where it would then become dormant, and a new volcano then would form over the fixed hot spot. A continuation of this process would build one volcano after another and would produce a linear chain of volcanoes parallel to the direction of plate motion.

Isolated volcanoes may result from small hot spots that do not endure long enough to produce a volcanic chain, or they may develop from the occasional tapping of minor pockets of magma carried with the moving asthenosphere.

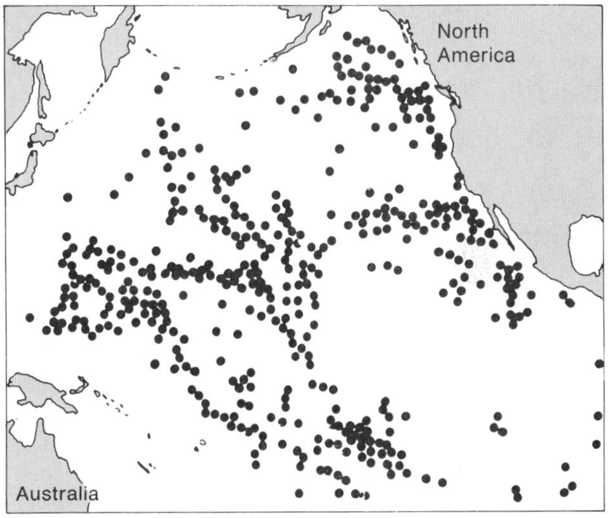

Figure 19.12 *Map showing intraplate volcanism in the Pacific.*

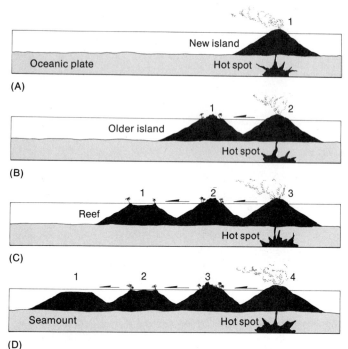

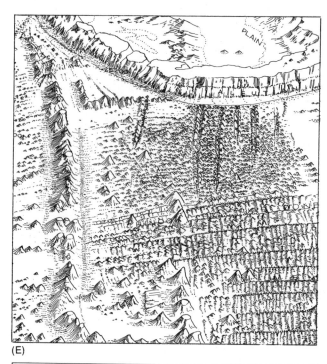

(A)

New island

Oceanic plate   Hot spot

(B)

Older island

Hot spot

(C)

Reef

Hot spot

(D)

Seamount   Hot spot

(E)

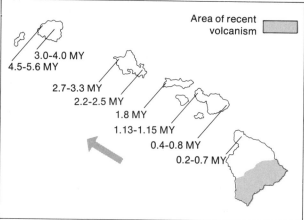

3.0-4.0 MY
4.5-5.6 MY

2.7-3.3 MY

2.2-2.5 MY

1.8 MY

1.13-1.15 MY

0.4-0.8 MY

0.2-0.7 MY

Area of recent volcanism

Figure 19.13 The origin of linear chains of volcanic islands and seamounts according to the plate tectonic theory. (A) A volcanic island or a seamount is built up by extrusions from a fixed hot spot, or source of magma, in the mantle. (B) As the plate moves, the volcano passes over the source of magma and becomes dormant. The surface of the island then may be eroded to sea level, and reefs may grow to form an atoll. A new island is formed over the hot spot. (C) Continued plate movement produces a chain of islands. (D) The islands of the chain become progressively older away from the hot spot. (E) An abrupt change in the direction of plate movement is indicated by a change in direction of a chain of islands in the Pacific. The Emperor Seamount chain began to form more than 40 million years ago when the plate was moving northward. About 25 million years ago, the plate moved northwestward and started to form the Midway-Hawaiian chain.

The type of volcanic activity in the interior of plates is similar, in most cases, to that produced along diverging plates. The products are largely basaltic lava extruded by quiet fissure eruptions. The basaltic lava thus is believed to be a derivative from the mantle that is like that found along the oceanic ridges.

Summary

Most volcanic activity coincides with active seismic zones and is clearly associated with the plate boundaries. The type of volcanism depends upon the type of plate boundary. At diverging plate margins, basaltic magma is generated by the partial melting of the asthenosphere and is extruded largely as fissure eruptions. Pillow lavas form on the seafloor, and plateau basalts are extruded where continents overlie rift zones.

At converging plate margins, andesitic or granitic magma is generated in the subduction zone by the partial melting of the oceanic crust, and it is intruded in the

upper plate as batholiths or it is extruded as composite volcanoes in island arcs or mountain belts. Andesitic magma is viscous and erupts violently, commonly in the form of an ash flow.

Minor intraplate volcanic activity probably indicates local hot spots in the mantle. The area of greatest intraplate volcanism is in the Pacific, where numerous shield volcanoes form islands and seamounts.

Additional Readings

Bullard, F. 1962. Volcanoes: In History, in Theory, in Eruption. Austin, Tex.: University of Texas Press.

Francis, P. 1976. Volcanoes. New York: Penguin Books.

Green, J., and N. Short. 1971. Volcanic Landforms and Surface Features: A Photographic Atlas and Glossary. New York: Springer-Verlag.

Stearns, H. T. 1946. "Geology of the Hawaiian Islands." Terr. Hawaii Div. Hydrogr. Bull. (8).

Tazieff, H. 1974. The Making of the Earth. Westmead, Farnborough, Hants, England: Saxon House.

Williams, H. 1951. "Volcanoes." Sci. Amer. 185 (6): 45-53.

Crustal Deformation

From the first geologic studies that began over 150 years ago, geologists have found that rocks in certain parts of the continents have been folded and fractured on a scale so large that, at first, it is difficult to comprehend. Deformation of the crust is most intense throughout the great mountain belts of the world where sedimentary rocks—which were originally horizontal—are now folded, contorted, fractured, and, in places, completely overturned. In some mountains, large bodies of rock have been thrust several tens of kilometers over younger strata. In the deeper parts of mountain belts, deformation (and elevated temperature) are great enough to cause metamorphism (see chapter 7). It is apparent, therefore, that many rock bodies contain an important record of the earth's movements and the mobility of the earth's crust during past geologic ages. It is this record of the earth's movements that will be examined in this chapter.

Major Concepts

1. *Deformation of the earth's crust is well documented by historical movement along faults, raised beach terraces, and deformed rock bodies.*
2. *Folds in rock strata range in size from microscopic wrinkles to large flexures hundreds of kilometers long. The major types are domes and basins, plunging anticlines and synclines, and complex flexures.*
3. *Faults are fractures in the crust along which slippage or displacement has occurred. The three basic types of faults are normal, thrust, and strike-slip.*
4. *Joints are fractures in rocks without displacement.*
5. *Unconformities are major discontinuities in rock sequence that indicate interruptions in rock-forming processes and result from crustal movement.*

Evidence of Crustal Deformation

Statement

Although the casual observer may feel that the crust of the earth is permanent and fixed, there is a great deal of evidence, both direct and indirect, indicating that the crust is in continuous motion and that it has moved on a vast scale throughout all of geologic time.

Discussion

Perhaps the most convincing evidence that the crust is moving is seen in the effects of historical earthquakes, during which movement of the crust is experienced and deformation can be seen in displacements of roads, fences, and buildings. One of the most impressive examples is the movement along the San Andreas fault during the 1906 San Francisco earthquake that offset fences and roads by as much as 7 m. Other examples include the 1899 earthquake at Yakutat Bay, Alaska, during which a beach along the coast was uplifted 15 m above sea level. Similar displacements are well docu-

Figure 20.1 *Evidences of crustal deformation produced during historical earthquakes.*

mented with photographs from the Good Friday earthquake in Alaska in 1964, the Hebgen Lake earthquake in Montana in 1959, and the Dixie Valley earthquake in Nevada in 1954 (*figure 20.1*).

Various topographic features also testify to earth movements in prehistorical times. One of the most striking examples is the raised beach terraces along the coast of southern California, where a series of ancient wave-cut cliffs and terraces containing remnants of beaches with barnacles, shells, and clean sand rise in steplike forms above the present shore (*figure 20.2*).

The rocks themselves contain additional evidence of deformation on a much larger scale. One of the most striking examples, seen in practically every mountain range, is the deformation of sedimentary strata that originally were deposited in a horizontal position below sea level. This evidence, preserved in tilted and contorted rock layers, is present in all types of sedimentary rocks,

Figure 20.2   *Evidence of recent uplift along the southern coast of California is well documented by a series of elevated wave-cut platforms such as these, on the Palos Verdes Hills.*

both young and old, and testifies to the continuing motion of the lithosphere and the ceaseless deformation it produces (*figure 20.3*).

Orogenesis.   Most of the deformation in the earth's crust is the result of the convergence of lithospheric plates. Where plates converge along continental margins, the sedimentary rocks (which originally were deposited in the ocean surrounding the continental platform) are deformed into an elongated mountain belt consisting of strongly folded and faulted rocks. This type of deformation is called **orogenic** (Greek: mountain genesis), and the event is referred to as an **orogeny.** The folded rocks in the major mountain belts of the world (Appalachians, Rockies, Himalayas, Urals, and Alps) are all excellent examples of this type of deformation (*figure 20.4*).

On the stable platforms of the continents, another type of deformation is present and is restricted largely to broad upwarps and downwarps of the crust. Coal beds now being mined at great depths below the surface are evidence of subsidence inasmuch as the bedding planes of coal were once at the surface where vegetation could

grow and now are buried thousands of meters below the surface. Deposits of sedimentary rocks, now buried thousands of meters below the surface, also show evidence of slow subsidence during sedimentation, and, in areas such as the Colorado Plateau, the marine sedimentary rocks that exist 1000 m above sea level indicate gradual uplift.

In summary, the evidence of a mobile crust can be seen almost everywhere. Movement is experienced during earthquakes, and displacement along associated faults usually is expressed in an impressive way. Evidence of earlier movement can be seen in features of the landscape, and ancient crustal movement is recorded in a variety of ways in the rocks themselves. All of the lines of evidence point to the same conclusion: the crust of the earth is in motion and has been in motion throughout geologic time. More will be said about crustal movements in later chapters, but first the types of structures produced by this movement will be considered.

Dip and Strike

Statement

Many structural features of the crust are too large to be seen from any given point on the ground and are recognized only after the geometry of the rock body is

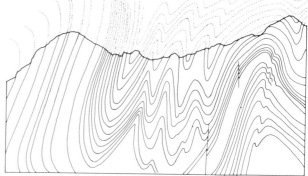

Figure 20.3 *The structure of a mountain. Highly deformed layers of sedimentary rock such as these show the degree of deformation that occurs in a mountain belt.*

determined from geologic mapping. At the outcrop, two fundamental observations can be made that describe the positions of bedding planes, fault planes, joints and other planar features in the rock. These are **dip** and **strike.** The dip of a plane is the angle and direction of inclination from the horizontal. The strike is the direction, or trend, of the planar feature; more precisely, it is defined as the compass direction of a horizontal line on the planar feature. These two measurements together define the position of the planar surface in space.

Figure 20.4 Folded rocks of a mountain belt as seen from space (see also plates 7, 26, and 28).

Figure 20.5 The strike and dip of an inclined bed. The strike of a bed is the compass direction of a horizontal line drawn on the bedding plane. This can be readily established by reference to the horizontal waterline shown in the illustration. The dip is the angle and direction that the bed is inclined.

Discussion

The concept of dip and strike can be easily understood with reference to the roof of a building. The dip is the direction and amount of inclination of the roof, and the strike is the trend of the ridge.

In rock bodies, the planar elements and the angles are a little more subtle, but they can be recognized with a little experience. In *figure 20.5,* the dip and strike of an inclined bed is illustrated. For easy reference, the water level is a horizontal line, and the trend of the shore along the bedding plane of the inclined strata would be the strike direction. The angle between the water level and the bedding plane is the angle of dip. The photograph shown in *figure 20.6* shows a sequence of beds striking south (to the top of the picture and dipping to the left at 40°).

Measurements of dip and strike are made in the field by a geological compass that is designed for measuring angles. The symbol used for recording the dip and strike of bedding planes on a map is ⟋₃₀. The long crossbar shows the strike, the short line perpendicular to it is the direction of dip, and the number is the angle of dip. The symbol shown here would represent beds striking N 45° E and dipping 30° to the southeast.

Figure 20.6 *Sequence of inclined beds striking south (top of the photo) and dipping 40° to the east.*

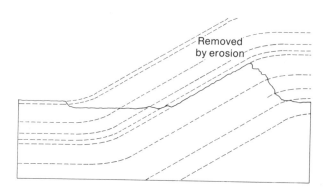

Folds

Statement

Folds are warps (wavelike contortions) in rock bodies that indicate that rocks, like most other solids, can be deformed by plastic flow. They are three-dimensional structures ranging in size from microscopic crinkles to large domes and basins hundreds of kilometers wide. Small flexures in sedimentary rocks are abundant and can be seen in mountain sides, road cuts, and even in small hand specimens; but large folds cannot be seen in their entirety from any one point on the ground and are recognized only when extensive geologic mapping has been done. Folds are important, because they constitute one of the clearest and most obvious records of the mobility of the earth's crust. The complexity of folds is proportionate to the degree of deformation. Broad, open folds form in the stable interiors of continents, where the rocks are only mildly warped. Complex folds develop in mountain belts where deformation has been more intense.

Discussion

Fold Nomenclature. In order to characterize and interpret the significance of folds, you must first become familiar with the terms used to describe them. Three general types of folds are recognized and illustrated in the block diagram in *figure 20.7.* A **monocline** is a fold in which horizontal or gently dipping beds are modified by a simple steplike bend. **Anticlines,** in their simplest form, are uparched strata with the sides of the fold dipping away from the crest. Rocks in an eroded anticline are progressively *older* toward the interior of the fold. **Synclines,** in their simplest form, are downfolds or troughs with the sides (limbs) toward the center. Rocks in an eroded syncline are progressively *younger* toward the center of the fold.

For purposes of description and analysis, it is useful to divide the folds into two equal parts by an imaginary plane called the **axial plane.** The line formed by the intersection of the axial plane and a bedding plane is called the axis, and the downward inclination of the axis is referred to as the **plunge** (*figure 20.8*). In *figure 20.9,* the folds plunge toward the back of the block diagram at an angle of approximately 20° (see also *figures 12.22* and *12.23*).

Domes and Basins. The sedimentary rocks that cover much of the interior of the continents have been mildly warped into broad domes and basins many kilometers in diameter. One large basin covers practically all of the state of Michigan (see *figure 14.30*); another underlies the state of Illinois. Similarly, an elongate dome constitutes the bedrock structure of central Tennessee, central Kentucky, and southwestern Ohio.

Figure 20.7 *Diagram showing three types of folds: a monocline, an anticline, and a syncline.*

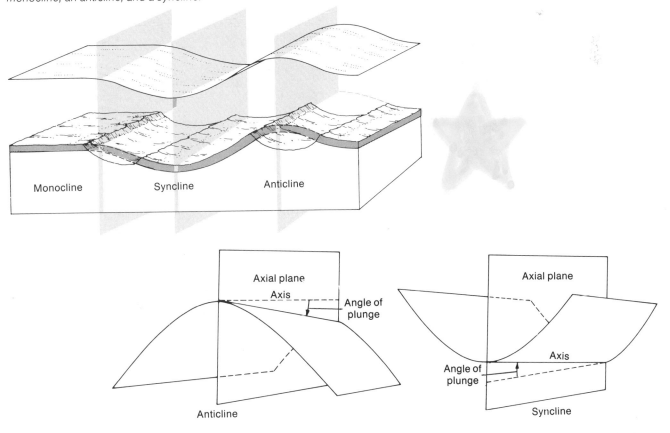

Monocline   Syncline   Anticline

Axial plane
Axis
Angle of plunge
Anticline

Axial plane
Axis
Angle of plunge
Syncline

Figure 20.8 *Schematic diagrams showing the terminology of folds.*

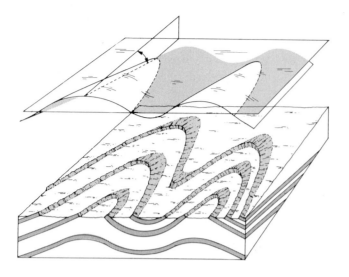

Figure 20.9 *Outcrop pattern of a plunging fold.*

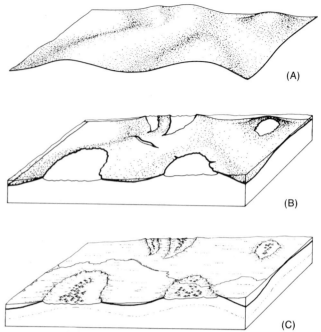

Figure 20.10   *The geometry of domes and basins and the outcrop patterns they produce when they are eroded.*

Although these warps of flexures in the sedimentary strata are extremely large, the configuration of the folds is known from geologic mapping and from information gained through drilling. The nature of these flexures and their topographic expression are diagrammed in *figure 20.10* (see also *plates 5* and *6*).

In diagram *A*, the configuration of a single bed that has been warped into broad domes and basins is shown in perspective. Such a layer is large enough to extend across several states. If erosion cuts down to this layer of rock, the tops of the domes are eroded first so that the layer would look like the one shown in diagram *B*. In diagram *C*, the outcrop pattern of the eroded domes and basins is shown. The exposed rocks of both domes and basins typically have a circular or elliptical outcrop pattern, with the oldest rocks exposed in the central parts of the domes and the youngest rocks exposed in the centers of the basins.

A classic example of a broad fold in the continental interior is the large dome that forms the Black Hills of South Dakota (see *figure 12.21*). The resistant rock units form ridges that can be traced completely around the core of the dome, and the nonresistant formations make up the intervening valleys. A small dome in southern Wyoming, similar to that of the Black Hills, is shown in the photograph in *figure 20.11*. It clearly illustrates the typical elliptical outcrop pattern of domes that commonly is expressed by alternating ridges and valleys that circle the structure.

Plunging Folds.   Where deformation is more intense, such as is typical in mountain belts, the rock layers are deformed into a series of tight folds (see *plate 7*). The geometry of folds is not exceedingly complex and, in many ways, it closely resembles the wrinkles in a rug. The diagrams in *figure 20.12* illustrate the general configuration of folds and their surface expressions after being eroded. The basic form of the fold is shown in diagram *A*. Diagram *B* shows the fold system after the upper part has been removed by erosion. Diagram *C* shows the typical surface expression of the folds after differential erosion. The outcrop of the eroded plunging

anticlines and synclines forms a characteristic zigzag pattern. An anticline forms a V-shaped pattern that closes in the direction of plunge, and the *oldest* rocks are in the center of the fold. Plunging synclines form V-shaped patterns that open in the direction of plunge, and the *youngest* rocks are in the central part of the fold. On the basis of the outcrop pattern, together with the age relationship of the rocks in the center of the fold, it is possible to determine the subsurface configuration of the structure.

Complex Folds.   Intense deformation in some mountain ranges produces complex folding like that illustrated in *figure 20.13*. The size of the structure commonly is over 100 km wide, so complex folds can extend through a large part of a mountain range. Details of such intensely deformed structures are extremely difficult to work out because of the complexities of outcrop patterns.

Figure 20.13 demonstrates the geometry and surface expression of complex folds. In diagram *A*, a single bed in a typical complex fold is shown in perspective. The fold consists of a huge, overturned anticlinal structure, with numerous minor anticlines and synclines that form digits on the larger fold. Diagram *B* shows the fold after it has been subjected to considerable erosion and most of the upper limb has been removed. Note the crop section of the structure on the mountain front and the outcrop pattern on the top of the diagram. The topographic expression of complex folds (diagram *C*) is variable, usually consisting of a complex of mountain ranges, unless the area has been eroded to a surface of low relief.

Complex folds are common in the Swiss Alps but were recognized only after detailed geologic studies were conducted in the area for more than half a century.

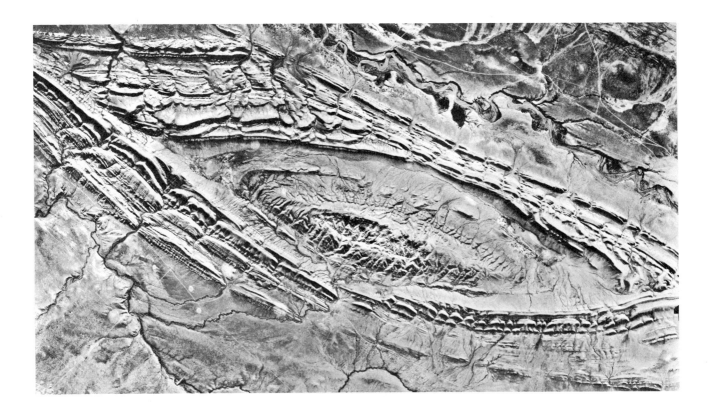

Figure 20.11 The surface expression of an elongate structural dome in western Wyoming.

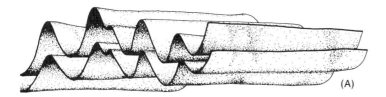

(A)

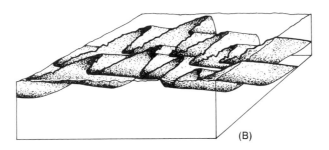

(B)

Figure 20.12 The geometry and surface expression of folded rocks. (A) The basic form of folded strata is similar to that of a wrinkled rug. In this diagram, the strata are compressed and inclined to the north. (B) If the top of the folded sequence of rock is eroded away, the map pattern of the individual layers forms a zigzag pattern at the surface. (C) Rock units that are resistant to erosion form ridges, and nonresistant layers erode into linear valleys. Compare this diagram with plates 26 and 28. The surface expression of a plunging fold commonly is a series of alternating ridges and valleys. In a plunging anticline, the surface map pattern of the beds forms a V that points in the direction of plunge.

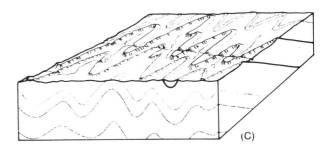

(C)

(A) *The configuration of a large, tightly overturned fold with minor folds on the limbs.*

(B) *The surface outcrop of the fold in diagram A after erosion has removed the upper surface.*

(C) *The topographic expression of the fold in diagram A can be a series of mountains or a flat, eroded surface.*

Figure 20.13 *The geometry and surface expression of complex folds.*

Faults

Statement

Faults are fractures in the earth's crust along which there has been slippage or displacement. In the walls of a canyon, the location of the fault plane may be obvious from the offset plane, and, in many areas, detailed geologic mapping may be necessary before the precise location of faults can be established.

Three basic types of faults are recognized.

1. **Normal faults** are inclined faults in which rocks above the fault plane have been moved down relative to those beneath the fault plane.
2. **Thrust faults** typically are low-angle faults in which the rocks above the fracture surface move up and over those below.
3. **Strike-slip faults** are high-angle faults that have displacement in a horizontal direction parallel to the strike or trend of the fault plane, rather than up or down.

Most major faults of the continents are associated with the folding and deformation of mountain ranges.

Some normal and thrust faults have displacements of many thousands of meters and can be traced for several hundred kilometers. Strike-slip faults commonly show evidence of much greater movement, with horizontal displacements measured in tens or hundreds of kilometers.

There is little or no evidence to suggest that displacement along a fault occurs as a single catastrophic event. On the contrary, all available data indicate that movement along faults is rarely more than a few meters at a time and that displacement occurs in a series of small increments occurring thousands of years apart. This is supported by displacement along faults observed during earthquakes that have occurred in historical times.

Discussion

The three basic types of faults are shown in the block diagrams in *figure 20.14*. Other types can be recognized, but those shown in the diagrams are the most common.

Normal Faults. Movement in normal faults is predominantly in a vertical direction, so that a **scarp** (cliff)

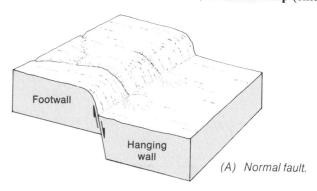

(A) *Normal fault.*

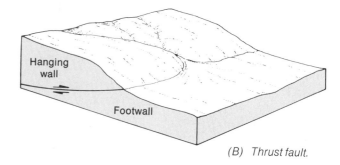

(B) *Thrust fault.*

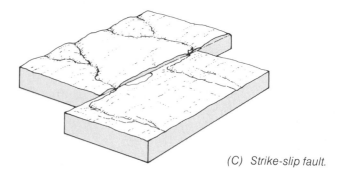

(C) *Strike-slip fault.*

Figure 20.14 *Diagrams showing relative displacement on the major types of fault.*

commonly is produced by the displacement. Most normal faults are steeply inclined, commonly between 65° and 90°. They rarely are isolated fractures but typically occur in a group or a series of parallel faults that may develop a steplike arrangement or a series of fault blocks. A narrow block dropped down between two normal faults is known as a **graben,** and an upraised block is known as a **horst** (*figure 20.15*). Grabens typically form a conspicuous fault valley or basin marked by relatively straight, parallel wells. Horsts form blocklike plateaus or mountains.

Normal faulting on a large scale is the result of tensional stresses that tend to stretch the crust. Normal faults are the major structures in the several rift systems of the continents and are some of the more prominent structures of the oceanic ridge. In the Basin and Range Province of western North America, normal faulting has produced a series of north-south-trending fault blocks that extend from central Mexico northward to Oregon and Idaho (see *plate 24*). The horsts form mountain ranges 2000 to 4000 m high that are dissected considerably by erosion. Grabens form topographic basins partly filled with erosional debris derived from the adjacent ranges (*figure 20.16*). Estimates are that the earth's crust,

between the Wasatch Range in central Utah and the Sierra Nevada Mountains on the Nevada-California border, has been extended some 60 km during the last 15 million years.

The great rift valleys of Africa offer another example of a zone of tension in the earth's crust that has produced large-scale normal faulting (see *plate 23*). There, large

Figure 20.15 *Normal faults resulting in downdropped blocks (grabens), upraised blocks (horsts), and titled fault blocks in Canyonlands National Park, Utah.*

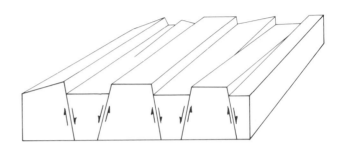

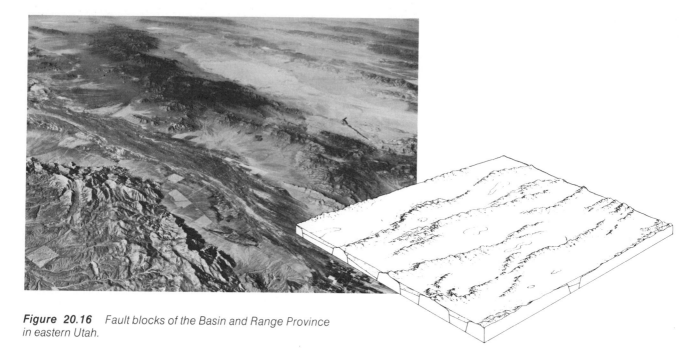

Figure 20.16 *Fault blocks of the Basin and Range Province in eastern Utah.*

grabens have been formed by a system of normal faults extending from the Zambesi River to northern Ethiopia, a distance of 2900 km. That distance is almost doubled when the rift system of the Red Sea and the Jordan Valley (which continues northward into Syria) are added. The rift valleys of Africa are remarkably uniform in width, ranging from 30 to 45 km. Thus, they are about the same size as the grabens of the Basin and Range in North America and the rift valleys of other continents.

Normal faulting on a grand scale, with displacements of several thousand meters, produces fault scarps that soon are attacked by erosion. Thus, the scarps evolve through a series of stages until they are obliterated and leave only obscure fault lines to mark the location of displacement (see *figure 12.25*). In the early stages of faulting, streams begin to erode through the scarp and form spurs that end abruptly in a triangular face at the fault surface. With continued erosion, the fault block recedes back from the fault line and becomes highly dissected. Deposition of erosional debris derived from the upthrown block accumulates in the adjacent basin. Ultimately, the topographic relief is destroyed, and the location of the fault line is largely obliterated.

Upwarping in regions of major normal faulting, such as those just described, could be produced by the rise of convection currents in the mantle that result in horizontal tensional stress where the convection currents move laterally and stretch the crust. Areas of extensive normal faulting (for example, the Basin and Range of the western United States and the African rift system) are areas where the continents are considered to be splitting apart, and they appear to be an extension of the oceanic ridge beneath the continents.

Thrust Faults. Although there are exceptions, most thrust faults with large displacements have low-angle dips. Some geologists apply the term *thrust* to faults dipping at angles less than 45° and refer to high-angle thrusts as **reverse faults.** Movement on a thrust is pre-dominantly horizontal, and displacement can be more than 50 km.

Thrust faults generally are associated with intense folding, which results from strong, compressive horizontal forces in the earth's crust, and are prominent in all the major folded mountain regions in the world. They commonly evolve from folds in the manner diagrammed in *figure 20.17*. Where resistant rocks are thrust over nonresistant strata, a scarp is eroded on the upper plate. The scarp is not straight or smooth as in normal faulting, but the outcrop of the fault surface typically is irregular in map view.

Strike-Slip Faults. As was mentioned in the statement, strike-slip faults are high-angle fractures in which displacement has been in a horizontal direction, rather than up or down. Since there is little or no vertical movement, high cliffs do not form from strike-slip faults. Instead, the faults are expressed on the surface by a straight line that extends across the surface, commonly marking a discontinuity in the topography (see *figure 12.26* and *plate 25*).

Some of the topographic features produced by strike-slip faulting and subsequent erosion are shown in *figure 12.26*. One of the most obvious is the offset of the drainage pattern. The relative movement often is shown by abrupt right-angle bends in the streams as they cross the fault line. The stream follows the fault for a short distance and then turns abruptly and continues down the regional slope. As the blocks move, some parts may be depressed to form sag ponds; others may bend and buckle to form low, linear ridges.

A strike-slip movement results in the juxtaposition of blocks with contrasting structures, rock types, and topographic forms (*figure 20.18*). Therefore, the fault line commonly marks the boundary between distinctly different surface features and rock types. This is expressed on the diagram in *figure 12.26* by the contrast in degree of dissection on the fault blocks. Also, the fault

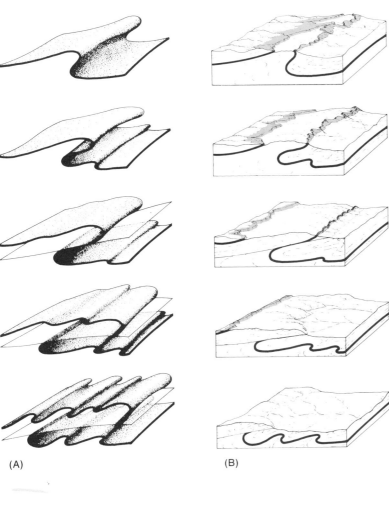

(A)

(B)

Figure 20.17 The evolution of thrust faults from folds. (A) The geometry of the rock units as deformation proceeds. (B) Common topographic expressions of thrust faults.

Figure 20.18 Photograph of a part of the San Andreas fault, a major strike-slip fault in California. The fault is delineated by the prominent straight valley. Note the discontinuity in the drainage on either side of the fault.

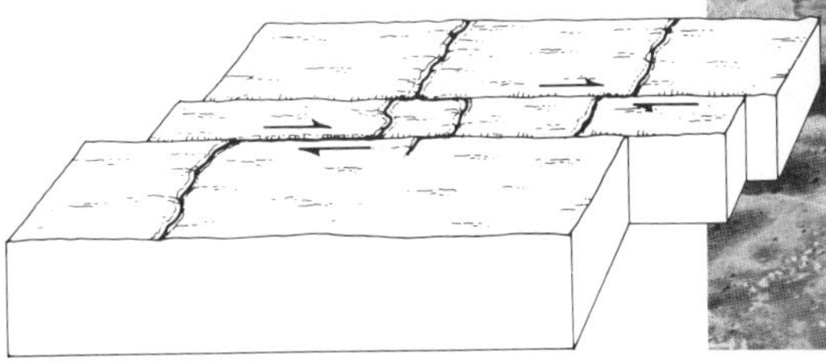

blocks disrupt patterns of ground-water movement, which is reflected by contrasts in vegetation, soils, and springs along the fault trace.

The topographic expressions of strike-slip faults generally are short-lived, since they do not form high cliffs and the low surface features are modified rapidly by erosion. They are best preserved in arid regions and are most obvious where there has been relatively recent movement along the fault.

Observed Movement on Faults. The movement along faults during earthquakes rarely exceeds a few meters. However, in the great San Francisco earthquake of 1906, the crust slipped horizontally as much as 9 m along the San Andreas fault, offsetting roads, fence lines, and

Figure 20.19 *Recent displacement along the Wasatch fault in central Utah has produced the fresh cliff at the base of the mountain front. Cumulative movement on the fault produced the mountain range.*

orchards. Recent vertical faults in Nevada and Utah have produced fresh scarps 3 to 6 m high (*figure 20.19*). The Good Friday earthquake in Alaska in 1964 was accompanied by a 15 m uplift of Montague Island. The largest well-authenticated displacement during an earthquake appears to have been near Disenchantment Bay in 1899, where beaches were raised as much as 16 m above sea level.

Movement along faults is not restricted entirely to earthquakes, however. Precise surveys along the San Andreas fault in California show slow shifting (**creep**) along the fault plane, at an average rate of 4 cm per year. Such movements have broken buildings constructed across the fault line and eventually will cause considerable displacement. Total displacement on a fault can be several hundred kilometers.

The important point that should be emphasized is that the displacement on a fault does not occur in a single violent event but is the result of numerous periods of displacement and slow tectonic creep.

Joints

Statement

Joints are simple cracks or fractures in rocks along which there has been no appreciable displacement. They are very common and are found in almost every expo-

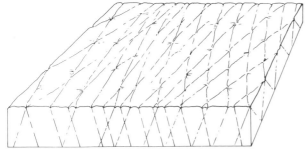

Figure 20.20 *Joint systems in resistant sandstones in Arches National Park, Utah.*

sure. Usually, they occur as two sets of fractures that intersect at angles ranging from 45° to 90°. Thus, they divide the rock body into large, rectangular blocks. Joints commonly are related to major faults and can form remarkably persistent patterns over hundreds of square kilometers.

Discussion

The best areas to study joints are those where brittle rocks, such as thick sandstone, have been fractured and the joint planes have been accentuated by erosion. The massive sandstones of the Colorado Plateau provide an excellent example, because the joint plane commonly is accentuated by selective erosion. Thus, the joint is expressed by deep, parallel cracks, which are most impressive when seen from the air (*figure 20.20;* see also *figure 20.15*). In places, joints control the development of stream courses, especially the secondary tributaries and areas of solution activity.

Most joints result from broad regional upwarps or from compression or tension associated with faults and folds. Some tensional joints result from erosional unloading and expansion of rock. Columnar joints in volcanic rocks are produced from stresses set up as the lava cools and contracts.

Joints can be of economic importance. They can provide the necessary porosity for ground water and for petroleum migration and accumulation. Therefore, analyses of jointing have been important in both the exploration and the development of these resources. Joints also control ore deposits of copper, lead, zinc, mercury, silver, gold, and tungsten; since hydrothermal solutions, associated with igneous intrusions, can migrate along joint systems and precipitate minerals along the joint walls, thus forming mineral veins. Modern prospecting techniques include detailed fracture analyses. Major construction projects (such as dams) are especially affected by jointing systems within rocks, and allowance must be made for them in the projects' planning. For example, the Flaming Gorge Dam on the Green River in northeastern Utah was constructed nearly parallel to the major vertical joint system in a small bend of the river so that stresses caused by water storage would tend to close the fractures. Had the dam site been selected in the east-west-trending part of the river, the joint planes would have presented major structural weaknesses in the bedrock foundation.

Joint systems can be either an asset or an obstacle to quarry operations. Closely spaced joints seriously limit the sizes of blocks that can be removed. On the other hand, when the quarryman utilizes one orientation of intersecting joints, the expense of removing the building blocks is greatly reduced and waste is held to a minimum.

Figure 20.21 Angular unconformity exposed in the wall of the Grand Canyon. The lower sequence of strata was tilted and eroded prior to the deposition of the younger horizontal sequence.

Unconformities

Statement

In most sequences of sedimentary rocks, there are many major discontinuities that indicate significant interruptions in the rock-forming process, interruptions caused by movement of the earth's crust. If, for example, an area where sedimentary rocks are being deposited is uplifted and subjected to erosion, an **unconformity** (a physical discontinuity in the succession of strata) is produced. The presence of an unconformity is of paramount importance for the interpretation of geologic events, especially earth movements.

Unconformities are classified into three main categories on the basis of the structural relationships between the underlying and overlying rocks.
1. **Angular unconformities.**
2. **Disconformities.**
3. **Nonconformities.**

Each records three important geologic events.
1. Formation of the older rocks.
2. Uplift and erosion of the older rocks to a surface independent of its structure.
3. Subsidence and burial of this ancient erosional surface beneath younger strata.

Discussion

Angular Unconformities. The relationship among the rock bodies shown in *figure 20.21* is an example of an angular unconformity, the significance of which can be

appreciated only by careful consideration of what the angular discordance implies. The older strata were deposited first, then folded or tilted, and then eroded. Subsidence followed with the subsequent deposition of the overlying beds. A complete reversal of geologic processes occurred: the area ceased to be the site of deposition, was deformed by earth movements, was subjected to erosion, and subsequently became the site of deposition again.

Disconformities. A disconformity consists of an erosional surface carved on horizontal rocks without tilting or folding. The rocks above and below the erosional surface are parallel. Erosion can strip off the top of the older sequence and cut channels into the older beds, but there is no structural discordance between the two rock bodies.

Disconformities are formed by broad, gentle upwarps in which the layered sediments on the seafloor are elevated above sea level without folding. Subsequent subsidence then lowers the rocks below sea level so that deposition of the new sequence of beds, essentially, is parallel to the stratification of the older sequence. Generally, disconformities are more difficult to recognize, inasmuch as the beds above and below the erosional surface are parallel (*figure 20.22*).

Nonconformities. A nonconformity is an unconformity separating intrusive igneous rocks or metamorphic rocks from overlying sedimentary strata. A structure of this sort is shown in *figure 20.23*. Clearly, the light-colored granite dike in the center of the photograph did not invade the overlying horizontal sandstone, because the basal layers of the sandstone contain pebbles of granite and coarse sand of quartz and feldspar derived

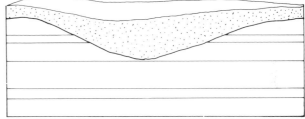

Figure 20.22 *A disconformity. Disconformities do not show an angular discordance, but an erosional surface separates the two rock bodies. In this exposure, it is clear from the channel in the central part of the photograph that the lower shale units were deposited and eroded before the upper units were deposited.*

from the weathering and disintegration of the granite below. Moreover, the sandstone is unaltered by heat, and a local soil profile can be seen above the granite and below the sandstone.

Why does sandstone deposited on a granite or a metamorphic rock indicate mobility of the crust and a major discontinuity in the rock-forming processes? The answer becomes readily apparent when the environment in which intrusive igneous rocks and metamorphic rocks form is considered. Both originate deep within the earth's crust, the granite cooling slowly at great depth and the metamorphic rock recrystallizing at high temperatures and pressures far below the surface. In order for these rocks to be weathered and eroded, the overlying cover of rock must be removed. Uplift must occur in order for these rocks to be uncovered and exposed. Thus, sedimentary rocks deposited upon a granitic or metamorphic rock imply three major events: (1) formation of

Figure 20.23 A nonconformity. The metamorphic rocks and the igneous dikes shown in this photograph were formed at great depths in the crust. The rocks that covered them when they were formed subsequently were eroded away so that the igneous and metamorphic rocks were exposed to the surface. The horizontal sedimentary rocks, lying on the igneous and metamorphic rocks, form this nonconformity.

an ancient sequence of rocks (the lower part of which is represented by the metamorphic rocks in *figure 20.23*), (2) uplift and erosion to remove the cover and expose the granites or metamorphic rocks to the surface, and (3) subsidence and deposition of the younger sedimentary rocks on the old erosional surface developed on the granite.

Studies of unconformities show the close relationship among crustal movements, erosion, and sedimentation and provide important insight into crustal movements during the earth's history.

Summary

Deformation of the earth's crust is well documented by historical movements along faults, raised beach terraces, and contorted rock layers. The major types of rock structures are folds, faults, and joints. Folds in stratified rocks range in size from small microscopic wrinkles to large flexures hundreds of kilometers long.

In the stable platform, sedimentary rocks are mildly warped into broad domes and basins. More intense deformation occurs in mountain belts, where rocks are folded into tight, plunging anticlines and synclines. Intense deformation in parts of some mountain ranges produces complex folding.

Faults are fractures in the earth's crust along which there has been slippage or displacement. Three basic types of faults are recognized: (1) normal faults, resulting from tension; (2) thrust faults, resulting from compression; and (3) strike-slip faults, resulting from shear stresses. Normal faults develop on a large scale in rift systems, thrust faults generally are associated with tight folds and occur at converging plate margins, and strike-slip faults are associated with shearing in fracture zones that offset spreading centers.

Joints are fractures with no noticeable displacement and are associated with all types of deformation.

Unconformities are major discontinuities in rock sequences and commonly indicate crustal deformation.

Deformation of the earth's crust constitutes an important record of plate movement and a history of the earth's dynamics.

Additional Readings

Billings, M. P. 1972. Structural Geology. 3rd ed. Englewood Cliffs, N.J.: Prentice-Hall.

Spencer, E. W. 1977. Introduction to the Structure of the Earth. New York: McGraw-Hill.

Evolution of the Ocean Basins

21

During the past few decades, while space exploration has held the attention of popular science, oceanographers have been quietly at work mapping the topography of the ocean floor. Through this work, in which many nations have participated, a great amount of bathymetric data has been compiled into regional physiographic maps showing in perspective the surface features of the ocean floor. In addition, special seismic-reflection techniques have been developed that show the thickness of the sediment that overlies the bedrock of the oceanic crust. The results are fascinating and truly spectacular. A great "mountain range" called the oceanic ridge extends as a continuous unit through all of the ocean basins, a distance of more than 64,000 km. The ridge is cut by long fracture zones, some of which can be traced for several thousand kilometers. Flanking the oceanic ridge is the abyssal floor, above which rise numerous submarine volcanoes called seamounts. The deepest parts of the ocean are the trenches, some of which extend down to depths of 11 km. In addition, there are a number of small ocean basins, commonly referred to as seas, that mostly are isolated from the major oceans.

As has been shown already, the new knowledge of the landforms on the ocean floor has helped to revolutionize geologic thinking, because, for the first time, we are able to consider all parts of the planet and develop the unifying theory of plate tectonics. No amount of research on the continents alone could reveal what we now know from studies of the ocean floor and the continents together.

From a variety of studies of the seafloor, we have learned that the oceanic crust is continually being created at the ridges and destroyed as it is consumed into the mantle at the deep-sea trenches. This is a major modification of our ideas about the seafloor, because, prior to the 1960s, it seemed logical that the ocean basins were permanent features of the planet and as the major reservoirs of water should contain the oldest and most complete record of sedimentary rocks on the earth.

In this chapter, the geology of the ocean basins will be discussed in some detail, because it provides some of the most important information we have concerning the dynamics of the earth.

Major Concepts

1. The major features of the ocean floor are (a) ridges, (b) abyssal hills, (c) abyssal plains, (d) trenches, (e) islands and seamounts, and (f) continental margins. Their origin is best explained by the theory of plate tectonics.
2. The oceanic crust is composed of four major layers: (a) a surface layer of marine sediment, (b) a layer of pillow basalt, (c) a zone of sheeted dikes, and (d) a layer of gabbro.
3. Although the ocean basins originated by seafloor spreading, they are quite different in size, shape, and topography. This results largely from differences in ages and stages of development.
4. During the last 200 million years, the continents have drifted apart to produce the present distribution of continents and oceans.

Seafloor Morphology

Statement

The surface features of the continents are familiar to everyone, but the landscape of the ocean floor remained largely a mystery until the development of modern oceanographic research. Today, the ocean floor can be seen through the use of photography and television cameras (see *figures 19.2* and *19.3*), but, in order to study the regional landscape of the seafloor, oceanographers must use a variety of new instruments, including echo sounders, seismic-reflection profilers, and magnetometers. These instruments provide maps and charts that show relief forms on the ocean floor in detail similar to that provided by early topographic maps of the land. Within the past two decades, our knowledge of the ocean floor has increased enormously. This provides us with our first real understanding of the rocks and landforms of the ocean floor, an area covering three-fourths of the planet.

The major landforms of the ocean basins are:

1. The oceanic ridge.
2. The abyssal floor.
3. The trenches.
4. The seamounts.
5. The continental margins.

Discussion

Seismic-Reflection Profiles. An example of the type of information about the seafloor obtained from seismic-reflection profiles is reproduced in *figure 21.1*. Pulses of sound are emitted from a ship and reflected back from the ocean floor. However, some of the energy penetrates the sediment and is reflected from the surface of the hard rock below. With the aid of electronics, a continuous profile of the seafloor and the configuration of the unconsolidated sediments are plotted automatically. The profile in *figure 21.1* shows a cross-sectional view of the geologic structure of the ocean floor flanking the mid-Atlantic ridge. From the profile, it can be seen that the sediments are relatively flat and undeformed while the bedrock consists of irregular mounds protruding through the sediment.

The Oceanic Ridge. The oceanic ridge is the most pronounced tectonic feature on the earth, and, if the ridge were not covered with water, it certainly would be visible from observation points as far away as the moon. It is, in essence, a broad, fractured swell, generally more than 1400 km wide, with the highest peaks rising as much as 3 km above the surrounding ocean floor. The remarkable characteristic about the ridge is that it extends as a continuous feature from the Arctic basin, down through the center of the Atlantic, into the Indian Ocean, and across the South Pacific, terminating in the Gulf of California. Its total length is more than 64,000 km, so, without question, it is the greatest "mountain" system on the earth. The "mountains" of the oceanic ridge, however, are nothing like the mountains of the continents, which were built largely of folded sedimentary rocks. By contrast, the ridge is composed entirely of basalt and is not deformed by folding.

Many of the characteristics of the ridge are apparent in the seismic-reflection profile reproduced in *figure 21.2*. On a regional basis, the ridge is a broad, uparched segment of the ocean floor broken by numerous fault blocks that form linear hills and valleys. The highest and most rugged topography is located along the axis, and a prominent rift valley marks the crest of the ridge throughout most of its length. As can be seen in *figure 21.1*, oceanic sediments are thickest down the flanks of the ridge but thin rapidly toward the crest and are absent where new crust has just been generated.

Throughout most of its length, the oceanic ridge is cut by a series of transform faults (*figure 21.3*). These

Figure 21.1 *A photograph of a continuous seismic-reflection-profile record showing the morphology of the ocean floor and the geologic structure of the oceanic sediments.*

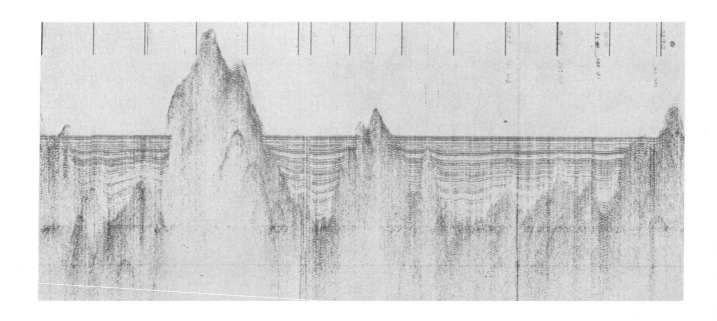

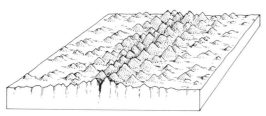

Figure 21.2   *Tracing of seismic-profiler records across the mid-Atlantic ridge at 44° north latitude. The crest of the ridge is marked by a deep rift valley that can be traced along its entire length. Sediment is thickest down the flanks of the ridge, but it thins rapidly near the crest.*

are the site of continued seismic activity that occurs as the blocks of oceanic crust slide past each other between ridge segments. Beyond the active transform fault, the fracture zone as it is expressed by an abrupt, steep cliff may be traceable for several thousand kilometers.

Detailed studies of the axial spreading zone of the mid-Atlantic ridge were made in 1974, when scientists in deep-diving vessels sampled, observed, and photographed the ridge for the first time. Without doubt, this project made some of the most remarkable submarine discoveries of modern times. The photographs showed extensive pillow basalts so recent that little or no sediment covered the fine details of the surface textures (*figures 19.2* and *19.3*). Numerous open fissures in the crust also were observed and mapped, and, in one small area of only 6 km², 400 open fissures were observed, some of which are as wide as 3 m. These are considered to be conclusive evidence that the oceanic crust is being pulled apart. The eruption of lava from these fractures, which are parallel to the rift valley, would tend to create long, narrow ridges—a feature that indeed seems to characterize the morphology of the oceanic ridge.

The general character of the oceanic ridge seems to

be a function of the rate of plate separation. Where the spreading rate is relatively low (less than 5 cm per year), the ridge stands higher and is more rugged and mountainous than where rates of spreading are more rapid. Moreover, rift valleys on slow-spreading ridges are prominent, whereas rift valleys in areas with high-spreading rates are more subdued.

The Abyssal Floor.   Vast areas of the deep ocean consist of broad, relatively smooth surfaces known as the **abyssal floor.** This type of seafloor topography was discovered in 1947 by oceanographic expeditions surveying the mid-Atlantic ridge and subsequently has been mapped in detail with the use of precision depth recorders capable of measuring elevations on the ocean floor with relief as small as 2 m. The abyssal floor extends from the flanks of the oceanic ridge to the continental margins and generally lies at depths of about 3600 m. In most ocean basins, the abyssal floor can be subdivided into two sections: (1) the **abyssal hills** and (2) the **abyssal plains.**

The abyssal hills are relatively small hills rising above the ocean floor 75 to 900 m (*figure 21.4*). They are circular or elliptical and range from 1 to 8 km wide

Figure 21.3   *Tracing of seismic-profiler records across the Murray fracture zone in the eastern Pacific Ocean. The fracture is expressed by a pronounced vertical cliff that separates areas of contrasting topography. On the left side of the fault, seamounts are abundant, whereas, to the right, the seafloor is relatively smooth and featureless.*

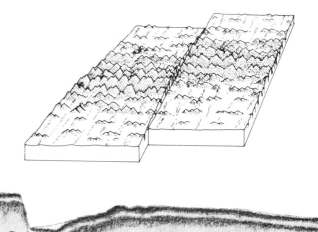

Figure 21.4   *Profile across the abyssal floor in the Atlantic Ocean, showing buried abyssal hills and the smooth abyssal plain covered with sediment.*

at their bases. The hills are found along the flanks of the oceanic ridge and occur in profusion in parts of the ocean floor separated from land by trenches. In the Pacific, they cover 80% to 85% of the ocean floor. Thus, abyssal hills can be considered to be the most widespread landform on earth.

The abyssal plains are exceptionally flat surfaces of the ocean floor where the abyssal hills are buried completely by sediment (*figure 21.7*). Commonly, they are located near the margins of a continent, where sediment from a continental mass is transported by turbidity currents and is spread over the adjacent ocean floor.

The origin of the abyssal hills and abyssal plains can be traced to the oceanic ridge and the development of new crust at the spreading center. The new lithosphere created at the ridge crest slowly recedes from the rift zone and is gradually modified in several important ways. The new crust cools and contracts, deepening the ocean basin as it moves away from the ridge. Fine-grained sediments, consisting of dust and shells of marine organisms, slowly but continually settle over all of the seafloor. The linear hills formed by volcanic activity, intrusions of magma, and block faulting become the foundations for the abyssal hills. As the plates move away from the spreading center, the original superficial features of the landforms created at the ridge are gradually modified and concealed and eventually can be buried completely to form flat abyssal plains. The outline of the buried rock surface can be traced by seismic-reflection profiling (as exemplified in *figure 21.4*).

The distribution of abyssal plains and abyssal hills throughout the oceans substantiates this explanation of their origin. Abyssal plains occur *only* where the topography of the seafloor does not inhibit turbidity currents from spreading out from the continents over the seafloor. Throughout the Atlantic Ocean, abyssal plains occur along the bases of the continents of North America, South America, Africa, and Europe. In the Pacific Ocean, by contrast, there are few abyssal plains, because turbidity currents cannot flow past the deep trenches that occur along most of the continental margins. The trenches trap the inflowing sediment, so most of the Pacific floor lacks abyssal plains and is covered with abyssal hills in-

stead. The largest abyssal plains in the Pacific are found in the northeast, off the coast of Alaska and western Canada. This is the only significant segment of a continental mass bordering the Pacific that is not marked by deep trenches. In the Indian Ocean, abyssal plains occur only along the margins of Africa and India and are absent in the eastern part of the ocean, where the deep Java trench along the continental margins acts as a sediment trap. Another major area of abyssal plains lies off the northern shore of Antarctica. In the North Pacific, between the Aleutian trench and the continental shelf, the Bering abyssal plain covers most of the deep-sea floor north of the Aleutian Islands. This deep basin is underlain by an abnormally thick section of layered sediment that, in places, can be as thick as 2 km. As the physiographic map (*figure 21.11*) shows, the Aleutian ridge has cut off this corner of the Pacific basin and has acted as a dam behind which sediments have accumulated rapidly.

Large cone- or fan-shaped deposits of sediment derived from the continents have been mapped on the abyssal plains offshore from most of the world's great rivers. These are referred to as **deep-sea fans** (see *figure 6.24*). They are similar to alluvial fans and deltas on land in that they are fan-shaped accumulations of sediment located at the mouth of a river. They are different, however, in their mode of transportation and deposition of sediment, which occur primarily by means of turbidity currents. Most large fans are located at the base of the continental slope, with their apexes at the mouth of a submarine canyon cut into the edge of the shelf. Intermittently, turbidity currents flush sediment through the canyon and build up depositional fans where they reach the lower gradient of the ocean floor.

Most, if not all, fans are marked by one or more deep-sea channels that usually are the extension of submarine canyons cut in the continental slope. As the slope flattens, the channels develop natural levees similar to low-gradient streams on land.

The submarine fan of the Ganges River in the northwestern Indian Ocean is by far the largest deep-sea fan in the world. It is over 2800 km long and covers an area of slightly more than 4 million km². This represents about 70% of the erosional debris derived from the Himalayas.

The remaining 30% goes to the flood plain and delta of the Ganges and to the Indus River to the west.

Other large fans are the Indus fan on the west side of the Indian peninsula, the Amazon fan and the Congo fan in the South Atlantic, and the Mississippi and the Laurentian fans in the Atlantic. A number of smaller fans have been mapped off the Pacific coast of North America. There are no large fans in the Pacific, because most major rivers drain into the Atlantic and Indian Oceans and because the deep-marine trenches trap most of the sediment that flows into the Pacific.

Trenches.   The subduction zone where two plates converge and one slab of lithosphere plunges down into the mantle is expressed topographically by a trench. As has been seen in previous chapters, a subduction zone is characterized by intensive volcanic activity, high seismicity, and large gravity anomaly.

Trenches, some of which reach nearly 11 km below sea level, are the deepest parts of the ocean. As can be seen from the seismic-reflection profile shown in *figure 21.5,* trenches typically are asymmetrical, with a relatively steep slope adjacent to the continental landmass and a more gentle slope on the side of the ocean basin. Individual trenches less than 100 km wide can extend as continuous features across the deep-ocean floor for several thousand kilometers.

The most striking trenches occur in the western Pacific, where a trench system extends from the vicinity of New Zealand to Indonesia and Japan and northward along the southern flank of the Aleutian Islands. Long trenches also occur along the western coast of Central and South America, in the Indian Ocean west of Australia, in the Atlantic off the tip of South America, and in the Caribbean Sea.

Islands and Seamounts.   Literally thousands of submarine volcanoes occur on the ocean floor, the greatest concentration of which are found in the eastern Pacific. Some rise above sea level and form islands, but most remain submerged and are referred to as seamounts (*figure 21.6*). They often occur in groups or chains, with individual volcanoes being as much as 100 km in diameter and 1000 m high. The term **guyot** refers to a special type of seamount, the top of which is a flat, mesalike surface, rather than a cone.

The chains of islands and seamounts are believed to develop by movement of the seafloor over a hot spot in the upper mantle, where a huge column of upwelling lava known as a **plume** lies in a fixed position under the lithosphere (see *figure 19.13;* see also *plate 9*).

The origin of guyots, however, has been the subject of much debate. There is little doubt that guyots represent ancient erosional platforms developed by stream and wave action on volcanic islands. Dredge samples show that some guyots are covered with obviously abraded volcanic debris and many are coated with coral, which can grow only in very shallow water. The big question raised by this fact is how the coral became submerged to depths of 1500 to 3000 m. In addition, many atolls are composed of thousands of meters of coral reef, all of which had to originate near sea level. These facts can be accounted for by theories of seafloor spreading and plate tectonics. The initial volcano forming on the oceanic ridge could rise above sea level near the ridge crest, because this segment of the ocean floor is arched up by ascending convection currents. Thus, wave action would be able to erode the top of the island to form a flat surface, and coral reefs could grow on this platform near sea level. As seafloor spreading continued, the flattopped island would migrate off the ridge and become submerged. If environmental factors were satisfactory, coral

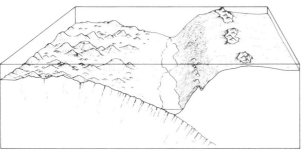

Figure 21.5   *Tracing of seismic-profiler records across the central part of the Aleutian trench. Note the steep flank adjacent to the Aleutian island arc (right) and the gentle slopes toward the ocean basin.*

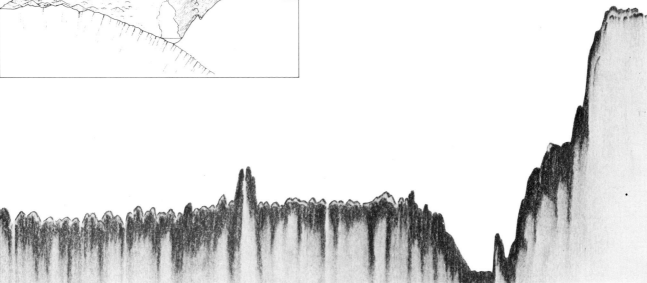

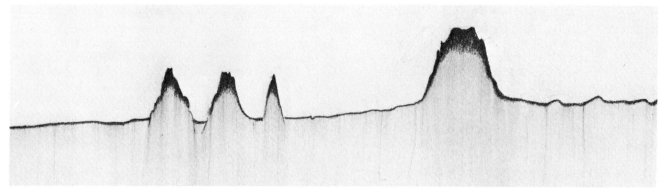

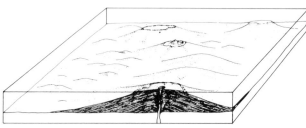

Figure 21.6 *Tracing of seismic-profiler records across seamounts in the central Pacific Ocean. Seamounts are submarine volcanoes that usually occur in groups or chains. Some rise above sea level to form islands.*

reefs would continue to grow upward as the seamount moved off the flanks of the ridge into deeper water. Thus, atolls such as the Bikini Atoll could develop a reef rock over a thousand meters thick as the island slowly moved into deeper water.

Continental Margins. The margins of the continents are covered by the ocean but are not geologically part of the oceanic crust. The submerged part of the continent is called the **continental shelf** and it is barely covered with water. It is a gently sloping platform extending from the shoreline to the start of the steep descent to the deep-ocean floor. The shelf ranges up to 1500 km in width, and, at its outer edge, the depth of the shelf ranges from 20 to 550 m. At present, the continental shelves are equal to 18% of the earth's total land area, but, at times in the geologic past, they were much larger because the oceans spread much farther over the continental platforms.

Characteristically, the continental shelf is smooth and flat, but the topography has been influenced greatly by changes in sea level. Large areas once exposed as dry land were subjected to submarine processes. Probably the most dramatic period in the history of the shelves has been the last million years during the glacial epoch, when sea level fluctuated with the advance and retreat of the continental glaciers.

The seafloor descends from the outer edge of the continental shelf as a long, continuous slope to the deep-ocean basin (*figure 21.7*). This slope is appropriately called the **continental slope,** because it marks the edge of the mass of continental granitic rock. Continental slopes are found around the margins of every continent and around smaller pieces of continental rock, such as Madagascar and New Zealand. If you study closely the continental slopes shown on the physiographic map (especially those surrounding North America, South

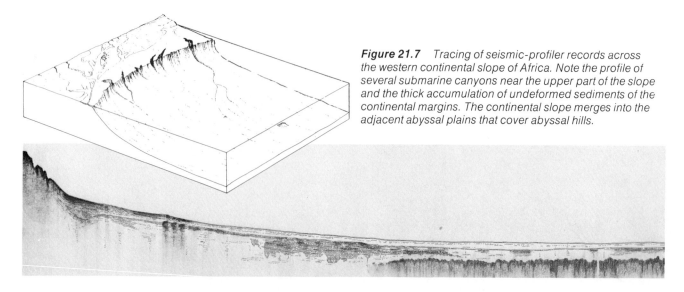

Figure 21.7 *Tracing of seismic-profiler records across the western continental slope of Africa. Note the profile of several submarine canyons near the upper part of the slope and the thick accumulation of undeformed sediments of the continental margins. The continental slope merges into the adjacent abyssal plains that cover abyssal hills.*

America, and Africa), you should note that the slopes form one of the major topographic features on the earth's surface. Obviously, they are by far the steepest, longest, and highest slopes on the earth. Within this zone, which is 20 to 40 km wide, the average relief above the seafloor is 4000 m; adjacent to the marginal trenches, relief is as great as 10,000 m. In contrast to the shorelines of the continents, the edges of continental slopes are relatively straight over distances measured in thousands of kilometers. This is the topographic expression of the geologic difference between the continental and oceanic crusts and reflects a fundamental difference in structure and rock type of the earth's crust.

Two types of continental slopes can be recognized. Each is subjected to different stresses and, as a result, develops different characteristics. The slopes on the trailing edge of the continent are relatively passive and are not subjected to strong compressive forces. The slopes of the Red Sea are an example of a newly formed continental margin caused by the separation of Africa from

Arabia. The continental slopes of North and South America bordering the Atlantic Ocean are similar and are considered to have originated in much the same way. The major difference is that the slopes of the Atlantic are much older and have been modified considerably by erosion and sedimentation.

In contrast to the passive trailing edge of the continent, the **leading edge** of the continent is subjected to much more stress, since it is converging on another moving plate. Crustal deformation results. Where the continent converges with an oceanic plate, the oceanic plate moves down and under the lighter continental crust and a trench develops along the continental margin. Such is

Figure 21.8 (A) The Monterey submarine canyon off the coast of central California shows many characteristics of canyons cut by rivers. (Constructed from bathymetric charts.) (B) A profile across the Monterey canyon is similar to a profile across the Grand Canyon constructed with the same number of control points.

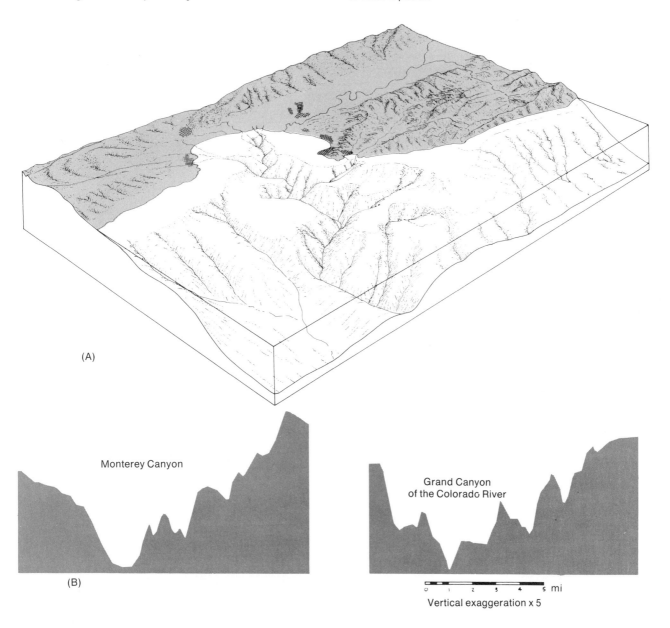

the case along the Pacific coast of South and Central America, where there is a tendency for the continental slopes to be straight and seismically active. Volcanic activity is prominent on the adjacent continental mass.

Submarine canyons are common along the continental slope and have been studied for many years, long before the ocean floor was mapped. As shown on the physiographic map (*figure 21.11*), the submarine canyons typically cut through the edge of the continental shelf and terminate on the deep abyssal floor some 5000 to 6000 m below sea level. The profile in *figure 21.7* cuts across three canyons near the upper part of the continental slope. The large submarine canyons have a V-shaped profile and a system of tributaries, so they closely resemble the great canyons cut by rivers on the continents (*figure 21.8*). Many pioneer researchers, therefore, suggested that the canyons were cut by rivers, but the problem of *how* remained unanswered. Some canyons are 3000 m below sea level, and it is difficult to conceive how sea level changed to the degree necessary for rivers to cut to that depth. Others suggested that the continents were uplifted thousands of meters, were dissected by streams, and then subsided so the canyons became submerged.

More recently, submarine canyons were thought to have been eroded by turbidity currents, but it is not known whether or not turbidity currents are capable of such erosion. It is clear, however, that much sediment has flowed through many of the canyons because of the great volumes of sand and mud occurring at their lower ends as deep-sea fans.

The theory of plate tectonics could explain the origin of canyons as a product of stream erosion during the initial phases of continental rifting (*figure 21.9*). When convection currents occur beneath a continent, they arch up the overlying continental crust and begin to pull it apart. The initial response is the formation of rift valleys like those in East Africa. The sides of the rift valleys soon are attacked by erosion, and many canyons and ravines are cut into them. As the continents separate

Figure 21.9 *Formation of submarine canyons by continental erosion along the initial rift zone. (A) The continent is arched and pulled apart by convecting currents in the asthenosphere. Erosion cuts canyons into the sides of the rift valley, the future continent's margins. (B) Continents move off the uparched area, and the margins become partly submerged. The canyons cut by erosion on the sides of the rift valley become submarine canyons. (C) As the continents drift farther from the spreading center, the canyons are submerged and modified by the activity of turbidity currents.*

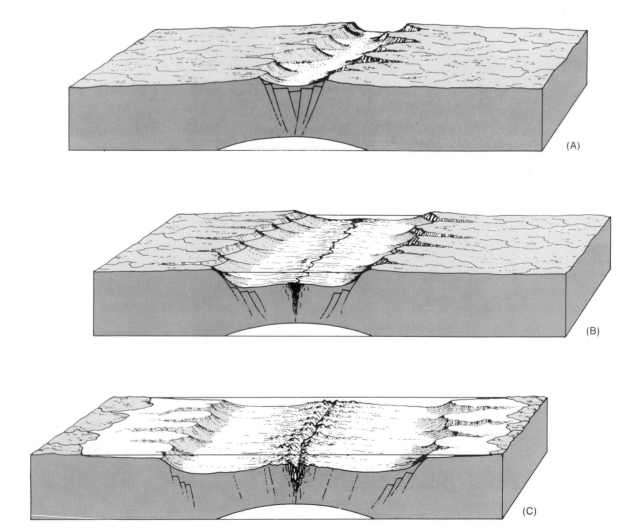

completely and move off the uparched spreading center, the margins become submerged and the canyons drowned. They then act as channels for turbidity flows and could be modified by them. The canyons, therefore, would be due to initial erosion by stream action and later modification by turbidity currents.

Composition and Structure of the Oceanic Crust

Statement

With expanded research on the ocean basins, considerable effort has been made to determine the composition and structure of the oceanic crust and how it differs from the continental crust. A variety of methods have been used, including seismic reflection, drilling, and studies of fragments of oceanic crust exposed at the surface. Among the most significant facts that have been learned about the oceanic crust are these.

1. The oceanic crust is composed of four major layers.
   (a) A surface layer of marine sediments.
   (b) A layer of pillow basalt.
   (c) A zone of sheeted dikes.
   (d) A layer of gabbro.
   The oceanic crust and topographic features are related in some way to igneous activity.

2. Rocks of the ocean floor have not been deformed by compression, so their structure stands out in marked contrast to that in the folded mountains and the shields of the continents.

3. Rocks of the ocean floor are young in a geologic time frame. (Most are less than 150 million years old, compared to the ancient rocks of the shield that are more than 700 million years old.)

Discussion

Important insight into the rock types and structure of the oceanic crust is also obtainable in certain areas where oceanic crust is exposed for direct observation. Although most of the oceanic crust descends back into the asthenosphere during subduction, locally, large slices of oceanic crust are scraped off the descending plate and are incorporated into the orogenic belt of the continental crust. Later, isostatic uplift and erosion expose the fragment of oceanic crust, and this makes them accessible for direct geologic studies. Many examples of fragments of oceanic crust are known around the world; most notable are exposures in Cyprus, Greece, New Guinea, Newfoundland, and California. These rocks display a sequence of structure and rock types shown in the diagram in *figure 21.10.* At the top of the sequence (*layer 1*) is a relatively thin layer of sedimentary rock consisting of carbonate and siliceous shells of microscopic marine organisms

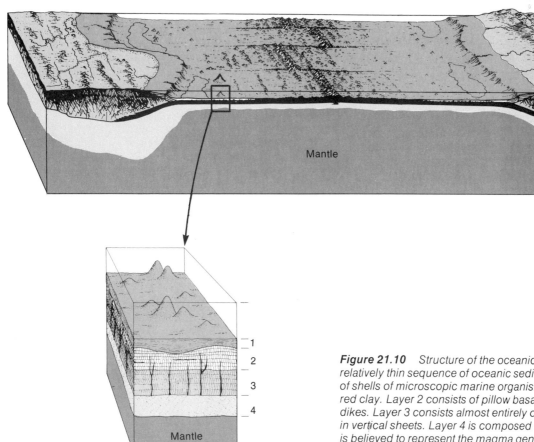

Figure 21.10 *Structure of the oceanic crust. Layer 1 is a relatively thin sequence of oceanic sediment consisting of shells of microscopic marine organisms together with red clay. Layer 2 consists of pillow basalt fed by numerous dikes. Layer 3 consists almost entirely of basalt dikes in vertical sheets. Layer 4 is composed of gabbro, which is believed to represent the magma generated at the spreading center but which cooled slowly at depth.*

together with red clay, all of which are typical of deep-marine sediments. *Layer 2* consists of pillow basalts fed by numerous dikes that originally were intruded into vertical fractures and that now form a series of parallel sheets of coarse-grained basalt or gabbro. The pillow lava is considered to represent the volcanic extrusions on the seafloor, and the dikes are the feeder vents produced as the plates split and move apart. Below the pillow lava, the rock consists almost entirely of the sheeted dikes (*layer 3*), and many younger dikes are intruded into portions of earlier dikes. Below the sheeted dike complex, the rocks consist of coarse-grained gabbro, which is believed to represent magma that was generated at the spreading center but cooled very slowly at some depth. Underlying the gabbro are peridotites composed almost entirely of olivine and pyroxene. This material is considered to be part of the mantle, and the Moho would be the boundary between the gabbro and the peridotite.

One of the most significant facts about the oceanic crust is that the samples thus far brought to the surface are all much younger than the rocks that form the bulk of the continents. The oldest known basalts retrieved from the ocean floor are only 50 million years old, and there is little reason to suspect that any basalt on the ocean floor is older than 160 million years. Extensive drilling in recent years has produced sediments ranging only from 125 to 150 million years old.

The Ocean Basins

Statement

Although all the ocean basins are believed to have originated in the same manner (that is, by seafloor spreading), they are quite different in size, shape, and topographic features. These differences are very significant, because they indicate a great deal about the ages, origins, and evolutions of the oceanic basins.

Discussion

The Atlantic Ocean. The regional topography of the Atlantic floor is basically very simple and shows remarkable symmetry in the distribution of the major features (*figure 21.11*). You should note from the physiographic map that the dominant feature is the mid-Atlantic ridge, which forms an S-shaped pattern down the exact center of the ocean basin. It separates the ocean floor into two long, north-south-trending, parallel subbasins characterized by abyssal plains. The abyssal hills occur along the margins of the ridge, and the plains occur along the margins of the continental platforms. The symmetry of the Atlantic basin extends to the continental margins, where the outlines of Africa and Europe fit those of South America and North America. This symmetrical distribution of the major features of the Atlantic has attracted scientific attention for many years and is one of the cornerstones for the theories about continental drift and seafloor spreading. The structure and topographic features of the Atlantic Ocean are simple and reflect the initial opening of this young ocean basin.

The Arctic Ocean. Although the Arctic Ocean has been considered by some to be simply an extension of the Atlantic Ocean, it is clear from recent exploration that it is unique in several respects. (1) It is nearly landlocked, and, when the basin beyond the continental slope is considered, there is only one inlet, the Lena trough between Spitzbergen and Greenland. (2) The continental shelf north of Siberia is the widest shelf in the world (over 1100 km across). Most of it is deeper than 100 fathoms, but very little exceeds 300 fathoms. (3) Two prominent submarine ridges trend roughly 140° east and divide the deep Arctic Ocean into three major basins.

The Indian Ocean. The Indian Ocean is the smallest of the three great oceans, and it connects with both the Atlantic and the Pacific through broad, open seas south of Africa and Australia. There, as in the Atlantic, the most conspicuous feature is the oceanic ridge, which continues from the Atlantic around South Africa and splits near the center of the Indian Ocean to form a pattern similar to a large inverted Y (*figure 21.12*). The northern segment of the ridge extends into the Gulf of Aden, where it apparently connects with the African rift valley and the rift of the Red Sea. Thus, the ridge divides the ocean basin into three major parts. Unlike the other oceans, the topography of the Indian Ocean floor is dominated by scattered blocks and some remarkably linear plateaus called **microcontinents.** Most are oriented in a north-south direction. Prominent parallel fracture zones are numerous. The striking northern trend of the parallel fracture zones, together with the trend of the linear microcontinents, imparts a remarkable linear structural fabric to the floor of the Indian Ocean.

The Pacific Ocean. The Pacific Ocean is somewhat different from the other ocean basins in that the oceanic ridge occurs near its eastern margin. The basin covers approximately half of the planet and is the largest single unit of oceanic crust. In addition, it is probably the oldest ocean basin, and it lacks the symmetry of the Atlantic and Indian basins.

The oceanic ridge continues in a broad sweep from the Indian Ocean between New Zealand and Antarctica and then northward along the American side of the ocean. The crest of the ridge disappears at the head of the Gulf of California but reappears off the coast of Oregon.

The floor of the western Pacific is studded with more seamounts, guyots, and atolls than all other oceans combined (*figure 21.13*). As is apparent on the physiographic map, many of the seamounts occur in linear chains that extend for considerable distances. The margins of the Pacific are also different in that they generally are marked by a line of deep arcuate trenches. On the eastern side of the Pacific, the trenches lie adjacent to the margins of Central and South America and are parallel to the great mountain systems of the Andes and the Rockies. The local relief from the top of the Andes to the bottom of the trench is 14,500 m (nearly nine times the depth of the Grand Canyon). On the western side of the Pacific, a nearly continuous line of trenches extends from the margins of the Gulf of Alaska along the margins of the Aleutians, Japan, and the Philippines and down to New Zealand.

Figure 21.13 *Physiographic map of the western Pacific Ocean.*

Small Ocean Basins. A number of small deep-ocean basins are not directly connected to the major oceanic plates but are, nevertheless, considered to be a direct consequence of plate tectonic processes. Excellent examples are the Mediterranean Sea, the Red Sea, and the Sea of Japan. Most of these are nearly landlocked and are connected to the main ocean by narrow gaps (*figure 21.14*). These small ocean basins originate in three ways: (1) by the growth of island arcs, (2) by initial rifting of continental plates, and (3) by convergence of continental plates. The small ocean basins are temporary features and, like the major oceanic plates, can grow but ultimately are destroyed.

The growth of island arcs has created the small ocean basins of the Pacific. The basins are formed because the subduction zones are not directly adjacent to a continental plate and the associated volcanic arcs isolate segments of oceanic crust to form small independent basins not directly attached to the major oceanic basins. Several such basins extend from the Aleutian Islands to New Guinea. Most of these basins are shallower than the adjacent open ocean. That is due, in part, to the fact that they trap the sediment washed in from adjacent continents. Sediment in the Bering basin even overflows through gaps in the Aleutian arc and spills over into the adjacent trench.

Figure 21.14 *Small ocean basins.*
(A) Basins of the western Pacific.
(B) Basins of the western Atlantic.
(C) Basins of the Mediterranean area.

Small ocean basins develop in the initial rift zone where continents split and begin to spread apart. Basalt from the upwelling mantle commonly fills the new opening during its early stages, and sediment eroded from the adjacent continents can help keep the floor of the basin filled to near sea level. When spreading is rapid, a long, narrow oceanic basin develops. The Red Sea and the Gulf of California are both excellent examples. (An interesting and potentially important feature of these new ocean basins is the circulation of hot water through the fresh magma and up through the sediments on the ocean floor to develop pools of hot brine with concentrations of rare metals.)

As the seafloor continues to spread, the intervening sea develops the characteristics of a major ocean basin, that is, an oceanic ridge, abyssal plains, and continental slopes. The Arctic Ocean is apparently in this transitional stage from sea to ocean.

The Mediterranean and Black Seas, in contrast to those previously described, are basins formed by the collision of Africa and Eurasia. These basins are remnants of the once-vast Tethys Ocean basin, which extended in an east-west direction and separated Laurasia from Gondwanaland 200 million years ago (*figure 21.15*). As Africa and India moved northward, the Tethys Sea gradually closed, although it remained a connecting link between the Atlantic and Indian Oceans for many millions of years. The compression from the northward movement of Africa and India developed the sinuous mountain belts of the Alps and the Himalayas, much of which are composed of folded and up-

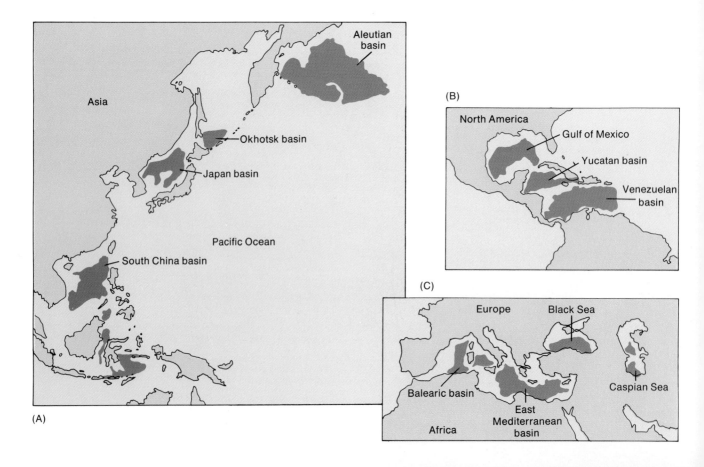

(A)

(B)

(C)

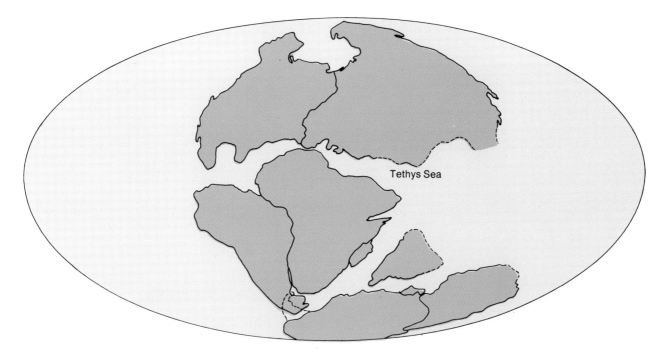

Figure 21.15 *Sketch map of the Tethys Sea.*

lifted deep-ocean sediments. Eventually, if the present plate motion continues, the Mediterranean basin will be closed completely and Africa will be tucked under the Alps in much the same way as India is thrust under the Himalayas.

Recent drilling in the Mediterranean basin indicates a fascinating history during the last 20 million years, the period when it became an isolated basin. As can be seen in *figure 21.14C,* the Mediterranean is almost landlocked as it is connected with the Atlantic by a narrow gap through the Straits of Gibraltar. The deep basin, however, is completely isolated. This produces some very interesting hydrologic conditions. Evaporation removes more than 4000 km³ of water each year, but less than 500 km³ is replaced by rain and surface runoff from Europe and Africa. An influx of approximately 3500 km³ of water each year flows from the Atlantic through the Straits of Gibraltar to maintain the Mediterranean at sea level. If the straits were closed, the Mediterranean Sea would evaporate completely in about 1000 years.

Deep drilling in the Mediterranean basin by the *Glomar Challenger* indicated that this actually happened 5 to 8 million years ago and that the Mediterranean basin was a deep, desolate, dry ocean basin 3 km below sea level. Core samples of windblown sand and salt deposits indicate that the basin was a desert region, with huge volcanoes (now islands) rising above the basin floor. The continental platforms of Africa and Europe surrounded the basin as huge plateaus. The area was undoubtedly void of life since temperatures would have reached 150 °F.

The salt deposits and windblown sand that formed in the dry basin are covered throughout the basin with deep-sea organic oozes that indicate that the Mediterranean basin was flooded almost instantaneously about

5.5 million years ago. This probably resulted from erosion of the barrier at the Straits of Gibraltar, which permitted water from the Atlantic to flow into the Mediterranean basin over an enormous waterfall. Estimates based on fossils preserved in the sediments indicate that the flow through the straits was 1000 times greater than the present-day flow over Niagara Falls and that the basin filled in about 100 years.

The History of Plate Movement during the Last 200 Million Years

Statement

The tectonic system probably has been operational during much of the earth's history, and it is believed to be responsible for the origin and evolution of continents as well as for the growth and destruction of ocean basins. Our understanding of the early history of plate movements comes from evidence preserved mostly in continental rocks. Ocean basins come and go, because the ancient oceanic crust is consumed at subduction zones and replaced by newer ocean basins at spreading centers. Continents have been drifting, with tectonic plates splitting and colliding a number of times, but details on the patterns of plate movements are very scanty. There are, however, considerable amounts of data on plate motion during the last 200 million years, and it is possible to reconstruct the position of continents and trace plate movement with considerable certainty. The variety of geologic data discussed on pp. 290-92 all suggest the same basic pattern of plate motion and agree that a large continental mass (Wegener's Pangaea) began to break up and drift apart about 200 million years ago. Dispersal and collision of the fragments have continued to the present time. Reconstruction of Pangaea in terms of absolute coordinates now is possible, and the direction and rate of plate movement have been determined.

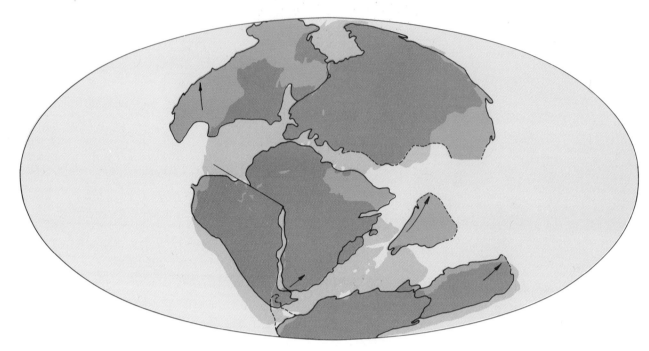

(A) Plate movement 200-100 million years ago.

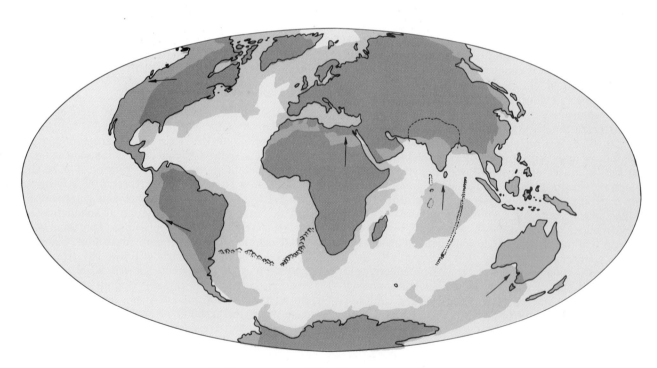

(B) Plate movement 100 million years ago to present.

Figure 21.16 History of plate movement during the last 200 million years.

Discussion

The history of relative plate movement during the last 200 million years is shown in *figure 21.16*. These maps are adapted from those prepared by Dietz and Holden in 1970, which were constructed with cartographic precision using all available evidence, including the geologic fit of continents, paleomagnetic pole positions, and patterns of seafloor spreading.

The initial event in the splitting of the continents was the extrusion of large volumes of basalt along the initial rift zone. Remnants of these basalts are found in the Triassic basins of the eastern United States and the plateau basalts of southwestern Africa, western India, and eastern Brazil. A northern rift split Pangaea along an east-west line slightly north of the equator and separated Laurasia from Gondwanaland, rotating Laurasia clockwise. A southern rift split South America and Africa away from the remaining Gondwanaland. Soon afterwards, India was severed from Antarctica and moved rapidly northward. The plate containing Africa converged toward Eurasia, forming a subduction zone in the Tethyan Sea. By the end of the Cretaceous period 65 million years ago, after 135 million years of movement, the separation of South America from Africa was complete and the South Atlantic Ocean had widened to at least 3000 km. All of the major continents were blocked out by this time, except for the connection between Greenland and Europe and between Australia and Antarctica. A new rift separated Madagascar from Africa, and India continued moving northward.

A north-south trench system must have existed in the Pacific (the South American-Central American trench)

and consumed oceanic crust on the western edge of the rapidly westward-moving plates carrying North and South America. North America probably encountered this trench in late Jurassic time. The trench eventually was overridden with the continued westward drift of North America, causing the deformation of the Rocky Mountains. About the same time, the same trench was encountered by South America. This encounter developed the early Andean folded mountain belt.

Throughout the Cenozoic period, the mid-Atlantic ridge extended into the Arctic and finally detached Greenland from Europe. During that time, the two Americas were joined by the Isthmus of Panama, which was created by volcanism along the subduction zone. The Indian landmass completed its northward movement and collided with Asia, creating the Himalayas. Australia drifted northward from Antarctica.

Finally, a branch of the Indian rift system split Arabia away from Africa, creating the Gulf of Aden and the Red Sea. Then, a spur of the rift meandered west and south to create the East African rift valleys. Less pronounced changes induced the partial closing of the Caribbean region and the continued widening of the South Atlantic.

Figure 21.17 shows the anticipated position of plates 50 million years from now. Through the extrapolation of present directions and rates of plate movement, certain changes seem likely. The Atlantic and Indian Oceans will continue to grow at the expense of the Pacific. Australia will move northward and encounter the Eurasian plate. The eastern part of Africa will split off along the rift valleys, but the northern part of the continent of Africa probably will move northward and close the Mediterranean Sea. New land will be created in the Caribbean by compressional uplift. Baja, California, and part of western California will be severed from North America and drift to the northwest.

Figure 21.17 Anticipated plate movement during the next 50 million years.

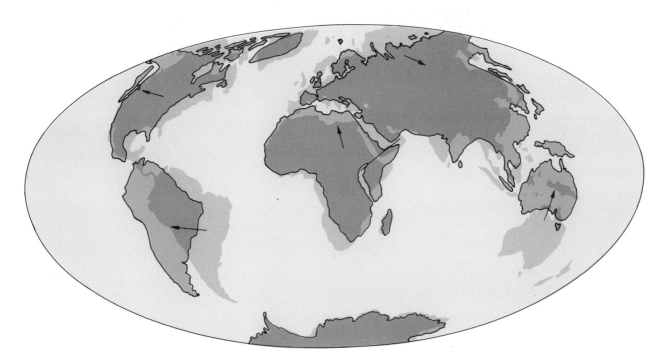

Summary

The major topographic features of the ocean floor have been mapped with precision-depth recorders and have been found to be related in some way to volcanic activity.

The oceanic ridge is a broad, fractured swell extending as a continuous feature for a distance of more than 64,000 km. It has a narrow axial rift valley and is cut by numerous transverse fractures.

Abyssal hills are formed at, or near, the ridge and move away from the ridge with the drifting plates. As they move, they become covered with sediment and ultimately can be buried to form abyssal plains.

Trenches form at subduction zones where two plates converge and one is thrust down into the mantle. Typically, they are bordered by island arcs or young mountain ranges.

Volcanic islands, or seamounts, can originate near the spreading center and move with the spreading plates. They also can form over local hot spots in the mantle and form linear chains.

The continental shelf and slope are not parts of the oceanic crust but are parts of the continental platform, which is submerged slightly below sea level. The continental slope is cut by many deep submarine canyons, which are similar in size and depth to the great canyons cut on the continents by rivers.

The oceanic crust is relatively thin (7 to 8 km) and is composed mostly of basalt. It is composed of four major layers: (1) a surface layer of marine sediment, (2) a layer of pillow basalt, (3) a zone of sheeted dikes of basalt, and (4) a basal layer of gabbro. The oceanic crust is *not* deformed by tight folding or thrust faulting and is geologically young, with most of it having been formed during the last 50 million years.

The three major ocean basins—the Atlantic, Indian, and Pacific—are believed to have originated in the same way; that is, by seafloor spreading; but each is different in size, shape, seafloor topography, and age. The small ocean basins in the western Pacific, Red Sea, Caribbean, and Mediterranean areas result from growth of island arcs, initial rifting, and by converging continental plates.

The history of the ocean basins during the last 200 million years can be outlined in some detail. The Atlantic, Indian, and Arctic Oceans were formed by the rifting of a large super-continent (Pangaea). The major fragments of the super-continent moved with the spreading plates, and, as the ocean basins enlarged, the continents eventually arrived at their present positions.

Additional Readings

Bullard, E. 1969. "The Origin of the Oceans." Sci. Amer. 221(3): 66-75. (Reprint no. 880.) San Francisco: W. H. Freeman and Company.

Heezen, B. C. 1960. "The Rift in the Ocean Floor." Sci. Amer. 203(4): 98-110.

Heirtzler, J. R. 1968. "Sea-Floor Spreading." Sci. Amer. 219(6): 60-70. (Reprint no. 875.) San Francisco: W. H. Freeman and Company.

Menard, H. W. 1969a. "The Deep-Ocean Floor." Sci. Amer. 221(3): 126-42. (Reprint no. 883.) San Francisco: W. H. Freeman and Company.

Menard, H. W. 1969b. "Growth of Drifting Volcanoes." J. Geophys. Res. 74(20): 4827-37.

Moore, J. R., ed. 1971. Oceanography: Readings from Scientific American. San Francisco: W. H. Freeman and Company.

The Ocean. 1969. A Scientific American Book. San Francisco: W. H. Freeman and Company.

Phinney, R. A., ed. 1968. The History of the Earth's Crust. Princeton, N.J.: Princeton University Press.

Evolution of the Continents

22

Although the theory of plate tectonics was developed largely from new studies of the seafloor, a record of plate motion throughout most of geologic time is preserved only in the rocks of the continents. This is because the oceanic crust is a temporary feature that is continually being created along the oceanic ridge and destroyed along the subduction zones. Plate movement can be studied and measured in the relatively young oceanic rocks, but all of the oceanic crust is younger than about 200 million years.

By contrast, rocks on the continents are as old as 3.8 billion years, and a record of plate movement during the early history of the earth is preserved only in the ancient metamorphic and igneous rocks of the continental shields. On the continental shields, geologists find a long and complex history of earth dynamics, one which suggests that the tectonic system has been operating during most of the time since the earth was formed.

This chapter will consider how new continental crust is made by a combination of deformation, metamorphism, igneous activity, erosion, sedimentation, and isostatic adjustment. The continents are products of the tectonic system. They grow by accretion but also can split and drift apart. All of these processes of continental evolution are involved with the fundamental process of segregating and differentiating the materials of the earth.

Major Concepts

1. The continents are made up of three basic structural components: (a) shields, (b) stable platforms, and (c) young mountain belts.
2. Mountain-building (orogenesis) occurs at converging plate margins and results in differentiation of the oceanic crust and concentration of silica-rich, low-density material in folded mountain belts.
3. The major features of orogenesis are (a) intensive crustal deformation, (b) metamorphism, and (c) igneous activity.
4. The characteristics of a mountain belt and the sequence of events that produce it depend upon the rock types involved and the types of interactions at converging plate boundaries.
5. Three different plate interactions are recognized: (a) convergence of two oceanic plates, (b) convergence of a continental and an oceanic plate, and (c) convergence of two continental plates.
6. Continents grow by accretion as new crustal material is formed in an orogenic belt.

The Continental Crust

Statement

The continental crust has been studied for more than 200 years, but only recently have we been able to synthesize the vast amount of data concerning continental geology and develop a reasonably clear concept of what continents really are. A summary of the more important geologic facts about continents is as follows.

1. Continents are composed of a huge, flat slab of granitic rock. This does not mean that all the rocks of the continents are igneous; much of the continental crust consists of metamorphic rocks composed of roughly the same material as granite. Seismic data suggests that there is a gradation from granitic composition observed near the surface of a continent to a more basaltic composition near its base.
2. Continents range in thickness from 30 to 50 km. The thickest portions of continents are beneath mountain ranges.
3. Continents cover roughly one-third of the planet and appear to have grown in size throughout most of geologic time.
4. The structure of continents is extremely complicated and consists mostly of a highly deformed sequence of metamorphosed sediments and volcanic rocks, together with large volumes of granitic intrusions and granitic gneiss.
5. The continents contain the oldest rocks on earth, which range in age up to 3.8 billion years. This stands out in contrast to the age of the oceanic crust, which is all less than 200 million years old.
6. Preliminary studies indicate that the earth is the only terrestrial planet of the solar system with continental crust and that continental crust results from planetary differentiation.
7. Although each continent may appear to be unique, they all have three basic components.
   (a) A large area of very old and highly deformed metamorphic and igneous rocks known as **shields.**
   (b) Broad, flat **stable platforms,** where the shield is covered with a veneer of sedimentary rocks.
   (c) Young, **folded mountain belts** located along the continental margins.
8. Geologic differences among continents are mostly in the size, shape, and proportions of these components.

Discussion

The major features of continents were described briefly in chapter 1, but, in summing up, the main ideas

Figure 22.1   An oblique view of the Canadian Shield. The linear lakes emphasize the structural trends of the metamorphic rocks that are etched out by differential erosion into a relief of a few tens of meters. On a regional basis, the surface of the shield is a broad, flat surface (compare with plate 4 and figure 22.3).

concerning their origin will be repeated for the sake of clarity.

Shields.   In order to understand modern theories of the origin and evolution of continents, you first must become better acquainted with the characteristic features of the continental shields, because the shields are the best indications of the true nature of the continental crust. Perhaps the best way to do this is to study carefully *figures 22.1* and *22.2*, together with *plates 4* and *28*, and note some of the features that have been observed in the field. The most striking characteristic of the shield seen in these photographs is the vast expanse of its low, rela-

tively flat surface. Throughout an area of thousands of square kilometers, the surface of the shield is within a few hundred meters of sea level. The only features that stand out in relief are the resistant rock formations that rise a few tens of meters above the surrounding surface. On a regional basis, the shield is flat and almost feature-less. The structural complexities of the shield are shown by the patterns of erosion, the alignment of lakes, and

Figure 22.2   *Vertical aerial photograph of part of the Canadian Shield. The complex metamorphic rocks (dark tones) are intruded by granite (light tones).*

the differences in the tones of the photographs. Faults and joints are common and are expressed at the surface by linear depressions, some of which can be traced for hundreds of kilometers.

A second fundamental characteristic of the shields is their structure and composition. Shields are composed of a complex mixture of metamorphic rocks and large volumes of granite. Most, but not all, of the rocks are Precambrian and were formed under high temperature and high compressive stresses several kilometers below the surface. In *figure 22.2,* metamorphic rocks are shown in tones of dark gray and granitic intrusions appear in lighter tones and have a more massive texture. Evidence, such as shown in *figures 22.1* and *22.2,* shows that the shields have been subjected to intense deformation in which sedimentary and volcanic rocks were converted to complex metamorphic rocks and intruded by granitic magmas. Subsequently, erosion removed the upper cover of the sedimentary and metamorphic terrain to expose what is seen now at the surface.

On the basis of these facts, our present understanding of the shield could be summarized graphically, as in the diagram in *figure 22.3.* This basic structure of complex igneous and metamorphic rocks eroded to a flat surface near sea level forms the nucleus of every continent.

The belts of metamorphic rocks, together with igneous intrusions, indicate that the shields are composed of a series of zones that were once highly mobile and tectonically active. These facts have long been known, and, for many years, geologists have considered the shields to consist of a series of ancient mountain belts that have

Figure 22.3.   *Schematic block diagrams showing the general characteristics of a shield. (A) The shield consists of a complex of metamorphic rocks and a variety of igneous intrusions. The upper surface is nearly flat and is eroded down to sea level. (B) Throughout most of the interior of the United States, the shield is covered with a veneer of horizontal sedimentary rock thousands of meters thick. In some places, such as in the Grand Canyon, erosion has cut through the sedimentary veneer to expose the shield below.*

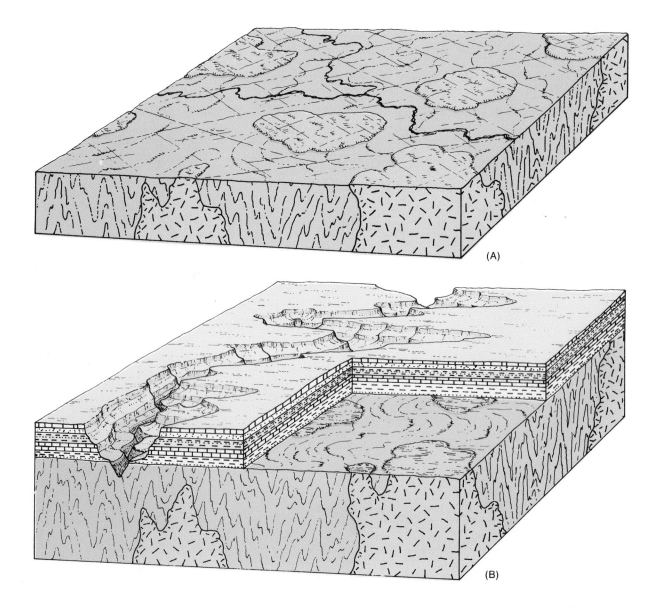

(A)

(B)

been eroded down to their roots and have since remained stable.

The Canadian Shield of North America is relatively well known, and the map in *figure 22.4* summarizes many of the structural details compiled from many years of fieldwork, geophysical studies, and radiometric age determinations. In a general way, North America is quite

Figure 22.4 *Map showing the major structural trends and radiometric dates of the rocks in the Canadian Shield. The shield consists of a number of geologic provinces, each representing major mountain-building events. The numbers refer to the ages, in billions of years, of the major granites, and the lines represent the trends of the folds. The shield appears to have grown by accretion as new mountain belts were formed along its margins.*

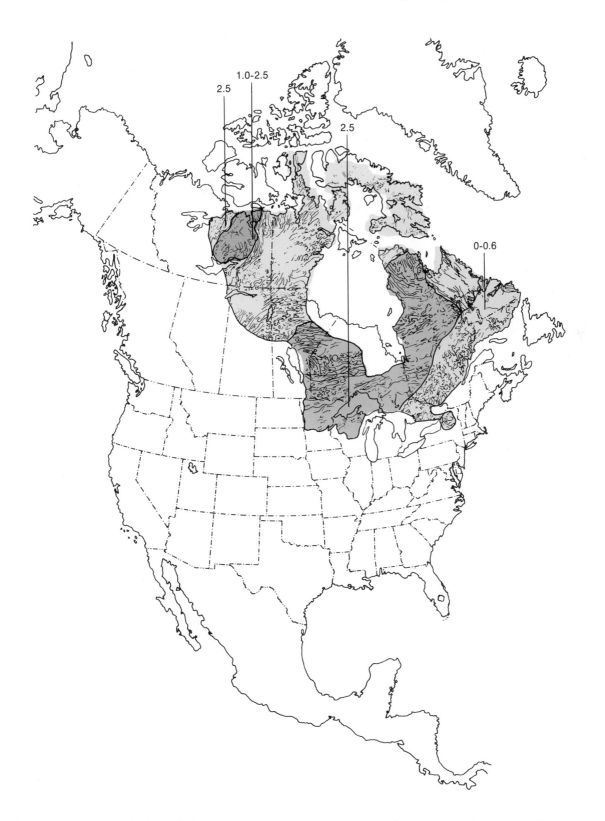

symmetrical. The oldest rocks, recording the first clearly recognizable geological events, are in the Superior-Wyoming and Slave Provinces. They consist of volcanic flows and volcanic-derived sediments similar to those forming today in the island arc areas of the Pacific. These sediments are metamorphosed and engulfed by granitic intrusions. The granite-forming event, as determined by radioactive dating, was 2.5 to 3.0 billion years ago. Since the earth has been estimated to be a little more than 4.5 billion years old, this segment of crust apparently formed 1.5 to 2.5 billion years after the origin of the earth. Generally speaking, the composition of this ancient conti-

nental crust, prior to the granitic intrusion, was closer to that of basalt than granite. The granitic intrusions now constitute three-fourths of the area. The metamorphosed sediments and volcanics have structural trends in a northward direction, as shown in *figure 22.4*. This segment of crust is significant, because it is thought to represent the development of the first stable and resistant granitic crust in North America. Since the time of its formation, the upper surface has been eroded down almost to sea level and has persisted as a stable unit, at times partly submerged beneath a shallow sea and at times slightly above sea level, but it was never again the

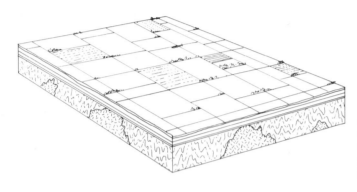

Figure 22.5 *Structure of the stable platform in the central United States. The stable platform is, in essence, part of the continental shield covered with a veneer of sedimentary rocks. It constitutes much of the broad, flat lowland area of the world known as plains, steppes, and low plateaus.*

site of mountain-building processes. In all probability, this block of crust was as much as 20% to 60% larger than it is presently, because the trends of the metamorphic structures are terminated abruptly by younger provinces. We can assume that the original continental mass was split and fragmented by plate movement, with each segment subsequently acting as a center for future continental growth.

Surrounding the Superior-Wyoming Province to the south and west is a vast area of gneiss and granite from 1.0 to 1.4 billion years old. The rocks in this younger province also are metamorphosed, but the original sediment was quite different from that now seen in the Superior Province. These younger rocks contain less lava and show a distinct increase in quartz-rich sandstone and limestone formed on the continental shelf. In addition, the structural trends are oriented in a different direction.

Beyond this province, the rocks are still younger and the granitic intrusions range in age from 0.8 to 1.1 billion years. They are mostly quartzites, marbles, and schists derived from well-sorted sandstone, shale, and limestone and volcanics richer in silica than the older volcanic rocks of the Superior Province.

The concentric pattern of the provinces in North America is considered to be strong evidence that the continent grew by accretion of material around the margins during a series of orogenic events. Each province represents a mountain-building event during which sediments were deformed into a mountain range by compressive forces. Subsequently, erosion removed the upper part of the deformed belt to expose only the deeper, highly deformed metamorphic rocks and associated granitic intrusions.

Stable Platforms.   Large areas of all the shields are covered with a series of horizontal sedimentary rocks in a manner similar to that shown in *figure 22.3*. These areas have been removed from the tectonically active plate margins since the shields were formed. They have, therefore, never been uplifted a great distance above sea level or submerged far below it. They have remained stable, hence the term *stable platform*.

The stable platforms form much of the broad, flat lowlands of the world (*figure 22.5*) and are known as plains, steppes, and low plateaus (see also *plate 6*). The relationship between the sedimentary rocks of the stable platform and the underlying igneous and metamorphic complex of the shield (also referred to as the **basement complex**) is known from the thousands of wells that penetrate the sedimentary cover and from seismic studies of rock structure beneath the surface. Although locally the rocks appear to be almost perfectly horizontal, on a regional basis they are warped into broad, shallow domes and basins (see *plates 5* and *6*).

The flat-lying sedimentary rocks that locally cover parts of the shield are predominantly sandstone, shale, and limestone deposited in shallow-marine environments. These flat-lying marine sediments, preserved on all continents, clearly indicate that large areas of the shields have been flooded periodically by shallow seas and then have reemerged as dry land. At the present time, more than 25% of the continental crust is covered with water (the continental shelf), but, at various periods in the past, shallow seas spread over a much greater percentage of the land surface.

There are several reasons why large areas of the shields were covered with extensive shallow seas during various periods of geologic time. In the first place, the shields are broad, flat areas eroded down to within a few tens of meters of sea level, so any uplifts or downwarps as the plates move across the globe would cause expansion or contraction of shallow seas over the continents. For example, a period of rapid convection in the mantle would cause the oceanic ridge to arch up and displace the water from the ocean basins onto the continents. When the rate of convection decreased, the oceanic ridge would subside and the seas would recede from the continents. Also, as a tectonic plate moves with the convecting asthenosphere, it can rise and fall slightly, causing expansion and contraction of shallow seas over the shield areas.

Mountain Belts.   The great linear belts of folded mountains that typically occur along the margins of continents constitute one of the most distinctive tectonic features of the earth. As was mentioned in several previous sections, the rocks within these relatively narrow zones are compressed, folded, and—in many places—broken by thrust faults with the result that the crust is shortened as much as 30% (*figure 22.6*; see also *plates 7* and *28*).

The young, active mountain belts on the earth today occur along convergent plate margins and coincide with zones of high seismicity and andesitic volcanism (*figure 22.7*). Two major mountain belts are still in the process of growing; they are (1) the Cordilleran belt, which includes the Rockies and the Andes; and (2) the Himalaya-Alpine belt, which extends across Asia, western Europe, and into North Africa. Older mountain ranges, in which deformation ceased long ago but which are still expressed by significant topographic relief, include the Appalachian Mountains of the eastern United States, the mountains of eastern Australia, and the Ural Mountains of Russia.

The location of young mountains in long, narrow belts along continental margins is significant, because it indicates that they could not result from a uniform worldwide force evenly distributed over the earth. Mountains must be the result of forces concentrated along the margins of continents. Another important aspect of the location of mountains is that many mountain belts extend to the ocean and abruptly terminate at the continental margin. The northern Appalachians, the Atlas Mountains of Africa, and the mountains of Great Britain are excellent examples. In these systems, the abrupt termination of the folded structures suggests that the mountain systems were once much more continuous and have been separated by continental rifting.

A Description of the Continents.   Shields, stable platforms, and young, folded mountain belts are the three basic structural components of all continents, and the major geologic differences among continents are mostly in the size, shape, and proportions of these three components. This section will briefly consider the structure of each continent and see how the continents are similar and how they are different. *Figure 22.8* shows the distribution of shields, platforms, and young, folded mountain

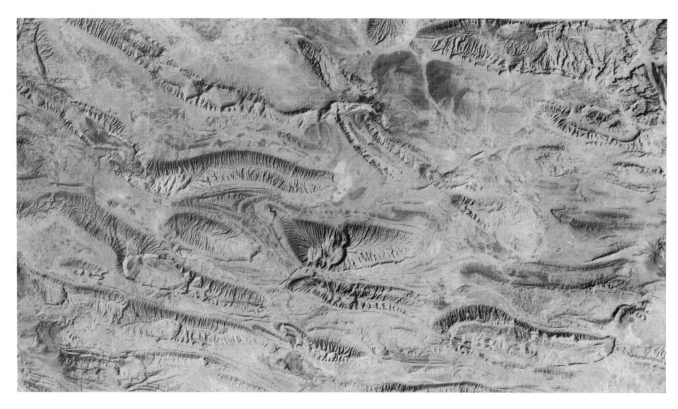

belts on each continent and will serve as a graphic summary. This is important background material for subsequent discussions on the origin and evolution of continents.

North America has a large shield, most of which is less than 300 m above sea level, in Canada. It extends from the Arctic islands southward to the Great Lakes area and westward to the plains of western Canada. The

Figure 22.6 *Structure and topography of a folded mountain belt in southern Iran.*

Figure 22.7 *Map showing the folded mountain belts and island arcs of the world.*

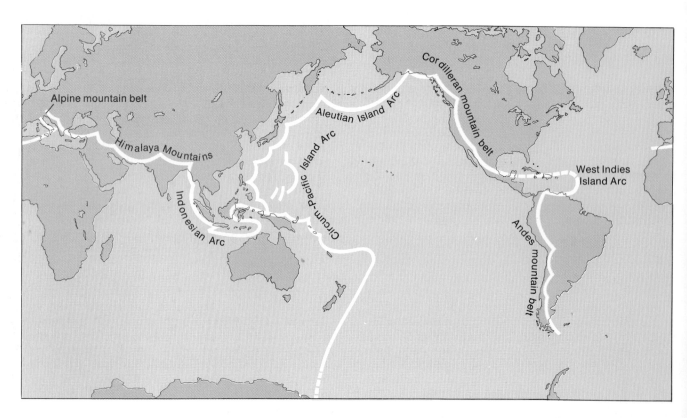

platform extends throughout the central United States and western Canada and is covered with sedimentary rocks that have been slightly warped into broad domes and basins.

South America consists of a broad shield, parts of which are covered in the south and west, in Brazil. The Andes Mountains are an extension of the Cordilleran folded mountain belt, which extends from Alaska to the southern tip of South America. There are no mountain belts on the eastern margin of South America.

Australia is much like South America. Most of the continent is a stable platform with a thin veneer of sedimentary rocks partly covering the shield. A single mountain range occurs along the eastern margin.

By contrast, Africa consists of an extensive shield covered only locally with a thin veneer of sediments preserved largely in circular downwarped basins. The only folded mountain belt occurs in narrow bands along the southern and northern margins of the continent.

Asia has a much larger area of folded mountains—the Himalayas—that forms a wide east-west range. Most

of the Asian Shield is covered with sedimentary rocks that form the vast, low areas of central Asia, but exposed igneous and metamorphic rocks of the shield occur in parts of Siberia. The Ural Mountains form the margin between the Asian and European continents, and the Himalaya Mountains form the margin between Asia and India.

The shield of Europe is exposed in the Baltic area but is covered throughout most of the central and southern parts of the continent to form a large stable platform. In central Europe, the sedimentary rocks have been warped into broad domes and basins. The eroded edges of these structures, such as the Paris basin, can be seen on a regional physiographic map. The Alps form the folded mountains along the southern margin of the European continent.

Most of India is a shield with a limited platform area of horizontal basalt flows and sedimentary rock.

The style of landforms developed on each major structural component of a continent is quite distinctive. The shields are flat lowlands on a regional basis, but, on a local scale, many structural details of the complex metamorphic and igneous rocks are etched out into relief (see *figures 22.1* and *22.2* and *plates 4* and *28*). The sedimentary rocks of the stable platform form flat terrain commonly dissected by stream erosion to form low, rolling hills. Where the sedimentary rocks are tilted slightly

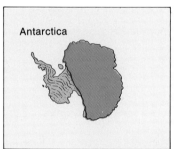

Antarctica

Figure 22.8 *Tectonic map of the world showing the shields, stable platforms, and folded mountain belts of each continent.*

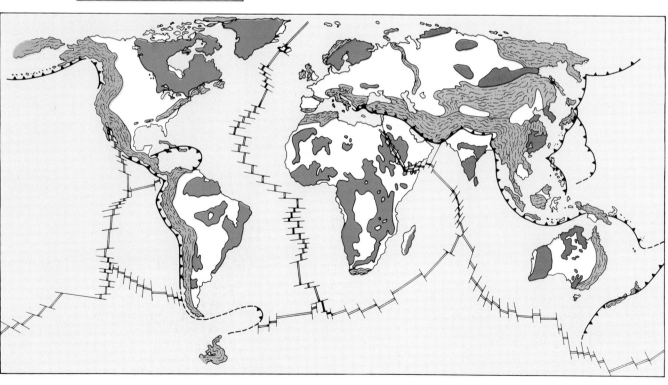

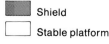 Shield
Stable platform

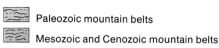 Paleozoic mountain belts
Mesozoic and Cenozoic mountain belts

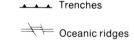 Trenches
Oceanic ridges

in broad domes and basins, the eroded edges of the strata form a series of ridges (see *plates 5* and *6*). Eroded mountain belts typically form alternating ridges and valleys (see *plate 7* and *figure 22.6*). Details of the style of landforms developed on each structural unit are, of course, influenced profoundly by climate (see *figure 8.15*).

Orogenesis (Mountain-building)

Statement

The highly deformed rocks in mountain belts have attracted the attention of geologists for over 150 years, but, until recently, the question of why the earth has mountains has escaped a satisfactory answer. The question becomes even more important when we recognize that folded mountain belts have not been formed on the moon, Mercury, or Mars. Only the earth has this distinctive structural feature.

We now know that mountain belts are formed at converging plate margins and that the compressive forces result from plate collision. Mountains, therefore, provide a record of ancient plate motion and a history of the earth's dynamics.

The process of mountain-building is called **orogenesis.** The factors that appear to be most important in this process are:
1. Rock sequences.
2. Structural deformation.
3. Metamorphism.
4. Igneous activity.

In addition, erosion and isostatic adjustment of the crust occur until the mountain belt is eroded to near sea level.

Discussion

Rock Sequences and Orogenesis. Over 100 years ago, James Hall, a noted American geologist, recognized that the total thickness of strata in the Appalachian Mountains is approximately 8 to 10 times as thick as sedimentary rocks of equivalent age in the stable platform. He also noted that, in both areas, most of the rocks throughout the entire sequence had been deposited in shallow water, as indicated by shallow-marine fossils, mud cracks, and interbedded layers of coal. From these observations, he concluded that the present mountain range was once an elongate trough, or a segment of the crust, that gradually subsided much more than the rest of the continent. The gradual subsidence of the trough permitted a great thickness of shallow-marine sediments to accumulate, the rates of sedimentation being roughly equivalent to the rates of subsidence. The elongate, subsiding trough was called a **geosyncline.** After receiving a critical thickness of sediments (approximately 14 km), it was compressed into a folded mountain range.

The geosynclinal theory of mountain-building was further developed by James Dana, another American geologist, who also worked in the Appalachians. He proposed that mountain-building involved a three-phase cycle consisting of (1) geosynclinal sedimentation and contemporaneous subsidence, (2) compression and deformation, and (3) uplift and erosion.

Studies of mountain belts in different parts of the world indicated that the Hall-Dana concept was too simple and that parallel belts of different types of geosynclinal sedimentation existed. One belt adjacent to the stable platform was called the **miogeosyncline** and was underlain by continental crust. It consisted of clean, well-sorted, shallow-water sandstone, limestone, and shale, with no volcanic rocks. The other, called a **eugeosyncline,** consisted of sediments deposited in deep-marine environments and typically included volcanic rocks. Modern thought concerning geosynclinal sedimentation and orogenesis utilizes much new information about sedimentation along continental margins and in the adjacent deep-marine basins and relates the formation of geosynclines to the plate tectonic theory.

The theory of plate tectonics proposes that mountain-building occurs along converging plate margins, but, unlike the geosynclinal theory, it does not require a specific sequence of events to occur in the same order in all orogenic belts. Moreover, a variety of rock sequences deposited along continental margins and on the ocean floor are known to be deformed at converging plate margins to produce different styles of mountain belts. Modern oceanographic studies have determined that the classical

Figure 22.9 *Schematic diagram showing the two types of rock sequences that accumulate along continental margins. Clean, well-sorted sandstone, shale, and limestone deposited in shallow water accumulate on the continental shelf. Poorly sorted, dirty sandstone and shale are deposited by turbidity currents in the deep water beyond the continental margins.*

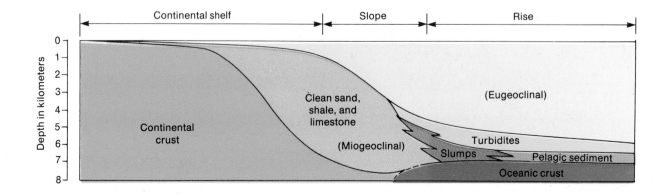

miogeosynclinal sequence of sediments accumulates along the continental margins and that the eugeosynclinal suites of rocks form in the deep water beyond. Probably the best known example is the present Atlantic continental margins of the eastern United States (*figure 22.9*). The sequence of thick sediments has recently been called a **geocline** by some geologists, because it is not a two-sided trough but is open toward the ocean.

Two distinctly different sequences of rock are clearly recognized: (1) a rock sequence consisting of well-sorted, clean sandstone, shale, and limestone derived from erosion of the continents and deposited in shallow water along the continental margins and (2) poorly sorted sandstone and shale deposited in deep water off the continental margins by turbidity currents together with submarine slump blocks and landslides. The clean, well-sorted, shallow-water sediments are sometimes referred to as miogeoclinal, and the dirty, deep-water sediments are called eugeoclinal. As illustrated in *figure 22.9*, the turbidities of the eugeoclinal sequence grade seaward into deep-marine organic oozes.

Shallow-water miogeoclinal sediments also can form behind an island arc, and deep-marine eugeoclinal assemblages can form in trenches and on the seaward side of arcs.

A third rock assemblage common in some mountain belts is called the ophiolite sequence. It consists of peridotite, gabbro, pillow basalt, and deep-marine sediments that form the oceanic crust. These rocks are scraped off the subducting plate and plastered against the upper plate as a chaotic melange, with complex structures such as folding and thrust faulting.

One of the features of geoclinal sequences that always has been difficult to explain is the great thickness of shallow-marine sediments. In order for 10 km of shallow-water sediments to accumulate, the crust had to subside, essentially, at the same rate as the sediment accumulated. What caused the gradual subsidence? This question has not been completely answered, but the subsidence may be related to the vertical movements of the crust following extension and rifting of the continent. When a continent begins to split apart and a new ocean basin begins to form, the continental crust is uparched, extended, and split. The edges of the rift zone, which will become the new continental margins and the site of geoclinal sedimentation, soon are leveled by erosion. As the continent moves off from the uparched spreading center, it begins to subside, and sediment derived from erosion of the continent accumulates along the margins as the continent moves. Gradual subsidence of the continental margins occurs for two reasons: (1) the continent moves off from the rising mantle that underlies the ridge, and (2) the weight of the sediment deposited causes the crust to be depressed.

To sum up, sedimentary rocks that commonly are involved in orogenesis are (1) shallow-marine miogeoclinal sediments, (2) deep-marine eugeoclinal sediments, and (3) ophiolite sequences of the oceanic crust. In studying the history of mountain-building and the plate movements that produce it, geologists not only study the direct results of orogenesis (such as deformation, metamorphism, and igneous activity) but pay particular attention to details of sedimentary rock assemblages that were deposited prior to plate collision, because these assemblages provide important insight into the nature of plate interactions.

Structural Deformation. The single most distinctive feature of a mountain belt is the structural deformation of the rocks. The nature of this deformation was discussed in chapter 20 (refer again to *figures 20.3, 20.4*, and *20.13*, as well as *plates 7, 26*, and *28*). The structures result from compression, and the range of deformation goes from small, deformed grains or fossils in the rock to large-scale folds tens of kilometers wide.

In the figures cited in the preceding paragraph, only a small part of a mountain belt can be seen, but, from detailed field mapping, the structure of an entire mountain range can be determined and illustrated by geologic maps and cross sections. *Figure 22.10* includes three well-known examples.

In the Canadian Rockies (*figure 22.10A*), a major type of deformation is thrust faulting, in which large slices of rock have been thrust over another in a belt 0.60 km wide. It is apparent from the orientation of the faults and the direction of displaced beds that the rocks were thrust from the margin of the continent toward the interior.

A cross section of the Appalachian Mountains (*figure 22.10B*) shows a different style and magnitude of deformation. The major structural feature is a series of tight folds. Deformation is most intense near the continental margin, and it dies out toward the continental interior. Both the Rockies and the Appalachians show compression from the continental margins as a result of plate collision.

The structure of the Alps is even more complicated (*figure 22.10C*), with great overturned folds called **nappes** (French: sheets) showing an enormous amount of crustal shortening. The rocks involved are so intensely deformed that originally spherical pebbles are stretched out into rods as much as 30 times longer than the original diameters of the pebbles. Most of these structures can be explained only in terms of compressive forces.

The significant conclusion from these and other examples that could be cited is that, in each mountain belt, the internal structures show the results of strong, horizontal compressive forces. Similar deformation in older rocks, in places where the topographic relief has been eroded down, constitutes one of the basic evidences for concluding that regions such as the shields are roots of ancient mountain systems.

Metamorphism. In the deeper parts of an orogenic belt, intense plastic deformation and recrystallization at elevated temperature and pressure cause the original sedimentary and volcanic rocks to be metamorphosed into schists and gneisses. The horizontal stress generated by the converging plates causes recrystallization of many minerals and develops foliation perpendicular to the stress. Thus, slaty cleavage, schistosity, and gneissic layering in the deeper parts of a mountain range are characteristically vertical, or they dip at a high angle. Abnormally high heat in certain areas within a mountain belt can produce a system of concentric metamorphic zones around a thermal center. In the deeper parts of a mountain belt, metamorphism can become intense enough to

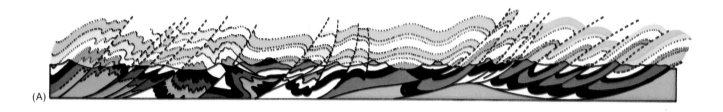

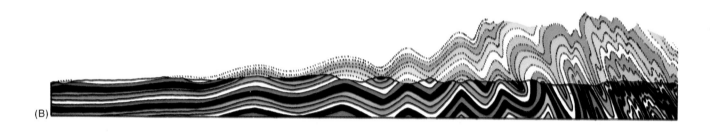

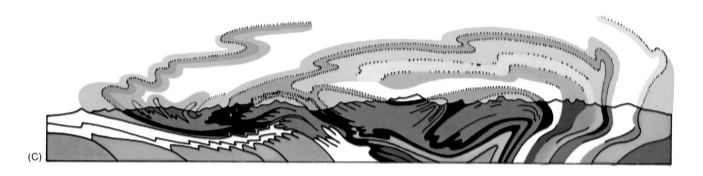

Figure 22.10   *The structure of mountain ranges in cross section. (A) The Canadian Rockies contain both folds and thrust faults. (B) The Appalachian Mountains consist of tight folds that have been eroded down to within two or three thousand meters of sea level. Resistant sandstones form the mountain ridges. (C) The Alps are complex folds, many of which are overturned.*

produce granitic migmatite complexes. (*Migmatite* is a term applied to a complex mixture of thin layers of granitic material between sheets of schist or gneiss). The migmatite develops largely from partial melting of the preexisting rocks in the immediate vicinity, and, apparently, the magma does not migrate far. In these zones, temperatures and pressures are high enough to soften the entire rock body, which behaves as a highly viscous liquid when stressed. Consequently, metamorphic rocks in deeper parts of orogenic belts exhibit complex flow structures (see *figures 7.2, 7.3,* and *7.4*).

Igneous Activity.   We now know that mountain-building is much more significant than simple deformation of the crust; it is part of a fundamental process by which the earth's materials are separated and concentrated into layers according to density—a process known as **differentiation.**

The process of concentrating the lighter material of the earth in the continental crust occurs in two steps. The first phase begins at the spreading center, where partial melting of the peridotite in the upper mantle generates a basaltic magma that rises to form the oceanic crust. The basalt is richer than peridotite in the lighter elements, particularly silicon and oxygen. The second phase occurs at the subduction zone, where partial melting of the oceanic crust forms a silica-rich magma that then is emplaced in the mountain belt as granitic intrusions or andesitic volcanic products. This process further separates the lighter elements and concentrates them in the continental crust. The granitic continental crust is less dense than the mantle and the oceanic crust, and its buoyancy is what prevents it from being consumed at subduction zones. Once formed, the continental crust remains on the outer surface of the earth.

The generation of silica-rich magma, conceivably, can result from partial melting in three different regions in the subduction zone: (1) in the subducting oceanic crust, (2) in the overlying mantle as fluids percolate upward, and (3) near the base of the continental crust as upward-migrating magma increases temperature (*figure 22.11*). The presence of water in the pore spaces of the rock and in chemical combination in many minerals of

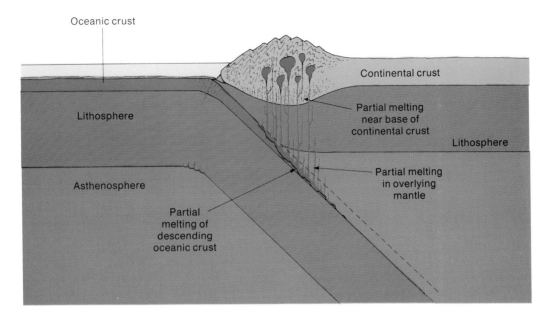

Oceanic crust

Continental crust

Lithosphere

Partial melting
near base of
continental crust

Lithosphere

Asthenosphere

Partial melting
in overlying
mantle

Partial
melting of
descending
oceanic crust

Figure 22.11 *Schematic diagram showing the generation of magma in a subduction zone.*

the oceanic crust plays an important role in the type of igneous activity at converging plate margins. As the crust is subducted, this water is driven out with increasing temperature, and, as it percolates upward, it enhances melting.

The Evolution of a Mountain Belt.   The general concepts of the relationships among metamorphism, igneous activity, and structural deformation in a mountain belt are shown in *figure 22.12.* Folding and thrusting occur at relatively shallow depths, metamorphism occurs deeper, and partial melting at still greater depth. Granitic magma initially forms deep within the crust at a large number of points where the rock begins to melt. The magma then is injected into the foliation of the adjacent metamorphic rock to form a migmatite. Much of the magma can migrate upward, because it is less dense than the solid rock. The boundaries of the body of magma are largely parallel with the broad zones of foliated metamorphic rock. As the granitic magma rises, it cuts across the upper folded strata, which are not metamorphosed but are deformed only by folding and faulting. The magma can rise high in the crust and cool within a few kilometers of the surface to form a batholith, or it can be extruded as andesitic volcanic material.

Erosion and Isostatic Adjustment.   The history of a mountain belt does not end with deformation, metamorphism, and igneous activity. After the orogenic activity is terminated—presumably by shifting convection cells in the mantle—deformation ceases, but erosion and isostatic adjustment combine to modify the orogenic belt. As illustrated in *figure 22.12,* the entire crust is deformed by the orogeny so that a mountain root composed of the most intensely deformed rocks extends down into the mantle. The high mountain range and its deep roots are in isostatic balance. As erosion vigorously attacks the high peaks and the sediment is carried to the continental margins, this balance is destroyed and the vertical uplift of the mountain belt occurs to reestablish an isostatic

balance. In the early stages of erosion, the removal of 500 m of rock is accompanied by an isostatic uplift of approximately 400 m, with a net lowering of the surface of only 100 m. This rate is believed to decrease as the mountain belt is eroded down and approaches isostatic equilibrium. Uplift continues as long as erosion removes material from the mountain range, but, eventually, a balance is reached when the mountainous topography is eroded to near sea level and the mountains' roots are removed.

An important point to note here is that the rocks exposed at the surface when isostatic balance is established are the metamorphic and igneous rocks formed at great depths in the orogenic belt. These rocks then are tectonically stable and become part of the shield.

Types of Orogenic Activity

Statement

In the preceding section, it was shown that mountain-building is the result of plate convergence and involves intensive deformation, metamorphism, and igneous activity. The characteristics of a mountain belt and the sequence of events involved in its development can vary, depending upon the types of interactions that occur at converging plate margins and the types of rock sequences involved in the deformation. Three fundamentally different types of converging plate boundaries exist, each of which develops a mountain belt with distinctive characteristics. They are:
1. Boundaries involving only oceanic crust.
2. Boundaries involving both oceanic and continental crust.
3. Boundaries involving only continental crust.

When both plates contain oceanic crust, one will override the other and the lower plate will descend down into the mantle. A volcanic island arc will be generated along the margin of the overriding plate. Deformation, metamorphism, and igneous activity will all be involved with island arc rocks.

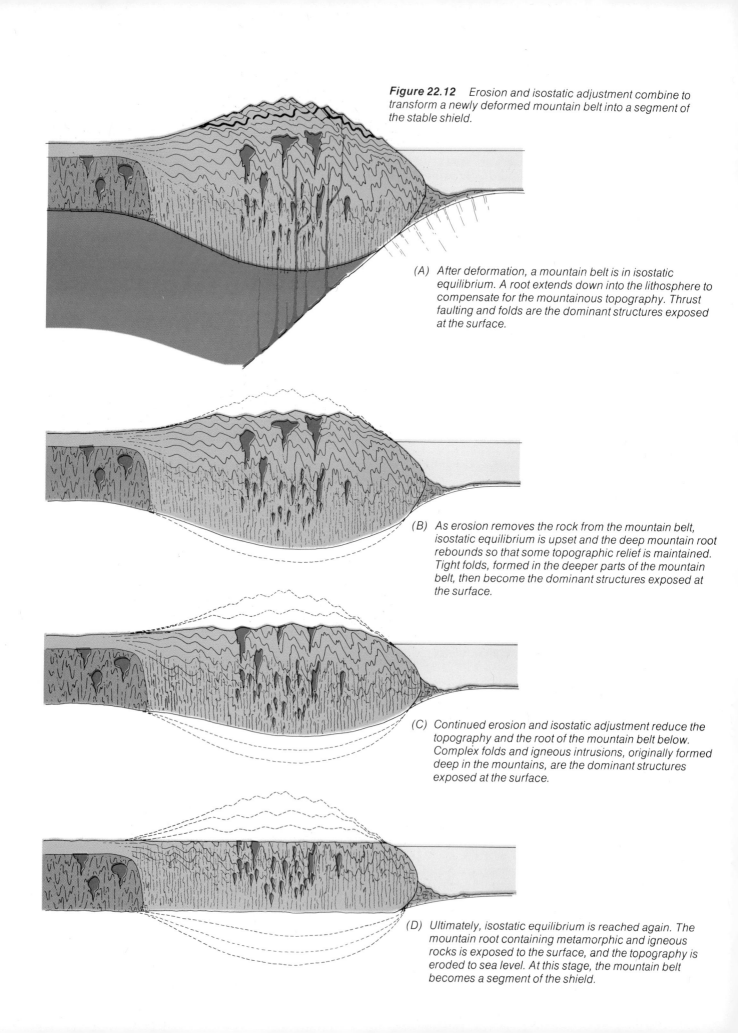

Figure 22.12 *Erosion and isostatic adjustment combine to transform a newly deformed mountain belt into a segment of the stable shield.*

(A) *After deformation, a mountain belt is in isostatic equilibrium. A root extends down into the lithosphere to compensate for the mountainous topography. Thrust faulting and folds are the dominant structures exposed at the surface.*

(B) *As erosion removes the rock from the mountain belt, isostatic equilibrium is upset and the deep mountain root rebounds so that some topographic relief is maintained. Tight folds, formed in the deeper parts of the mountain belt, then become the dominant structures exposed at the surface.*

(C) *Continued erosion and isostatic adjustment reduce the topography and the root of the mountain belt below. Complex folds and igneous intrusions, originally formed deep in the mountains, are the dominant structures exposed at the surface.*

(D) *Ultimately, isostatic equilibrium is reached again. The mountain root containing metamorphic and igneous rocks is exposed to the surface, and the topography is eroded to sea level. At this stage, the mountain belt becomes a segment of the shield.*

When one plate contains continental crust and the other oceanic crust, the buoyant granitic mass of the continental crust always will override the adjacent oceanic plate. Thick sequences of sediments eroded from the continent and deposited along its margin will be deformed by the converging plates, and magma generated by the descending plate will be placed within, or upon, the continental crust.

When both converging plates contain continental crust, significant subduction cannot occur. The oceanic basin between the continents will close, and the bounding prisms of sediments will be deformed. The buoyant, low-density continental masses cannot descend into the mantle, although one may override the other for a distance and create a double thickness of continental crust with exceptionally high topographic relief. Subduction in such a case would be deactivated and the continents would be welded together.

Discussion

Convergence of Two Oceanic Plates. The major result of two converging plates containing only oceanic crust is the development of a subduction zone, where one plate is thrust under the other and is consumed in the mantle. In the first stages of convergence, the major process may be restricted to volcanic activity and the development of an andesitic volcanic arc on the overriding plate. This situation may not involve widespread metamorphism or granitic intrusion. An example of a simple island arc in the present oceans is Tonga Island (*figure 22.13*).

In more complex island arcs (such as Japan), crustal deformation, metamorphism, and igneous activity combine to produce distinctive rock associations and deformational patterns. These are illustrated in *figure 22.14*. The major processes illustrated in this diagram are these.

1. Rock sequences consisting of oceanic sediments and pillow basalts of the oceanic crust develop, in addition to sediment derived from erosion of the volcanic arc.
2. Partial melting of rocks in the subduction zone produces a silica-rich magma that rises to form granitic intrusions and volcanic products in the island arc.
3. A zone of high-pressure/low-temperature metamorphism develops at the outer margins of the overriding plate.
4. An inner zone of higher temperature metamorphism develops with associated granitic intrusions.
5. Crustal deformation results from the converging plates and the intrusion of granitic magma.

The zone of high-pressure/low-temperature meta-

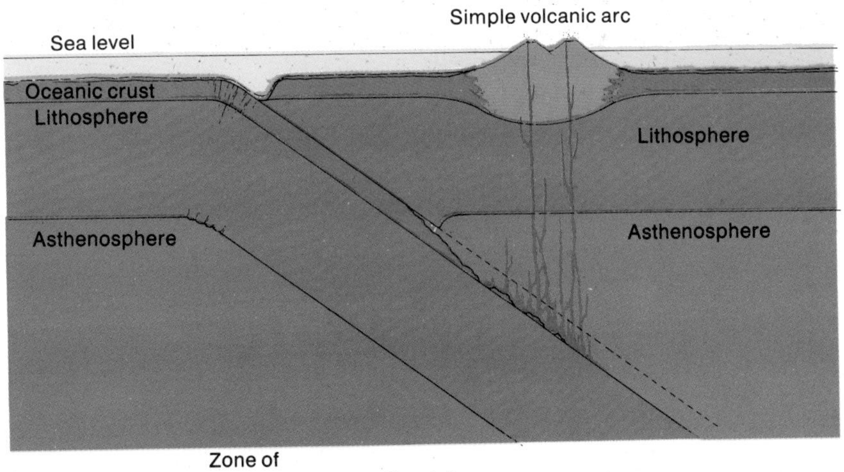

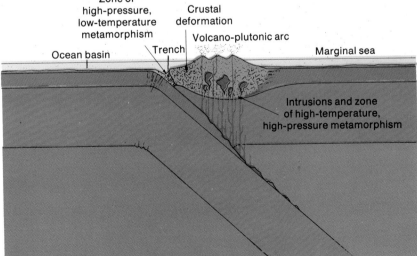

Figure 22.13 *Simple ocean-to-ocean orogenesis. This simple initial phase of orogenesis is restricted largely to volcanic activity and does not involve widespread metamorphism or granitic intrusions. An example is the Tonga arc in the South Pacific.*

Figure 22.14 *Complex ocean-to-ocean orogenesis. In complex island arcs (such as Japan), crustal deformation, metamorphism, and granitic intrusions are produced in addition to andesitic volcanism.*

morphism that develops on the outer margin of the over-riding plate is easily understood in light of its dynamic setting. High pressure is produced because this is the focal point of the converging plates, and low temperature results because the cold oceanic plate descends at rates between 5 and 15 cm per year—rates in excess of the rates at which heat can be conducted from the mantle through the rock in the descending lithosphere. The rocks that are metamorphosed consist of sediment derived from the erosion of the adjacent volcanic arc, in addition to oceanic sediments and pillow basalts of the oceanic crust. This complex mixture is squeezed downward along the top of the descending plate at temperatures generally less than 300 °C. The mixture of pillow basalts, oceanic sediments (ophiolite complex), and erosional debris from the island arc is metamorphosed into a distinctive, fine-grained schistose rock, characterized by high-pressure/low-temperature mineral assemblages. The rock typically develops a bluish color due to the presence of the mineral glaucophane and is referred to as a blueschist. There are no contemporaneous igneous intrusions in this zone, be-cause magma is generated only where the descending plate reaches a greater depth.

Blueschists are found in Japan, California, and New Zealand and have been recognized in the Alps, but they represent only a small fraction of metamorphic rocks exposed on the continents. This may be explained by the fact that these low-temperature/high-pressure assem-blages would lose their distinctive character if they were heated subsequently above 300 °C., which would occur from granitic intrusions that develop in continental oro-genic belts.

In the deeper parts of the subduction zone, magma is generated by partial melting and rises to form the large granitic batholiths that make up the deeper parts of the island arc. These intrusions develop a surrounding zone of high temperature that develops a very different assem-blage of metamorphic minerals characterized by rela-tively high temperature associations. At shallow depths, low-pressure/high-temperature metamorphism occurs; at deeper levels, high-pressure/high-temperature meta-morphism occurs. The original rocks are the volcanic flows and sediments that formed within the arc system proper, so the resulting metamorphic rocks are quite distinct from the blueschist formed near the trench.

As a result of the converging plates, the older rocks in the volcanic arc are compressed, folded, and broken by thrust faults, in addition to the deformation resulting from the emplacement of granitic batholiths. The fold axes and thrust faults trend parallel to the long axis of the arc and the linear belts of granitic batholiths.

The significance of orogenesis in island arcs is that the deformed andesitic volcanic rocks, and sediments derived from them, are a differentiate from oceanic basalt and are the initial phase in development of new conti-nental crust in areas where only oceanic crust previously existed. This provides important insight into the origin and evolution of continents.

Convergence of Continental and Oceanic Plates. The style of mountain-building produced by the collision of continental and oceanic plates is similar in some respects to that produced by the convergence of two oceanic plates, but it is distinctive for two main reasons.

1. Thick sequences of geoclinal sediments, derived from the continent, commonly occur along the continental margins and are quite different from rock sequences

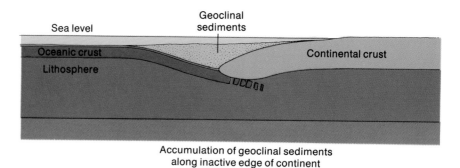

Figure 22.15 *Continent-to-ocean orogenesis. The thick geoclinal sequences of sedimentary rock that accumulate along the margin of a continent are deformed into a folded mountain belt. Granitic batholiths intrude the continental crust. High topography is formed, and andesitic volcanism occurs in the mountain belt.*

deposited around a volcanic arc. This produces a mountain belt characterized by folded sedimentary rocks with roots of granitic intrusions and metasediments.

2. Silica-rich magma generated by the partial melting of the descending oceanic plate is placed upon, or within, the continental crust, rather than upon the ocean floor. This produces a high topography, with the resulting rapid erosion. Volcanic material soon is stripped off the mountain range, and granitic batholiths commonly are exposed.

A schematic cross section showing the major features produced by the collision of continental and oceanic plates is shown in *figure 22.15*. Before the arrival of the continent at the subduction zone, a considerable thickness of sandstone, shale, and limestone may accumulate in the miogeocline along the continental margins. During this same period, deep-marine sediments accumulate on the oceanic plate. As the continental plate approaches the subduction zone, some of the deep-marine sediments on the oceanic plate can be crumpled and deformed and slabs of oceanic crust are sheared off and incorporated into the chaotic mass. This material, like that developed at the convergence of two oceanic plates, is subjected to high-pressure/low-temperature metamorphism.

The thick sequence of geoclinal sediments along the continental margin then is compressed and deformed. Thrust faults occur in the shallow zone of the mountain belt, where the rocks are brittle and fail by rupture. At intermediate depths, plastic flow occurs, developing tight folds and nappes. Intense metamorphism occurs within the deeper zones, where relatively high temperatures and high pressures exist. The partial melting of the descending lithosphere generates silica-rich magma that rises to form granitic intrusions in the deformed sediments in the upper plate or is extruded as volcanic material at the surface.

With continued compression, deformation of the geoclinal sediments and the resulting crustal thickening progress landward, and the edge of the deformed continental margin rises above sea level. Erosion of the mountain belt begins and continues contemporaneously with subsequent deformation. Sediments eroded from the growing mountain range can be shed off in both directions, and thick sequences of coarse conglomerates and sandstone can be deposited as alluvial fans, deltas, et cetera, in front of the rising mountain belt.

As the orogenic belt evolves, resistance builds up and convergence may stop, or the subduction zone may step seaward to form an ocean-to-ocean orogenic belt. After compression terminates, erosion continues to wear down the mountain range. Isostatic rebound occurs, and, ultimately, the mountain roots are exposed at the surface and the topography is eroded to near sea level.

Excellent examples of orogenic belts formed by convergence of continental and oceanic plates are the Rocky Mountain belt of western North America, which was deformed during late Mesozoic and early Tertiary time; the Andes Mountains of South America, deformed from the Tertiary to the present; and the Appalachian Mountains of the eastern United States, deformed during the late Paleozoic.

Convergence of Continental Plates. In the tectonic system, virtually all oceanic crust is destined to descend into the mantle by the subduction processes at converging plate margins. The continents, carried by the plates, move toward the subduction zone, and the oceanic basin continually is reduced in size and is eliminated altogether when the two continents collide. The collision of two continental plates generates an orogenic belt with a number of characteristics quite different from those previously described. Although, in detail, mountain belts generated by continental collision are exceedingly complex, the major events are similar to those outlined in *figure 22.16*.

1. Before continents actually collide, the wedge of sediments along their margins above the subduction zone are deformed and plastered against the leading edge of the continental crust. The oceanic lithosphere is consumed at the subduction zone, and the ocean basin decreases in size (*figure 22.16A*).

2. As the continents approach collision, segments of the remaining oceanic crust are deformed by overthrusting and eventually are squeezed between the converging plates (*figure 22.16B*).

3. When the continental crust moves into the subduction zone, its buoyance prevents it from descending down into the mantle more than perhaps 40 km below its normal depth. It can be thrust under the overriding plate, however, creating a double layer of low-density continental crust, which rises buoyantly to create a wide belt of deformed rock plus an adjacent high plateau (*figure 22.16C*).

4. An alternative response is that the continental masses become welded together and, between them, fragments of ophiolite assemblage (oceanic crust) can be caught and squeezed upward.

5. The oceanic slab of lithosphere descending down into the mantle ultimately becomes detached and sinks independently. As the slab is consumed, volcanic activity and earthquakes generated by it are terminated.

6. Eventually, convergence stops as resisting forces build up and the mountain belt is eroded and adjusts isostatically.

7. As a result of the welding of two continents together, one large continental mass with an internal mountain range is produced.

The Himalaya Mountains are, of course, an example of continental collision. They were formed as the Indian plate moved northward during the last 100 million years, destroying the oceanic lithosphere that formally separated India and Asia. As the two continents collided, India was thrust under the Asian plate to form the Himalaya Mountains, as well as the extensive highlands of the Tibetan Plateau. Earthquakes are frequent, but they are shallow and occur in a broad, diffuse zone because there is no descending oceanic plate.

The Alps and Ural Mountains are other examples. The Alps of Europe result from the convergence of the African and Eurasian plates. In many ways, Africa, moving northward against Eurasia, is like India, but it has not evolved to the point where the oceanic crust (the Mediterranean Sea) is completely consumed. The Urals were formed much earlier by the collision of the Siberian

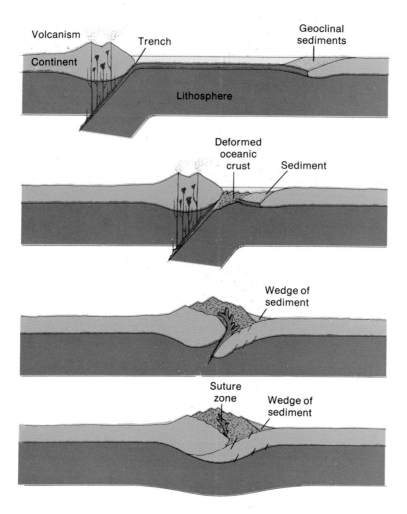

Figure 22.16 *Continent-to-continent orogenesis. As two continents converge, the oceanic crust between them is caught and deformed. A double layer of continental crust may be produced, resulting in abnormally high topography. Continents are welded together. The descending oceanic plate becomes detached in the subduction zone and sinks independently. As this slab is consumed, volcanic activity and deep earthquakes terminate.*

continental mass with Europe during late Paleozoic time and are in a later stage of development in which erosion and isostatic adjustments are the main processes.

In summary, the theory of plate tectonics proposes that mountain-building is a fundamental process in the differentiation of the earth. Orogenesis occurs along convergent plate boundaries but, unlike the activity outlined by the geosynclinal theory, does not require a specific sequence of events. Compressional deformation, metamorphism, and igneous activity are the processes in mountain-building; but the style of the orogenesis and the specific events can vary significantly, depending upon the various interactions that may occur along converging plate boundaries. As was emphasized in the statement, three fundamental types of converging plate margins exist (ocean to ocean, ocean to continent, and continent to continent). Each produces its own distinctive style of mountain-building.

Origin and Evolution of Continents

Statement

The origin and the evolution of continents are believed to be the results of plate tectonic processes operating along convergent plate boundaries. A continent originates as an island arc when silica-rich volcanic material is produced by the partial melting of a descending plate. Once formed, this low-density material resists subduction because of its buoyancy, and it remains on the surface of the lithosphere and continually drifts with the moving plates. The small, embryonic continent grows by accretion of new granitic material with each subsequent mountain-building event. Erosion and isostatic adjustment reduce the mountainous topography to a surface of low relief and develop a stable shield. Subsequent rifting may split and separate a continent, but each segment then would act as a separate center for future continental growth.

Discussion

An idealized graphic model showing the origin and evolution of continents is shown in the series of schematic diagrams in *figure 22.17*. The initial step in the development of a continent is the formation of a subduction zone (diagram *A*). Magmatic differentiation occurs by the partial melting of the descending plate. The silica-rich

Figure 22.17 *Evolution of a continent.*

(A) Partial melting in the subduction zone produces an island arc composed of andesitic volcanic material that is too light to be pulled down into the mantle. Erosion of the islands further concentrated light material consisting of sediment deposited on the islands' flanks.

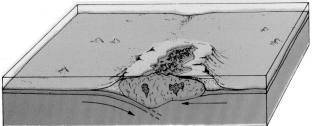

(B) The sediment and andesitic volcanoes are deformed by compression at converging plates to produce mountain belts and metamorphic rocks that formed an embryonic continent.

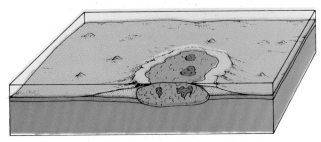

(C) Erosion of mountains concentrates light minerals (quartz, clay, and calcite) as geoclinal sediments along margins of the small continent.

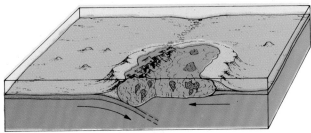

(D) The geoclinal and orogenic cycle is repeated, with new material added to the continental mass as granitic batholiths and andesitic flows.

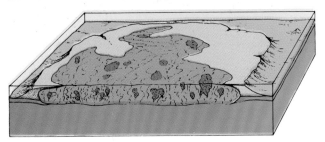

(E) The continent continues to grow by accretion.

minerals melt first and produce a silica-rich magma that erupts as andesitic volcanoes or forms granitic intrusions. Sediment derived from erosion of the island arc further separates the low-density material and concentrates it by sedimentary differentiation. The new igneous and sedimentary rocks produced in the island arc are less dense than the basaltic oceanic crust, and their buoyancy prevents any significant consumption in a subduction zone.

The volcanic material and sediment in the island arc are deformed into an orogenic belt by subsequent collision with another plate. Metamorphism occurs in the mountain roots, and granitic intrusions result from partial melting in the subduction zone (diagram *B*). Erosion and isostatic adjustment produce a small, stable shield

in which igneous and metamorphic rocks formed in the mountain roots are exposed at the surface (diagram *C*). Sediment derived from the erosion of the mountains is deposited along the margins of the new continent. Repetition of this process would deform the geoclinal sediments along the continental margins into a new orogenic belt, with new igneous material being added from the partial melting of the oceanic crust in the subduction zone (diagram *D*). The continent would continue to grow by accretion of material along its margins during each subsequent orogenic event (diagram *E*).

The major processes involved in continental growth are (1) mountain-building along convergent plate margins involving deformation and metamorphism of geo-

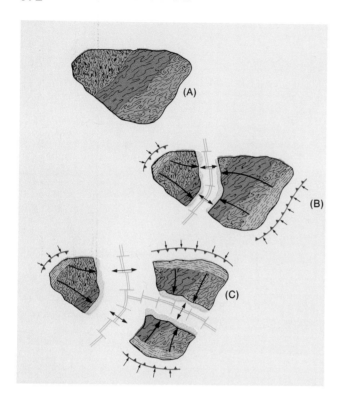

Figure 22.18 *Fragmentation of continents can occur at any time during the evolution of a continental mass.*

(A) Original continental mass composed of orogenic belts welded together into a stable shield.

(B) A spreading center splits the continent, and the fragments drift apart. A subduction zone ultimately will develop along the leading edge of the plate, and each fragment may grow as an independent continental mass.

(C) Further fragmentation may occur, or the continents may collide and become welded together.

clinal sediments, (2) igneous activity and emplacement of new material in the orogenic belt, and (3) erosion and isostatic adjustment to produce a shield. Broad upwarps or slight vertical movements of the shield would permit the ocean to expand and contract over the continent and deposit a sediment cover to form the stable platform.

At some stage, rifts may form within the stable continental block as a new spreading center develops, presumably as a result of shifting convection currents in the mantle. The continent then splits, and each continental fragment drifts apart with the spreading plates (*figure 22.18*). Each continental fragment acts as a separate center for future continental growth.

Since the earth is a sphere, the moving plates carrying continental blocks converge, and continents may collide and become welded together.

Summary

Extensive geologic mapping shows that continents consist of three major components: (1) shields composed of complexly deformed and recrystallized metamorphic and granitic rocks generally eroded to a flat surface near sea level; (2) stable platforms, or areas of the shield that are covered with a veneer of essentially horizontal sedimentary rocks; and (3) young, folded mountain belts. Continents also can be split by rift systems.

The origin of continents is intimately involved with the mountain-building at convergent plate boundaries and the concentration of silica-rich, low-density material on the outer part of the crust through the partial melting of the oceanic plate in the subduction zone. The characteristics of a mountain belt and the sequence of events that produce it depend largely upon the type of crust on the converging plates and the type of sediments involved in the deformation. Three distinctive types of converging plate boundaries are recognized: (1) convergence of two oceanic plates, (2) convergence of a continental plate and an oceanic plate, and (3) convergence of two continental plates. Convergence of two oceanic plates produces island arcs with andesitic volcanism, metamorphism of ophiolite rock sequences, and emplacement of granitic batholiths in the deeper parts of the orogenic belt. Convergence of a continental plate and an oceanic plate involves deformation of a thick geoclinal sequence of sediments, andesitic volcanism, and emplacement of granitic batholiths in the deformed mountain belt. Convergence of two continental plates produces continental welding, or a double layer of continental crust with high mountainous topography, and limited igneous activity.

After orogenesis, erosion and associated isostatic adjustment reduce the mountain topography to a flat surface near sea level and expose the metamorphic rock of the deep mountain roots as a new segment of the shield. Thus, continents grow by accretion, as new, silica-rich material is added to an orogenic belt through igneous activity at converging plate margins. This low-density crust resists any further subduction and remains on the outer lithosphere. Subsequent rifting can split and separate continental blocks, but each block then will act as a separate center for future continental growth.

Continental growth is simply part of the process of planetary differentiation in which the lighter elements are separated and concentrated in the outer layers of the earth. The process of differentiation has produced the ocean basins, the continental platforms, and the surface fluids (air and water). Thus, differentiation is directly or indirectly responsible for the origin of all surface features and the internal structure of the planet. The next chapter will consider how the processes operating in, and on, the earth affect the environment and produce the natural resources used by man.

Additional Readings

Condie, K. C. 1976. Plate Tectonics and Crustal Evolution. New York: Pergamon Press.

Engel, A. E. J., and C. G. Engel. 1964. "Continental Accretion and the Evolution of North America." *In* A. P. Subramanian and S. Balakrishna, eds., Advancing Frontiers in Geology and Geophysics: A Volume in Honour of M. S. Krishman, pp. 17-37e. Hyderabad: Indian Geophysical Union. (Reprinted in P. E. Cloud, ed. 1970. Adventures in Earth History, pp. 293-312. San Francisco: W. H. Freeman and Company.)

Hallam, A. 1973. A Revolution in the Earth Sciences. Oxford: Clarendon Press.

Wyllie, P. J. 1976. The Way the Earth Works. New York: John Wiley and Sons.

Environment, Resources, and Energy

The preceding chapters have studied the earth as a dynamic system that produces a constantly changing environment. Some rapid changes such as earthquakes, volcanoes, and floods produce catastrophic effects, but most changes are so slow by human standards that they are barely noticed, even during a human lifetime. Perhaps there is comfort in realizing that most geologic processes are extremely slow and are of little concern to men as individuals, but, in the last 10,000 to 20,000 years, many changes that significantly affect man's history have occurred. For example, the present sites of many northern cities, including Chicago, Detroit, Montreal, and Toronto, were buried beneath thousands of meters of glacial ice as recently as 15,000 to 20,000 years ago, while, at the same time, the sites of most major port cities, such as New York, were located many kilometers inland. San Francisco Bay did not exist. The Missouri and Ohio Rivers had not been established completely at that time, and the valley where Salt Lake City is located contained a large freshwater lake 300 m deep. These changes were largely the results of worldwide climatic changes associated with the ice age, and they took place over long periods of time.

In contrast, man is capable of simulating drastic geologic changes within a few years and can alter radically the environment in which he lives. When this is done, the balance of the natural systems established over thousands of years is upset instantaneously, and adjustments to the changes cannot always be anticipated. One of the greatest goals of twentieth-century man, therefore, should be to live in harmony with the natural environment. It is the only environment there is.

Not only is the environment of great concern, but men are beginning to realize that resources and energy are finite and are being depleted. During the next 10 years, men will use more oil, gas, iron, aluminum, and other mineral resources than they have consumed throughout all of history. This is because the growth rate of the world's population has been exponential. It grew slowly at first but doubled from 1830 to 1930. By 1970, it nearly doubled again, and, by the year 2000, it is expected to double once more. In addition, individuals consume mineral resources at an extremely rapid rate compared to rates of usage in earlier times. As a result of technological advances, Americans, on a per-capita basis, use four times more steel, a hundred times more aluminum, and a million times more uranium than they used in 1900.

Only a moment's reflection is necessary to realize that the affluence of American society depends upon mineral resources for food, shelter, transportation, and the extras and that it is impossible to match exponential population growth with finite mineral resources.

In this chapter, the geologic basis for the environment, resources, and energy will be discussed.

Major Concepts

1. The environment is the product of natural systems that have operated on the earth for a long period of time, in a delicate balance.
2. Man is capable of manipulating many natural systems, including (a) rivers, (b) slopes, (c) ground water, and (d) shorelines.
3. Natural systems can adjust to changes induced by man in ways that cannot always be anticipated.
4. Waste disposal involves the natural systems, and problems include the handling of (a) solid, (b) liquid, (c) gaseous, and (d) radioactive wastes.
5. Mineral resources are concentrated by geologic processes operating in the hydrologic and tectonic systems. The resources are finite and nonrenewable.
6. Renewable energy sources include solar energy, hydropower, and tidal energy.
7. The nonrenewable energy sources, the fossil fuels, will reach their peak development in a few decades.
8. There are limits to growth in the earth's system.

Environmental Geology

Statement

The Problem. Man as a part of nature is one of the many species of animals adapted to the present natural environment. He is so well adapted that he dominates all other species, and his numbers are so great that he has become an effective agent of physical, chemical, and biological change. Presently, he is capable of modifying significantly the natural systems which operate on the earth. Many of his modifications, unfortunately, conflict with the normal evolution of the earth's environment. For example, since man has developed an agrarian culture, he has changed drastically large areas of the earth's surface from natural wilderness to controlled, cultivated land. This has put a great strain on the food supply of many animal populations that previously had established a degree of balance. As a result, many species have become extinct. The advancements of science and technology have telescoped thousands, if not millions, of years of normal evolutionary change into days.

Drainage systems are diverted and modified; the quality and flow of water, both in the surface and subsurface, are altered; the atmosphere is changed; and large parts of the ocean are affected. The changes instituted by man, almost without exception, are intended to produce improvements and advantages for his societies, but frequently the results are quite the opposite. At best, some are detrimental on a short-term basis; others are catastrophic and irreversible in the long run. Therefore, man is forced to adapt to a rapidly changing environment, one that he in part created. Changes are necessary, and, at times, the undesirable side effects (anticipated or unanticipated) must be tolerated as the "necessary price" for change.

Since man is capable of changing his environment, the decision to do so should be made carefully and only after enough geologic and biologic facts are available to enable him to see how the natural systems operate and how they will be affected. If man decides to dam a river, he should understand the side effects of altering part of the natural river system. He should question and know how the damming will affect erosion and sedimentation downstream, how the ground-water system will respond, how lakes will be altered, how the beaches along the shore far removed from the dam will be affected, and how marine and terrestrial life will be changed. If he develops a great volume of waste, he should know how it will be assimilated into the earth's system. How will it affect the quality of surface and ground water, atmosphere and oceans? If metropolitan areas are built, he should have full understanding of how this development will affect the terrain conditions, the atmosphere, and the hydrosphere. In too many cases, in the process of spending money to correct or alter a natural condition to better suit man's needs, he upsets a natural balance, which requires that more money be spent to reestablish the balance.

The ability to change and alter the environment is limited mostly to surface processes. Some natural earth processes operate regardless of man's activities. For ex-

ample, earthquakes and volcanism are the results of internal processes and cannot be stopped effectively or altered. Indeed, their occurrence cannot even be predicted with any confidence. Thus, there exist a number of geologic hazards that should be understood in any attempt to direct growth and development.

Man exists by geologic consent, because the basis of his entire environment is the geologic system. Therefore, it is incumbent upon man to understand the system and to live within it.

Discussion

Throughout many of the preceding chapters, there have been examples of man modifying and changing natural systems. These examples now will be reviewed in the general perspective of environmental geology.

Modifying River Systems. Water is perhaps the most important natural resource, and, as population has grown, man has become involved increasingly in changing river systems so that he can better utilize this finite resource. Man's attempts to modify river systems have been largely through the construction of dams and canals and by urbanization. The important point to be considered in all of these examples is that a river is a system that has approached equilibrium with a large number of variables over a long period of time. When this equilibrium is upset, a number of rapid adjustments occur. You should recall the example of the Aswan Dam in Egypt (see pp. 148-49) and the adjustments of the river to the dam. When a dam is built, the major geologic adjustments are these.

1. Sediment is trapped in the reservoir, and the clear water discharged downstream is capable of accelerated erosion.
2. Deltas are deprived of their source of sediment, and coastal erosion accelerates.
3. Ground-water systems are modified by creation of artificial lakes behind dams.
4. A more universal problem is the subtle modification of a river system by urbanization (pp. 149-50). As cities are constructed, the surface runoff is modified greatly because the streets, sidewalks, parking lots, and roofs of buildings make large areas of the surface impermeable. This produces two major effects. (1) The volume of surface runoff is greatly increased and is channeled into gutters and storm drains. As a result, flooding increases in intensity and frequency. (2) Ground-water systems are altered by reduced infiltration.

Modifying Slope Systems. Slopes are also dynamic systems in which there is some degree of balance (pp. 163-64). Construction on hills and slopes modifies the system, resulting in an increase in the magnitude and frequency of mass movement. The landslide in northern Italy that resulted from slope modification associated with the Vaiont Reservoir was cited earlier in the text as an example, but the problem can be seen wherever uncontrolled hillside construction occurs, especially where the slope is underlain by unconsolidated rock material.

Modifying Ground-Water Systems. The use of ground-water resources is increasing constantly. As people drill wells and use ground water, they, in essence, create a new and unnatural ground-water discharge. One of the more important problems resulting from this is subsidence as the ground water is removed. Mexico City is an excellent example: many of its buildings have subsided more than 3 m (pp. 210-11). Also, the encroachment of salt water into the lens-shaped body of fresh ground water on an island or a peninsula is commonly a result of excessive pumping (p. 208).

Alteration of the ground-water table can occur from the modification of surface infiltration, such as by the construction of canals in southern Florida that caused a variety of problems in the Everglades (pp. 208-9). In contrast, the irrigation of arid lands causes the water table to rise, a condition that can produce unstable slopes and accelerated mass movement such as occurred in the Pasco basin of Washington (p. 210).

Modifying Shorelines. Perhaps even more dramatic are the results of modifying shorelines. Santa Barbara, California (pp. 251-52), is an example in which the process of longshore drift was manipulated by constructing a breakwater to form a harbor. When longshore drift is disrupted, the supply of sand is eliminated or reduced and beach erosion results.

The great central theme that these and other examples illustrate is that the geologic systems are complex dynamic systems in which some type of balance has been established with a number of variables. Manipulating these systems often results in many serious side effects —some detrimental on a short term, others catastrophic and irreversible in the long run. Man's decision to change any aspect of his environment should be made carefully and only after enough geologic and biologic facts are available to enable him to understand fully what he is doing.

Geologic Problems with Waste Disposal. Man's industrialized society produces an ever-increasing variety and quantity of toxic and noxious waste materials. Most people are aware that, traditionally, man has used fresh water to remove and dilute solid waste and the atmosphere to dilute the gaseous waste products of combustion. But, until recently, people generally have been unaware of the fact that a local natural system of waste disposal can become saturated and create an unhealthy environment. People cannot take their waste "away" as some politicians have suggested. The waste products are in the earth's natural systems and will remain in them. The problem is particularly acute because of business practices based on such things as planned obsolescence, throwaway containers, and the hard sell of new models of old products. In addition, high labor costs often make it uneconomical to repair, reclaim, and recycle used items, so the volume of waste grows unnecessarily, at a staggering rate. This, of course, greatly reduces natural resources. Unfortunately, waste is not just a by-product of production; eventually, it is a product itself. The American economy has reached a point where thrift is too expensive. Since society has created a volume of waste that cannot be assimilated and diluted in the river systems, oceans, and atmosphere, people must live with it by converting it into one state or another (solid, liquid, or gas).

Waste disposal has many geologic ramifications. When waste is buried, the quality of the ground water is threatened. When it is dumped into streams and rivers, it is not all carried away and lost; it accumulates on beaches and in estuaries, where it alters the environment of the oceans. Previous methods of elimination have not been "waste disposal," they have been "waste dispersal." Any significant solution to the elimination problem must involve a consideration of what kind of waste disposal and dispersal methods the given geologic environment of an area will accommodate without the geo-biologic environment being altered critically.

Most people are aware of the great variety of waste products produced by modern culture. Commonly, waste is labeled as either municipal, agricultural, or industrial, but it is probably more important to consider waste according to its composition rather than its source. The information listed in *table 23.1* should be considered man's contribution to the hydrologic system.

1. *Solid Wastes.* Solid wastes are disposed of in many ways, including landfilling, incinerating, composting, open dumping, animal feeding, fertilizing, and disposing in oceans. The geologic consequences include changes in the surface of the land where the waste is deposited and changes in the environment (rivers, lakes, oceans, ground water) where the mass of waste is concentrated. The major problems associated with solid waste disposal are related to the hydrologic characteristics of the site, including the porosity and permeability of the rock in which the fill is located and whether or not the waste deposit intersects the water table. The altered topography associated with dumps and landfill is also critical because it can change the drainage and ground-water conditions. Perhaps the most critical contamination problem is created by the leachate formed as circulating water passes through the landfill and dissolves the organic and inorganic compounds that enter ground-water reservoirs.

2. *Liquid Wastes.* Traditionally, liquid wastes simply are discharged into the surface drainage systems and diluted. Ultimately, they accumulate in lakes and oceans, where they are stored. As the volume of liquid waste increases, the capacity for the natural water system to dilute it is overwhelmed and the drainage system becomes a system of moving waste.

One very subtle type of liquid pollutant is hot water created by cooling systems in power and industrial plants. Although the water itself is not contaminated, the temperature alone is enough to alter the biologic conditions of the streams and lakes into which it flows. Such pollution is called thermal pollution.

3. *Gaseous Wastes (Air Pollution).* The average United States citizen is well aware of air pollution because the problem is heavily publicized by the news media. At sea level, the natural gaseous environment—the atmosphere—is composed of approximately 78% nitrogen, 21% oxygen, 1% argon, and about 0.03% carbon dioxide. Water vapor content varies from nearly nothing to as much as 4%.

The population explosion, with its consequent

Table 23.1 Wastes That Enter the Hydrologic Cycle

Solid Wastes

Garbage
  Waste from preparation of
    food
  Market refuse—handling,
    storage of produce and
    meat
Rubbish (household)
  Paper—boxes, cartons
  Plastics
  Rags
  Grass, leaves
  Metal cans
  Dirt
  Stones, brick, ceramics
  Glass
Ash
  Fly ash
  Residue from combustion
    of coal
Bulky waste
  Large auto parts
  Appliances
  Furniture
  Trees, stumps, branches
Dead animals
Construction and demolition
  Lumber and sheeting
    scraps
  Broken concrete, plaster,
    etc.
  Pipes, wire
  Asphalt paving fragments
Industrial waste
  Food processing (slaughter-
    house, etc.)
  Wood, plastic scraps
  Slag
Agricultural waste
  Manures
  Crop residues
  Pesticide residues

Solid Wastes (continued)

Sewage and sewage
  treatment residues
Sludge
Mining wastes
  Concentrations of various
    minerals
  Strip mine tailings
  Placer mine tailings
  Quarries, rock piles

Liquid Wastes

Domestic sewage
  Raw sewage
  Treated residues
Industrial
  Liquid chemical residues
  Saline brines
  Paint sludge

Gaseous Wastes

Carbon monoxide
  Internal combustion of
    automobiles
Sulfur dioxide and sulfuric
acid
  Sulfur-bearing fuels—
    power plants, smelters
Hydrogen fluoride
  Production of fluorine ores
Particulate matter
  Dust, soot, ash
  Metals
  Insecticides, herbicides
  Lead
  Jet exhaust

Radioactive Wastes

Explosions of nuclear devices
  Radioactive fallout
Atomic energy plants
  Radioactive waste

From Gazda and Malina, 1969

industrial expansion, has produced, along with gaseous pollutants, pollutants in the form of minute liquid and solid particles that are suspended in the atmosphere. In the past, pollutants were expelled into the air with the reasonable assurance that normal atmospheric processes would disperse and dilute the waste to a harmless unnoticeable level. In many areas of heavy industrialization, the atmosphere's capacity for absorption and dispersal has been exceeded, and the composition of the air has been altered severely. If the troposphere (that part of the atmosphere involved in most human activities) extended indefinitely into space, there would be little air pollution problem. However, few pollutants move out of the troposphere (the lower 10 to 15 km of the atmosphere) into the overlying stratosphere for any great length of time; thus, a steadily increasing volume of pollutants is concentrated mostly in the lower part of the troposphere.

4. *Radioactive Wastes.* All industries are faced with problems of disposing of waste products, but none are greater than the problem of disposing of waste from nuclear energy plants. As nuclear energy is generated, a variety of radioactive isotopes are created, some with short half-lives and others with very long half-lives. Nuclear waste is extremely hazardous and generates large amounts of heat. Thus, any disposal system must be capable of removing heat and, at the same time, completely isolating the waste from the biologic environment. In addition, containment must be maintained for exceptionally long periods of time. Compared to that produced by many other industries, the volume of radioactive waste is not large, but the hazards and heat generated are considerable. By 1980, an estimated 9 million liters of radioactive waste will be produced, and, by the year 2000, the amount will increase to 144 million liters.

In many respects, the problems associated with radioactive waste disposal are similar to those with other pollutants produced by American society. What long-range effect will it have on the environment? Unlike other pollutants, radioactive waste can induce mutations in living species and, since society is looking more and more to nuclear energy to solve many of its resource shortages, the long-term effect of radioactive waste could be catastrophic.

One of the more promising methods of radioactive waste disposal involves the storage of waste in thick salt formations. Salt deposits are desirable because they are essentially impermeable and are isolated from circulating ground water. In addition, salt yields to plastic flow so that it is unlikely to fracture and make contact with leaching solutions over extended periods of time. Salt also has a high thermal conductivity for removing heat from waste and has approximately the same shielding properties as concrete.

In theory, the wastes would be solidified and sealed in containers 15 to 60 cm in diameter and up to 3 m in length. The containers then would be shipped to salt mines in the stable interior of the continent, where seismic activity is minimal. There, they would be placed into holes drilled into a salt formation in a deep salt mine. When filled with waste, the hole would be closed and the room itself would be filled with crushed salt. Since the plastic flow of salt causes consolidation and recrystallization, the salt formation would return to its original state within a few decades.

5. *Mining Wastes.* The waste products from mining operations include (1) tailings and dumps, (2) altered terrain (open-pit mining), (3) changes in the composition of the surface, and (4) solid, liquid, and gaseous wastes produced by refining.

In the United States, approximately 2.7 billion metric tons of rock are mined each year. Approximately 85% comes from open-pits and strip mines, which require the removal of an additional 5.4 billion metric tons of rock as overburden. In these operations, about 12,000 km² of land in the United States have been affected by surface mining. The principal geologic problem arises from the alteration of the terrain by the creation of open-pits and from the creation of artificial mounds and hills from tailings.

Most mine dumps are unstable and are highly susceptible to mass movement unless they accumulate under proper engineering supervision. In 1966, the mudflow from coal mine dumps in Aberfan, Wales, completely destroyed a school and killed and injured many children.

An additional problem arises when the mine tailings enter the drainage system, since they can choke a stream channel and increase the flood hazards. Alteration of a stream system also can be produced from placer mining, where the movement of large quantities of sediment upsets the balance of the stream.

Resources

Statement

The important minerals upon which modern civilization depends constitute an infinitesimally small part of the earth's crust. Whereas the rock-forming minerals (such as feldspar, quartz, calcite, and clay) are abundant and distributed widely, copper, tin, gold, and other metallic minerals occur in such small quantities that their occurrence is measured in parts per million (and in most cases a very few parts per million). The important question then is: how are these very small quantities of important minerals concentrated into deposits large enough to be used? The answer is: by the various geologic processes operating in the earth's system. It may come as a surprise to some people, but essentially every geologic process—including igneous activity, metamorphism, sedimentation, weathering, and deformation of the crust—plays a part in the genesis of some valuable mineral deposit. The occurrence or absence of most mineral deposits, therefore, is controlled by the specific geologic conditions of a region. Furthermore, it is critical to understand that the processes by which these minerals are formed operate so slowly (by human standards) that the rates of replenishment are infinitesimally small compared to rates of consumption. People must clearly understand that mineral deposits have finite dimensions and, therefore, are exhaustible and nonrenewable. If the approximate extent of a deposit and the rate of its consumption are known, how long it will last can be predicted. Once these resources are consumed, they cannot be replenished. Moreover, today, there are relatively few unexplored areas of potential mineral deposits. Most of the continents have been mapped and studied extensively, so that the inventory of natural resources is nearly complete.

Discussion

Perhaps the best way to appreciate the fact that most natural resources are finite and nonrenewable is to consider some of the principles that govern the concentration of rare minerals into ore deposits, as well as the origin of some nonmetallic resources. These principles are complex and diversified, but the origin of most mineral deposits is related in some way to the various processes in the tectonic and hydrologic systems. Therefore, it is possible to recognize major groups of mineral deposits formed by (1) igneous processes, (2) metamorphic processes, (3) sedimentary processes, and (4) weathering processes.

Igneous Processes.   Many metallic ores are concentrated by magmatic processes in much the same way as are silicate minerals, that is, as a result of differences in crystal structure, order of crystallization, density, et cetera. Magmatic intrusions themselves are concentrations, on a regional scale, of silicate minerals and metallic ores and characteristically are generated at plate boundaries. On a local scale, metallic minerals are concentrated in specific areas of intrusion in a variety of ways. One method is by direct segregation, in which heavy mineral grains that crystallize early sink down through the fluid magma and accumulate in layers near the base of the igneous body. Deposits of chromite, nickel, and magnetite are good examples of this type of concentration (*figure 23.1*).

Another method of concentrating rare minerals results from late crystallization. Many elements that occur in amounts of only a few parts per million in the original magma do not fit readily into the crystalline structure of the silicate rock-forming minerals and are concentrated in the residual liquid as the feldspar, amphibole, mica, et cetera crystallize. Through this process, much of the silicon, oxygen, and aluminum is incorporated into the crystal structure of rock-forming minerals, so that, as a magma cools, rare elements such as gold, silver, copper, lead, and zinc become concentrated in the last remaining fluid. These "late-stage" metal-rich solutions then can be squeezed into the fractures of the surrounding rock; and, as they cool, the rare elements are precipitated, often with quartz, as veins of metallic minerals that differ little from ordinary dikes and sills except in their chemistry and mineralogy. The residual hot solutions of the magma contain much water (because of the low temperature at which water crystallizes), so the deposits are known as hydrothermal deposits (*figure 23.2*).

Hydrothermal solutions are not only injected into the surrounding rock to form veins of mineral deposits, but they also can permeate through the early-formed crystals in the granitic rock and disseminate metallic minerals

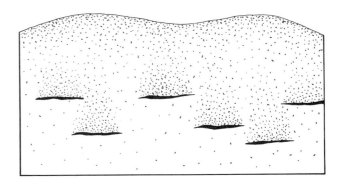

Figure 23.1   *Concentration of chromite as early-formed crystals settle to form layers near the base of a magma.*

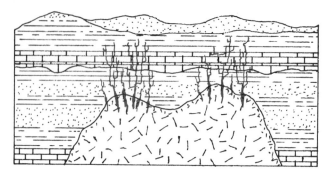

Figure 23.2 *Schematic cross section through an intrusive body, showing the location of hydrothermal vein deposits.*

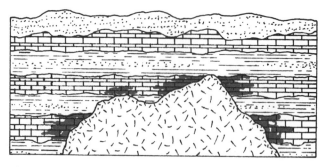

Figure 23.3 *Schematic diagram showing the concentration of ore deposits by contact metamorphism.*

throughout the body. This can produce large deposits of low-grade ore, because practically the entire igneous body would be mineralized. When the intrusion is exposed at the surface, these low-grade deposits can be mined profitably by using special equipment and open-pit-mining techniques. An important example of this type of deposit is the disseminated copper in porphyritic intrusions. These deposits are called porphyry coppers and at present account for over 50% of the world's copper production.

The porphyry copper and many other metallic minerals are associated with granitic intrusions that, in turn, are associated with convergent plate boundaries. The metals are believed to have been derived from partial melting of the oceanic plate in the subduction zone, with final concentration and emplacement resulting from differentiation of the magma as it cooled. The mineral deposits are, therefore, a result of the tectonic system, and the plate tectonic theory promises to be an important key to better understanding the genesis of minerals and to future exploration of minerals.

Metamorphic Processes. Ore deposits are formed also by metamorphism along the contact between the igneous intrusion and the surrounding rocks (contact metamorphism). In this process, the interactions among heat, pressure, and chemically active fluids of the cooling magma alter the adjacent rock by adding or replacing certain components. Limestone surrounding granitic intrusions is particularly susceptible to alteration and replacement by the chemicals in the hot, acid solutions of the magma. For example, large volumes of calcium can be replaced by iron to form a valuable ore deposit (*figure 23.3*).

Regional metamorphism changes the texture and mineralogy of a rock and in doing so produces deposits of nonmetallic minerals, such as asbestos, talc, graphite, et cetera.

In the generation of mineral deposits by metamorphism, again the close association of ore genesis with the dynamics of converging plate margins can be seen.

Sedimentary Processes. One of the most significant results of the processes of erosion, transportation, and deposition of sediments is the segregation and concentration of material according to size and density. As was emphasized in chapter 6, the soluble minerals are transported in solution, silt and clay-size particles are trans-

ported in suspension, and sand and gravel are moved mostly as bed load by the stronger currents. As a result, sand and gravel are concentrated in river bars and beaches. These deposits, both modern and ancient, constitute a very valuable resource for the construction industry. In the United States alone, over $1 billion worth of sand and gravel is mined each year, making this the largest industry not associated with fuel production in the country. In areas where sand and gravel have not been concentrated by natural processes, they must be made by crushing and screening, an operation requiring considerable energy and expense.

Sedimentary processes also concentrate other valuable materials such as gold, diamonds, and tin. Originally formed in veins, volcanic pipes, and intrusions, these minerals are eroded and transported by streams, and, because they are heavy, they are deposited and concentrated where current action is weak (such as on the inside of meander bends or on protected beaches and bars [*figure 23.4*]). The layers and lenses of valuable minerals are known as **placer** deposits, and large operations are mining both modern and ancient river and beach deposits.

Sedimentary processes also were responsible for the concentration of the great bulk of iron ore mined today. The banded iron formations of Precambrian age are especially significant because of their abundance. These deposits consist of alternating layers of iron oxide and chert that were formed during a unique period of earth's history, 1.8 to 2.2 billion years ago. The combination of conditions necessary to produce these deposits has never been repeated.

Another way sedimentary processes concentrate valuable minerals is by evaporation of saline waters in large lakes and restricted embayments of the ocean. As evaporation proceeds, dissolved minerals are concentrated, and eventually they are precipitated. These **evaporite** deposits include a variety of minerals of important commercial value, including potassium, sodium, and magnesium salts, gypsum, sulfates, borate, and nitrates. Many marine evaporites occur in extensive layers interbedded with shale and limestone and have been mined for thousands of years. Thick salt deposits are especially significant, because, when they are buried to sufficient depths, the pressure from the overlying beds is commonly sufficient to cause the salt to flow and rise like an intrusive

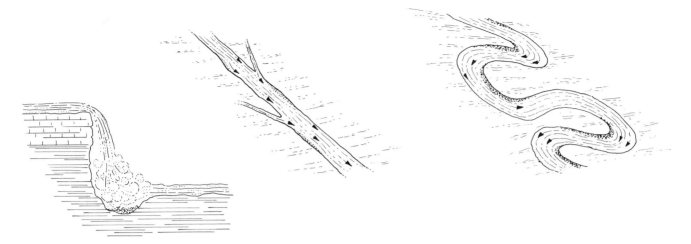

Figure 23.4 *Schematic diagrams showing several ways in which ore can be concentrated into placer deposits.*

body of magma into the overlying strata. These intrusions are known as **salt domes** and are common in the Gulf Coast of the United States, Germany, and Iran. Salt domes are economically important not only because they constitute a source of nearly pure salt but because they also constitute important traps for petroleum.

Weathering Processes. The simple process of weathering also concentrates minerals into deposits as it removes the soluble material from a rock and leaves the insoluble material as a residue. Weathering, therefore, can serve to concentrate and enrich ore deposits originally formed by other processes (*figure 23.5*) or it can concentrate in regolith material originally dispersed throughout a rock body. For example, extensive weathering of a granite in tropical and semitropical zones commonly concentrates the relatively insoluble oxides of iron and aluminum in thick regolith while removing the more soluble material. Deposits of aluminum, nickel, iron, and cobalt are formed in this way.

In summary, the few examples cited here clearly illustrate that the valuable mineral deposits of the world were formed in a systematic way by the major geologic processes. They were formed during various periods in the geologic past over a long period of time under specific geologic conditions. Some deposits were formed in very restricted geologic settings that approach uniqueness. For example, 40% of the world's reserve of molybdenum is in one igneous intrusion in Colorado, 77% of the tungsten reserves are in China, more than 50% of the world's tin is in southeast Asia, and 75% of the chromium reserves are in South Africa. If the resources are depleted, people cannot "just go out and find some more." There simply are no more. Fortunately, most metals, unlike fossils fuels, can be recycled. The process of recycling must be emphasized in the near future.

Energy

Statement

The technological progress and standard of living of our modern society are intimately related to energy consumption, and, until recently, energy resources and capacity for growth seemed to be unlimited. Now, for the first time, people all over the world are experiencing a serious energy shortage, one that is sure to become more acute in the future. Understanding the sources of energy and how they can be utilized most effectively is one of the most pressing problems of the twentieth century.

When the entire dynamic system of the earth is

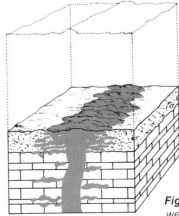

Figure 23.5 *Concentrations of ore by solution and weathering of surrounding rock. The insoluble minerals accumulates as a residual deposit.*

considered, it is found that sources of energy usable by man are found in both renewable and nonrenewable forms. A renewable energy source is one that, for all practical purposes, is either available in almost infinite amounts or one that will not be appreciably diminished in the foreseeable future. Solar energy, tidal energy, and geothermal energy are the most important examples. Nonrenewable energy sources, like mineral resources, are finite and exhaustible and cannot be replaced once they are consumed. The fossil fuels, coal and gas, upon which modern culture relies so much, are nonrenewable. These forms of energy have been concentrated and preserved by geologic processes that operated during past periods of geologic time, and, although the same processes may function today, they operate too slowly to be considered a potential energy source for man.

An important fact in today's energy picture is that, at present, over 90% of the energy used by man is produced by the nonrenewable fossil fuels. Estimates are that the peak production of petroleum will occur in either 1990 or 2000 and will decline to near exhaustion by the year 2070. Coal is expected to replace petroleum as the main hydrocarbon resource, with peak production by the year 2100 or 2200. It is quite clear, therefore, that in the very near future man's reliance on fossil fuels will have to be curtailed greatly and that fossil fuels will have to be replaced by other energy sources.

Discussion

Renewable Energy Sources. Solar energy is the most important renewable, or sustained-yield, energy source there is, and it has the added benefits of being clean, constant, and reliable. The major problem, of course, is that solar energy is distributed over a broad area, and, to be used in society, it must be concentrated into a small control center where it can be converted to electricity and be distributed. Panels for collecting solar energy have been constructed on buildings to provide energy for heating or cooling, but large-scale use of solar energy has not yet occurred. Large systems for collecting solar energy are technically feasible, but they are not at present economical. For example, to produce the present electrical energy needs of the United States, a collecting system covering 26,000 km² (one-tenth the size of the state of Nevada) would have to be built. Large-scale use of solar energy is, therefore, a long way in the future, although local use in homes and buildings would help alleviate the need for other energy forms.

Hydropower is another source of sustained energy. It has been developed in the United States to approximately 25% of its maximum capacity. With full development of hydropower by the year 2000, it would still only provide 10% to 15% of America's energy needs. A problem involved with hydropower is that, when a river system is modified by a dam, a number of unforeseen side effects can occur. Moreover, the reservoir behind the dam is a temporary feature and is destined to become filled with sediment. The useful life expectancy for most large reservoirs is only 100 to 200 years, so this source of energy is, in reality, limited.

Another form of sustained energy can be derived from the ocean's tides. To develop tidal power, a dam must be built across the entrance of a bay. The rise and fall of the tides then can be used to develop a strong current flow that is used to turn turbines to generate electricity. With maximum development, tidal power, however, could produce only 1% of America's energy needs.

Geothermal Energy. The earth has its own source of internal heat that is expressed on the surface by hot springs, geysers, and active volcanoes. In general, there is a systematic increase in temperature with depth at the rate of approximately 3 °C per 100 m. At the base of the continental crust, temperatures may range from 200 to 1000 °C, and, at the center of the earth, temperatures are perhaps as high as 4500 °C. Unfortunately, most of the earth's heat is far too deep ever to be tapped by man and that which could be reached by drilling, in most places, is too diffuse to be of economic value. Geothermal energy, like ore deposits, however, can be concentrated in local areas and has been used for years in Iceland, Italy, New Zealand, and the United States. The concentration of geothermal energy is along active plate margins or in mantle plumes, and the regions of greatest potential are where magma is relatively close to the surface.

Estimates are that, at its maximum world development, geothermal energy would yield only 10 times more than that produced in 1969, which is a small fraction of the world's total energy requirements. Locally, however, geothermal energy could be significant.

Fossil Fuels. Coal, petroleum, and natural gas commonly are referred to as **fossil fuels** because they contain solar energy preserved from past geologic ages. The idea that people presently are using energy released by the sun over 200 million years ago may seem remarkable, but the basic process of storing solar energy is quite simple. Heat from the sun is converted by biological processes into combustible, carbon-rich substances (plant and animal tissues) that subsequently are buried with sediment and preserved.

Coal. Coal originates from plant material that flourished in ancient swamps typically found in low-lying coastal plains. A modern example of such a swamp is the present Dismal Swamp along the coast of Virginia and North Carolina (*plate 20*). In this area, the lush growth of vegetation has produced a layer of peat over 2 m thick, covering an area of more than 5000 km². (Peat is an accumulation of partly decomposed plant material containing approximately 60% carbon and 30% oxygen.) In such an environment, the layer of peat may be covered with sand mud from the adjacent lagoon and beach as sea level slowly rises (*figure 23.6*). Under pressure from the overlying sediment, water and organic gases (volatiles) are squeezed out and the percentage of carbon increases. By this process, peat eventually is compressed and transformed into coal.

If the rise and fall of sea level occurs repeatedly, a series of coal beds may develop, that would be interbedded with beach sand and nearshore mud. The essential factors in developing coal deposits, therefore, are a luxuriant growth of vegetation and relatively rapid burial by a sediment cover to prevent decomposition. Coal deposits, therefore, are restricted to the latter part of the geologic record when plant life became plentiful. The most important coal-forming period in the earth's

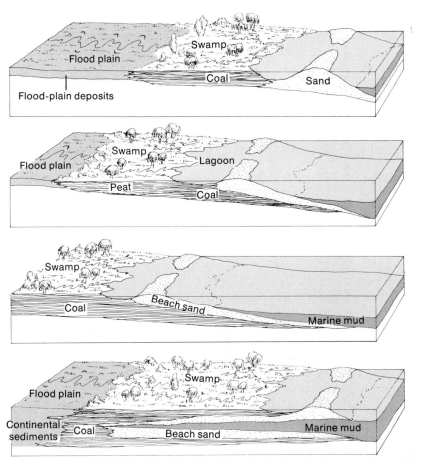

(A) *The sequence of environments is from flood plain, to swamp and lagoon, to beach, to offshore mud.*

(B) *As the sea expands inland, swamp is deposited over the flood plain, beach is deposited over the previous swamp (peat) muck and marine mud is deposited over the beach.*

(C) *Continued expansion of the sea superposes coal over flood-plain sediments, beach over coal, and mud over beach.*

(D) *As the sea recedes, the sequence is reversed; that is, beach sand is deposited over offshore mud, coal is deposited over beach sand, and flood-plain sediments are deposited over coal. Thus, by expansion and contraction of the sea, layers (lenses) of sediments and coal are deposited in an orderly sequence.*

Figure 23.6   *Schematic diagram showing how a swamp can be buried to form a coal deposit.*

history was during the Pennsylvania and Permian periods, when great swamps and forests covered large parts of most of the continents. It was from these coals that the great industrial nations such as Great Britain, Germany, and the United States developed. Other important periods of coal formation include the Cretaceous and Tertiary periods.

With the completion of at least a reconnaissance geologic mapping of most of the continents, all of the major coal fields are believed to have been discovered, and a reasonably accurate inventory of the world's coal reserves can be made. Most of the reserves are found in the United States and Russia, so the developing nations are not likely to develop by using vast amounts of coal.

The importance of coal lies in the fact that there are reasonably large reserves, and, with the depletion of oil and gas, people are already witnessing an important shift to a reliance on coal. The real pinch on oil and gas is still a few years away, but, when it comes, it will put tremendous pressure on the coal industry, which undoubtedly will bring serious environmental problems.

Petroleum and Natural Gas.   The energy sources of petroleum and natural gas are hydrocarbons (molecules composed primarily of hydrogen and carbon combined in various ways). In contrast to coal, the hydrocarbons that form oil and gas deposits originate largely from micro-

scopic organisms that once lived in the ocean or in large lakes. The remains of these organisms accumulated with mud on the seafloor and escaped complete decomposition by rapid burial. In order for deposits of oil and gas to form, three basic conditions must be met: (1) there must be sufficient organic material in the fine-grained sediments (source beds); (2) the source beds must be buried to sufficient depths (usually at least 500 m) in order for heat and pressure to compress the rock and cause the chemical transformations that change the organic debris into hydrocarbons; (3) once formed, the hydrocarbons must migrate upward from the source beds into more porous and permeable rock (usually sandstone or porous limestone or dolomite) called reservoir beds; and (4) as the oil and gas migrate through the reservoir beds, they must encounter a barrier (a trap) that causes them to accumulate. If the reservoir beds provide an unobstructed path to the surface, the oil and gas will seep out and be lost. This is one of the reasons why most oil and gas deposits are found in relatively young rocks, since there has been greater time for erosion and earth movements to provide a means for oil and gas to escape in older reservoir rocks. The barriers, or traps, can result from a variety of geologic conditions, such as those shown in *figure 23.7*. The exploration for oil and gas, therefore, is based on finding sequences of sedimentary rocks that would provide good source and reservoir beds and then locating an effective trap.

In some instances, hydrocarbons can remain as solids in the shale in which the organic debris originally ac-

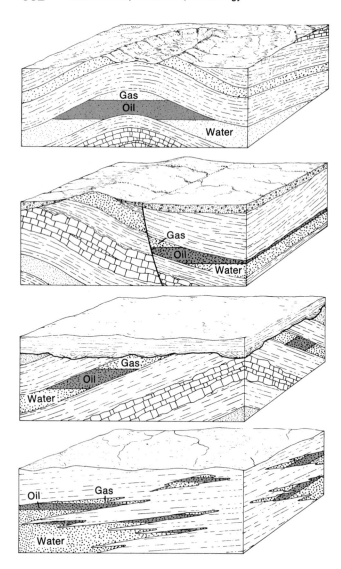

Figure 23.7 *Cross sections showing various geologic conditions that can produce an oil trap.*

(A) Anticline—oil, being lighter than water, migrates up the dip of permeable beds and can be trapped beneath an impermeable shale bed in the crust of an anticline.

(B) Fault trap—impermeable beds can be displaced against a permeable stratum and trap the oil as it migrates updip.

(C) Unconformity—an impermeable layer caps inclined strata to form a seal that traps the upward-migrating oil.

(D) Shale surrounding a sandstone lens forms an impermeable barrier and prevents the oil from escaping.

cumulated. These deposits are known as **oil shales.** They constitute reservoirs of oil that may become important in the future. The problem with oil shale, however, is that the shale must be mined and heated to extract the oil. This process requires considerable energy and is not economically feasible at present.

Oil and gas are convenient forms of energy in that they are exceptionally easy to handle and transport. Unfortunately, at the present rate of consumption, the presently known reserves will soon be depleted, and, unless other large accumulations are found, people will be forced to go to large-scale gasification and liquification of our coal and oil shale deposits and rely more on nuclear and solar energy. It is clear from present trends that people can expect to pay a great deal more for petroleum in the future and that they should utilize alternate sources of energy wherever possible.

Nuclear Energy. The ever-increasing demands for energy and the decreasing supply of fossil fuels naturally puts the light on nuclear energy as the answer to society's energy requirements. The technology necessary to utilize nuclear energy is well developed, but modern society has hesitated to move toward large-scale nuclear energy pro-

duction because of the serious long-range environmental problems that are possible. Radiation hazards, problems of waste disposal, and thermal pollution of fresh and marine waters, as well as potential terrorist activities, are among the greatest concerns.

The key element in the development of nuclear energy is uranium. The average uranium content in the rocks of the earth's crust is only two parts per million, but it is concentrated into deposits by late magmatic segregation and occurs in veins associated with intrusive igneous rocks (see *figure 23.2*). As the veins are oxidized, the water-soluble uranium is leached out and transported by surface and ground water and later may be deposited in permeable sedimentary rocks as it is absorbed by clay minerals and organic matter. In the Colorado Plateau, the rich uranium deposits are concentrated in ancient stream channels, especially where fossil wood and bones are found.

Important uranium deposits occur in Canada, the United States, South Africa, and Russia, and, if the hazards and environmental problems can be solved, nuclear energy may indeed provide a source of almost unlimited cheap energy.

Limits to Growth

Statement

The consumption of natural resources is proceeding at a phenomenal rate. Only a moment's reflection is necessary to understand that the period of rapid population and industrial growth that has prevailed during the last few hundred years is not normal but is one of the most abnormal phases of human history. Present growth and the associated consumption of natural resources present one of the most serious problems facing man today. The problem is basically one of changing from a period of growth to nongrowth, and it will entail a fundamental revision of those aspects of current economic and social thinking that are based on the assumption that growth can be permanent.

Discussion

The preceding section emphasized the fact that mineral and energy resources are formed by geologic processes in a systematic manner. They do not occur in a haphazard or random way, nor are they distributed evenly throughout the continents. This creates continents and nations that have resources and those that do not. Iceland and Hawaii, for example, having been built up exclusively of basaltic lava, cannot be considered as good potential sources of petroleum and natural gas no matter how long those areas are explored.

In addition, mineral deposits are a nonrenewable resource. Once a deposit is mined out and used, it is gone and cannot be replaced. It is important, therefore, to consider how resources are used and the rates at which they are being depleted. Some of the more important mineral and energy resources used in today's major industries are listed in *table 23.2*. The number in column 3 is the number of years the presently known reserves will last at the current rate of consumption. These figures assume no growth and that the usage rate will remain constant. It is called the *static index* and normally is used to express resources available in the future.

However, the rate of resource consumption is not static but is increasing exponentially. This is a result of the growth in population plus the fact that the average consumption per person per year is increasing each year. In other words, the exponential increase in resource consumption is driven by growth in both the population and capital and the desire for a higher standard of living. A more accurate estimate of the lifetime of a given mineral resource is the *exponential index,* or the number of years the known reserves will last at an exponential consumption rate. For the resources listed in *table 23.2,* this

Table 23.2 Nonrenewable Natural Resources

1 Resource	2 Known Global Reserves	3 Static Index (years)	4 Exponential Index (years)	5 Exponential Index Calculated Using Five Times Known Reserves (years)
Aluminum	1.17×10^9 tons	100	31	55
Chromium	7.75×10^8 tons	420	95	154
Coal	5×10^{12} tons	2300	111	150
Cobalt	4.8×10^9 lbs	110	60	148
Copper	308×10^6 tons	36	21	48
Gold	353×10^8 troy oz	11	9	29
Iron	1×10^{11} tons	240	93	173
Lead	91×10^8 tons	26	21	64
Manganese	8×10^8 tons	97	46	94
Mercury	3.34×10^6 flasks	13	13	41
Molybdenum	10.8×10^9 lbs	79	34	65
Natural gas	1.14×10^{15} ft³	38	22	49
Nickel	147×10^9 lbs	150	53	96
Petroleum	455×10^9 bbls	31	20	50
Platinum group	429×10^6 troy oz	130	47	85
Silver	5.5×10^9 troy oz	16	13	42
Tin	4.3×10^8 lg tons	17	15	61
Tungsten	2.9×10^9 lbs	40	28	72
Zinc	123×10^6 tons	23	18	50

estimate is shown in column 4. For many minerals, the usage rate is growing even faster. Exponential rates of consumption drastically reduce the estimated lifetime of a reserve. Coal will be depleted in 111 years rather than in 2300. Chromium will be gone in 95 years rather than in 420, and so on.

It can be assumed that the known reserves can be expanded to five times the presently known reserves by exploration and new discoveries. This will not extend the lifetime of a deposit five times, because the effect of exponential growth is to consume resources at an even faster rate. Column 5 shows the number of years that five times the known global reserves will last with consumption growing exponentially at the average annual rate of growth.

Figure 23.8 illustrates the effect of exponentially increasing consumption of a nonrenewable resource. Chromium is taken as an example, because it has one of the largest static indices of all the resources listed in *table 23.2*. At the current rate of usage, the known reserves would last about 420 years. The actual reserves of chromium, however, are being consumed at a rate increasing at 2.6% annually. If it is assumed that reserves yet undiscovered could increase the known reserves five times, this five-fold increase would extend the lifetime of the reserves only from 95 to 154 years. Even with 100% recycling of chromium, so that none of the initial reserves were lost, the demand would exceed supply in 235 years, because people simply would want more than there is.

A similar graph for each resource could be drawn from the data in *table 23.2*. The time scale would vary, but the general shape of the curves would be the same. The exponential index shown in column 5 is the critical figure. At the present rate of growth in consumption, copper will be depleted in 21 years, and in only 48 years, if reserves are multiplied by five. Gold has a projected lifetime of 9 years; natural gas, 22; tin, 15; et cetera.

It is obvious from these data that, with the present resource consumption rates and the projected increase in these rates, the great majority of the currently important nonrenewable resources will be extremely costly 100 years from now, regardless of the most optimistic assumptions about undiscovered reserves. The limit of growth in the world system probably will not be pollution; it will be depletion of natural resources. The interaction of some of the major variables involved as the world system grows to its ultimate limits is shown in *figure 23.9*, which is a computer model of growth and the consumption of resources. Assuming no major change in the physical, economic, and social relationships that have governed the world system historically, depletion of natural resources will be the main factor in limiting the growth of the industrial complex and population. According to *figure 23.9*, food, industrial output, and population will continue to grow exponentially until the rapidly diminishing resources force a sharp decline in industrial growth.

Summary

The natural systems that operate upon the earth constitute the basis of the ecology. They are complexly interrelated and are in a delicate balance. Man is capable of modifying and manipulating most of the natural systems, including rivers, slopes, ground water, and shorelines. When this is done, the systems adjust to the changes with many possibly unanticipated side effects.

Waste disposal is a major environmental problem. Because matter cannot be created or destroyed, it can only change from one state to another or from one chemical combination to another. Problems with waste disposal

Figure 23.8 *The lifetime of known chromium reserves depends on the future usage rate of the mineral. If usage remains constant, reserves will be depleted linearly (dashed line) and will last 420 years. If usage increases exponentially at its present rate of 2.6% per year, reserves will be depleted in just 95 years. If actual reserves are five times present proven reserves, chromium ore will be available for 154 years (dotted line), assuming exponential growth in usage. Even if all chromium is perfectly recycled, starting in 1970, exponentially growing demand will exceed the supply after 235 years (horizontal line).*

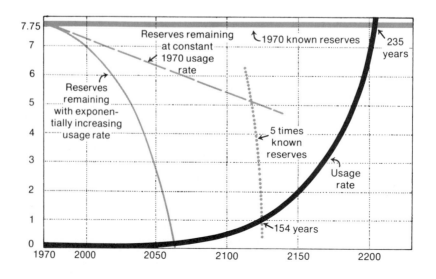

involve (a) solid wastes, (b) liquid wastes, (c) gaseous wastes, and (d) radioactive wastes.

Mineral resources have been concentrated by a variety of geologic processes operating in the earth and are related to the plate tectonic and hydrologic systems. Most, therefore, are finite and nonrenewable. Igneous processes concentrated valuable minerals by (a) early heavy crystals settling to the base of the magma chamber and (b) hydrothermal activity during the late stages

of cooling. Metamorphic processes concentrated minerals by (a) contact metamorphism around igneous intrusions (replacement) and (b) regional metamorphism that changes the texture and mineral composition of rock bodies. Sedimentary processes concentrated minerals by (a) sedimentary differentiation (sand and gravel), (b) placers (gold, diamonds, et cetera), and (c) evaporation. Weathering concentrates minerals by leaching and secondary enrichment.

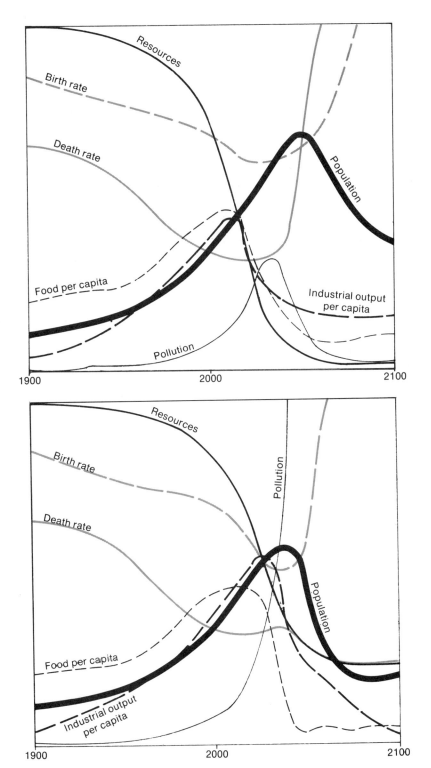

Figure 23.9 The computer model of resource consumption and its influence on other variables assumes no major changes in the physical, economic, and social relationships that historically have governed the development of the world system. All variables plotted here follow historical values from 1900 to 1970. In the upper diagram, food, industrial output, and population grow exponentially until the rapidly diminishing resource base forces a slowing down in industrial growth. Because of natural delays in the system, both population and pollution continue to increase for some time after the peak of industrialization. Population growth finally is halted by a rise in the death rate, due to decreased food and medical services. The lower diagram represents the same assumptions, with double resource reserves. Industrialization can reach a higher level, but pollution rises much more rapidly, causing an immediate increase in death rate and a decline in food production.

Energy resources are classified as renewable and nonrenewable forms. Renewable energy sources include (a) solar energy, (b) hydropower, (c) tidal energy, and (d) geothermal energy, none of which can be expected to supply a large percentage of worldwide energy requirements in the near future. Nonrenewable energy sources include fossil fuels and nuclear energy. Most industry is based on fossil fuels—oil and gas particularly. Fossil fuels are essentially solar energy from past geologic ages stored in the form of organic matter. Coal originates from vegetation produced principally in coastal swamps that later became buried by other sediments. Oil and gas originate from microscopic organisms that accumulated with sediment in the ocean or in large lakes. In order for oil and gas deposits to form, the organic matter involved must migrate from the source bed to more permeable rocks and must become trapped by some geologic structure.

Nuclear energy certainly will become a major energy source in the future, although there are some serious environmental concerns, such as radiation hazards, waste disposal, and misuse by irresponsible persons or groups.

The impact of rapid population growth and the associated industrial expansion are increasing consumption of natural resources at an exponential rate. Man is finding that there are limits to growth, and these limits probably will be reached through the depletion of natural resources.

Additional Readings

Cloud, P., chairman. 1969. Resources and Man. (A study by the Committee on Resources and Man of the National Academy of Sciences—National Research Council.) San Francisco: W. H. Freeman and Company.

Flawn, P. 1970. Environmental Geology. New York: Harper and Row.

Hubbert, M. K. 1962. Energy Resources: A Report to the Committee on Natural Resources. Washington, D.C.: National Academy of Sciences—National Research Council.

Hubbert, M. K. 1971. "The Energy Resources of the Earth." Sci. Amer. 224(3):60-84. (Reprint no. 663.) San Francisco: W. H. Freeman and Company.

Keller, E. A. 1976. Environmental Geology. Columbus, Ohio: Charles C. Merrill.

Matthews, W. H., ed. 1970. Man's Impact on the Global Environment: Assessments and Recommendations for Action. (Report of the Study of Critical Environmental Problems [SCEP]). Cambridge, Mass.: The MIT Press.

Meadows, D. H., D. L. Meadows, J. Randers, and W. Behrens, III. 1972. The Limits to Growth. New York: Universe.

Office of Emergency Preparedness. 1972. Disaster Preparedness 1(3).

Park, C., Jr. 1968. Affluence in Jeopardy. San Francisco: Freeman, Cooper and Company.

Skinner, N. 1969. Earth Resources. Englewood Cliffs, N.J.: Prentice-Hall.

Strahler, A. N., and A. H. Strahler. 1973. Environmental Geosciences. Santa Barbara, Calif.: Hamilton Publishing.

U.S. Geological Survey. Professional paper no. 820. United States Mineral Resources.

Geology of the Inner Planets

The exploration of the inner planets of the solar system (the moon, Mercury, and Mars) has been one of the most exciting adventures ever experienced by geologists, because for the first time entire new planets can be compared on a geologic basis with the earth. This does much more than merely satisfy scientific curiosity; for, as we are able to compare the details of the geologic nature and history of planets, we can better recognize those principles and processes fundamental to the geology of the earth and those that are of secondary importance.

The moon is a record of early planetary development, with some features nearly as old as the solar system itself. Mercury has surface features much like those of the moon, but its density is like that of the earth. Mars is a controlled experiment in planetary evolution with limited hydrologic and tectonic systems. We wonder why the moon has no atmosphere, no folded mountain belts, and no recent volcanism. Why does Mars have such large shield volcanoes and the great canyon? How did stream channels and catastrophic flooding develop on Mars, when water cannot exist as a liquid on the planet today? Why did Mercury, the moon, and Mars all have major thermal events and outpourings of basalt early in their history? How important is the earth's atmosphere to its surface features? To gain an insight into these and other questions about the four planets, this chapter will venture into a brief study of planetology in an effort to understand their similarities and differences.

Major Concepts

1. Cratering has been the dominant geologic process in the early history of the moon, Mercury, and Mars, and in all probability it was also important during the early history of the earth.
2. The superposition of crater ejecta and crater frequency on a surface provide a basis for establishing sequences of events on the moon, Mercury, and Mars.
3. The moon, Mercury, and Mars form a family of related planets that experienced similar sequences of events in their early histories. Each planet records (1) an early period of intense bombardment, (2) a thermal event and the extrusion of fluid lava flows that formed extensive plains, and (3) a subsequent period of light bombardment.
4. Both the moon and Mercury are primitive bodies, and their surfaces have not been modified by a hydrologic system or a tectonic system. Mars, in contrast, has evolved beyond an impact-dominated planet and has had a varied and complex history involving changes resulting from its own hydrologic and tectonic systems.
5. The earth, in contrast to the other terrestrial planets, has an active tectonic system and has continued to change throughout its history by crustal deformation, volcanism, and activity in the hydrologic system.

The Geology of the Moon

Statement

Before the space program, many scientists thought of the moon as a geologic Rosetta stone: an airless, waterless body untouched by erosion, containing clues to events that occurred in the early years of the solar system that would help reveal details of its origin and history and would provide new insight about the evolution of the earth. Now, with the Apollo missions completed, many of their fondest hopes have been realized. The thousands of satellite photographs beamed back from the moon have permitted us to map its surface with greater accuracy than we could map the surface of the earth a few decades ago. We now have over 380 kg of rocks from eight places on the moon, plus data from a variety of experimental packages that continue to send back new information. As it turns out, the moon is truly a whole new world, and the rocks (analyzed by hundreds of scientists from all over the world) and surface features provide a record of events that occurred during the first billion years of the solar system.

We have found that the moon is a planet without hydrologic and tectonic systems. Cratering has been the most important surface process and is responsible for the formation and degradation of most of the lunar landscape. A major thermal event did occur on the moon, however, when basaltic lava was extruded in a series of eruptions to form the lunar maria. Only minor structural features have been found. This indicates that the crust has not been deformed by a tectonic system.

A geologic time scale has been developed for the moon with the use of the same principles of superposition originated in the early nineteenth century by geologists studying the earth. Radiometric dates of lunar rock samples provide a base of absolute time for events in lunar history.

Perhaps the most important aspect of the moon's geologic evolution is that most of it occurred during the early history of the solar system, before the oldest rock on the earth was formed. Thus, the moon provides important insight into planetary evolution unobtainable from studies of the earth.

Discussion

Major Physiographic Divisions.   When Galileo first observed the moon through a telescope, he discovered that the dark areas were fairly smooth and the bright areas were rugged and densely pockmarked with craters. He called the dark areas **maria,** the Latin word for seas, and the bright areas **terrae** (lands). These terms are still used today, although we know that the maria are not seas of water and the terrae are not geologically similar to the earth's continents.

The maria and terrae are easily visible from the earth with the use of a small telescope or binoculars. As shown in *figure 24.1,* the maria on the near side of the moon appear to be flat, with only a few large craters. Some occur within the walls of large, circular basins; others occupy a much larger, irregular depression. We know from the rock specimens and surface features that the maria are vast floods of lava that flowed into low depressions on the lunar surface. The terrae, or highlands, constitute about two-thirds of the visible surface of the moon and exhibit a wide range of relief form. This is the highest and most rugged topography on the moon, with a local relief in many areas up to 5000 m. The single most important characteristic of the lunar highlands is that they are completely saturated with craters, many of which range from 50 to 100 km in diameter. The far side of the moon is composed almost entirely of densely cratered highlands (see *plate 1*).

The maria and highlands not only represent different types of terrain, but they broadly represent two different periods in the history of the moon. For the most part, the highlands, which occupy about 80% of the entire lunar surface, are old surfaces that became extensively pockmarked with craters early in the moon's history. After the highlands were formed, the mare basins were formed as large impact structures and subsequently were filled with lava, which in places overflowed the basins and spread over parts of the lunar highlands. Thus, the maria are relatively young features of the lunar surface, although they were formed 3 to 4 billion years ago.

Craters.   Although cratering is a rare event on the earth today, it is a fundamental and universal process in planetary development. The moon is pockmarked with literally billions of craters, which range in size from microscopic pits on the surface of a rock specimen to huge, circular basins hundreds of kilometers in diameter. The same is true for the surfaces of Mercury and Mars (and probably for Venus, the asteroids, and the satellites of the outer planets, too). Indeed, cratering was undoubtedly the dominant geologic process on the earth during the early stages of its evolution, and, in all probability, its surface once looked much like that of the moon today.

The Mechanism of Crater Formation.   Impact processes are nearly instantaneous, but they can be studied in the laboratory with the use of high-speed motion pictures. Conceptually, the process is quite simple, as is illustrated in *figure 24.2.* As a meteorite strikes the surface, the kinetic energy is almost instantaneously transferred to the ground as a shock wave, which moves downward and outward from the point of impact. This initial compression is followed by a rebound (a rarefaction wave that moves in the opposite direction) that causes material to be ejected from the surface and thrown out along ballistic trajectories. This fragmented material then accumulates around the crater to form an ejecta blanket and a system of splashlike rays. A central peak on the crater floor can result from the rebound, and crater rims can be overturned.

It is important to emphasize that the impacts of meteorites produce landforms (craters) and new rock bodies (ejecta blankets) and are, therefore, similar to other rock-forming processes (such as volcanism and sedimentation) that operate at the surface of a planet. In each process, there is a transfer of energy, material is transported and deposited to form a new rock body, and a new landform (crater, volcanic cone or flow, or delta) is formed in the process. The rock-forming processes associated with impact include fragmentation, transportation, and deposition of rock particles and shock meta-

Figure 24.1 *The major surface features of the moon. The densely cratered highlands occur in the Southern Hemisphere and on the far side of the moon, occupying about 80% of the lunar surface. It is, for the most part, an old surface that was formed during an early period of intense bombardment. The dark areas were formed from vast floods of basalt that filled large multiringed basins and lapped upon the older highlands. Little impact occurred after the lava flows (these are indicated by the few bright-rayed craters)(see also plate 1).*

morphism, which results from the force of impact in which new high-pressure minerals are created and partial melting and vaporization take place.

Types of Craters. The surface of the moon has been described as a "forest of craters," and at first glance all

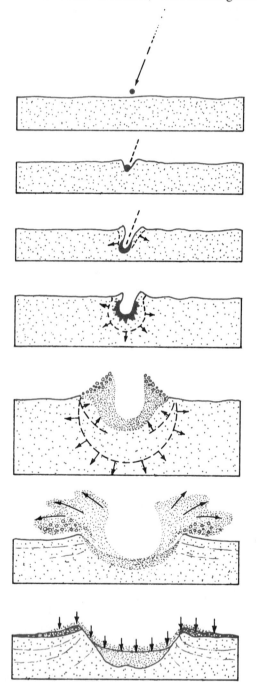

Figure 24.2 *Hypothetical stages in the formation of a meteorite impact crater. The kinetic energy of the meteorite is almost instantly transferred to the ground as a shock wave, which moves out, compressing the rock. The rock, at the point of impact, is intensely fractured, fused, and partly vaporized by shock metamorphism. The shock wave is reflected back as a rarefaction wave, which throws out large amounts of fragmental debris. Also, the solid bedrock is forced upward; this creates the crater rim. A large amount of fragmental material falls back into the crater.*

craters may look the same. After a few moments of thoughtful study, however, one is able to identify various "trees in the forest" and recognize various types of craters by certain characteristics of their size and form. Detailed study of lunar craters shows that many morphologic differences are associated with size, so it is useful to classify craters into groups that are similar in size and shape and that probably originated from the impact of similar sizes of meteorites.

Craters less than 20 km in diameter are almost perfectly circular and typically are bowl shaped. The raised rim is well defined, and a prominent ejecta blanket extends out from the rim at a distance equal to approximately the diameter of the crater (*figure 24.3A*). The surface of the ejecta blanket consists of a series of hummocky ridges that are covered with large boulders.

Craters 20 to 200 km in diameter are the most common craters on the moon. They typically have a series of terraces on their inner walls that become well developed with increasing size (*figure 24.3B*). In contrast to the smaller craters, many craters in the 20-to-200-km size range have flat floors, apparently filled with landslide debris or by lava flows. The floors of larger craters, in fact, can be arched upward, presumably as a result of isostatic adjustment and slumping. A central peak or a cluster of promontories is another conspicuous feature of craters larger than 40 km.

Craters 200 to 300 km in diameter are transitional in morphology to still larger, multiringed basins. The terraced walls are retained, but the central peaks change progressively from single promontories through clusters of peaks to a ring of peaks halfway between the center and the rim (*figure 24.3C*).

Craters larger than 300 km are called **basins** (*figure 24.3D*). These are the largest structures on the moon, and most are fringed by a series of concentric ridges and depressions and are called **multiringed basins.**

The youngest and best-preserved multiringed basin is the Orientale basin (*figure 24.4*). The basin is largely hidden from telescopic view, but, in orbiting satellite photography, it stands out as one of the most spectacular features on the lunar surface. The basin resembles a gigantic bull's-eye consisting of three concentric ridges and intervening lowlands. Beyond the outermost ridge most topographic features are ridges oriented radially to the basin. Mare basalt covers much of the lowlands in the center of the basin, but only isolated parts of the outer lowlands are flooded with lava.

Multiringed basins are believed to have been produced by impacts of asteroid-size bodies. One opinion maintains that the impact of such a large body generated shock waves that could be considered analogous to the ripples formed when a stone is thrown into a pool of water. The waves were large enough to exceed the elastic limits of the lunar crust and became "frozen" (the crust remained permanently deformed). Another considers the rings to be the result of inward-dipping, circular fault systems (*figure 24.5*). The radial ridges and valleys are believed to be ejecta with secondary craters. As can be seen in *figure 24.4*, the ejecta blanket from the Orientale basin extends out a distance roughly equal to the diameter of the basin.

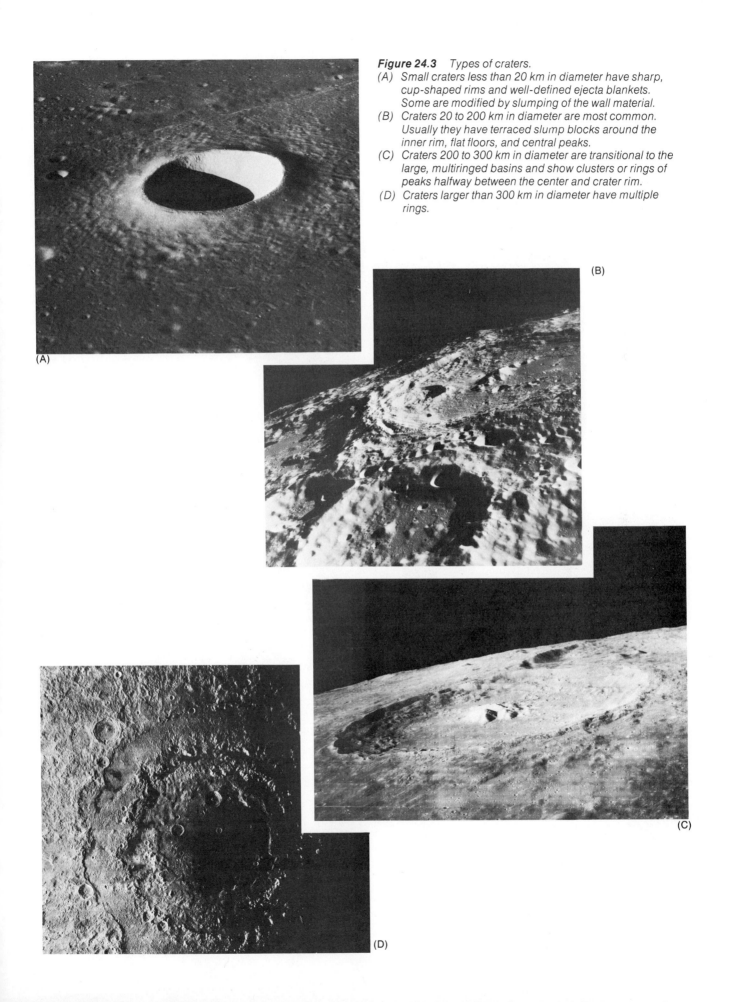

Figure 24.3 *Types of craters.*
(A) *Small craters less than 20 km in diameter have sharp, cup-shaped rims and well-defined ejecta blankets. Some are modified by slumping of the wall material.*
(B) *Craters 20 to 200 km in diameter are most common. Usually they have terraced slump blocks around the inner rim, flat floors, and central peaks.*
(C) *Craters 200 to 300 km in diameter are transitional to the large, multiringed basins and show clusters or rings of peaks halfway between the center and crater rim.*
(D) *Craters larger than 300 km in diameter have multiple rings.*

(A)

(B)

(C)

(D)

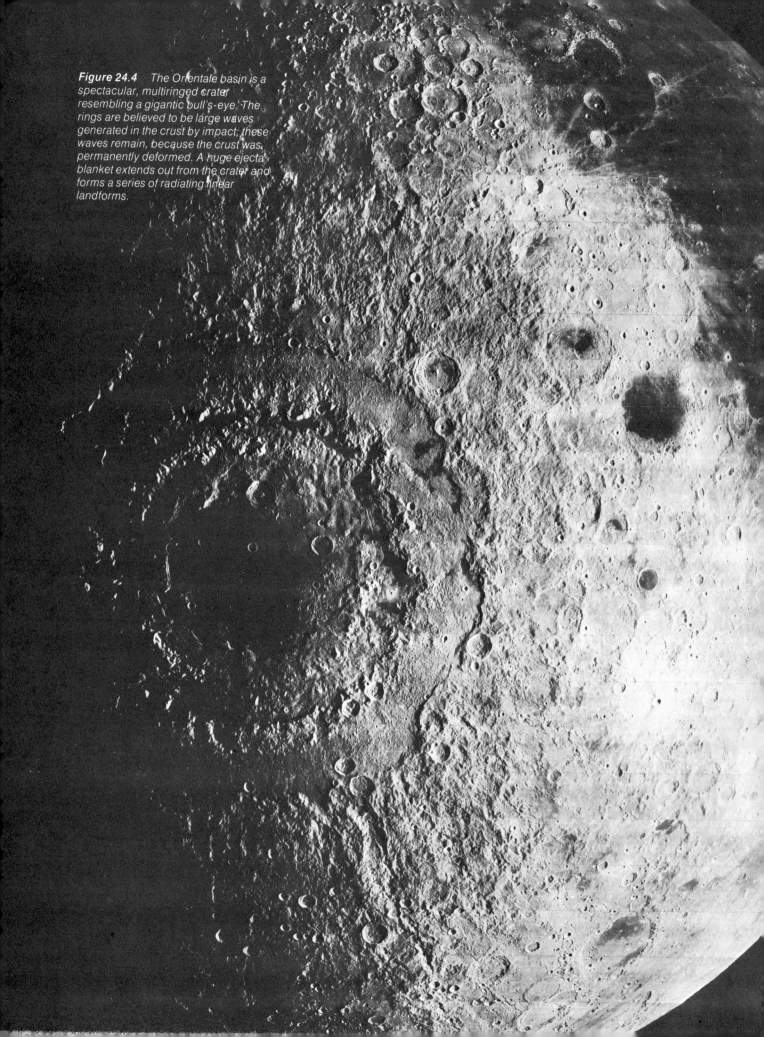

Figure 24.4 The Orientale basin is a spectacular, multiringed crater resembling a gigantic bull's-eye. The rings are believed to be large waves generated in the crust by impact; these waves remain, because the crust was permanently deformed. A huge ejecta blanket extends out from the crater and forms a series of radiating linear landforms.

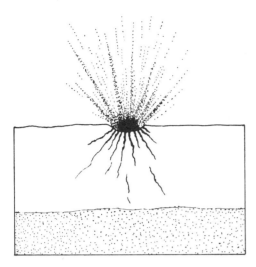

(A) Impact of asteroid-size body blasts out debris.

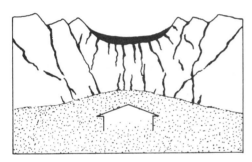

(B) Shock waves exceed the elastic limits of the rocks, so the crust is permanently deformed.

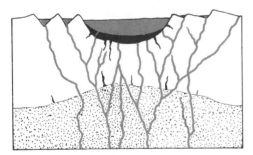

(C) Subsequent volcanic activity may partly fill the basin with numerous layers of basalt.

Figure 24.5 *The origin of multiringed basins.*

Crater Degradation and Modification. The morphology of the craters described above refers to the original size and shape produced by impact. After a crater is formed, however, it is subjected to certain types of modification. The steep crater walls soon begin to slump, resulting in partial filling of the depression and the formation of concentric terraces (slump blocks) inside the crater rim. Isostatic adjustment can arch up the crater flow to compensate for material removed in the excavation. Further modification results from three main processes: (1) subsequent impact can partly destroy or obliterate the crater and ejecta blanket; (2) the crater can be

(A)

(B)

(C)

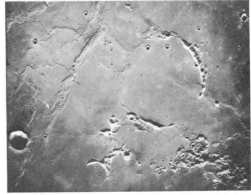

(D)

Figure 24.6 *Examples of crater degradation and modification in the size range of 20 to 45 km.*
(A) Slumping of crater walls and isostatic adjustment.
(B) Modification of crater rims by impact.
(C) Crater partly covered by ejecta from younger craters.
(D) Crater partly covered by lava flows.

covered with ejecta from the formation of younger craters; and (3) the crater and ejecta can be buried by younger lava flows. Examples of crater degradation and modification are shown in *figure 24.6*.

Volcanic Features. The nature and distribution of volcanic activity are especially important in studies of planetary dynamics, because volcanic products constitute a type of window into a planet's interior and provide valuable insight into how a planet operates. Although craters dominate the lunar landscape, a variety of volcanic features have been discovered that show that the moon has had a thermal history. The most important volcanic features, of course, are the maria—the vast floods of lava that fill the large, multiringed basins on the near side of the moon and commonly overflow, spilling into the surrounding areas. From distant views, it may appear that the maria represent one huge flood during which the basins all filled to the same level, as the oceans on the earth did; but, upon closer examination, we find that the maria lava flows are not at the same level and were extruded over a period of nearly a billion years (*figure 24.7*). They were formed, however, during a specific interval of time extending from 3.9 to 3.1 billion years ago and record a major thermal event in lunar history.

It is interesting to note that some rims of old craters project above the mare material and show that the lava is relatively thin. The basalt in the interior of the mare basins completely covers the old craters and must be in excess of 1500 to 2000 meters thick, but the "ghost" craters along the margins show that a considerable part of the margins of the basins is covered with only 200 to 400 m of basalt (*figure 24.8*).

Looking at the extensive lava fields of the maria, one

Figure 24.7 *Lava flows in Mare Imbrium. These flows were extruded from fissures near the edge of the basin and may extend downslope for 1200 km. They are approximately 35 m high and 10 to 25 km wide.*

Figure 24.8 *Many craters, near the margins of the maria, are almost completely covered, so that only parts of their rims can be detected. This indicates that the basalt flows are only 200 to 400 m thick.*

might expect to find spectacular stratavolcanoes like Vesuvius, Shasta, and Fuji, together with numerous cinder cones. These volcanic features did not develop on the moon for several reasons. First, the basaltic lava on the moon was extremely fluid and was extruded by fissure eruptions. As has been shown, fissure eruptions produce extensive lava plains but few volcanoes. Second, the physical environment on the moon would have had important effects on the development of volcanic features. If a typical cinder cone eruption occurred on the moon, the lack of atmospheric drag on the ejected particles and the lower gravitational field would tend to form a thin, widespread blanket of ash, rather than a cone. Some distinctive patches of dark material surrounding low-rim craters may very well be the lunar equivalent of cinder cones.

A number of smooth, low domes occur in the lunar maria and are considered by most geologists to be volcanic in origin. Foremost among these are the Marius Hills on the western side of the earth-side disk just north of the equator. There, the array of domes and cones is similar in many respects to volcanic features on the earth (*figure 24.9*). These are interpreted as a type of shield volcano or possibly laccoliths.

Figure 24.9 *An oblique view of the Marius Hills, showing domes believed to be of volcanic origin.*

In summary, the basalts of the lunar maria represent a major thermal period in lunar history that extended over a period from 3.9 to 3.1 billion years ago. There are a number of volcanic features associated with the flood basalts, but there are no spectacular volcanic mountains like those found on the earth and Mars. The extrusions occurred as quiet fissure eruptions that terminated 3 billion years ago. Since then, the moon has not experienced significant volcanic activity.

Tectonic Activity. There has been almost no tectonic activity on the moon during the last 2.5 billion years, and very little during its entire history. We know this because the millions of craters that cover the lunar surface provide an excellent reference system for even the most subtle structural deformations. Just as lines on graph paper, the circular craters would show the effects of deformation by compression, tension, or shear and would record even the slightest disturbances. The network of craters, however, is essentially undeformed, and the lunar crust appears to have been fixed throughout time. There is no evidence of intense folding or thrust faulting and no indication of major rifts. The major features that can be attributed to structural deformation are linear rilles, lineaments, and wrinkle ridges.

Many linear rilles are sharp, linear depressions that generally take the form of flat-floored, steep-walled troughs ranging up to several kilometers in width and hundreds of kilometers in length (*figure 24.10*). The val-

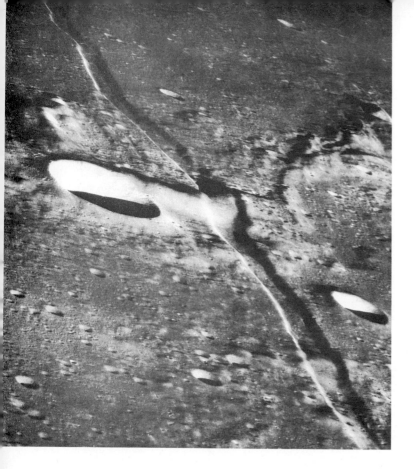

ley walls are straight or arcuate and stand at the same elevation as though they had been pulled apart and the floor had subsided as a graben. Straight rilles are parallel or arranged in an echelon pattern. Some intersect, while others form a zigzag pattern similar to normal faults on the earth.

Other linear features are common on the surface of the moon, although their origins are a matter of conjecture. Some are radial to multiringed basins and are undoubtedly faults related to impact. Others are believed to be surface expressions of fracture systems in the old rocks beneath the surface cover.

The only expressions of compression in the lunar crust are found in the wrinkle ridges of the maria (*figure 24.11*). These structures are long, narrow anticlines that have a sinuous outline and extend for considerable distances parallel to the margins of the mare basins. Some geologists consider them to be compressional features produced as the maria subsided. Others feel that they represent differential compaction.

When compared to the deformation seen in the rocks of the earth, these structural features on the moon are minor indeed. There are no folded mountain belts and no great rift valleys on the moon. The moon's crust has been rigid and essentially undeformed by tectonic activity throughout all of its recorded history.

Figure 24.10 *Linear fault blocks resulting from tensional stresses in the lunar crust.*

Figure 24.11 *Wrinkle ridges in the lunar maria are long and straight or sinuous anticlines and are probably the only compressional features in the lunar crust.*

Lunar Rocks. To many people, the rocks returned from the moon were a great disappointment. Moon rocks are not exotic and mysterious but are much like basalts and other common rock types found on the earth. Yet, the lunar rocks hold a key to understanding the chronology of events in the history of planetary development and a record of the sun's activity. They were sampled from a variety of structural settings (*figure 24.12*) and have been studied by approximately 1000 scientists from 19 countries. Lunar rocks can be classified easily into the following types:

1. Basalt.
2. Anorthosite and other associated crystalline igneous rock.
3. Breccia—angular, fragmental particles formed by fragmentation of surface material by meteorite impact and subsequently compacted into a coherent rock by shock compression during cratering.

Figure 24.12 *Schematic block diagrams showing the structure of the strata at the Apollo landing sites.*

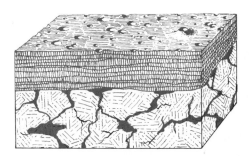

(A) *Apollo 11 landed on the flood basalts of Mare Tranquillitatis. The lunar maria consist of numerous thin layers of basalt flows with an aggregate thickness of several thousand meters extruded 3.7 billion years ago. The mare basalts rest on older rocks of the cratered highlands.*

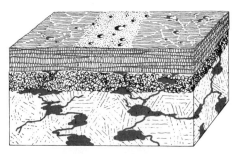

(B) *Apollo 12 landed on a ray of Copernicus, approximately 400 km south of the crater. The ray material rests on a sequence of basalts, which forms Oceanus Procellarum. Below the basalts is a layer of ejecta, which was formed during the early stages of impact and development of the lunar highlands.*

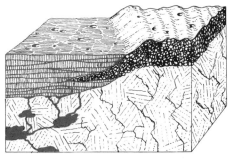

(C) *Apollo 14 landed on the Fra Mauro Formation, which is composed of material ejected by the Imbrium impact event. Basalts from Oceanus Procellarum lap up against the ejecta, proving that the maria lavas are younger than the highlands. The rocks of the Fra Mauro Formation were thrown out of the Imbrium basin about 4 billion years ago.*

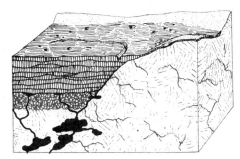

(D) *Apollo 15 landed near Hadley Rille at the base of the Apennine Mountains. The mare basalts lap up against the ancient rocks of the lunar highlands, which have been dated as more than 4 billion years old.*

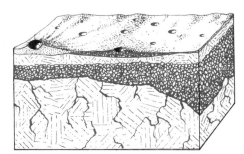

(E) *Apollo 16 landed in the highlands of the Descartes region. The surface material is composed of debris churned up by North Ray crater and South Ray crater and overlies layers of breccia, which were formed by more ancient meteorite impact.*

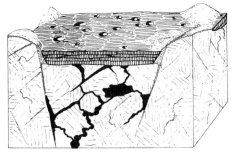

(F) *Apollo 17 landed in the Tarus-Littrow valley, which was formed by ejecta from the Serenitates basin.*

4. Regolith, or soil—a mixture of crystalline and glassy fragments formed by repeated fragmentation of surface material by meteorite impact.

Lunar Basalt. Most of the igneous rocks collected from the moon are very similar to terrestrial basalt, the most common rock in the earth's crust. These rocks were once totally molten, as is indicated by their vesicles, interlocking crystalline texture, and composition (*figure 24.13*). The principal minerals found in lunar basalt are plagioclase, pyroxene, ilmenite, and olivine, all found in terrestrial basalt. Only minor amounts of a few minerals unknown on the earth were found. The most significant difference between lunar and terrestrial basalts is that the former contain a greater amount of heat-resistant elements (titanium, zirconium, and chromium). Lunar basalts are devoid of water and have much lower amounts of relatively volatile elements, such as sodium and potassium, than do terrestrial basalts. These chemical characteristics are significant in that they suggest that the material that forms the moon was, at one time, heated to higher temperatures than the material from which the earth was formed.

Anorthosite. Anorthosite is a coarse-grained, crystalline igneous rock composed almost entirely of the mineral plagioclase (*figure 24.14*). This rock type was collected from the lunar highlands and is an important constituent in many samples of soil and breccia. The significance of anorthosite is that it records a thermal event very early in the moon's history, before the development of the mare basalts and before the period of intense bombardment that formed the craters of the lunar highlands.

Breccia. Breccia is a fragmental rock in which the individual particles are angular, in contrast to the rounded particles of sand and gravel. Lunar breccias consist of fragments of rock and glass from a variety of sources (*figure 24.15*). They result from fragmentation from impact. Some are consolidated ejecta blanket deposits that show various degrees of recrystallization.

Others are consolidated regolith and soil that are characterized by a high proportion of glassy fragments and cinders.

Regolith. The surface of the moon is mantled in most places by a thin layer of relatively loose, uncon-

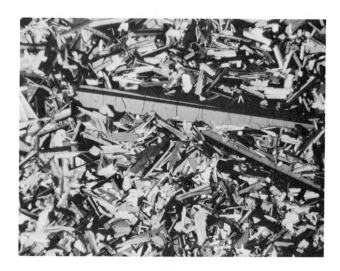

Figure 24.14 *Photomicrograph of a lunar anorthosite. The width of field of view is approximately 4 mm. This rock consists of a meshwork of plagioclase crystals, which are characteristically lath shaped, and some pyroxene occupying the interstitial spaces between the plagioclase crystals. Olivine occurs in amounts up to 1%, with small traces of opaque minerals and glass.*

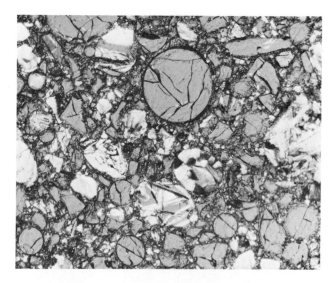

Figure 24.15 *Photomicrograph of a thin section of lunar breccia. The width of the field of view is approximately 4 mm. The lunar breccia constitutes a major rock type on the lunar surface and consists of angular fragments of broken rocks from a variety of sources. Some lunar breccia contain large amounts of glass, with particles that are remarkably spherical. Typically, the fragments are angular and show essentially no evidence of modification by abrasion. Glass particles are the dark black material.*

Figure 24.13 *Photomicrograph of a sample of lunar basalt. The width of field of view is approximately 4 mm.*

solidated fragments of rock, crystals, and glass, similar in many respects to a fine powder. This layer is called the lunar regolith. It results from the churning action of impact in contrast to the chemical and mechanical processes that produce regolith on the earth.

Glass particles are abundant in nearly all samples of lunar regolith, and glass is one of the features that distinguishes lunar soils from terrestrial soils. The glass occurs as beautifully formed spheres and teardrops and as spatter on other fragments (*figure 24.16*). The lunar glass is distributed with ejecta and is believed to have formed by shock melting of rock debris during the process of impact.

Under high magnification, the glass beads show pits, grooves, and spatter that resulted from micrometeorites striking the glass surface after the particles had cooled and accumulated in the regolith (*figure 24.17*). Thus, practically every rock fragment in the regolith appears to have been involved in one or more impact events.

In summary, the rock samples returned from the moon tell exactly what one would expect from studies of the surface features. They record the details of the major lunar processes (impact and volcanism). Their greatest values, however, are the information they provide about the absolute dates of major lunar events and the establishment of a radiometric time scale for other planets.

The Structure of the Moon. Our present understanding of the internal structure of the moon is based on a variety of physical observations, including density, magnetism, and seismicity. Much remains uncertain, but several facts place significant constraint on what the internal structure may or may not be. First, the bulk density of the moon is 3.34 g/cm³, and the mean density of lunar surface rocks is about 3.3 g/cm³. Thus, the density of the surface material is only slightly less than that of the moon as a whole, so there is little possibility for a significant increase in density with depth. The earth, in contrast, has a mean density of 5.5 g/cm³, with the density of surface rocks being only 2.7 g/cm³. This clearly indicates that the interior of the earth is much denser than the crustal material. However, it is also clear from studies of lunar rocks, surface features, and lunar seismicity that the moon is layered and the composition of the interior is different from the surface material. The

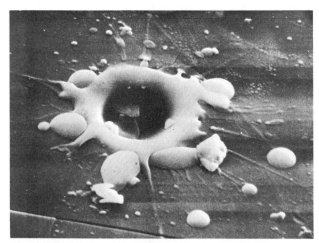

Figure 24.17 *Micrometeorite crater on the surface of a crystal fragment in lunar breccia.*

favored model of the moon's interior is shown in *figure 24.18*. The major units are (1) a crust, (2) a rigid mantle (lithosphere), and (3) a central core (asthenosphere).

The Crust. Data from Apollo seismic stations shows that the moon's crust varies in thickness from approximately 60 km on the near side to 100 km on the far side.

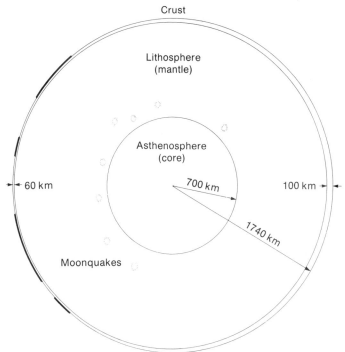

Figure 24.18 *Cross section showing our present understanding of the internal structure of the moon. The solid black represents mare basalts. The thickness of the crust ranges from 60 to 100 km. The rigid mantle (lithosphere) extends down to a depth of about 1000 km. Most moonquakes originate from the region near the base of the lithosphere. A possible central core (asthenosphere) is localized beneath the moonquake zone.*

Figure 24.16 *Glass beads from lunar regolith.*

Variations in seismic velocities with depth show that the crust is layered (*figure 24.19*). The low velocity near the surface that gradually increases with depth most likely indicates a layer of fine breccia and broken rock fragments 1 to 2 km thick. Below this, to a depth of 25 km, the seismic velocities have values similar to basalt. The rapid increase in the velocity from the surface to approximately the 25-km depth is believed to be the result of increasing load pressure on fractured basalts of the maria. From 25 km to 60 km, the velocities are similar to those determined for anorthosite.

The Mantle (Lithosphere).   At a depth of 60 to 100 km, a sharp increase in velocity occurs that marks the contact between the lunar crust and the mantle. The mantle rocks show a higher seismic velocity than most earth rocks and are believed to be rich in olivine. The thick mantle is rigid and could be called the lithosphere. It is believed to be composed of olivine-rich rock. Most moonquakes occur between 600 and 950 km and are apparently triggered by earth tides (the gravitational attraction of the earth on the moon).

The Core (Asthenosphere).   An inner zone, or core, possibly exists beneath the moonquake zone (*figure 24.18*). S waves traversing the deep interior are strongly attenuated; this suggests that the rocks are partly molten and perhaps somewhat like the earth's asthenosphere. As yet, there is no evidence indicating an iron core, either solid or liquid.

Mascons.   Studies of the orbits of Apollo and Lunar Orbiter spacecraft close to the moon show that there are significant irregularities in the lunar gravitational field. The areas of high gravitational attraction are attributed to mass concentrations and are called **mascons.** Most of the mascons are found in the circular mare basins and do not occur with any prominence in the highland regions. The origin of mascons is quite controversial. Some of the explanations proposed thus far include these. (1) They represent large buried meteorites of nickel and iron. (2) They represent disk-shaped bodies of high-density lava that fill the mare basins. (3) They represent accumulations of heavy minerals at the bases of lava lakes. More refined gravity mapping is needed before this question can be resolved.

Developing a Lunar Geologic Time Scale.   The deposits of ejecta from the craters, together with lava flows and other volcanic deposits, form a complex sequence of overlapping strata that covers most of the lunar surface. The individual deposits can be recognized by their distinctive topographic characteristics and by their physical properties (color, tone, and thermal and electrical properties) determined from measurements made with optical and radio telescopes.

In 1962, geologists from the U.S. Geological Survey developed a geologic time scale for the moon so that the major geologic events could be arranged in their proper chronologic order. The basic principles used by the survey to interpret lunar history are essentially the same as those used to study the history of terrestrial events on the earth; the most important of these principles are the law of superposition and the law of crosscutting relations. These principles for determining relative ages are, of course, as valid on the moon or Mars as they are on the earth. In addition, a second method of determining relative ages of lunar features was based on the abundance of craters (crater frequencies). The development of a lunar geologic time scale was a major advancement in the study of the moon, because, for the first time, the sequence of events in the history of another planet was firmly established.

The first lunar chronology was developed in 1962 by Shoemaker and Hackman, who interpreted the sequence

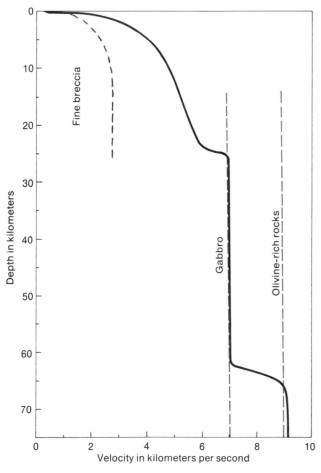

Figure 24.19   *Graph showing variations in seismic velocities and interpretation of the structure of the lunar crust. Three major discontinuities in seismic velocities indicate that the moon has a layered crust. Seismic velocities increase very rapidly at shallow depths (to about 10 km). A very sharp increase in velocities occurs at about a depth of 25 km, and, between 25 and 60 km, the seismic velocity is nearly constant. A significant discontinuity, or increase in velocity, occurs at 65 km. This is interpreted as marking the base of the crust. Comparing these seismic velocities with those of major rock types, it is interpreted that, near the surface to a depth of about 1 to 2 km, the composition of the rock is that of fine breccia and broken rock fragments. Below this, to a depth of about 25 km, is a layer composed largely of basalt. A second layer of lunar crust occurs between 25 and 65 km and appears to be composed of gabbro and anorthosite. Below 65 km, the mantle of the moon is interpreted (tentatively) to be composed of magnesium-rich olivines.*

of events in the vicinity of the crater Copernicus. In many ways, what Shoemaker and Hackman did in providing a rationale for interpreting moon history is comparable to what Smith, Lyell, and their contemporaries did in establishing the geologic time scale on earth during the early 1800s. The planet is different, and the nomenclature is different, but the logic remains the same. Let us carefully study the ejecta from the major craters studied by Shoemaker and Hackman and consider their reasons for recognizing the sequence of events they represent. As you read the following discussion, study the physiographic map of the moon and the selected illustrations, for it is only by recognizing the physical relationships among the features discussed that you can gain an appreciation for the relative time involved in their formation.

Copernicus. One of the most outstanding features seen on the near side of the moon is the crater Copernicus and its spectacular system of bright rays that ex-

tend outward in all directions. The rays and ejecta blanket surrounding Copernicus are superposed on essentially every feature in their path (*figure 24.20*). *From this superposition, it is clear that Copernicus and its associated system of ejecta and ray material are younger than the mare basalts and are younger than rayless craters such as Eratosthenes.* The ejecta from rayed craters

Figure 24.20 *Relative ages of lunar features in the vicinity of the crater Copernicus. Ejecta from Copernicus are superposed on all other features and are the youngest materials. The ejecta from Eratosthenes rest on the mare basalts and are, therefore, younger than the mare material but older than Copernicus. The ejecta from the Imbrium basin that contain the mare lava rest on the densely cratered highlands and are thus younger than the complex ejecta in that area.*

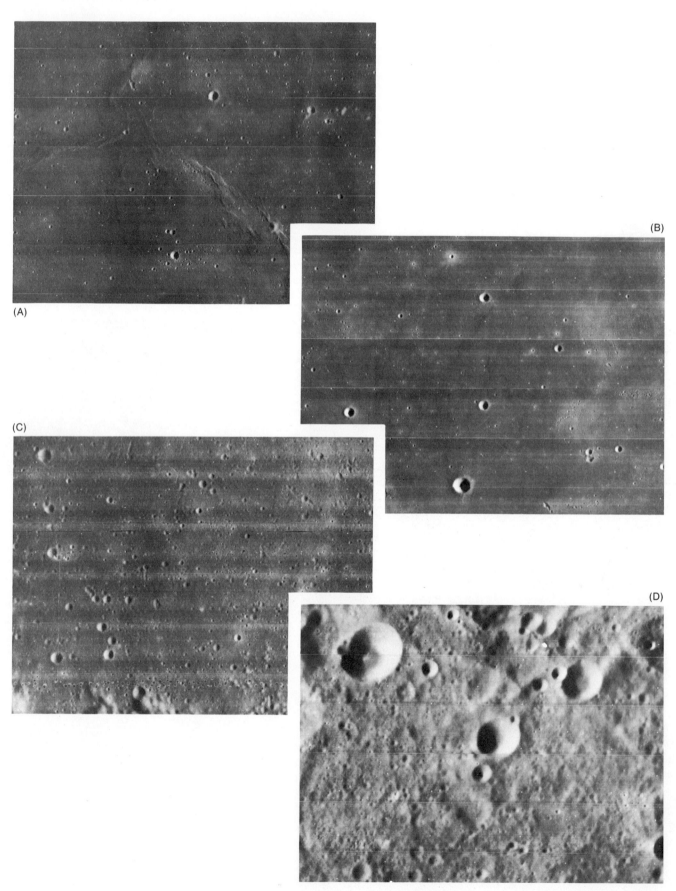

Figure 24.21 *Relative ages of surfaces determined by the frequency of craters. A is the youngest, D is the oldest.*

form the uppermost and youngest system of strata on the moon and are called the **Copernican system.** The period of time during which rayed craters and their associated rim deposits were formed has been called the **Copernican period.**

Eratosthenes. About half of the craters larger than 10 km in diameter that occur on the maria are rayed craters and belong to the Copernican system. Most other craters of this size found on the maria are similar, but their ejecta blankets are dark and they lack rays. An example is the crater Eratosthenes, just northeast of Copernicus (*figure 24.20*). It has terraced walls, a roughly circular floor with a central peak, a hummocky rim, and a distinctive pattern of secondary craters. However, unlike Copernicus, it does not have a visible ray system. Eratosthenes and similar craters, together with their ejecta, are superposed on the maria and are, therefore, younger than the mare lava flows on which they are formed. However, they are older than the rayed craters of the Copernican system. *The deposits of the dark-rimmed craters are called the Eratosthenian system of strata, and the period of time during which they were formed is referred to as the Eratosthenian period.*

The Imbrium Basin. In the northwest part of the near side of the moon is an enormous, multiringed crater, largely filled with lava flows, called the Imbrium basin. The Imbrium basin is surrounded by ejecta deposits similar to those formed by smaller craters, the best exposures being the Apennine Mountains extending outward from the southwestern rim.

The ejecta from the Imbrium basin are partly covered with lava, as is most of the interior of the basin (Mare Imbrium). *The lava plus the ejecta deposits constitute the Imbrian system of strata, and the time during which they were formed is called the Imbrian period.*

The Ancient Terrae. The ejecta from the Imbrium basin partly overlie the complex sequence of ejecta formed in the lunar highlands. *These strata are collectively called the Pre-Imbrian system of strata, and the time during which they were formed is called the Pre-Imbrian period.* The Pre-Imbrian deposits constitute the oldest material on the lunar surface, and their stratigraphic and structural relationships are very complex. Large craters are closely spaced and modified by impact. Apparently, the surfaces of the terrae have been churned by repeated formation of large craters early in lunar history.

Crater Frequencies as a Method of Determining Relative Age. Cratering has been the most universal process in the solar system and provides a wealth of information concerning the history of a planet and the relative age of its surface features. Cratering can be used to determine the relative age of a surface because of the simple fact that a larger number of craters will have developed on an older surface than on a younger one. This holds true regardless of whether the rate of cratering is constant, steadily decreasing, or erratic. A simple example illustrates this point. Assume that it has been snowing for several days and that the snow is 1 m deep on undisturbed lawns throughout the neighborhood. If the snow is 1 m deep on the sidewalks of a house, you can conclude that nobody has shoveled the snow since the storm began. If the snow is about half a meter deep in front of another

house, it would be obvious that the walk had been shoveled, possibly midway through the storm. If another walk has only a few centimeters of snow, it would be obvious that it had been shoveled only an hour or two before. A walk with only a few scattered snowflakes apparently would have just been shoveled.

To understand the basis for using cratering to determine relative ages, one needs only to substitute a planet's surface for the sidewalk and cratering projectiles for the snowflakes. An example of how the frequency of craters has been used in studies of the moon is shown in *figure 24.21.*

Crater counts on the rims of the Orientale, Imbrium, and Humorum basins show that Humorum has the largest number of craters and Orientale the least and they are, respectively, the oldest and youngest basins.

Radiometric Dates for Lunar Events. Some of the most critical information about the geology of the moon has been obtained from isotopic age determinations of lunar rocks. Many geologists expected that the lunar surface was old, but the fresh lava of the maria and bright-rayed craters appeared as though they could have formed as late as the earth's ice age. The first samples to be dated were basalts from one of the maria. They gave an age of 3.65 billion years, which is older than any rock found on the earth. Additional radiometric dates of other mare basalts indicate that the extrusion of lava to form the maria began about 3.9 billion years ago and continued for about 700 million years. The lunar highlands, of course, are older, and samples collected from Apollo 17 show that crystallization of the anorthosite occurred 4.6 billion years ago. This is the age of the oldest meteorites and is believed to represent development of the lunar crust close to the time of the formation of the moon.

The age of the crater Copernicus was determined from ray material collected from Apollo 12 landing sites. The date of this event, one of the most recent in lunar history, was 0.8 to 0.9 billion years ago.

By integrating these and other radiometric dates into the relative geologic time scale of the moon determined from superposition and crater counts, it is possible to construct an absolute time scale for the moon and a graph showing rates of cratering (*figure 24.22*). This shows that, early in lunar history, the rate of cratering was hundreds or even thousands of times greater than today. Moreover, the rate of decline in impact was very rapid until about 3 billion years ago. Since then, the rate of cratering has been relatively constant.

It is believed that other planets and satellites experienced similar variations in cratering. If so, cratering may be a means of correlating interplanetary events.

The Geologic History of the Moon. By utilizing data obtained from studies of the relative ages of lunar landforms and rock bodies, absolute ages of lunar rock samples, and the chemistry of lunar rocks and soils, it is possible to construct a sequence of events outlining the geologic evolution of the moon. Five major stages are recognized, each containing several distinct but overlapping events or processes (*figure 24.23*).

Stage I (4.7-4.6 billion years ago). The moon, like the other planets of the solar system, is believed to have been formed by accretion of many small objects in a sur-

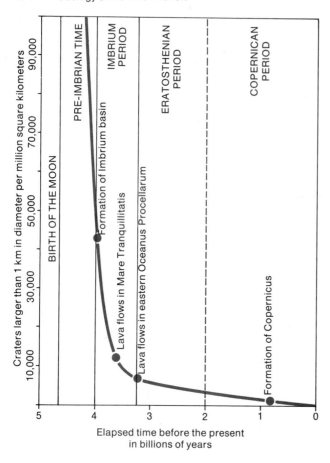

Figure 24.22   *Graph showing the variations in the numbers of craters formed on the moon's surface during different periods of time.*

prisingly short period of time. The processes of formation may have taken about 10 million years, but some scientists believe it took less than 1000 years. It is uncertain whether the moon formed as a hot or cold body, but most scientists believe that at least the outer 500 km melted very early and underwent extensive differentiation and the formation of a crust composed of aluminum-rich basalt, gabbro, and anorthosite. From a geologic point of view, the development of a global crust that still makes up most of the lunar highlands was the first important event in the evolution of the moon.

Stage II (4.6-3.9 billion years ago) (figure 24.23A). The next major event was a period of intense bombardment from large and small bodies orbiting around the planet or following trajectories in the solar system that brought them into collision with the moon. The impact of these bodies formed a densely cratered terrain over the entire surface of the moon. This terrain is now preserved in the lunar highlands.

Stage III (3.9 billion years ago) (figure 24.23B). A distinct event in the moon's history was the impact of a number of large, asteroid-size bodies that produced the multiringed basins. This was something of a discontinuity in the moon's early evolution, because at the time the moon was undergoing the early intense bombardment by many small objects and was hit by several large ones

(50 to 200 km in diameter) in a relatively short period of time. These large objects might have been small proto-moons formed at the same time as the moon, and their infall might have been a result of the larger gravitational pull of the moon. In any event, their impact significantly modified the surface of the moon. After the formation of the multiringed craters, additional impact of "normal" meteorites formed craters on the basins, including Plato, Archimedes, and Sinus Iridum. These craters show that there was a significant time lapse between the formation of the multiringed basins and their infilling with lava.

Stage IV (3.9-3.2 billion years ago) (figure 24.23C). The next major event in lunar history was a thermal event that produced a second period of differentiation and the extrusion of the mare lava that filled the multiringed basins. The lava was extruded from a series of eruptions during an interval of about 700 million years. Each mare was formed by hundreds or thousands of flows. This was a major event in the moon's history. Similar features are found on Mercury and Mars. It is, therefore, possible that the formation of the maria represents a basic element in the evolution of all terrestrial planets.

Stage V (3.2 billion years ago to the present) (figure 24.23D). Most of the moon's present features were developed by the end of Stage III, and its differentiation was essentially complete. The most significant events to occur since that time were the impact of meteorites to form post-maria craters. The influx of meteorites was greatly reduced and most authorities consider that the craters formed after the mare lavas were extruded were formed by bodies from the asteroid belt or from comet nuclei.

Minor local volcanic activity probably has occurred since the maria were formed to create such features as the domes in the Marius Hills. But, if the moon is geologically active, it is so only at great depths, and the landscape evolution is largely complete.

The Geology of Mercury

Statement

The surface features on the planet Mercury, as observed by the Mariner 10 space probe, are strikingly similar to those on the moon. The early geologic history of the two planets appears to be remarkably similar, and the differences that do exist can be explained as the result of differences in mass and internal structure.

Discussion

Surface Features.   The similarities between Mercury and the moon are apparent from the photomosaic of Mercury reproduced in *figure 24.24.* From this distance (approximately 230,000 km), the entire surface appears to be heavily craterial, much like the far side of the moon. A number of bright craters, comparable with Copernicus, are apparently the youngest features on Mercury, since their bright haloes of ejecta and long, splashlike rays appear to be superposed on everything they come in contact with. Craters range in size from small pits (limits of resolution of the photographs) to large, multiringed basins. Differences in crater morphology with size are

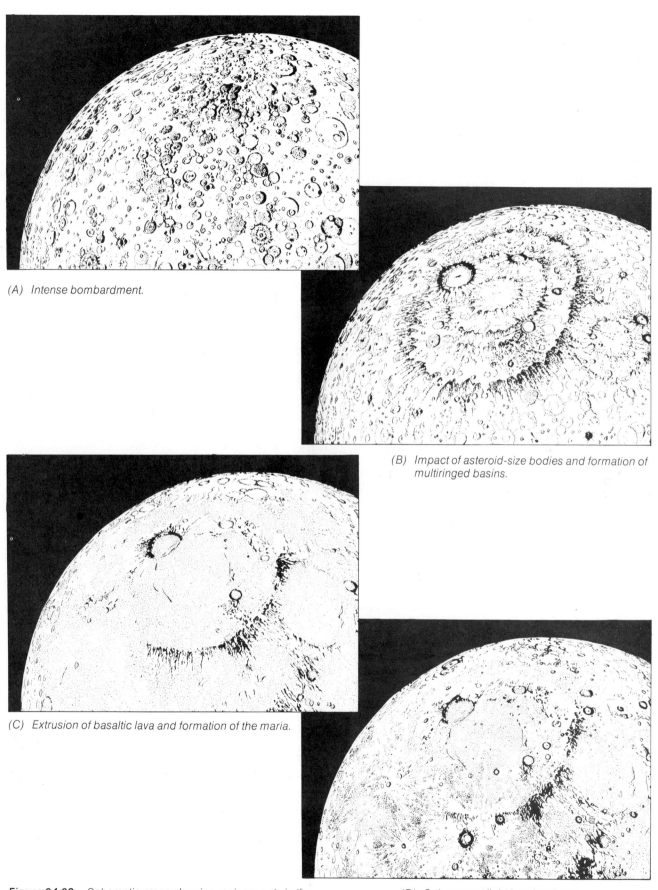

(A) Intense bombardment.

(B) Impact of asteroid-size bodies and formation of multiringed basins.

(C) Extrusion of basaltic lava and formation of the maria.

(D) Subsequent light bombardment.

Figure 24.23 Schematic maps showing major events in the history of the moon.

Figure 24.24 *Photomosaic of Mercury made from photographs taken at a distance of 234,000 km.*

similar to those observed on the moon. However, two important differences can be noticed: (1) the progression of changes in crater morphology occurs at small crater diameters on Mercury, and (2) the extent of the ejecta blanket and secondary craters is systematically smaller for a given crater size on Mercury than on the moon. The differences are the result of the greater gravitational pull of Mercury, which is over twice that of the moon. Although the style of cratering is slightly different, the process is fundamentally the same.

The largest structure photographed by Mariner 10 is the Caloris basin (*figure 24.25*), a large multiringed basin similar in size and form to the Imbrium basin on the moon. Ejecta from the basin radiate out from the main ring to form elongate hills and valleys that extend outward a distance equal roughly to the diameter of the basin. These are best expressed beyond the northeast part of the basin, which is possibly the most rugged topography on Mercury. Other multiringed basins are smaller (*figure 24.26*).

Smooth plains material covers the floor of the Caloris basin inside the main scarp, as well as the lowlands be-

yond. Although it cannot be said unequivocally that the plains material is volcanic, its similarity to the lunar maria suggests that it formed by a similar process, that is, by the extrusion of fluid basaltic lava. Evidence supporting this conclusion comes from the great volume of material that forms a level surface and fills older depressions. The volume is far too great to have been produced by mass movement, and there are no surface processes to transport that volume of sediment. In addition, the tone and texture of the plains material are different from the surrounding areas, a fact suggesting that it is composed of different material. The plains material is highly ridged and fractured. This appears to be unique, since similar features have not been found on the moon or Mars.

Careful mapping of the surface of Mercury shows that there are two major generations of plains material. The older type, referred to as intercrater plains, is the most extensive terrain type observed on the planet and forms rolling surfaces between, and around, areas of densely cratered terrain (*figure 24.27*). The distinguishing characteristic of the intercrater plains is their high density of small craters in the size range of 5 to 10 km. The small craters are commonly elliptical, partly superposed on one another to form irregular ridges between them. The intercrater terrain predates, at least in part, the period of intense bombardment. This older sequence of plains material is tentatively interpreted to be volcanic as a result of part of the early differentiation of the planet. Its definite origin will probably require more data than is available from Mariner 10.

Younger plains material forms smooth, nearly level surfaces covering approximately 15% of the planet viewed by Mariner 10. The surface is only sparsely cratered and closely resembles the lunar maria. The density of the relatively small, superposed craters is approximately the same wherever the unit occurs, indicating that most of it is approximately the same age. A notable feature of the plains material in many large craters is the presence of large scarps that terminate at the enclosing crater wall. These may be flow fronts. If they are, their heights would be enormous by earth standards and would indicate magmas of high viscosity (*figure 24.28*). Volcanic cones and domes have not been found on Mercury, so it appears that the only probable volcanic features are the two generations of plains material.

Figure 24.25   *The Caloris basin, a large, multiringed basin on Mercury that closely resembles basins on the moon and Mars. Radiating valleys and ridges are formed in the ejecta beyond the outer rim. The interior of the basin is completely flooded with plains material, which spreads out over the surrounding lowlands.*

Tectonic Features.   Although most surface features on the moon and Mercury are quite similar, Mercury appears to have a tectonic framework not found on other terrestrial planets. In contrast to the earth, Mars, and to a lesser extent the moon, evidence of tensional stress on Mercury is found in only two localized areas, both apparently related to the Caloris basin. Mercury appears to be unique in that the dominant tectonic feature is a series of lobate scarps that are often more than 500 km long and rise more than 3 km above the surrounding area. The general nature of these scarps can be seen in *figure 24.29*. The scarps are lobate in outline and have rounded crests that stand out in contrast to the sharp crests and straight ridges formed by normal faults and grabens on the moon and Mars. These characteristics seem to indicate reverse and thrust faults resulting from compression in the Mercurian crust. The scarps transect both the old intercrater plains and some craters of the densely cratered terrain, while some old craters and smooth plains interrupt the scarp. This would indicate that the scarps were formed sometime around the final phase of heavy bombardment and continued to be formed some time after the smooth plains were formed.

Figure 24.26   *Small, multiringed basins on the planet Mercury.*

Figure 24.27   Old intercrater plains. Prominent scarp formed by a thrust fault cuts both older and younger craters.

Figure 24.28   Flow fronts on the floor of craters that are about 85 km in diameter.

Figure 24.29   Fault scarps cutting cratered surface of Mercury. The scarp is about 550 km long and 3 km high. Note deformed crater rim on the crest of the scarp. (See also figure 24.27.)

An important characteristic of the scarps is their apparent planetwide distribution. Preliminary maps show that they extend from pole to pole throughout most of the visible surface. Therefore, the entire planet seems to have been subjected to the compressive forces that caused the crustal shortening. The relatively uniform global distribution of the scarps suggests that the entire planet was subjected to compressive forces resulting from minor crustal shortening, as the planet contracted during the early period of differentiation.

The Geologic History of Mercury. Only part of the surface of Mercury has been photographed, but geologists, utilizing the same methods and techniques used to study the moon, have been able to establish a preliminary geologic time scale and develop a working hypothesis for the geologic evolution of Mercury. The surface features of Mercury record a sequence of events broadly similar to those recorded on the moon and indicate five major events.

Stage I—Accretion and Differentiation. The early differentiation of Mercury is indicated by the planet's mean density of 5.44 g/cm^3 and inferences from its moderate magnetic field (about 1% that of the earth). By analogy with the materials on the surface of the earth and the moon, volcanism on Mercury would imply the existence of a silicate-rich mantle. For a planet the size of Mercury with a density of 5.44 g/cm^3, the silicate shell could only be 500 to 600 km thick and an iron core would extend out to 75% to 80% of the planet's radius. The existence of an iron core is inferred also from Mercury's magnetic field. Thus, the conclusion that Mercury is a differentiated planet is based on two separate observations.

The first period of differentiation must have been completed well before the period of intense bombardment, because melting and development of a liquid surface, or at least a plastic crust, on the planet would obliterate or greatly modify all preexisting landforms. Since the craters from the period of intense bombardment are well preserved, they must have been formed after the first period of differentiation. The intercrater plains, which are older than the highly cratered terrain, are probably the oldest feature on Mercury and may be the most primitive surface in the solar system.

Stage II—Heavy Bombardment. A period of intense bombardment is recorded by the clusters of densely packed, large craters. The old surface of Mercury is not saturated with large craters (as the moon is), and it appears that a period of heavy bombardment was an episode separate from the bombardment that formed the intercrater plains. Crustal shortening on a planetwide scale was taking place during the period of heavy bombardment as a result of contraction from minor core shrinkage.

Stage III—Formation of Caloris Basin. The formation of the large, multiringed Caloris basin was a major event in the geologic development of Mercury. The excavation of the basin modified the landscape over much of the visible surface and formed a large ejecta deposit and radial ridges and valleys far beyond the outer ring of mountains. Hilly and lineated terrain beyond the Caloris ejecta suggests the possibility of other large, multiringed basins not seen by Mariner 10 photography.

Stage IV—Basin Flooding. The smooth plains material that fills the Caloris basin and parts of the surrounding areas represents the flooding of the earlier basins at a time when heavy bombardment had greatly decreased. This material is probably of volcanic origin and the flooding occurred over a period of time (as the flooding of the lunar maria did), but the total time interval involved in the formation of the plains may have been relatively brief. If the smooth plains material were volcanic, it would represent a second differentiation of Mercury similar to the second period of differentiation on the moon.

Stage V—Light Cratering. After the period of basin flooding, the surface of Mercury was subjected to light cratering, which formed the bright-rayed craters. The density distribution and morphology of these craters resembles the cratering that took place on the moon after the maria were formed.

The absence of subsequent modification of the surface of Mercury by tectonic or volcanic activity or by atmospheric processes is significant, because it indicates that, after the period of basin flooding, the geochemical and tectonic evolution of Mercury was essentially completed. There is no observable deformation of the outer silicate shell of Mercury arising from any postulated fluid core. The extrusion of the plains material was apparently the end of Mercury's planetary dynamics. As with the moon, the only processes that could modify Mercury after the end of its final period of volcanic activity would be degradation by occasional objects (small meteorites, micrometeorites, and cosmic particles). However, the unexplored 50% of Mercury's surface could reveal more evidence of internal activity.

The Geology of Mars

Statement

In November 1971, Mariner 9 obtained an orbit around Mars and became the first man-made satellite of another planet. During its period of operation, which lasted nearly a year, it provided some 7300 spectacular photographs and a wealth of other scientific information that constitute one of the most significant scientific advancements of the space program. The Viking mission was equally successful, with Viking I landing on Mars on July 20, 1976, and Viking II on September 3, 1976. These spacecraft have not only sent back beautifully detailed photographs taken from the surface of Mars, but the Viking orbiting satellites have provided additional detailed photographic coverage of selected areas with much greater resolution than the Mariner 9 mission.

With this data, photomosaics of the entire planet have been made, shaded relief maps of the surface published, and geologic maps compiled in a variety of scales. This newly acquired information has changed drastically man's understanding of the planet that generations of astronomers have thought to be most like the earth. Mars is an extremely exciting planet of enormous geologic interest. Almost every geologic feature on Mars is not just big, it is gigantic. There are huge volcanoes 21 km high, enormous canyons, and giant landslides. There is evidence not only of stream action but of catastrophic flooding that modified large areas of the planet. Also, there are a variety of terrain features that have no counterpart here on the earth. Wind action also appears to be an important process and has modified most surface features. The polar regions are covered with alternating layers of ice and windblown sediments, and details of the advancing and retreating polar caps have been photographed with a type of time-lapse photography. It is also apparent that the geologic agents operating on Mars have not only varied from place to place but have varied throughout the planet's history.

The importance of Mars is that, in a study of physical geology, it provides a fourth reference point (the earth, the moon, and Mercury are the other three) for understanding planetary dynamics. This additional reference helps emphasize the fundamental principles of how planets evolve.

Discussion

Physiographic Provinces.   The whole surface of Mars was photographed during the epic flight of Mariner

9. Accurate terrain maps of the entire planet have been made, some at the scale of 1:5,000,000. From these maps and photographs, we see that Mars can be divided into two major regions, or physiographic provinces: (1) the old, densely cratered highland to the south and (2) the smooth sparsely cratered plains in the north (*figure 24.30*). The boundary forms a great circle inclined at approximately 35° to the equator.

The densely cratered terrain in the Southern Hemisphere is a highland somewhat like the cratered highlands of the moon. It rises abruptly above the low volcanic plains in the Northern Hemisphere and is separated from them throughout much of the planet by a prominent escarpment 2 to 3 km high. The scarp is well defined in the east but is masked in the west by the great volcanic field in the Tharsis region. It is dissected and modified by the unique type of stream erosion and mass movement on Mars that has produced a variety of terrain types referred to as chaotic terrain, hummocky terrain, and fretted terrain. The major channels, believed to have been formed by running water, appear to have evolved along the global escarpment draining the southern highland and emptying into the low northern plains.

The northern plains are believed to be the products of lava flows, and they resemble in many aspects the mare regions of the moon and the plains regions of Mercury.

Various sections within the two major regions have been produced by a variety of surface and internal processes.

Two large domal upwarps capped with huge volcanoes are located in the Northern Hemisphere near the escarpment. The largest is in the Tharsis region to the west, where the volcanic field appears to sit astride the global escarpment; the smaller—the Elysium field—is located in the northern plains to the east.

The large structural upwarp in the Tharsis region has

Figure 24.30   *Map showing the major physiographic provinces of Mars. The Southern Hemisphere is an old, densely cratered highland; the Northern Hemisphere is a relatively smooth lowland presumably formed by floods of basalt. A well-defined escarpment separates the two provinces, except where the escarpment is covered by younger volcanics in the Tharsis region. Most of the unique terrain types (chaotic, hummocky, and fretted) are formed by erosion and slope retreat along the global escarpment. Most river channels on Mars also are associated with drainage from the southern highlands across the scarp to the northern lowlands.*

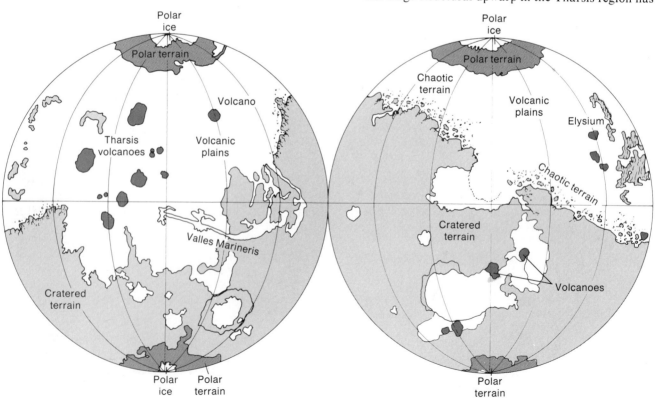

created a system of radial fractures that extends nearly halfway around the planet. Erosion and enlargement of an east-west segment of the fracture system has developed Valles Marineris, the Grand Canyon of Mars, which extends from the fractured bulge in the Tharsis area, across the cratered highland, and breaks up into a number of distributaries near the global escarpment.

Several large basins occur on the highly cratered plateau and are believed to be multiringed basins similar to those on the moon and Mercury. Wind action has been a major surface process and has played a role in modifying all the surface features.

The polar regions have several unique features resulting from processes restricted to those areas, including sections of pitted and etched terrain and regions of layered sedimentary deposits and regions of permanent ice.

Craters.   Although the photographs from Mariner 9 have revealed a variety of landforms on the surface of Mars, cratering still appears to have been a dominant process, especially during the early history of the planet. The craters on Mars reflect impact from a population of meteorites similar to that which produced the cratered surface on the moon and Mercury. The craters on Mars, however, have many distinctive features that reflect the particular gravitational attraction and surface processes and erosional history of the planet. Most of the craters are shallow, flat-floored depressions that show evidence

Figure 24.31   *The crater Yuty, an excellent example of ejecta debris that has flowed like a mudflow. The flowlike appearance of the ejecta blanket suggests the fluidization of ejected material, probably through the melting of near-surface ground ice.*

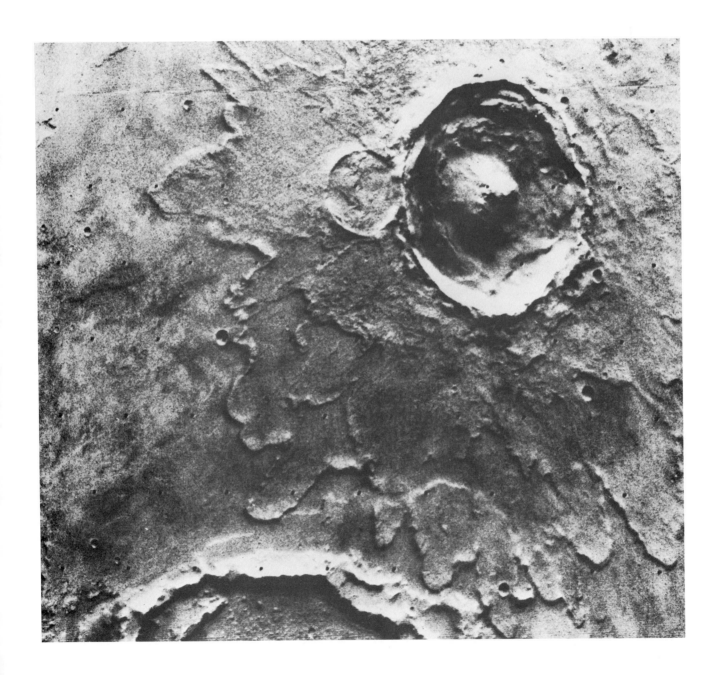

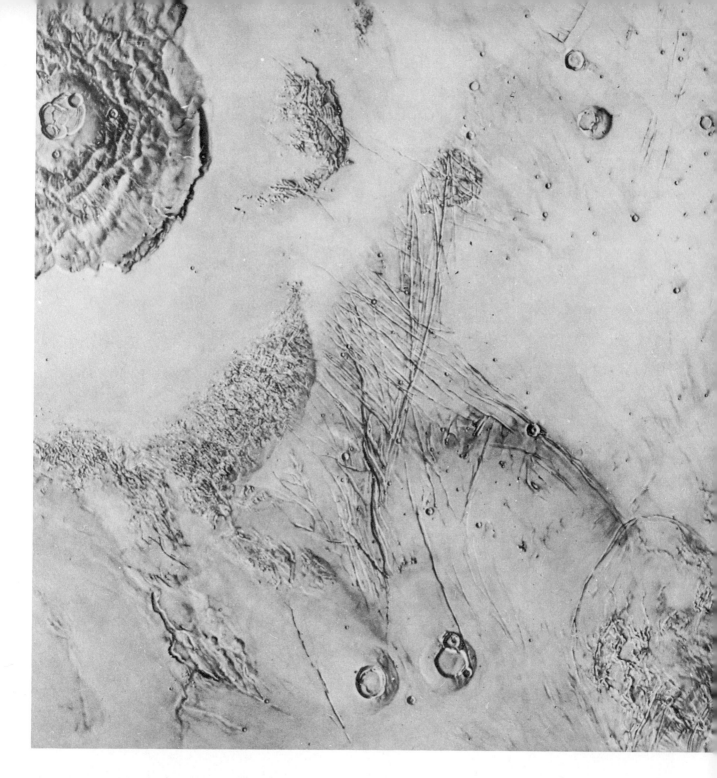

Figure 24.32 *Volcanoes in the Tharsis region.*

of much more erosion and modification by sedimentation than those on the moon and Mercury. Rayed craters are rare, because the winds on Mars can easily erode the loose ray material.

Photographs from the Viking mission (August 1976) with much higher resolution than those from Mariner 9 show that fresh craters do exist on Mars, but, to the surprise of most scientists, the ejecta blanket appears to have flowed over the surface as avalanches and mudflows do on the earth. This phenomenon is unique to Mars, not

having been observed on the moon or Mercury. One of the most spectacular examples is the crater Yuty, shown in *figure 24.31*. Tongues of ejecta debris like huge splashes of mud have flowed outward from the crater rim. To the bottom the ejecta flow overlaps the older eroded debris from the large adjacent crater and actually flows up and over the cliff formed on the older eroded ejecta blanket. The flow of ejecta material is believed to be the result of a planetwide distribution of ground ice, similar to the permafrost in the polar regions of the earth. Upon impact, the ground ice would melt so that the material ejected from the crater would move like great globs of mud.

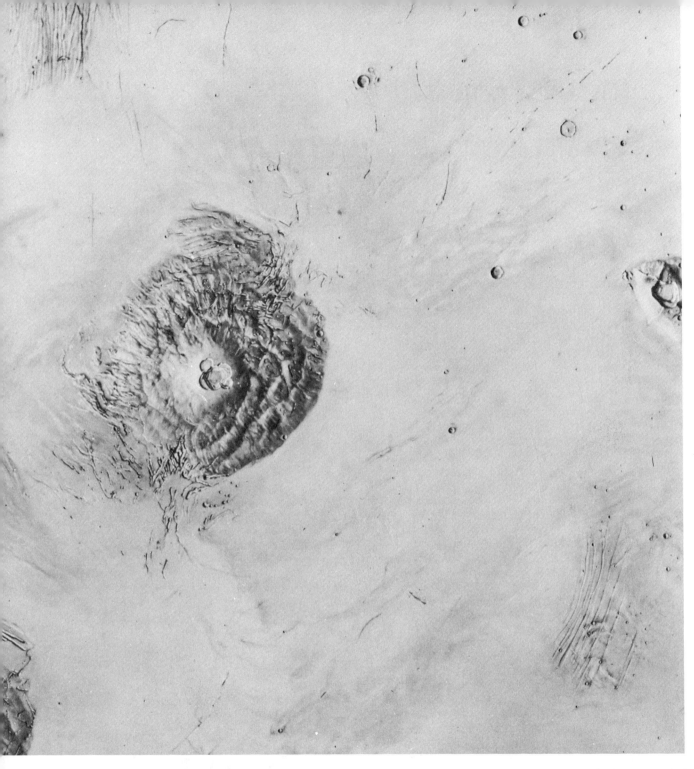

An important finding is that the cratering history on Mars parallels that of the moon and Mercury. Each planet experienced an early period of intense bombardment and a subsequent period of impact by large asteroid bodies that produced large, multiringed basins. Since then, the rate of impact has rapidly decreased to a minimum.

Volcanic Features. One of the most spectacular features observed on the surface of Mars is the series of giant volcanoes in the Northern Hemisphere. The volcanoes are much larger than anything that can be seen on the earth, and their freshness suggests that Mars has just been "turned on" volcanically. However, detailed studies of the available photography reveal that there are a num-

ber of older volcanoes, although they are highly eroded and are difficult to recognize. Therefore, it appears that Mars has had a long and very interesting volcanic history.

The most spectacular volcanic features on Mars are the enormous shield volcanoes, most of which occur in the Tharsis region and Elysium region (*figures 24.32 and 24.33*). Each of these volcanoes is at least twice the size of the largest volcano on the earth. Olympus Mons, the largest, is 500 to 600 km in diameter and rises 23 km above the surrounding plains. This is nearly half again the distance from the deepest spot on the earth—the Marianas trench—to the top of Mount Everest. At the crest is a huge, complex, collapsed caldera similar to

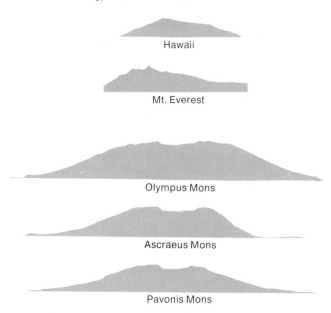

Hawaii

Mt. Everest

Olympus Mons

Ascraeus Mons

Pavonis Mons

Figure 24.33 *Profiles of some of the large Martian volcanoes compared to topographic features on the earth.*

those on the volcanoes of Hawaii. Other volcanoes resemble Olympus Mons but differ in size and detail.

The huge, young shield volcanoes on Mars are not scattered at random but are located high on the summit of large domes or upwarps in the Martian crust in the Tharsis and Elysium regions. Both upwarps are broken with major fracture systems that radiate out from the crest and extend across the Martian surface for great distances. Thus, the origin of the volcanoes is very likely closely associated with major upwarp of the crust.

Much older shield volcanoes have been discovered in the densely cratered terrain of the Southern Hemisphere and have been extensively modified by impact, wind erosion, and partial burial. Thus, volcanic activity has apparently extended over a large portion of the planet's history, going back as far as the period of intense bombardment.

In addition to the large volcanoes, Mars has a number of smaller volcanic domes and extensive lava plains that cover the Northern Hemisphere. These plains are similar to those on Mercury and the moon.

The volcanoes are important in that they indicate that Mars has been thermally active during much of its history. Whereas volcanic activity on the moon and Mercury terminated after the period of extensive flooding, it has continued on the large planets Mars and the earth.

Tectonic Features. The presence of undeformed craters across the surface of Mars clearly indicates that the crust has not been subjected to extensive horizontal compression. There are no folded mountains and no system of moving plates. Yet, there are some very important tectonic features on Mars that suggest tensional stress and extension of the rigid crust. These, together with the volcanic features, provide convincing evidence that significant tectonic activity has occurred and that, from a tectonic standpoint, Mars has a style of its own.

Two types of structures dominate the tectonic charac-

ter of Mars: (1) large domal upwarps and (2) grabens. Both are concentrated in the Northern Hemisphere and appear to be genetically related.

The large domal upwarps of the Tharsis and Elysium regions have been briefly mentioned with respect to their associations with the large volcanic shields. The Tharsis dome is a broad bulge 5000 km in diameter and 7 km high. It is asymmetrical, being much steeper on its northwestern flank. The Elysium dome is smaller, being only 2000 km across and 2 to 3 km high. Both, of course, are much larger than most domes on the stable platform of the earth.

The most significant tectonic feature associated with the Tharsis dome is a vast radial pattern of fractures that extends over about a third of the planet. The most extensive fracture system is northeast of Tharsis, where a fanlike array of grabens and normal faults converges toward the line of shield volcanoes. The fractures form magnificent sets of grabens that typically are 1 to 5 km wide and may be several thousand kilometers long (*figure 24.34*).

The western part of the fracture system deserves special attention because of its controlling influences on the development of the huge canyon system called the Valles Marineris (the Grand Canyon of Mars). The huge dimensions of this feature are vividly shown in *figure 24.35,* on which an outline map of the United States is superposed for scale. This great canyon system is more than 5000 km long and at least 6 km deep. A minor side canyon is similar in length and depth to the Grand Canyon in Arizona. The canyons of Valles Marineris consist of a series of parallel depressions characterized by steep walls and a sharp brink at the lip of each canyon. Each trough is highly irregular in detail, with sharp indentations, large embayments, and dendritic tributaries similar to the canyons cut by running water on the earth. The configuration of the canyon walls clearly indicates that considerable erosion has occurred to widen and modify the original rift valleys. The dominant processes were probably landsliding, debris flows, and, possibly, running water.

The features on the earth most comparable in size to Valles Marineris are the Red Sea and the rift valleys of East Africa. As with the rift system on the earth, Valles Marineris was initiated where the crust was pulled apart and the interior block subsided. Subsequently, the rim of the rift valley was sculptured by erosion. The fact that the head of Valles Marineris is near the crest of the upwarp nearly 9 km above the plains probably aided in accelerating erosion and widening of the canyon. Other fault systems might have developed similar canyons had they been favorably located.

The development of the domal upwarps and associated radial fracture systems clearly represents a major event in the history of Mars, and that event had some relationship with the focus of volcanic activity of the planet.

Fluvial Features. Perhaps the most startling result of the Mariner 9 project was the discovery of numerous channels on Mars that closely resemble dry riverbeds on the earth. These channels originate in the southern highlands near the erosional escarpment and empty into the low volcanic plains to the north. If these were viewed on the earth, no one would hesitate to call them dry river-

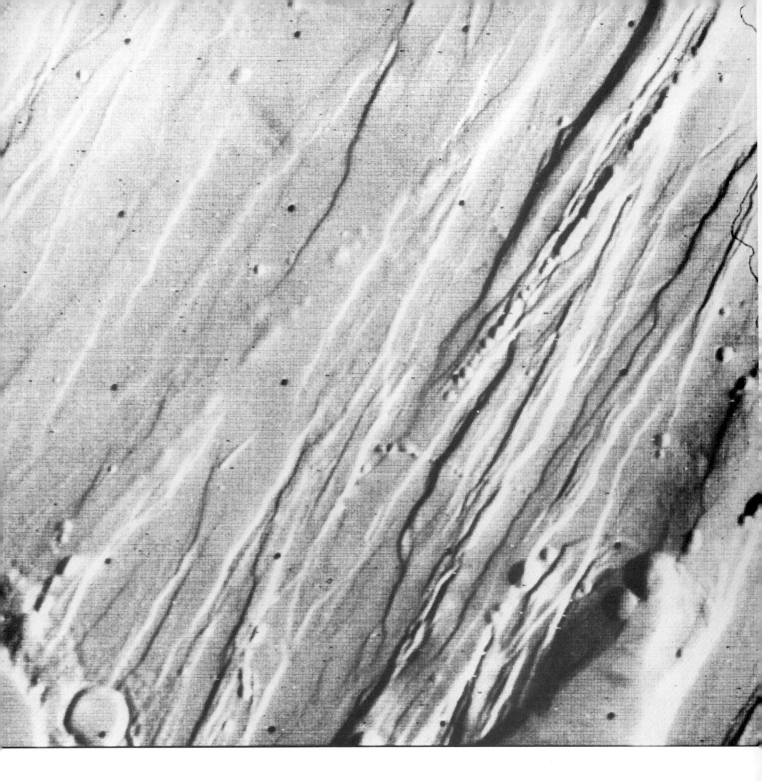

Figure 24.34 *Fracture system on Mars associated with the upwarps in the Tharsis area.*

beds, but their presence on Mars presents some of the most intriguing and perplexing questions about the planet. The existence of H_2O in a liquid state is dependent on the temperature and pressure regime at the planet's surface. The present atmospheric conditions on Mars are characterized by low pressures and low temperatures, so large amounts of water cannot exist at the surface. If present, water on Mars occurs in very small quantities for a short period of time when frost melts before reentering the atmosphere as vapor. Most water in the Martian system is locked up either as ice in the polar regions or as ground ice similar to permafrost on the earth. The presence of widespread river channels that represent not only running water but in places large-scale flooding leads us to the conclusion that either some unknown and presently unimaginable processes formed those features or at some time in the past geologic conditions on Mars permitted running water. The question is when did it (water) exist, where did it come from, and where has it gone? These questions are directly related to the question of whether there is, or has been, life on Mars. The discussion will continue for some time.

Figure 24.35 Valles Marineris, the Grand Canyon of Mars. This huge chasm is more than 5000 km of at least 6 km deep. The Grand Canyon of the Colorado River is equivalent to a minor tributary of the canyon on Mars.

Figure 24.36 Sketch map showing the major system of stream channels on Mars and their relationships to the global escarpment.

The system of channels along the eroded escarpment that separates the cratered highlands (source region of the channels) from the low volcanic plains supplies the best evidence for the former presence of rivers on Mars. *Figure 24.36* is a sketch map of the escarpment showing the size and distribution of the major channel system. The channels are wide and deep and show evidence of extensive flooding (*figures 24.37* and *24.38*), flooding that would be catastrophic by human standards.

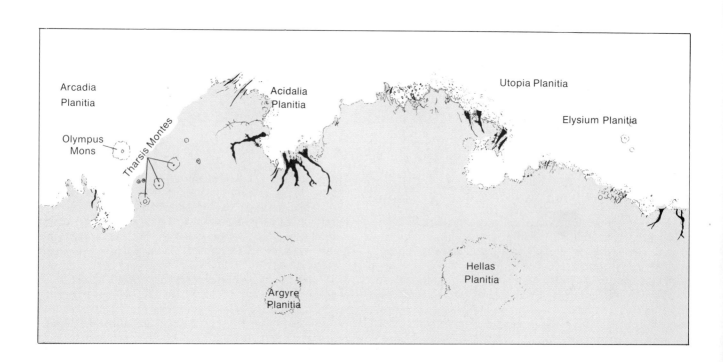

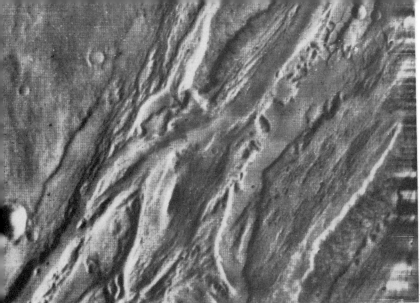

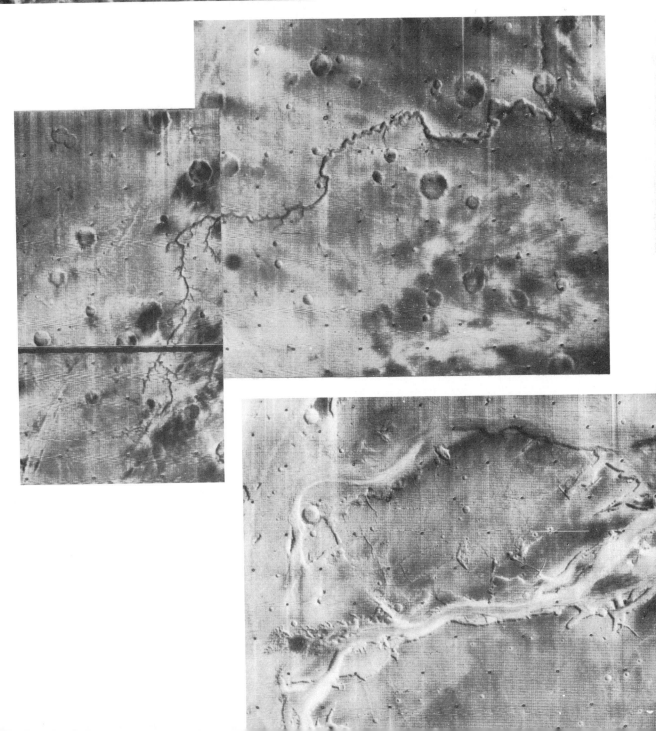

Figure 24.37 Stream channels on Mars.

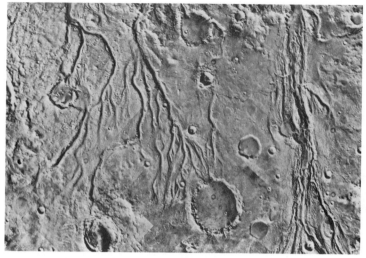

(A) *Tributary systems.*

(B) *Braided channels.*

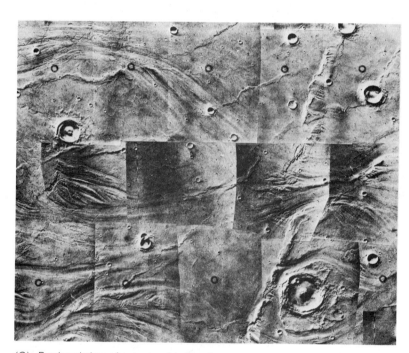

(C) *Regional view of catastrophic flooding.*

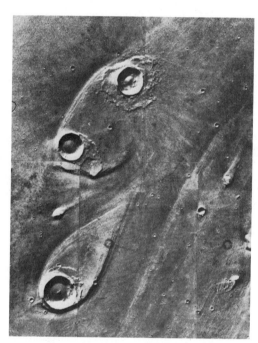

(D) *Detailed view of flooding.*

Figure 24.38   *Details of stream channels on Mars.*

Yet, the Martian channels are not like the river systems on the earth, which have delicate networks of tributaries, main trench streams, and large deltas where the rivers empty into the ocean. The tributaries to Martian channels are short and stubby, and many Martian channels have their headwaters in a highly fractured region or in a jumbled mass of angular blocks called chaotic terrain. Most of the chaotic terrain is located in the southern highlands near the boundary with the northern low plains.

At times in the past, great volumes of water were released resulting in catastrophic flooding. The closest thing on the earth comparable to this type of flooding is the Channeled Scablands of Washington, which were flooded when ice dams failed during the ice age and great sheets of water spread over what is now central Washington.

In summary, it is apparent that the surface of Mars has been modified by flowing liquid (presumably water) and that, at some earlier stage, fluvial processes operated on Mars. The change from ancient fluvial periods to the present dry period may be similar to the change from interglacial to glacial periods on the earth.

Mass Movement.   The Viking orbiter photography with its higher resolution and greater detail provided dramatic evidence that mass movement is an important process in the evolution of the landscape of Mars and is especially important in the enlargement of the canyons.

Figure 24.39 Mass movement along the walls of Valles Marineris. Multiple massive landslides and debris flows can be seen on both walls. The layers visible in the canyon walls probably represent alternating strata of basalt, ash, and windblown deposits.

Canyon rim

Older debris flow

Debris flow

Debris flows

Canyon rim

Figure 24.40 *Eolian features associated with craters in the Hesperia region (23°S 242°W). The streaks are believed to be fine, bright dust, which was transported into the craters in the waning stages of a duststorm and then was blown out by high-velocity winds having a constant direction.*

The oblique view in *figure 24.39* shows a section of the Valles Marineris Canyon and the important role that mass movement has played in slope retreat and enlargement of the canyon. In this photograph, the canyon is approximately 2 km deep and multiple massive landslides and debris flows are visible on both walls. Several resistant rock layers form a steep cliff at the top of the plateau and appear to break up in a series of large slump blocks, while the lower units are nonresistant and form a slope that appears to have moved as a debris flow. Some of the flows moved more than halfway across the canyon floor, at least 30 km from their point of origin. They have a lobate pattern similar to debris flows on the earth, and flow lines show direction of movement. These features show significant mobility and suggest the presence of a fluid (probably water) mixed with the finer debris. The low hills and knobs out on the flat canyon floor may be remnants of the more resistant rimrock that has slumped out into the canyon. The bright streaks behind the low hills are the result of wind shadows behind an obstruction and indicate that sediment is actively transported by wind after it slumps from the valley walls. The features shown in this photograph imply that the canyon is enlarged by mass movement and that the material subsequently is removed by the wind. Perhaps this is why we see so little evidence of stream deposits such as deltas and alluvial fans on Mars.

Eolian Features.   Wind is known to be a dominant surface process on Mars, and the scale of activity is much larger than in the great desert regions on the earth. When Mariner 9 entered orbit around Mars in 1971, a raging planetwide duststorm completely obscured the entire planet. It had begun two months before and was observed from earth-based telescopes, and it continued

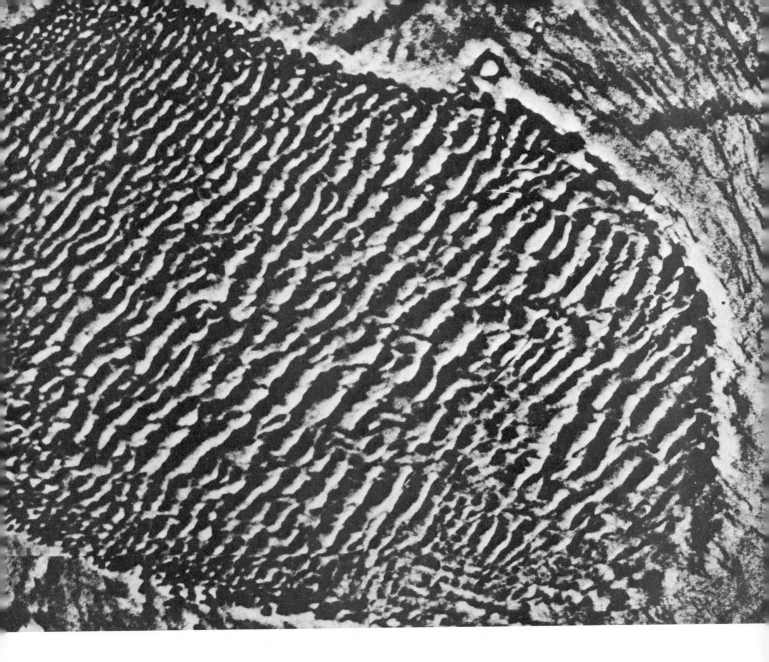

Figure 24.41 *Dune field in the Hellespontus region (near 75°S 331°W). The wavelike structure is highly suggestive of transverse sand dunes found on the earth. Note the tendency for the size of the dunes to decrease toward the margins of the field.*

for several months after the spacecraft was in orbit. The magnitude of a global storm on Mars is hard to imagine. Winds comparable to a strong hurricane on the earth raged for several months with gusts up to 300 mph. The storm blew dust high into the atmosphere so that only the tops of the highest volcanoes could be seen as dark spots rising above the global cloud of dust. When the storm cleared, a variety of features resulting from wind activity were photographed.

Dark streaks formed behind craters that created wind shadows occurred over broad areas (*figure 24.40*). Large dune fields were found in large craters and in the north polar regions and were remarkably similar to the dune fields on the earth (*figure 24.41*). In addition, surface photographs taken by the Viking Lander show an abundance of wind-generated features such as small dunes, wind-shaped pebbles, lag gravels, and sand streaks behind boulders (*figure 24.42*).

Polar Regions. The polar areas of Mars are of special interest, because they are the centers of contemporaneous geologic activity. The ice caps are believed to be very thin, probably only about 30 m thick, and, contrary to previous hypotheses, it now appears that the Martian ice caps are water ice, not carbon dioxide (CO_2).

Of particular interest are unique layered deposits that underlie the ice cap and are well exposed along its margins. These deposits are called laminated terrain and consist of alternating light and dark layers 20 to 30 m thick (*figure 24.43*). As many as 50 beds have been observed in a single exposure. These deposits are believed to be alternating layers of ice and windblown dust and sand formed during glacial and interglacial periods, respectively. At the present time, a great dune field of dark sand surrounds the north polar cap and will be covered by the next expansion of the ice.

(A)

(B)

Figure 24.42 *Views of the Martian surface. (A) The large boulder in the left foreground is approximately 2 m across. Many of the features in this photograph indicate the importance of wind action as an active surface process. The sand dunes show sediment transport into the area, and exposures of their internal cross-stratification indicate a certain amount of deflation. The gravel surface is probably a lag gravel produced as wind activity removed the finer sand and silt-size particles. Note the wind shadows behind most of the large boulders. (B) Lag gravels produced as wind removes fine-grained sand, silt, and dust.*

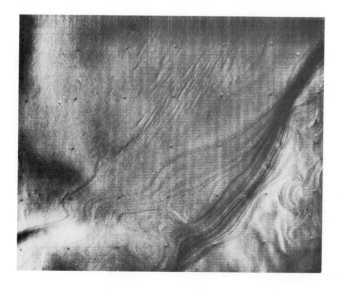

Figure 24.43 *Laminated terrain near the south pole of Mars. The oval tableland in the right half of the photograph is an eroded remnant of a series of layers of light and dark sediments that may represent layers containing dust or volcanic ash alternating (possibly) with carbon dioxide ice and water ice. The area covers about 57 by 60 km.*

The Moons of Mars. A number of photographs have been made of Phobos and Deimos, the two little moons of Mars. The photographs provide the first information concerning the moons' size, shape, and surface morphology. Both satellites appear to be irregular fragments of once larger bodies, and both are severely battered by repeated impact. On the basis of crater population (both appear to be nearly saturated), Phobos and Deimos appear to be at least 1.5 billion years old and may date back to the birth of the solar system some 4.5 billion years ago.

Typical close-up photographs of the Martian moons are reproduced in *figure 24.44*. Both are potato shaped and show a profusion of rimmed craters that range in size from about 5.3 km down to the limits of resolution.

The origin of the moons of Mars may be explained by one of two hypotheses: (1) they represent material left over from the period of intensive bombardment, or (2) they were captured, probably from the asteroid belt. Whatever the origin of the moons, the close-up photographs of Phobos and Deimos provided by Mariner 9 and the Viking missions give us our best views of the type of bodies that surely resemble asteriods in many details and that populate the asteroid belt in uncountable numbers.

Figure 24.44 The moons of Mars. The Martian moons are small and potato shaped and marked with numerous impact craters.

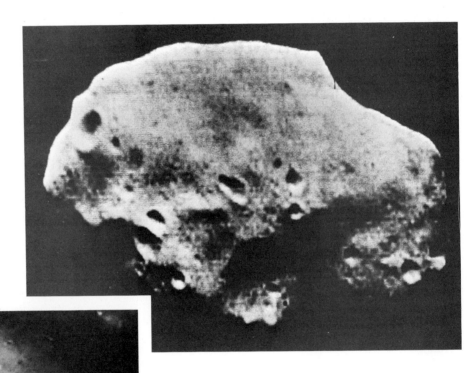

(A) Phobos, the larger of the two moons, is approximately 25 km long and 21 km wide.

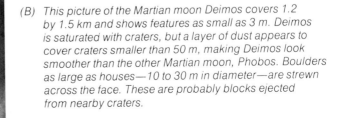

(B) This picture of the Martian moon Deimos covers 1.2 by 1.5 km and shows features as small as 3 m. Deimos is saturated with craters, but a layer of dust appears to cover craters smaller than 50 m, making Deimos look smoother than the other Martian moon, Phobos. Boulders as large as houses—10 to 30 m in diameter—are strewn across the face. These are probably blocks ejected from nearby craters.

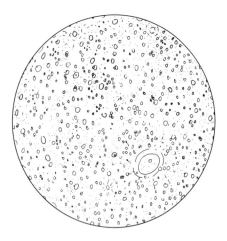

(A) Stage I: Period of intense bombardment.

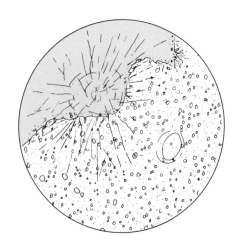

(B) Stage II: Uplift of Tharsis area. Development of a radial fracture system. Period of stream erosion.

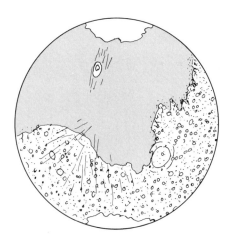

(C) Stage III: Widespread volcanism. Formation of plains in northern regions.

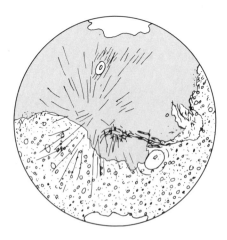

(D) Stage IV: Recurrent uplift in Tharsis region. Development of Valles Marineris and chaotic region.

(E) Stage V: Volcanism in Tharsis area and development of large shield volcanoes.

Figure 24.45 *Schematic maps showing the geologic evaluation of Mars.*

The Geologic History of Mars. The relative ages of the major rock bodies and terrain types on Mars can be determined by using the principles of superposition, crater densities, and degree of erosion in much the same way that they have been used in the study of the moon and Mercury. Although there are many uncertainties about the details, an outline of the major events in the geologic history of Mars may be constructed more or less as follows.

Stage I. Mars, like the other terrestrial planets, is believed to have formed by accretion of a myriad of smaller bodies over a relatively short period of time, perhaps only a few hundred thousand years. Planetary differentiation occurred shortly after the accretionary phase and formed a crust, mantle, and core. Following its formation as a planet, Mars was subjected to a period of intense bombardment. This produced a densely cratered surface like the one illustrated in *figure 24.45A*, which included several multiringed basins. The moon, Mercury, and, presumably, the earth experienced similar events in their early histories.

Stage II (figure 24.45B). Regional uplift in the Tharsis area with accompanying radial faulting occurred, presumably from plumelike convection. Much of the crust in the Northern Hemisphere was fractured at this time, possibly as a result of it being thinner than the crust in the Southern Hemisphere.

It is believed that Mars developed a dense atmosphere early in its history. Rainfall and flooding resulted in significant erosion at this time, with regional runoff to the northern lowlands. Channels were carved along the escarpment and locally in the cratered highlands. As surface temperature dropped, water became locked up in ground ice. Local melting, possibly as a result of volcanic activity, released its ground water and formed the present chaotic terrain. The period of great erosion probably occurred just before and during Stage III, the emplacement of the plains material in the Northern Hemisphere.

Stage III (figure 24.45C). Widespread volcanic activity occurred in the Northern Hemisphere and formed the great, low, sparsely cratered plains. Similar eruptions formed the lunar maria and plains of Mercury at this time, but on Mars volcanic activity has continued intermittently to the present.

Stage IV (figure 24.45D). Recurrent uplift in the Tharsis region developed additional radial faults. Valles Marineris developed along a structural trough of this system and was enlarged progressively by the release of subsurface water, slumping, and transport of sediment downslope to the northern lowlands. The chaotic terrain enlarged, and the escarpment continued to retreat southward.

Stage VI (figure 24.45E). Volcanic extrusions partly covered the Tharsis area with fresh lava, and the great shield volcanoes formed at this time. Eolian activity continued to modify the entire surface, with the polar regions forming the laminated terrain and surrounding dune fields. Eolian activity is presently the dominant active geologic process, although mass movement continues to cause slope retreat along the escarpment and along the canyon walls.

Geologic Contrasts of the Terrestrial Planets

Statement

The space probes to the moon, Mercury, and Mars provided an unprecedented store of scientific data about the origin and evolution of the terrestrial planets. The overwhelming impression gained from studying this data is that the earth, the moon, Mercury, and Mars constitute a family of planets in which the size and distance from the sun determine many characteristics of each body. All of the terrestrial planets appear to have originated at the same time and in much the same way and, as a result, have followed roughly the same evolutionary development during the first 1 to 2 billion years. Later development depended upon the development of an atmosphere (a hydrologic system) and an active tectonic system.

Discussion

Size and Density. One of the obvious and significant differences between the earth, the moon, Mars, and Mercury is the size and density of the four spheres (*figure 24.46*). The earth is the largest, with a diameter of 12,682 km and a mean density of 5.5 g/cm^3. Mars is slightly more than one-half the size of the earth, with a diameter of 6800 km and a density of 4.0 g/cm^3. Mercury is a small planet, having a diameter of 4848 km, less than half that of the earth, but its mean density is 5.5 g/cm^3. The moon is smaller than Mercury, with a diameter of 3456 km, but its mean density is only 3.3 g/cm^3.

These physical statistics at first may seem unimportant, but they are among the most significant facts underlying the geologic contrasts among the four planets. Most geologic differences among these planets can be explained by the differences in their size and density, which in turn would govern their internal heat, atmosphere, and type and degree of tectonic activity.

Internal Structure. Our knowledge about the internal structures of the terrestrial planets is based largely upon their densities, but other clues are supplied by the characteristics of surface features and magnetic fields. Seismic evidence has revealed much about the interior of the earth and the moon, but the structure of Mars and Mercury is less certain. *Figure 24.46* summarizes our present concepts about the internal structure of terrestrial planets. All are differentiated to some degree, with denser material being concentrated in the central part or core and the lighter elements in the mantle and crust.

The moon has a crust of igneous rock, mainly anorthosites and gabbro ranging in thickness from 60 to 100 km. The mantle, which is a rigid lithosphere, has a thickness of about 1000 km. The core is 1400 km in diameter and is a plastic asthenosphere consisting more or less of melted rock.

The high density of Mercury clearly indicates that it has a large core of dense material, probably iron. Such a core might contain 80% of the planet's mass and be 1800 km in diameter. The overlying mantle might be 640 km thick and probably is rigid because of the undeformed surface features.

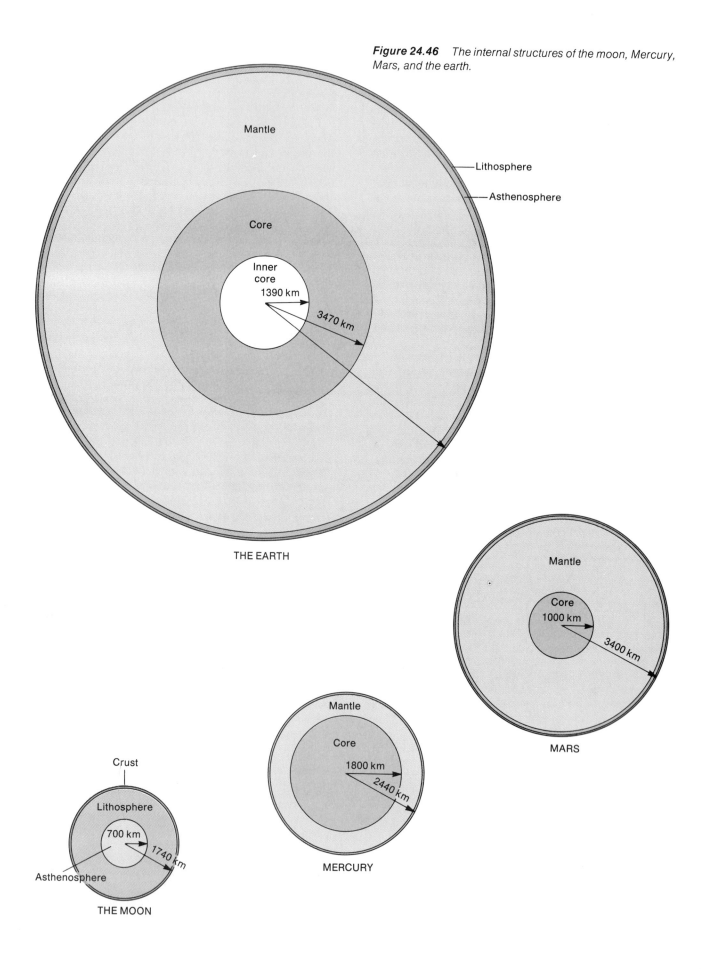

Figure 24.46 *The internal structures of the moon, Mercury, Mars, and the earth.*

The exact internal structure of Mars is less certain. It is differentiated into crust, mantle, and core similar to the other planets, but we have very little knowledge about the thickness of the structural units. The density of Mars (4.0 g/cm^3) suggests a core, relatively smaller than that of Mercury, and the uncompressed surface features indicate a relatively thick lithosphere.

The internal structure of the earth is, of course, best known and was discussed in chapter 1. The major structural units are a central core composed predominantly of iron and nickel; a thick surrounding mantle, composed of silicate minerals rich in iron and magnesium; a plastic, mobile asthenosphere; and a thin, rigid lithosphere, plus a cover of surface fluids of water and air.

Craters. One of the most significant results of the exploration of the planets has been the discovery that craters like those on the moon occur on all bodies in the inner solar system. Cratering is, therefore, an important key to understanding planetary evolution.

It is clear that the moon, Mercury, and Mars all recorded similar events in their early histories. All experienced an early period of intense bombardment during the first half-billion years of their histories, but the rate of impact declined rapidly and has remained very low for the last 3 billion years. On all three planets, large, asteroid-size bodies (probably at least 100 km in diameter) produced multiringed basins near the end of the period of intense bombardment.

The history of cratering on the earth is largely erased by tectonic and erosional processes, but the shields and platforms, which have been stable during the last 2 billion years, show a rate of impact during that time similar to that of the other planets. Therefore, it is quite likely that the earth, too, experienced an early impact history similar to that of the moon, Mercury, and Mars.

Atmospheric Differences. The earth's atmosphere and hydrosphere are two of its unique characteristics. These surface fluids have been in motion since the very early periods of the earth's history, and, as a consequence of its steady activity over such a long period of time, the earth has evolved through a series of stages quite unlike the other planets. The remarkable thing about terrestrial geologic processes is the gradual, but profound, changes that they effect. The entire surface of the earth appears to remain essentially constant from day to day and from year to year, yet changes occur in almost infinitely small increments and, in time, completely alter the landscape. Rocks at the earth's surface react with the elements in the atmosphere and are decomposed. Water transports the weathered debris to the ocean, where it is deposited to form, ultimately, parts of a new crust. The landscape evolves through a predictable series of stages, and high mountain ranges ultimately are worn down and can be covered with a blanket of younger sediment. This pattern of gradual change has resulted in a remarkably detailed record of the earth's history, in which segments of each interval of relatively recent geologic time are preserved. In contrast, the oldest features of the earth appear to have been destroyed completely.

The moon and Mercury do not have an atmosphere, so there is no change resulting from a hydrologic system. Their surface features are modified ever so slowly by churning action and fragmentation from impact and bombardment from cosmic rays. Mars with its thin atmosphere and some water has, at times, been modified by a hydrologic system, but the system has been local and sporadic and has operated much differently than the one on the earth. The water does not move in a continuous cycle but is locked up for long periods of time in ground ice and glaciers, to be released only under special conditions.

Why does the earth have a unique atmosphere? There are two major reasons, both related to its size. The gas in the atmosphere is believed to have been formed from thermal activity of the planet during which various volatile elements escaped to the surface during volcanic activity. If the planet's gravity is strong enough, the gases, (carbon dioxide, methane, water, et cetera) remain at the surface, while the lighter elements—hydrogen and helium—escape into space. The earth has had a long, continuous volcanic history and is large enough to retain the expelled gases to form the oceans and the atmosphere. The moon and Mercury have not been volcanically active and are much smaller, so what little atmospheric gas was produced during early periods of differentiation has escaped the gravitational field. Mars, being larger and more active volcanically, has a tenuous atmosphere that may have been denser during earlier periods of its history.

Tectonic Activity. The differences between the surface features of the earth, the moon, Mars, and Mercury go beyond the erosional and depositional features produced by running water and wind. If the earth had no atmosphere and were free from the effects of stream erosion, it would still not resemble these other planets. The earth's crust is extremely mobile and shows the effects of deformation by compressive forces. The most striking deformational features are the great folded mountain ranges in which crustal shortening on the order of 120 to 160 km can be demonstrated. Perhaps the best record of horizontal stresses is found in the shielded areas of the continents. There, all of the rocks have suffered extreme deformation, as is evident in the small-scale contortions in the metamorphic rocks (such as foliation in gneiss and schist). The great bulk of metamorphic rock has nearly vertical foliation, a structure produced only by horizontal stresses.

The crust of the seafloor is very young, and new ocean basins, such as the Red Sea, are just being born. Others have undoubtedly come and gone. The earth's crust has been mobile throughout its history and continually has been shaped and remolded by tectonic processes. The asthenosphere is in motion, giving rise to seafloor spreading, folded mountain ranges, metamorphic rocks, oceanic trenches, and a worldwide rift system.

Evidence of similar compression on the crust of other planets is completely absent. There are no linear mountain ranges and no suggestion of thrust faults and folds. The only deformation has been the impacts of meteorites, local wrinkles, and normal faulting along the basin margins. Considerable stability of the crust of the moon, Mars, and Mercury is also clearly indicated by the circular shape of craters. Regardless of their origin, their circular form indicates that they were formed by vertical forces. Therefore, the circular craters provide an excellent reference system that evidently has been fixed throughout time, not deformed by compression. The preservation of these structures shows that the crusts of the other inner planets have never been subjected to compressive forces similar to those that formed the folded mountains on the earth. Compressional structures such as folds, thrust faults, and strike-slip faults, if present, should be clearly expressed on these planets, because they would not be masked by sedimentary cover, water, or vegetation.

Volcanism. Although there are many things about the nature of terrestrial volcanism that are not completely understood, the concept of convection within the mantle can explain most aspects of volcanic activity on the earth. Volcanism in linear zones is related to mountain-building, island arcs, and oceanic trenches and can be explained by plate tectonics.

The rock record makes it abundantly clear that volcanism has occurred throughout geologic time. Some of the greatest outpourings of lava are the plateau basalts, now situated near the margins of continents, which appear to be related to the initial phases of continental drift. These flood basalts have occurred from Precambrian time through the present. Yet, these examples of volcanic activity are relatively minor when they are compared to the great volumes of lava extruded along the global rift zone that forms the oceanic crust. All of the present ocean basins have been formed during the last 200 million years, and there is good reason to believe that similar vigorous volcanic activity has occurred continually throughout most of the earth's history.

Volcanic activity on the moon and Mercury has been quite different. Both planets experienced an early thermal event during which large floods of lava were extruded over a period of possibly a billion years. During the last 3.2 billion years, there has been no clear evidence of volcanic activity at all. The exterior of Mercury resembles the moon not just in types of landforms but more surprisingly in surface history.

Volcanism on Mars might be considered as intermediate between that of the earth and the moon and Mercury. An early period of extensive floods of lava followed the period of intense bombardment and formed the sparsely cratered plains of the Northern Hemisphere. Periods of more recent volcanic activity also have formed the huge shield volcanoes, some of which may still be in a period of activity. Even so, the volume of volcanic material extruded on Mars is trivial compared to that extruded on the earth.

Similarities and Differences in Planetary Development. The theory of the origin of the planets favored by most people who have studied the subject visualizes the gradual condensation of small mineral grains from a nebular disk that is thought to have surrounded the infant sun. This material then was gathered together by gravitation, and the planets were formed. We can infer that the terrestrial planets (Mercury, Venus, the earth, the moon, and Mars) originated at the same time and in much the same way and as a result followed roughly the same plan of development during the first 1 to 2 billion years of their histories. Their early evolution involved three major

events. The first event was an early period of differentiation. As each planet grew larger by sweeping up material in its orbital path, it began to heat up as a result of gravitational infall of mass, impact and meteorites, and radiogenic heat from decay of uranium, thorium, and potassium. This resulted in a vast reorganization of the material in the entire planet. The heavier elements such as iron migrated toward the center of the planet, and the lighter elements such as silicon and aluminum floated toward the surface. The differentiation of the planets produced a zonation of material according to density into a core and mantle.

The second major event was a period of intense bombardment that occurred during the first half-billion years of each planet's history. This resulted in a surface saturated with craters, but, as the material in the planets' orbits was swept up, the rate of cratering rapidly declined. Then, large, multiringed basins were formed, possibly from a different population of meteorites of asteroid size.

During the third period of planetary development, vast floods of lava erupted and solidified in basins and lowlands.

These three early phases of planetary development are known to have occurred on the moon, Mercury, and Mars and probably on the earth as well, although the record of such events has been erased on the earth by erosion and tectonic activity.

After this sequence of events, the planets appear to have followed different paths in their development. Why? At least part of the answer may be found in the size and internal dynamics of the planets.

The moon and Mercury are both relatively small and appear to have a thick, rigid lithosphere. Forces generated in their interior would be unable to develop subsequent tectonic and volcanic activity. Any atmosphere generated by early volcanism would escape the small gravitational pull. The planet then would remain essentially unaltered.

The earth, being much larger, has a greater source of internal heat and has developed an active tectonic system that continues to operate. The size and density of the earth and its position in the solar system also may account for the atmosphere and the abundance of water. The gases leaked to the surface during the long volcanic history of the earth could not escape into space but were held in by the strong gravitational pull and formed the atmosphere and the oceans. If the earth had been closer to the sun and the surface temperatures had been higher, a greenhouse effect would have resulted, producing a dense atmosphere and high surface temperatures. This seems to be the case on Venus. If the earth were farther from the sun, the water reservoirs would be permanently frozen.

As it turns out, heat from the sun is just sufficient to provide the right amount of energy for the hydrologic system so that the water on the earth's surface is continually washing and eroding the surface of the continents at such a pace that the folded mountains are worn down almost as quickly as they are formed.

Mars is approximately twice the size of the moon but only half the size of the earth, so it is not surprising that Mars is a planet intermediate between the moon and the earth in its evolutionary sequence. It is large enough to develop a greater degree of differentiation than the moon and has, therefore, developed a more advanced style of tectonism and volcanism and a thin atmosphere. However, it has not developed a convecting mantle capable of forming folded mountains and continental masses as the earth has. All features of Mars indicate that it is a planet that has partly made the transition from a primitive, impact-dominated body (like the moon) to a tectonically mobile and volcanically active planet (like the earth).

When one compares the earth, the moon, Mercury, and Mars, it becomes obvious that the earth is different in two essential ways: it has a higher degree of internal energy, which results in an active tectonic system; and it has an abundance of surface fluids, which results in an active hydrologic system. These two features make the earth unique and inhabitable.

Summary

A vast amount of new knowledge has been obtained about the inner planets of the solar system as a result of the space program. This new knowledge has permitted geologists for the first time to study details of other planets and compare them with the earth. Some of the more important factual discoveries and some hypotheses that explain them are as follows.

Craters. Cratering has been the dominant geologic process on the moon, Mercury, and Mars and may have been the dominant process on the earth during the first half-billion years of its history. Crater populations on the moon, Mercury, and Mars are similar and represent similar sequences of events.

Volcanism. All three planets experienced a thermal event early in their histories after the period of intense bombardment. This event produced the maria on the moon and the plains regions of Mercury and Mars. Mars experienced more recent volcanic activity, as is indicated by its great shield volcanoes.

Tectonic Features. The circular craters on the moon, Mercury, and Mars indicate that the crust of these planets has not been deformed by strong compressive forces. Mercury has a global pattern of thrust faults, and Mars has upwarps and graben systems, but none of the planets have developed an active tectonic system like that operating on the earth.

Atmosphere. The moon and Mercury are primitive bodies, and their surfaces have not experienced changes from atmospheric processes. Mars, in contrast, has a thin atmosphere, and its surface has been modified by erosion and deposition by streams, mass movement, wind, ice caps, and ground ice.

Internal Structure. To various degrees, all of the inner planets appear to have experienced differentiation, but each has a unique, layered internal structure.

Geologic History. A geologic time scale for events on the moon has been established through the use of the principles of superposition and crosscutting relations; these provide a framework of relative time into which the major events of lunar history can be arranged in chronologic order. A geologic time scale has been developed for Mercury and Mars, and a tentative interplanetary correlation scheme has been established on the basis of crater populations. The early histories of the moon and Mercury are similar in that early intense bombardment developed highly cratered terrain, followed by development of large, multiringed basins and floods of basalt and a subsequent decline in cratering. Mars has had a more eventful geologic history involving (1) formation of densely cratered terrain, (2) uplift in the Tharsis area with radial fracturing, (3) widespread extrusion of lava to form the northern plains, (4) renewed uplift of the Tharsis region, (5) formation of great shield volcanoes and development of a rift valley, and (6) modification by surface erosion and deposition.

The moon and Mercury are primitive bodies, and their surfaces have not experienced change from atmospheric erosion or continuing convection in the planets' interiors.

Mars appears to represent an intermediate phase in planetary development that is transitional between an impact-dominated body (like the moon and Mercury) and a tectonically active, water-dominated planet (like the earth).

Additional Readings

Allen, J. P. 1972. "Apollo 15: Scientific Journey to Hadley-Apenine." Amer. Scientist 60:162-74.

Bowker, D. C., and J. K. Hughes. 1971. Lunar Orbiter Photographic Atlas of the Moon. Washington, D.C.: National Aeronautics and Space Administration.

Cross, C. A., and P. Moore. 1977. The Atlas of Mercury. New York: Crown Publishers.

Hartman, W. K. 1977. "Cratering in the Solar System." Sci. Amer. 236 (1): 84-99.

Hinners, N. W. 1971. "The New Moon: A View." Reviews of Geophysics and Space Physics 9:447-522.

Journal of Geophysical Research 78 (20):4009-439, July 10, 1973. (Entire issue devoted to the geology of Mars.)

Journal of Geophysical Research 80 (17):2342-514, June 10, 1975. (Entire issue is devoted to the geology of Mercury.)

Lowman, P. D. 1972. "The Geologic Evolution of the Moon." J. Geol. (80):125-66.

MacGauley, J. F., et al. 1972. "Preliminary Mariner 9 Report on the Geology of Mars." Icarus 17(2):289-327.

Mason, B. 1971. "The Lunar Rocks." Sci. Amer. 255(5):48-58. (A popular review of the first several groups of lunar samples.)

Mutch, T. A., et al. 1976. Geology of Mars. Princeton, N.J.: Princeton University Press.

Mutch, T. A. 1972. Geology of the Moon. Princeton, N.J.: Princeton University Press.

NASA. 1974. "The New Mars. The Discoveries of Mariner 9." NASA SP337.

NASA. 1976. "Mars as Viewed by Mariner 9." NASA SP329.

NASA. 1976. "Viking 1, Early Reports." NASA SP408.

Science 194 (4221), December 1977. (A large part of this issue is devoted to the Viking mission.)

Scientific American 233 (3):1-202, September 1975. (The entire issue is devoted to our new understanding of the solar system.)

Shoemaker, E. M. 1964. "The Geology of the Moon." Sci. Amer. 221(6):38-47.

Short, N. M. 1975. Planetary Geology. Englewood Cliffs, N.J.: Prentice-Hall.

Glossary

Aa flow. A lava flow whose surface is typified by angular, jagged blocks (see diagram).

Ablation. Reduction of a glacier by melting, evaporating, iceberg calving, or deflation.

Abrasion. The mechanical wearing away of a rock by friction, rubbing, scraping, or grinding.

Absolute time. Geologic time measured in a specific duration of years as contrasted to relative time, which signifies only the chronologic order of events.

Abyssal. Of, or pertaining to, the great depths of the ocean, generally below 1000 fathoms (2000 m).

Abyssal floor. The deep, relatively flat surface of the ocean floor located on both sides of the oceanic ridge. Includes the abyssal plains and the abyssal hills.

Abyssal hills. That part of the ocean floor consisting of hills rising to 1000 m above the surrounding floor. Abyssal hills are found seaward of most abyssal plains and occur in profusion in basins isolated from continents by ridges, rises, or trenches.

Abyssal plains. Flat areas of the ocean floor that slope less than 1:1000. Most abyssal plains lie at the base of the continental rise and are simply areas where abyssal hills are completely covered with sediment (see diagram).

Active margin (plate tectonics). The leading edge of a plate adjacent to the trench.

Aftershock. An earthquake that follows a larger earthquake. After a major earthquake, there are generally many aftershocks, which can be felt for many days or even months.

Agate. A variety of cryptocrystalline quartz in which colors occur in bands. Agate is commonly deposited in cavities.

Aggradation. The process of building up a surface by deposition of sediment.

A horizon. The topsoil layer in a soil profile.

Alcove. A large niche or recession formed in a precipitous cliff.

Alluvial fan. A fan-shaped deposit of sediment built by a stream where it emerges from an upland or a mountain range into a broad valley or a plain. Fans are typical of arid to semiarid climates but are not restricted to them (see diagram).

Alluvium. A general term for all sedimentary accumulations that occur as the result of the comparatively recent action of rivers. Alluvium thus includes sediment laid down in riverbeds, flood plains, and alluvial fans.

Alpine glacier. A glacier occupying a valley. Also called *mountain glaciers* and *valley glaciers*.

Amorphous solid. A solid in which atoms or ions are not arranged in a definite crystal structure. Examples: glass, amber, obsidian.

Amphibole. An important rock-forming mineral group of ferromagnesian silicates with a double chain of silicon-oxygen tetrahedra. Common example: hornblende.

Andesite. A fine-grained igneous rock composed mostly of plagioclase feldspar and 25% to 40% amphibole and biotite. It has no quartz or K-feldspar. It is abundant in the mountains around the borders of the Pacific Ocean, such as the Andes Mountains of South America, the area from which the name was derived. Andesitic magma is believed to originate from the fractionation of partially melted basalt.

Andesite line. The boundary in the Pacific Ocean separating volcanoes of the inner

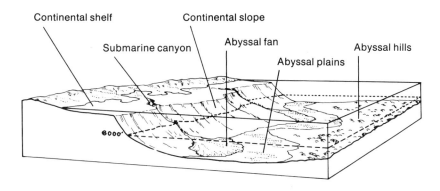

Continental shelf Continental slope

Submarine canyon Abyssal fan Abyssal hills

Abyssal plains

6000'

Pacific Ocean, which discharge only basalts, from those near the continental margins, which discharge both andesite and basalt.

Angular unconformity. An unconformity in which older strata dip at a different angle (generally steeper) than the younger strata (see diagram).

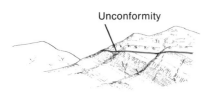

Unconformity

Anomaly. A deviation from the norm or the average.

Anorthosite. A coarse-grained igneous rock composed almost entirely of plagioclase.

Anticline. A fold in which the limbs dip away from the axis. When eroded, the oldest rocks are exposed in the central core (see diagram).

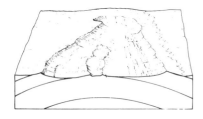

Aphanitic texture. A rock texture in which individual crystals are so small they cannot be identified without the aid of a microscope. The rock in a hand specimen appears to be dense and structureless.

Aquifer. A permeable stratum or zone below the earth's surface through which ground water moves (see diagram).

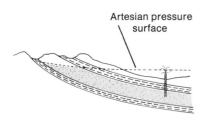

Artesian pressure surface

Arc-trench gap. The geographic area in an island arc and deep-sea trench system that separates the arc of volcanoes from the trench. In most cases the gap is 100 km or so wide.

Arête. A narrow, sharp ridge separating two adjacent glacial valleys.

Arkose. A sandstone containing at least 25% feldspar.

Artesian basin. A geologic structural feature in which ground water is confined and is under artesian pressure.

Artesian-pressure surface. The surface or level to which water in an artesian system would rise in a pipe high enough to stop the flow.

Artesian water. Ground water confined to an aquifer and under sufficient pressure so as to rise above the top of the aquifer when tapped by a well.

Ash. Volcanic fragments the size of dust.

Ash flow. A turbulent blend of unsorted pyroclastic material (mostly fine grained) mixed with high-temperature gas ejected explosively from fissures or craters.

Asteroid. A relatively small celestial body believed to be either the remains of an exploded planet or matter that never completed the planet-forming process.

Asthenosphere. A zone in the earth directly below the lithosphere 50 to 200 km below the surface. Seismic velocities are distinctly lower in the asthenosphere, suggesting that the material is soft and yields to plastic flow.

Astrogeology. The application of geologic methods and knowledge to the study of extraterrestrial bodies.

Asymmetric fold. A fold (anticline or syncline) in which one limb dips more steeply than the other (see diagram).

Atmosphere. The mixture of gases that surrounds a planet. The earth's atmosphere is chiefly oxygen and nitrogen, with minor amounts of other gases. Synonymous with *air*.

Atoll. A ring of low coral islands surrounding a central lagoon (see diagram).

Attitude. The three-dimensional orientation of a bed, fault, dike, or other geologic structure. Determined by the combined measurements of dip and strike.

Axial plane. As applied to folds, an im-

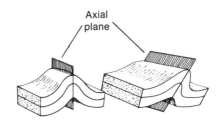

Axial plane

aginary plane through a fold that intersects the crest or trough in such a manner as to divide the folds as symmetrically as possible (see diagram).

Axis. 1. In crystallography, an imaginary line passing through a crystal, around which the parts are symmetrically arranged. 2. In folds, a line where folded beds show maximum curvature; also defined as a line formed by the intersection of the axial plane with the bedding surface (see diagram).

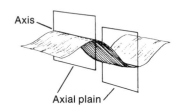

Axis

Axial plain

Backswamp. Marshy area of a flood plain at some distance beyond and lower than the natural levees that confine the river.

Backwash. The return sheet flow down a beach after the wave is spent.

Badlands. An area nearly devoid of vegetation and dissected by stream erosion into an intricate system of closely spaced, narrow ravines.

Bajada. The surface of a system of coalesced alluvial fans.

Bar. An offshore, submerged, elongate ridge of sand or gravel built on the seafloor by waves and currents.

Barchan dune. A crescent-shaped dune, the tips or horns of which point downwind. Formed in desert areas where sand is scarce.

Barrier island. An elongate island of sand or gravel built parallel to a coast (see diagram).

Barrier island

Barrier reef. An elongate coral reef that trends parallel to the shore of islands or continents, separated from them by lagoons (see diagram).

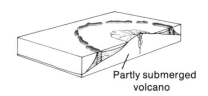

Partly submerged volcano

Basalt. A dark-colored, aphanitic (fine-grained) igneous rock composed of plagioclase (over 50%) and pyroxene. Olivine may or may not be present. Basalt and andesites represent 98% of all volcanic rocks.

Base level. The level below which a stream cannot effectively erode. Sea level is the ultimate base level, but lakes form temporary base levels for drainage systems inland.

Basement complex. A series of igneous and metamorphic rocks lying beneath the oldest stratified rocks of a region. The shield is the area where the basement complex is exposed over large areas (see diagram).

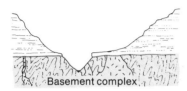

Basin. 1. Structural—a circular or elliptical downwarp with younger beds in the center after erosion exposes the structure. 2. Topographic—a depression into which the surrounding area drains.

Batholith. A large body of intrusive igneous rock at least 100 km² in areal extent (see diagram).

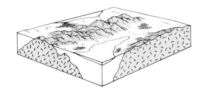

Bathymetric chart. A topographic map of the bottom of a body of water (such as the seafloor).

Bathymetry. The measurement of ocean depths and the charting of the topography of the ocean floor.

Bauxite. A mixture of various amorphous or crystalline hydrous aluminum oxides and aluminum hydroxides commonly formed as a residual clay deposit in tropical or subtropical regions. Bauxite is the principal commercial source of aluminum.

Bay (coast). A wide, curving recess or inlet between two capes or headlands.

Beach. A deposit of wave-washed sediment along the coast between the landward limit of wave action and the outermost breakers.

Beach drift. The migration of sediment along the beach caused by waves striking the shore at an oblique angle.

Bed. A layer of sediment 1 cm or more in thickness.

Bedding plane. A surface separating layers of sedimentary rock.

Bed load. Material transported along the bottom of a stream or river by rolling or sliding in contrast to material carried in suspension or solution.

Bedrock. The continuous solid rock that underlies the regolith everywhere and is exposed locally at the surface. An exposure of bedrock is called an outcrop.

Benioff zone. A zone of deep earthquakes that dips beneath the continents or island arcs from the deep-sea trenches.

Berm. A nearly horizontal portion of the beach or backshore formed by storm waves. Some beaches have no berms; others have several.

B horizon. The soil zone of accumulation where some of the material derived from leaching in the overlying A horizon is deposited.

Biologic material. A general term for material originating from organisms. Examples: fossils (shells, bones, leaves), peat, coal, et cetera.

Biosphere. A term that signifies the totality of living things on the earth's surface.

Biotite. "Black mica." A rock-forming ferromagnesian silicate with the tetrahedra arranged in sheets.

Birdfoot delta. A delta with distributaries extending seaward and resembling in plan the outstretched claws of a bird. The Mississippi River delta is an excellent example.

Block faulting. A type of normal faulting in which the crust is divided into fault blocks of different elevations and orientations.

Blowout. A basin excavated by the wind.

Boulder size. Fragments larger than 256 mm in diameter (approximately the size of a volleyball). Next size larger than a cobble.

Bracketed intrusion. An intrusive rock that has been exposed at the surface by erosion and subsequently covered by younger sediment. The relative age of the intrusive thus falls between (is bracketed by) the ages of the younger and the older sedimentary deposits.

Braided stream. A complex of converging and diverging stream channels separated by bars or islands. Forms where more sediment is available than can be removed by the discharge of the stream.

Breaker. A collapsing water wave.

Breccia. A rock composed of cemented angular fragments.

Butte. A somewhat isolated hill usually

capped with a resistant layer of rock and bordered by talus. Represents an erosional remnant of a formerly more extensive slope (see diagram).

Calcite. A mineral composed of calcium carbonate (CaCO₃).

Caldera. A large, more or less circular depression or basin having a diameter many times greater than that of the included volcanic vents. Calderas are believed to result from subsidence or collapse and may or may not be related to explosive eruptions.

Calving. The process by which large blocks of ice break off from glaciers that terminate in a body of water.

Capacity. The potential quantity of sediment a stream, a glacier, or the wind can carry under a given set of conditions.

Capillary. A small, tubular opening with a diameter about the size of a human hair.

Capillary action. The action by which a fluid (such as water) is drawn up into small openings (such as pore spaces in rocks) as a result of surface tension.

Capillary fringe. A zone above the water table where water is lifted by surface tension into openings of capillary size.

Carbon-14. Radioisotope of carbon that has a half-life of 5730 years.

Carbonaceous. Containing carbon (C).

Carbonate minerals. Minerals formed by the combination of the complex ion [CO₃]²⁻ with a positive ion. Examples: Calcite [CaCO₃], dolomite [CaMg (CO₃)₂].

Carbonate rock. Rocks composed chiefly of carbonate minerals. Examples: limestone, dolomite.

Catastrophism. The belief that geologic history consists of major catastrophic events far beyond any process we see today. (Contrast with *uniformitarianism.*)

Cave. A naturally formed subterranean open area or chamber or series of chambers commonly formed in limestone by solution activity.

Cement. Minerals precipitated from ground-water solution in the pore spaces of sedimentary rock that bind the particles together.

Chalcedony. A general term referring to fibrous cryptocrystalline quartz.

Chalk. A variety of limestone composed of the shells of microscopic oceanic organisms.

Chemical decomposition. In weathering, the chemical breakdown of a rock resulting from chemical reactions of elements in the atmosphere with those in the rocks.

Chert. Granular cryptocrystalline silica.

C horizon. The zone of soil consisting of partly decomposed bedrock, which lies directly beneath the B horizon and grades downward into fresh, unweathered bedrock.

Cinder cone. A cone-shaped hill that is composed of loose volcanic fragments.

Cinders. Fragments of volcanic ejecta 0.5 to 2.5 cm in diameter.

Cirque. An amphitheater-shaped depression at the head of a glacial valley; excavated mainly by ice plucking and frost wedging (see diagram).

Clastic. Fragments such as mud, sand, and gravel produced by the mechanical breakdown of rocks.

Clastic texture. A texture of sedimentary rocks resulting from physical transportation and deposition of broken particles of minerals, rock fragments, and organic skeletal remains (see diagram).

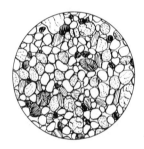

Clay minerals. Finely crystalline hydrous silicates formed by the weathering of minerals such as feldspar, pyroxene, and amphibole.

Clay size. Particles less than 1/256 mm in diameter.

Cleavage. The tendency for a mineral to break in a preferred direction along smooth planes.

Cobble size. Fragments having a diameter between 64 mm (size of a tennis ball) and 256 mm (size of a volleyball).

Columnar jointing. A system of fractures that split a rock body into long prisms or columns. Characteristic of lava flows and shallow igneous intrusion flows (see diagram).

Competence. The maximum potential size of particles that a stream, a glacier, or the wind can move at a given velocity.

Composite volcano. A large volcanic cone built by extrusion of alternating layers of ash and lava (see diagram). Synonymous with *strato-volcano*.

Compression. A system of stresses that tends to reduce the volume of, or shorten, a substance.

Conchoidal fracture. A type of fracture that produces a smooth, curved surface. Characteristic of quartz and obsidian.

Concretions. Spherical to elliptical nodules formed by the accumulation of mineral matter after deposition of sediment.

Cone of depression. A conical depression in the water table surrounding a well that results from heavy pumping (see diagram).

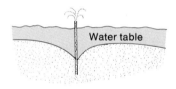
Water table

Conglomerate. A sedimentary rock composed of rounded fragments of pebbles, cobbles, or boulders.

Consequent stream. A stream whose course is a direct consequence of the original slope on which it developed.

Contact. The surface separating two different rock bodies.

Contact metamorphism. Metamorphism near the contact with a magma.

Continent. A large landmass, 20 to 60 km thick, composed mostly of granitic rock, that rises abruptly above the deep-ocean floor; includes marginal areas submerged beneath sea level. Examples: African continent, Australian continent, Asian continent, South American continent.

Continental accretion. The theory that continents have grown by the incorporation of deformed sediments along their margins.

Continental crust. The type of the earth's crust that underlies the continents (including the continental shelves). It is sometimes referred to as *sial*. The continental crust is about 35 km thick with a maximum of 60 km beneath mountain ranges. The density of the continental crust is 2.7 g/m³, and the velocities of primary seismic waves through it are less than 6.2 km/sec. (Contrast with *oceanic crust*.)

Continental drift. The concept that continents have moved relative to one another.

Continental glacier. A thick ice sheet that covers large parts of a continent. Examples are existing glaciers in Greenland and Antarctica.

Continental margins. The zone of transition from the continental mass to the ocean basins. Generally includes continental shelf, continental slope, and continental rise.

Continental rise. The gently sloping surface located at the base of the continental slope (see diagram for abyssal plains).

Continental shelf. Submerged margins of the continental mass extending from the shores to the first prominent break in slope at about 120 m depth.

Continental slope. The slope that extends from the continental shelf down to the ocean deep. In some areas, such as off eastern North America, the continental slope grades into a gentler slope of the continental rise.

Convection. A movement of portions of any fluid medium resulting from density differences produced by heating (see diagram).

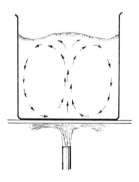

Convection cell. Space occupied by a single convection current.

Convection current. The transfer of material within a closed system as a result of thermal convection. Convection currents are characteristic of the atmosphere and bodies of water and are believed to be generated within the interior of the earth. Convection within the mantle is thought to be the process responsible for plate tectonics.

Convergent plate boundaries. The zone where the leading edges of two spreading plates meet or converge. This is the site of considerable geologic activity and is characterized by volcanism, earthquakes, and crustal deformation. (See also *subduction zone*.)

Copernican period. The most recent period of the lunar geologic time scale, the period during which the Copernican system of rocks was developed.

Copernican system. The youngest rocks and topographic features on the moon. It is characterized by ray craters and ejecta such as the crater Copernicus.

Coquina. A limestone composed of an aggregation of shells and shell fragments (see diagram).

Coral. A bottom-dwelling marine invertebrate organism belonging to the Class Anthozoa.

Core. The central part of the earth, 3380 to 3540 km in diameter, that is surrounded by the mantle.

Coriolis effect. The effect produced by the *Coriolis force*, namely, the tendency of all particles of matter in motion on the earth's surface to be deflected to the right in the Northern Hemisphere and to the left in the Southern Hemisphere.

Coriolis force. The force caused by the earth's rotation that serves to deflect a moving body on the earth's surface to the east in the Northern Hemisphere and to the west in the Southern Hemisphere.

Country rock. A general term for rock surrounding an igneous intrusion.

Covalent bond. A type of chemical bond in which electrons are shared between different atoms such that neither atom has a net charge.

Crater. 1. Volcanic—a circular depression at the summit of a volcano with a diameter less than three times its depth. (Compare with *caldera*.) 2. Lunar—a circular depression characteristic of the moon's surface. Most lunar craters are believed to have been caused by impact from meteorites (see diagram).

Creep. The imperceptibly slow movement of material downslope (see diagram).

Crevasse. 1. Glaciers—a deep crack in the upper surface of a glacier. 2. Natural levees—a break in a natural levee.

Crossbedding. Stratification inclined to the original horizontal surface upon which the sediment accumulated. Produced by deposition on the slope of a dune or sand wave (see diagram).

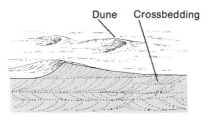

Dune   Crossbedding

Cross cutting relations, law of. A rock is younger than any rock across which it cuts.

Crust (earth's structure). The outermost layer or shell of the earth—generally defined as that part of the earth above the Mohorovičić discontinuity. It represents less than 1% of the earth's total volume (see diagram). (See also *continental crust; oceanic crust*.)

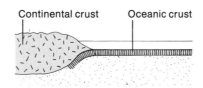

Continental crust   Oceanic crust

Crustal warping. Gentle upwarp or downwarp bending of sedimentary rocks in the continental interior.

Cryptocrystalline. A texture in which crystals are too small to be identified with an ordinary microscope.

Crystal. A solid, polyhedral form bounded by natural plane surfaces resulting from growth of a crystal lattice.

Crystal face. A natural, smooth plane surface of a crystal.

Crystal form. The geometric shape of a crystal (see diagram).

Crystal lattice. A systematic symmetrical network of atoms within a crystal.

Crystalline texture. The texture of a rock resulting from the simultaneous growth of crystals.

Crystallization. The process of crystal growth. May occur as a result of condensation from a gaseous state, precipitation from a solution, or cooling of a melt.

Crystal structure. The orderly arrangement of atoms in a crystal.

Cuesta. An elongate ridge formed on the tilted and eroded edges of gently dipping strata.

Daughter isotope. The isotope created by the radioactive decay of a parent isotope. The amount of a daughter isotope continually increases with time.

Debris flow. The rapid downslope movement of debris—rock, soil, and mud.

Debris slide. A type of landslide involving slow-to-rapid downslope movement of comparatively dry rock fragments and soil. The mass of debris does not show backward rotation (as in a slump) but slides and rolls forward.

Declination, magnetic. The horizontal angle between true north and magnetic north.

Decomposition. Weathering by chemical processes.

Deep-focus earthquakes. Earthquakes that originate at depths greater than 300 km.

Deep-sea fan. A cone- or fan-shaped deposit of land-derived sediment located seaward of large rivers or submarine canyons. Synonymous with *submarine cone, abyssal cone, abyssal fan*.

Deep-sea trench. See *trench*.

Deflation. Erosion of loose rock particles by the wind.

Degradation. The general lowering of the surface of the land by processes of *erosion*.

Delta. A deposit of sediment, roughly triangular, at the mouth of a river (see diagram).

Dendritic. A branching stream pattern that resembles the branching habit of certain trees, such as oaks and maples.

Density. The measure of the concentration of matter in a substance expressed in grams per cubic centimeter (g/cm^3), weight per unit volume.

Density current. A current that flows as a result of differences in density. In the ocean, density currents can be produced by differences in temperature, salinity, and turbidity (material held in suspension) (see diagram).

Denudation. The total action of all processes that result in the wearing away and lowering of the land. Includes weathering, mass wasting, stream action, groundwater processes, et cetera.

Deranged drainage. A distinctively disordered drainage pattern formed in recently glaciated areas. It is characterized by irregular direction of stream flow, few short tributaries, swampy areas, and many lakes (see diagram).

Desert pavement. A veneer of pebbles left in place where wind has removed the finer material.

Desiccation. Drying out. Used in discussions of sedimentation to refer to loss of water from pore spaces through evaporation or compaction.

Detrital. Fragment, clastic.

Diastrophism. Large-scale deformation involving mountain-building and metamorphism.

Differential erosion. Variations in rates of erosion on different rockmasses. As a result of differential erosion, resistant rocks form steep cliffs, whereas nonresistant rocks form gentle slopes.

Differentiated planet. A planet in which the various elements and minerals are separated according to density and concentrated at various levels. The earth, for example, is differentiated, with the heavy metals, iron and nickel, concentrated in the core; lighter minerals in the mantle; and still lighter material in the crust, hydrosphere, and atmosphere.

Differentiation. 1. Sedimentary—the progressive separation (by weathering, erosion, and transportation) of a well-defined rockmass into a variety of more stable components such as clay, quartz sand, and calcium carbonate. 2. Magmatic—processes that separate and segregate certain components of a magma.

Commonly, early-crystallized ferromagnesian minerals are separated by gravitational settling, leaving the parent magma enriched in silica, sodium, and potassium.

Dike. A tabular intrusive rock that occurs across strata or other structural features of the surrounding rock (see diagram).

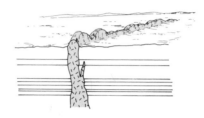

Dike swarm. A group of associated dikes.

Dip. The angle between a horizontal plane and the surface of bedding (joints, faults, foliation, et cetera).

Disappearing stream. A stream that disappears into an underground channel and does not reappear in the same, or even in an adjacent, drainage basin. Typical of karst regions, where streams disappear into sinkholes and follow channels through caves.

Discharge. Volume of flow moving through a given cross section of a stream in a given unit of time.

Disconformity. An unconformity in which beds above and below are parallel (see diagram).

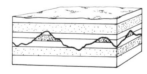

Discontinuity. A sudden or rapid change in the physical properties of the earth. Discontinuities are recognized by seismic data. (See also *Mohorovičić discontinuity.*)

Disintegration. Weathering by mechanical processes. Synonymous with *mechanical weathering.*

Dissolution. The process by which materials are dissolved.

Dissolved load. The part of the total stream load that is carried in solution.

Distributaries. Branches of a stream into which a river divides when it reaches its delta.

Divergent plate boundaries. The margins of a plate that form as the lithosphere splits and drifts apart. It is an area under tension, and new crust is generated by igneous activity. Also referred to as

spreading center, oceanic ridge, and *accreting boundary.*

Divide. A ridge separating two adjacent drainage basins.

Dolomite. 1. A mineral composed of $CaMa(CO_3)_2$. 2. A rock composed primarily of dolomite.

Dome. 1. Structural—a circular or elliptical uplift whose beds dip away in all directions from a central area (see diagram). 2. Topographic—a general term for any dome-shaped rockmass.

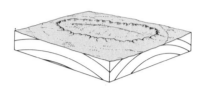

Downwarp. A downward bend or subsidence of a part of the earth's crust.

Drainage basin. The total area that contributes water to a given river (see diagram).

Drainage system. An integrated system of tributaries and a trunk stream that collect and funnel surface water to the sea.

Drift (glacial). A general term referring to any sediment deposited directly by ice or deposited in lakes, oceans, or streams as a result of glaciation.

Drip curtain (caves). A thin sheet of dripstone hanging from the ceiling or wall of a cave.

Dripstone. A cave deposit formed by the precipitation of calcium carbonate in ground water entering an underground cavern.

Drumlin. A smooth, glacially streamlined hill elongate in the direction of ice movement. Drumlins are generally composed of till (see diagram).

Dune. A low mound of fine-grained material that accumulates in response to sediment transport in a current system.

A dune has geometric form that is maintained as it migrates. Sand dunes are commonly classified according to shape. (See also *barchan, longitudinal, parabolic, seif, star,* and *transverse dunes.*)

Earthquake. Groups of elastic waves propagating in the earth, initiated where elastic limits are exceeded along a fault.

Ecology. The study of the relationship between organisms and their environments.

Ejecta (crater). Rock fragments, glass, and other material thrown out of a crater or volcano.

Ejecta blanket. A deposit of material thrown out of a crater that accumulates in the area beyond the crater's rim. In lunar craters, the material is chiefly crushed rock. Typically, the ejecta blanket forms a hummocky terrain extending a distance roughly equal to the diameter of the crater. Rays are also ejecta but do not form a continuous blanket deposit.

Elastic deformation. Temporary deformation after which a rock returns to its original size and shape. Example: bending mica flakes.

Elastic limit. The maximum stress any specimen can stand without undergoing permanent deformation either by solid flow or rupture.

Elastic rebound theory. The theory that explains earthquakes as a result of energy released by faulting. Earthquake waves are caused by the sudden release of stored strain.

End moraine. A ridge of till marking the former front of a glacier.

Entrenched meander. A meander cut into the underlying rock as a result of regional uplift or lowering of regional base level (see diagram).

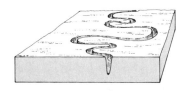

Environment of sedimentation. The physical, chemical, and biologic conditions at the site where sediment accumulates.

Eolian. Pertaining to the wind.

Epicenter. The area on the surface directly above the focus of an earthquake (see diagram).

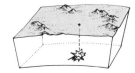

Epoch. A division of geologic time; a subdivision of a period. Example: Pleistocene epoch.

Erosion. The process that loosens and moves sediment to another place on the earth's surface. Agents of erosion include water, ice, wind, and gravity.

Erratic. A large boulder carried by ice to an area far removed from its point of origin (see diagram).

Escarpment. A cliff or very steep slope.

Esker. A long, narrow, sinuous ridge of stratified glacial drift deposited in a former tunnel or streambed beneath a glacier.

Estuary. A bay at the mouth of a river formed by subsidence of land or rise of sea level. Fresh water from the river mixes with, and dilutes, the seawater.

Eugeocline. A geocline in which volcanism is associated with clastic sedimentation—usually associated with an island arc.

Eustatic change of sea level. Worldwide change in sea level resulting from changes in volume of water or capacity of ocean basins.

Evaporites. Rocks that result from evaporation of mineralized water. Common examples are rock salt and gypsum.

Exfoliation. The process by which concentric shells, slabs, sheets, or flakes are successively broken loose and stripped away (see diagram).

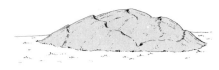

Exposure. Bedrock not covered with soil or regolith; outcrop.

Extrusive rock. Rock originating from a mass of magma that flowed out onto the surface of the earth. Example: basalt.

Faceted spur. A spur or ridge that has been beveled or truncated by faulting, erosion, or glaciation.

Facies. A distinctive group of characteristics within part of a rock, such as composition, grain size, and fossils, that differ as a group from those elsewhere in the same unit. Examples: conglomerate facies, shale facies, brachiopod facies, et cetera.

Fan. A fan-shaped deposit of sediment. (See *alluvial fan; deep-sea fan.*)

Fault. A surface along which a rock body has broken and been displaced.

Fault block. A mass of rock bounded by faults on at least two sides (see diagram).

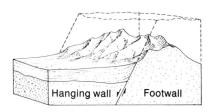

Hanging wall / Footwall

Fault scarp. A cliff produced by faulting.

Faunal succession. A law discovered by William Smith that states that fossils in a stratigraphic sequence succeed one another in a definite, recognizable order.

Feldspar. A mineral group consisting of silicate of aluminum and one or more of the metals K, Na, or Ca. Examples: K-feldspar, Ca-plagioclase, Na-plagioclase.

Felsite. A general term for light-colored aphanitic (fine-grained) igneous rocks. Example: rhyolite.

Ferromagnesian minerals. Silicate minerals containing abundant iron and magnesium. Olivine, pyroxene, and amphibole are common examples.

Fiord. A glaciated valley flooded by the sea to form a long, narrow, steep-walled inlet.

Firn. Granule of recrystallized snow intermediate between snow and glacial ice. Sometimes referred to as névé.

Fissure. An open fracture in a rock.

Fissure eruption. Extrusion of lava along a fissure (see diagram).

Flint. A popular name for dark-colored chert (cryptocrystalline quartz).

Flood basalt. Extensive volumes of basalt erupted largely along fissures. Synonymous with *plateau basalt.*

Flood plain. Flat area bordering a stream that is occasionally flooded.

Focus. The actual location of earthquake origin.

Fold. A bend or flexure in a rock (see diagram).

Folded mountain belt. A long, linear zone of the earth's crust where rocks have been intensely deformed by horizontal stresses and generally intruded by igneous rocks. The great folded mountains of the world (Appalachians, Himalayas, Rockies, Alps, et cetera, are believed to have been formed at converging plate margins).

Foliation. A planar feature in metamorphic rocks produced by the secondary growth of minerals. Three major types are recognized: (1) slaty cleavage, (2) schistosity, and (3) gneissic banding.

Footwall. The block beneath a dipping fault surface.

Foreshore. The seaward part of the shore or the beach lying between high tide and low tide.

Formation. A distinctive body of rock which is a convenient unit for study and mapping.

Fossil. The naturally preserved evidence of past life, such as bones, shells, casts, impressions, and trails.

Fossil fuel. Fuel containing solar energy preserved in chemical compounds of plants and animals of former ages. Includes petroleum, natural gas, and coal.

Fracture zone. 1. Field geology—a zone where the bedrock is cracked and fractured. 2. Oceanography—long linear fractures on the ocean floor expressed topographically by ridges and troughs. The topographic expression of transform faults.

Fringing reef. A reef marginal to the shore of a landmass (see diagram).

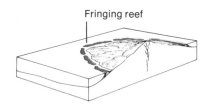

Fringing reef

Frost heaving. The lifting of unconsolidated material produced by the freezing of subsurface water.

Frost wedging. The mechanism of forcing rocks apart by the growth of ice in fractures and pore spaces.

Gabbro. A dark, coarse-grained rock composed of Ca-plagioclase, pyroxene, and possibly olivine, but no quartz.

Gas. The state of matter that has neither independent shape nor volume. It can be compressed readily and tends to expand indefinitely.

Geocline. An elongate prism of sedimentary rock deposited in a subsided part of the earth's crust. A modern example is the continental margins of the eastern United States. (See *eugeocline*.)

Geode. A cavity lined with crystals; when weathered from the rock body, it appears as a partly hollow, rounded rock (see diagram).

Geologic column. A composite diagram showing in a single column the subdivisions of geologic time and the rock units formed during each major period.

Geologic cross section. A diagram showing the structure and arrangement of rocks as they would appear in a vertical plane below the surface.

Geologic map. A map showing the distribution of rocks at the surface.

Geologic time scale. The time interpreted from the geologic column and radiometric dates.

Geosyncline. A subsiding part of the lithosphere in which thousands of meters of sediment accumulate.

Geothermal. Pertaining to heat of the interior of the earth.

Geothermal gradient. The rate of increase of temperature with depth. The approximate average in the earth's crust is about 24 °C/km.

Geyser. A thermal spring that intermittently erupts steam and boiling water.

Glacier. A mass of ice formed from recrystallized snow and thick enough to flow plastically.

Glass. A form of matter that has many properties of a solid but lacks crystalline structure.

Glassy texture. A texture of an igneous rock in which the material is not crystal but is in the form of natural glass.

Global tectonics. A study of the characteristics and origin of structural features of the earth that have regional or global significance.

Glossopteris flora. An assemblage of late Paleozoic fossil plants named for the seed fern *Glossopteris*, one of the plants in the assemblage. This flora is widespread in South America, Africa, Australia, India, and Antarctica and has been important in the development of concepts of continental drift.

Gneiss. A coarse-grained metamorphic rock containing a foliation consisting of alternating layers of light- and dark-colored minerals. Its composition is generally similar to that of a granite.

Gondwanaland. The name of the southern continental landmass thought to have split apart in Mesozoic time to form the present-day continents of South America, Africa, India, Australia, and Antarctica (see diagram).

Graben. An elongate fault block that has been lowered relative to the blocks on either side (see diagram).

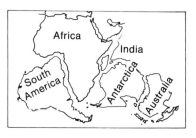

Gradation. Leveling of the land resulting from agents of erosion such as river systems, ground water, glaciers, wind, and waves.

Graded bedding. A type of bedding in which there is a characteristic increase in grain size from bottom to top (see diagram).

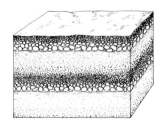

Graded stream. A stream that has attained a delicate adjustment or balance between erosion and deposition so that there is just the velocity necessary to transport the sediment load supplied from the drainage basin. A graded stream is in equilibrium, so neither erosion nor deposition occurs.

Gradient (stream). The slope of the stream channel measured along the course of the stream.

Grain. A particle of mineral or rock generally lacking well-developed crystal faces.

Granite. A coarse-grained igneous rock composed of K-feldspar, plagioclase, and quartz, with minor ferromagnesian minerals.

Granitization. The formation of granite from a metamorphic rock without complete melting.

Gravity anomaly. An abnormal gravitational attraction within the earth.

Graywacke. An impure sandstone consisting of rock fragments, quartz, feldspar, and a matrix of clay-size particles.

Groundmass. The matrix of finer grains in a porphyritic rock.

Ground moraine. Glacial deposits that cover an area formerly occupied by a glacier. Ground moraine typically produces a landscape of low, gently rolling hills.

Ground water. Water below the earth's surface. Generally occurs in pore spaces of rock and soil.

Guyot. A seamount with a flat top (see diagram).

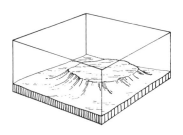

Half-life. The amount of time required for half of a radioactive isotope to decay to another isotope of less mass.

Hanging valley. A tributary valley whose floor lies ("hangs") above the valley of the main stream or shore to which it flows. Hanging valleys commonly are created by the deepening of the main valley by glaciation, but they also can be produced by faulting or rapid retreat of a sea cliff (see diagram).

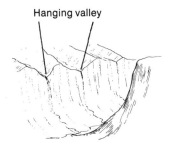

Hanging valley

Hanging wall. The surface or block of rock above an inclined fault plane (see diagram).

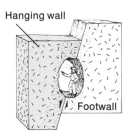

Hanging wall

Footwall

Hardness. 1. Minerals—a measure of the resistance of a mineral to scratching or abrasion. 2. Water—the amount of calcium carbonate and magnesium carbonate in solution.

Headland. An extension of land seaward out from the general trend of the coast. A promontory, cape, or peninsula.

Headward erosion. Extension of a stream headward up the regional slope of erosion.

Heat flow. The flow of heat from the interior of the earth.

High-grade metamorphism. Metamorphism that is accomplished under high temperature and pressure.

Hogback. A narrow, sharp ridge formed on steeply inclined, resistant rock.

Horizon. 1. Geologic—a plane of stratification assumed to have been originally horizontal; 2. Soil—a layer of soil distinguished by characteristic physical properties generally designated by letters, for example, A horizon, B horizon, et cetera.

Horn. A sharp peak formed by the intersecting headwalls of three or more cirques (see diagram).

Hornblende. A variety of the amphibole mineral group.

Horst. An elongate fault block that has been uplifted relative to the adjacent rocks (see diagram).

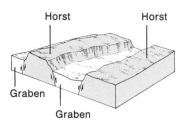

Horst   Horst

Graben

Graben

Hummock. A rounded or conical knoll, mound, or hillock or a surface of other small, irregular shapes. A surface not equidimensional or ridgelike.

Hydraulic. Pertaining to fluids in motion.

Hydraulic head. The pressure of a fluid on a given area caused by the height of the fluid surface above the area.

Hydrologic system. The system of moving water at the earth's surface (see diagram).

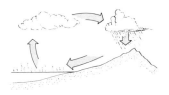

Hydrolysis. The chemical combination of water with other substances.

Hydrosphere. The waters of the earth as distinguished from the rocks (lithosphere), air (atmosphere), and living things (biosphere).

Hydrostatic pressure. The pressure exerted by the water at any given point in a body of water.

Ice sheet. A large body of ice of considerable extent and not confined to valleys. Localized ice sheets are sometimes called *ice caps.*

Igneous rocks. Rocks that originate from cooling and solidification of molten silicate minerals (magma), including volcanic rocks and plutonic rocks.

Imbrium period. Period of lunar geologic time marking the formation of Mare Imbrium.

Inclination (magnetic). The angle between a horizontal plane and the magnetic line of force.

Inclusion. A fragment of an older rock incorporated into an igneous rock.

Intermittent stream. A stream through which water flows only part of the time.

Internal drainage. A drainage system that does not extend to the ocean.

Interstitial. Material that occurs in the pore spaces of a rock. Petroleum and ground water are interstitial fluids. Minerals deposited by ground water in a sandstone are interstitial minerals.

Intrusion. The process of injecting a magma into a preexisting rock. The term is also used for the body of rock resulting from the process.

Intrusive rock. Igneous rock that, while fluid, penetrated into or between other rocks and solidified. It may be exposed later at the surface as a result of erosion.

Inverted valley. A valley filled with lava or some other resistant material that subsequently is eroded into an elongate ridge.

Ion. An atom or group of atoms that has gained or lost electrons and as a result has a net electric charge.

Ionic bonding. A type of bond between atoms formed by electrostatic attraction between oppositely charged ions.

Island. A body of land that is smaller than a continent and is surrounded completely by water.

Island arc. A chain of volcanic islands generally convex toward the open ocean. Example: the Aleutians.

Isostasy. The tendency for all parts of the earth's outer layers to establish a condition of flotational balance as they rest on the soft, denser layers beneath. Layers that have low density and layers that are thicker rise higher than those that are denser or thinner (see diagram).

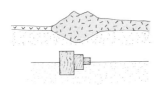

Isotopes. One of several forms of the same chemical element all having the same number of protons in the nucleus but differing in their number of neutrons and, thus, in atomic weight.

Joint. A fracture in a rock along which there has been no appreciable displacement.

Kame. A body of stratified glacial sediment. A mount, knob, or irregular ridge deposited by a subglacial stream as an alluvial fan or a delta.

Karst topography. A landscape characterized by sinks, solution valleys, and other features resulting from groundwater activity (see diagram).

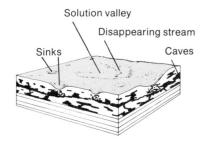

Solution valley
Disappearing stream
Sinks
Caves

Kettle. A closed depression in glacial drift created by the melting of a buried or partly buried block of ice.

Laccolith. A concordant igneous intrusion that has arched up the strata into which it was injected so it forms a pod- or lens-shaped body. The laccolith has a

floor that is generally horizontal (see diagram).

Lag deposits. A residual accumulation of coarse fragments remaining on the surface after the finer material has been removed.

Lagoon. A body of shallow water separated from the ocean by a barrier island or reef.

Lamina. A layer of sediment less than 1 cm thick.

Laminar flow. Flow in which the fluid moves in parallel lines. (Contrast with *turbulent flow.*)

Landform. Any feature of the earth's surface having a characteristic shape as the product of natural processes. Includes major features such as continents and ocean basins, plains, plateaus, and mountain ranges, and minor features such as hills, valleys, slopes, drumlins, and dunes. Taken together, landforms make up the entire surface configuration of the earth.

Landslide. A general term applied to relatively rapid mass movement such as debris flows, slumps, rockslides, et cetera.

Lateral moraine. A deposit of till along the margins of a valley glacier. Accumulates as a result of mass movement of debris onto the sides of the glacier.

Laterite. A soil rich in oxides of iron and aluminum formed by deep weathering in tropical and subtropical areas.

Laurasia. The original continental landmass comprising what is now Europe, Asia, North America, and Greenland.

Lava. Magma that reaches the earth's surface.

Leach. To separate and remove by dissolving the soluble constituents of a rock or soil.

Leachate. A solution obtained by leaching. Example: water percolating through a waste disposal site would dissolve certain soluble substances and would contain these substances in solution.

Leading edge (plate tectonics). The margin of a plate that is located farthest from the spreading center, as compared to the trailing edge, which is the plate margin at the spreading center.

Lee slope. That part of a hill, dune, or rock that is sheltered or turned away from the wind. Also called *slip face.*

Levee (natural). A broad, low embankment built up along the sides of a river channel during floods.

Limbs. The flanks or sides of a fold.

Limestone. A sedimentary rock composed principally of calcium carbonate.

Liquid. A state of matter that flows freely and lacks a crystal structure. Unlike a gas, a liquid retains its independent volume.

Lithification. The processes by which sediment is converted into sedimentary rock. Includes cementation and compaction.

Lithosphere. The relatively rigid outer zone of the earth. Includes the continent, the oceanic crust, and the part of the mantle above the softer asthenosphere.

Load. The total amount of sediment carried at a given time by a stream, a glacier, or the wind.

Loess. Unconsolidated, wind-deposited silt and dust.

Longitudinal dune. An elongate sand dune oriented in the direction of the prevailing wind.

Longitudinal profile. The profile of a stream or valley drawn along its length from the source to the mouth.

Longitudinal wave. A seismic body wave in which particles oscillate along lines in the direction the wave travels. Also called a *primary* or *P wave.*

Longshore current. A current in the surf zone moving parallel to the shore. Longshore currents result where waves strike the shore at an angle and push water and sediment obliquely up the beach. The backwash is straight down the beach face. The water and sediment thus follow a zigzag pattern, with a net movement that is parallel to the shore.

Low-grade metamorphism. Metamorphism that is accomplished under conditions of low to moderate temperature and pressure.

Mafic rock. An igneous rock containing more than 50% ferromagnesian minerals.

Magma. A mobile silicate melt that can contain suspended crystals and dissolved gases in addition to liquid.

Magmatic differentiation. A general term for the various processes by which early-formed crystals or early-formed liquids are separated and removed to form a rock with a composition different from that of the original magma.

Magnetic reversal. The complete 180° reversal of the polarity of the earth's magnetic field.

Mantle. The zone of the earth's interior between the Moho discontinuity and the core.

Mantle plume. A buoyant mass of hot mantle material that rises to the base of the lithosphere and commonly produces volcanic activity and structural deformation in the central part of lithospheric plates.

Marble. A metamorphic rock that originates from limestone or dolomite.

Mare (sea), pl. maria. The relatively smooth, low, dark areas of the moon that were formed by the extrusion of lava.

Mascons. Concentrations of mass in local areas beneath the maria of the moon.

Mass movement. The transfer of rock and earth material downslope through the direct action of gravity without a flowing medium such as a river or glacial ice. Also referred to as *mass wasting.*

Matrix. Small particles of a rock that occupy the space between larger particles. (See *Groundmass.*)

Meander. A broad, looplike bend in the course of a river (see diagram).

Mechanical weathering. The breaking down of rock by physical processes such as frost wedging. Synonymous with *disintegration.*

Medial moraine. A ridge of till in the middle of a valley glacier formed by the junction of two lateral moraines where two valley glaciers meet (see diagram).

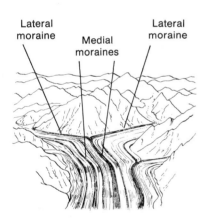

Lateral moraine   Lateral moraine
Medial moraines

Melt. A substance altered from a solid to a liquid state.

Metaconglomerate. A metamorphosed conglomerate.

Metamorphic rock. A major class of rocks formed from preexisting rocks within the earth's crust by changes in temperature, pressure, and the chemical action of fluids.

Meteorites. Particles of solid matter that have fallen to the earth, the moon, or other planets from space.

Micas. A group of silicate minerals exhibiting perfect cleavage in one direction.

Microcontinent. A relatively small, isolated fragment of continental crust. Example: Madagascar.

Mid-Atlantic ridge. The mountain range extending down the central part of the Atlantic Ocean.

Migmatite. A mixture of igneous and metamorphic rocks in which thin dikes and stringers of granitic material interfinger with metamorphic rocks.

Mineral. A naturally occurring inorganic solid having a definite internal structure and a definite chemical composition that varies only within strict limits. The chemical composition and internal structure produce specific physical properties, including the tendency to assume a specific geometric form (crystal).

Mobile belts. Long, narrow belts of the continents that have been subjected to mountain-building processes.

Mohorovičić discontinuity. The first global seismic discontinuity below the surface. Commonly referred to as the *Moho.* Depth varies from about 5 to 10 km beneath the ocean floor to about 35 km below the continents.

Monadnock. An erosional remnant rising above a peneplain (see diagram).

Monocline. A type of fold in which the beds on either side of the fold dip uniformly at low angles (see diagram).

Moraine. A general term for landforms composed of till.

Mountain. A general term for any landmass that stands above its surroundings. In the stricter geologic sense, a mountain belt is a highly deformed part of the earth's crust that has been injected with igneous intrusions and that has had deeper parts metamorphosed. The topography of young mountains is high, but old mountains may be eroded down to flat lowlands.

Mud crack. A crack in a deposit of mud or silt resulting from the contraction that accompanies drying.

Mudflow. A flow of a mixture of mud and water (see diagram).

Nappe. Faulted and overturned folds.

Névé. Granular ice formed by recrystallization of snow. Synonymous with *firn.*

Nodule. An irregular, knobby, or rounded body that is generally harder than the surrounding rock.

Nonconformity. An unconformity in which stratified rocks rest upon granitic or metamorphic rocks (see diagram).

Normal fault. A steeply inclined fault in which the hanging wall has moved down relative to the footwall. Also referred to as *gravity fault* (see diagram).

Footwall

Hanging wall

Nuée ardente. A hot cloud of volcanic fragments and superheated gases that flows as a mass, because it is denser than air. Upon cooling, it forms a rock called an ash flow tuff or a welded tuff (see diagram).

Obsidian. A glassy igneous rock with a composition equivalent to that of granite.

Ocean basin. The low part of the lithosphere that lies between continental masses. The rocks are mostly basalt with a veneer of oceanic sediment.

Oceanic crust. The type of crust that underlies the ocean basin. It is about 5 km thick, composed predominantly of basalt. It has a density of 3.0 g/cm³, and compressional seismic-wave velocities traveling through it exceed 6.2 km/sec. (Compare with *continental crust*.)

Oceanic ridge. The continuous ridge or broad fractured topographic swell which extends through the central part of the Arctic, Atlantic, Indian, and South Pacific Oceans. It is several hundred kilometers wide and has a relief of 600 m or more. It is thus a major structural and topographic feature of the earth.

Offshore. The area from low tide seaward.

Oil shale. Shale rich in hydrocarbon derivatives. In the United States, the chief oil shale is the Green River Formation of the Rocky Mountain region.

Oolite. A limestone consisting largely of spherical grains of calcium carbonate having concentric spherical layers (see diagram).

Ooze. Marine sediment consisting of more than 30% shell fragments of microscopic organisms.

Orogenesis. The processes of mountain-building.

Orogenic. Pertaining to deformation of the continental margins to the extent that mountain ranges are formed.

Orogenic belt. A mountain belt.

Orogeny. A term for a major episode of mountain-building.

Outcrop. An exposure of bedrock.

Outlet glacier. A tonguelike stream of ice resembling a valley glacier formed when a continental glacier encounters a mountain system and the ice moves through the mountain passes in large streams.

Outwash. Stratified sediment washed out from a glacier by melt-water streams and deposited in front of the end moraine.

Outwash plain. The area beyond the mar-

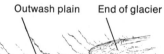
Outwash plain   End of glacier

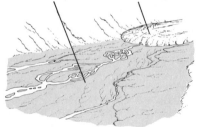

gin of a glacier where meltwater deposits glacier-derived sand, gravel, and mud (see diagram).

Overturned fold. A fold in which at least one limb has been rotated through more than 90° (see diagram).

Oxbow lake. A lake formed in the channel of an abandoned meander.

Oxidation. The chemical combination of oxygen with a mineral.

Pahoehoe flow. Lava with a billowy or ropy surface. (Contrast with *aa flow*.)

Paleocurrent. An ancient current that existed in the geologic past and whose direction can be inferred from cross-bedding, ripple marks, and other sedimentary structures.

Paleogeography. The study of geography in the geologic past, including the patterns of the earth's surface, distribution of land and ocean, ancient mountains, et cetera.

Paleomagnetism. The study of the earth's magnetic field during geologic time.

Paleontology. The study of ancient life.

Paleowind. An ancient wind in the geologic past. Its direction can be inferred from patterns of ancient ashfalls, orientation of crossbedding, and growth rates of colonial corals.

Pangaea. A hypothetical continent from which the present continents originated through processes of drifting from the Mesozoic era to the present.

Parabolic dune. A dune shaped like a parabola with the concave side toward the wind.

Partial melting. The process by which a rock becomes partly liquid as a result of an increase in temperature and/or a decrease in pressure that causes the minerals with low melting points to become liquid. If the liquid is removed a magma can be generated with a composition quite different from that of the parent rock. Partial melting is believed to be an important process by which basaltic magma is generated from peridotite at the spreading centers and granitic magma is generated from basaltic crust at the subduction zone.

Passive plate margins (tectonics). Margins of lithospheric plates in which crust is neither created nor destroyed; generally marked by transform faults.

Peat. An accumulation of partly semicarbonized plant material containing approximately 60% carbon and 30% oxy-

gen. Peat is considered an early stage or rank in the development of coal.

Pebble size. Sediment particles with a diameter ranging from 2 mm to 64 mm (about the size of a match head to the size of a tennis ball).

Pediment. A gently sloping erosional surface developed as a mountain front or a cliff recedes. A pediment surface cuts across bedrock and may be covered with a veneer of sediment. Forms in arid to semiarid climates (see diagram).

Pediment

Peneplain. An extensive erosional surface worn down almost to sea level. Later tectonic activity may lift a peneplain to high elevations (see diagram).

Peninsula. An elongate body of land extending out into the water.

Perched water table. A local zone of saturation above the regional water table (see diagram).

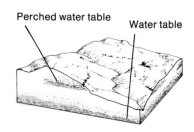
Perched water table   Water table

Peridotite. A dark, coarse-grained igneous rock composed of olivine, pyroxene, and some other ferromagnesian minerals, but with essentially no feldspar and no quartz.

Permafrost. Permanently frozen ground.

Permeability. The ability of a material to transmit fluids.

Phaneritic texture. A texture of igneous rock in which the interlocking crystals are large enough to be seen without aid of magnification.

Phenocryst. A crystal that is significantly larger than those that surround it. It forms during an early slow-cooling stage of the magma (see diagram).

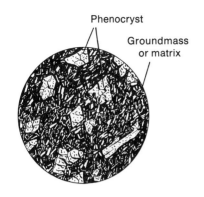

Phenocryst

Groundmass or matrix

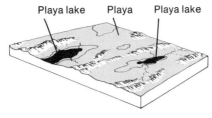

Playa lake   Playa   Playa lake

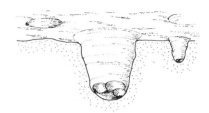

Pothole. A hole in a streambed formed by sand and gravel swirled around in one spot by eddy action (see diagram).

Physiographic map. A map showing surface features of the earth.

Physiography. A study of the surface features and landforms of the earth.

Pillar. A landform shaped like a pillar.

Pillow lava. An ellipsoidal mass of igneous rock formed by extrusion of lava underwater.

Pinnacle. A tall tower or spire-shaped pillar of rock.

Placer. A mineral deposit formed by the sorting or washing action of water. Usually a heavy mineral such as gold.

Plagioclase. A group of feldspar minerals having a composition ranging from $NaAlSi_3O_8$ to $CaAl_2Si_2O_8$.

Plastic deformation. Permanent change in the shape or the volume that does not involve failure by rupture.

Plateau. An extensive upland region.

Plateau basalt. Extensive layers of nearly horizontal basalt that tend to erode into great plateaus (see diagram). Synonymous with *flood basalt.*

Plates (tectonic). Broad segments of the lithosphere (includes the rigid upper mantle, plus oceanic and/or continental crust) that "float" on the underlying asthenosphere and move independently of other plates.

Plate tectonics. The theory of global dynamics in which the lithosphere is believed to be broken into individual plates that move in response to convection in the upper mantle. The margins of the plates are the sites of considerable geologic activity.

Platform reef. An organic reef with a flat upper surface developed on submerged segments of a continental platform. Platform reefs are common off the coast of Australia.

Playa. A depression in the center of a desert basin, the site of occasional playa lakes (see diagram).

Playa lake. A shallow temporary lake formed in a desert basin after a rain.

Plucking (glacial). The process of glacial erosion by which large rock fragments are loosened, detached, and transported by freezing of the meltwater along fractures and bedding planes with the resulting removal of the rock as it is frozen to the moving glacier.

Plunge. The inclination, with respect to the horizontal, of any linear structural element of a rock. The plunge of a fold is the inclination of the axes of the fold.

Plunging folds. Folds whose axes are inclined.

Plutonic rock. Igneous rock formed deep beneath the surface.

Pluvial lake. A lake that formed during a former climate when rainfall in the region was higher than at present. Such lakes were common in the arid regions during the period of Pleistocene glaciation.

Point bar. A crescent-shaped accumulation of sand and gravel deposited on the inside of a meander bend (see diagram).

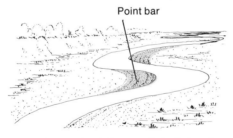

Point bar

Polarity epoch. A long period of time in which the earth's magnetic field has been oriented in either a normal or a reverse direction.

Polarity event. A shorter interval of opposite polarity within a polarity epoch.

Polar wandering. Apparent movement of the magnetic pole relative to the continents.

Pore fluid. Fluids in the pore spaces of a rock; can be ground water or liquid rock resulting from partial melting.

Pore spaces. The spaces in a rock body unoccupied by solid material. Can be the spaces between grains or fractures or can be voids formed by solution.

Porosity. The percentage of pore space within a rock or a sediment.

Porphyritic texture. A texture of igneous rock in which some crystals are distinctly larger than others.

Pre-Imbrian period. Period of lunar geologic time before the Imbrian period.

Pressure ridge. An elongate uplift of the congealing crust of a lava flow resulting from pressure of the underlying and still fluid lava.

Primary coast. Coasts shaped by subaerial erosion, deposition, volcanism, or tectonic activity.

Primary sedimentary structure. A structure such as crossbedding, ripple marks, or mud cracks that originated contemporaneously with the deposition of the sediment. Contrasts with secondary structures such as joints or faults that originate after the rock is formed.

Pumice. A rock consisting of frothy natural glass.

P waves (primary seismic waves). Waves in the earth which are propagated like sound waves. The material involved in the wave motion is alternately compressed and expanded.

Pyroclastic. Pertaining to fragmental rock material formed by volcanic explosions.

Pyroclastic texture. A rock texture consisting of fragments of ash, rock, and glass produced by volcanic explosion.

Pyroxene. A group of silicate minerals with a single chain of silica tetrahedra. Compare with *amphibole,* which has a double chain.

Quartz. An important rock-forming silicate mineral composed of silicon-oxygen tetrahedra joined in a three-dimensional network. Distinguished by its hardness, glassy luster, and conchoidal fracture.

Quartzite. A metamorphosed sandstone.

Radioactive dating. Calculating the age in years for minerals by measuring the ratios of the original radioactive material to their decayed product. Synonymous with *radiometric dating.*

Radioactivity. The spontaneous disintegration of an atomic nucleus with the emission of energy.

Radiocarbon. The radioactive isotope of carbon, C^{14}, which is formed in the atmosphere and is circulated throughout living matter.

Radiogenic heat. Heat generated by the process of radioactivity.

Ray craters. Lunar craters that have a system of rays extending like splash marks from the crater rim (see diagram).

Recessional end moraine. A ridge of till deposited at the end of a glacier during a period of temporary stability in its general recession.

Recharge. Addition of water to the ground-water reservoir.

Recrystallization. Reorganization of elements of the original minerals in a rock as a result of heat, pressure, and pore fluid.

Reef. A solid structure built by shells and other secretions of marine organisms, particularly coral.

Regolith. Soil and loose rock fragments that overlie the bedrock.

Rejuvenated stream. A change in a regimen in which more active erosion occurs.

Relative age. The age of rock or an event as compared to some other event.

Relative time. Dating by means of arranging events in their proper chronologic order. (Compare with *absolute time*.)

Relief. The difference in altitude between the high and low parts of an area.

Reverse fault. A fault in which the hanging wall has moved up relative to the footwall. A high-angle thrust fault (see diagram).

Rift system. A system of faults in the earth resulting from extension.

Rift valley. A valley formed by block faulting in which tensional stresses tend to pull the crust apart. Synonymous with *graben*. Also refers to the downdropped block along the crust of the oceanic ridge.

Rill. A very small stream of trickling water.

Rille. An elongate trench or a crack-like valley on the moon's surface. Rilles may be extremely irregular and meandering or relatively straight structural depressions.

Rip current. A current formed on a water surface resulting from convergence of currents flowing in opposite directions. Common along coasts where longshore currents move in opposite directions.

Ripple marks. Small waves produced in sand or mud by the effects of drag of moving wind or water.

River system. A river with all its tributaries.

Roche moutonnée. An abraded knob of bedrock formed by an overriding glacier. Typically, they are striated and have a gentle slope facing the upstream direction of ice movement (see diagram).

Direction of ice movement

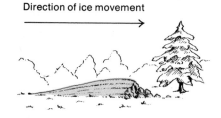

Rock. An aggregate of minerals that form an appreciable part of the lithosphere.

Rockfall. The most rapid type of landslide, ranging from large masses of rock to small fragments loosened from the face of a cliff.

Rock flour. Fine-grained rock particles pulverized while being eroded and transported by a glacier.

Rock glacier. A mass of poorly sorted, angular boulders cemented with interstitial ice. It moves slowly through the action of gravity.

Rockslide. A landslide involving a sudden and rapid movement of a newly detached segment of bedrock sliding over an inclined surface of weakness (such as a joint or a bedding plane).

Runoff. Water that flows over the surface.

Saltation. The transportation of particles in a current of wind or water by movement through a series of bounces (see diagram).

Direction of wind

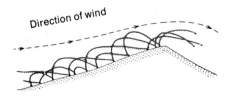

Salt dome. A dome in sedimentary rock produced by the upward movement of a body of salt (see diagram).

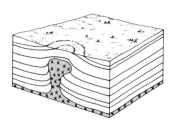

Sand. Fragments ranging in size from 0.0625 to 2 mm in diameter. Much sand is composed of quartz grains, because quartz is abundant and resists chemical and mechanical disintegration. Other materials, such as shell fragments and rock fragments, can form sand grains.

Sandstone. A sedimentary rock composed mostly of sand-size particles cemented usually by calcite, silica, or iron oxide.

Sand wave. A wave of sand created by the effects of drag of air or water moving over the surface. Includes dunes and ripple marks.

Scarp. A cliff produced by faulting or erosion.

Schist. A medium-to coarse-grained metamorphic rock containing strong foliation as a result of parallel orientation of platy minerals.

Schistosity. A type of foliation resulting from the parallel arrangement of coarse-grained platy minerals such as mica, chlorite, and talc.

Scoria. An igneous rock containing abundant vesicles.

Sea arch. An arch cut by wave erosion through a headland (see diagram).

Sea cave. A cave formed by wave erosion (see diagram).

Sea cliff. A cliff produced by wave erosion.

Seafloor spreading. The theory that the seafloors spread laterally away from the oceanic ridge as new lithosphere is created along the crest of the ridge by igneous activity.

Seamount. An isolated, conical mound rising more than 1000 m above the floor of the ocean. Seamounts are probably submerged shield volcanoes (see diagram).

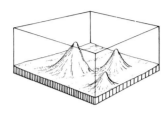

Sea stack. An isolated, pillarlike, rocky island formed by erosion through a headland near a sea cliff (see diagram).

Secondary coast. A coast formed by marine processes or the growth of organisms.

Sediment. Material, such as gravel, sand, mud, or lime, that has been transported and deposited by wind, water, ice, or gravity. Also includes materials precipitated from solution and deposits of organic origin such as coal and coral reefs.

Sedimentary differentiation. The progressive separation (by weathering, erosion, transportation, and deposition) of a well-defined rockmass into distinctive sedimentary products such as sand, shale, and limestone; the segregation and dispersal of the components of an igneous rock into sand, shale, lime, et cetera.

Sedimentary rock. Rock formed from the accumulation and consolidation of sediment.

Seep. A spot where water or other fluids ooze from the earth.

Seif dune. A longitudinal dune of great height and length.

Seismic. Pertaining to waves produced by natural or artificial earthquakes.

Seismic discontinuity. The surface within the earth at which seismic wave velocities abruptly change.

Seismic ray. The path along which seismic waves travel (the direction of a seismic ray is perpendicular to the wave crest).

Seismograph. An instrument that records seismic waves.

Settling velocity. The rate at which suspended solids subside and are deposited.

Shale. A fine-grained, clastic (fragmented) sedimentary rock formed by the consolidation of clay and mud.

Sheeting. A set of joints essentially parallel to the surface that allows layers to spall off when the weight of overlying rock is removed by erosion; especially well developed in granitic rock.

Shield. Large areas where igneous and metamorphic rocks are exposed and have approached equilibrium with respect to erosion and isostasy. Rocks of the shield are usually very old, i.e., more than 600 million years.

Shield volcano. A large volcano in the shape of a flattened dome built up almost entirely of numerous flows of fluid basaltic lava. Slopes seldom exceed 10°, so the profile resembles a shield or a broad dome.

Shore. The zone between high and low tide; a narrow strip of land immediately bordering any body of water, especially lakes or oceans.

Sial. A general term for the silica-rich rocks that form the continental masses.

Silicates. Minerals containing silicon and oxygen atoms linked in units of four oxygen atoms surrounding each silicon atom.

Silicon-oxygen tetrahedron. The arrangement of four oxygen atoms around a silicon atom to form a four-sided pyramid or tetrahedron.

Sill. A tubular body of intrusive rock injected between layers of the enclosing rock.

Siltstone. A fine-grained, clastic (fragmental) rock in which particles range from 1/16 to 1/256 mm in diameter.

Sima. A term referring to the magnesium-rich igneous rocks (basalt, gabbro, and peridotite) of the ocean basins.

Sinkhole. A depression formed by the collapse of a cavern roof (see diagram).

Slate. A fine-grained metamorphic rock with foliation resulting from the parallel arrangement of microscopic platy minerals such as mica.

Slaty cleavage. A type of foliation formed by the parallel arrangement of microscopic platy minerals such as mica and chlorite. The slaty cleavage forms distinct zones of weakness within the rock, causing it to split into slabs.

Slip face. See *lee slope.*

Slope retreat. The progressive recession of a scarp or a side of a hill or mountain by mass movement and stream erosion.

Slump. A type of mass movement in which material moves along a curved surface of rupture.

Soil. The surface material of the continents, produced by rock disintegration and decomposition as well as organic processes. Regolith that has undergone chemical weathering in place.

Soil profile. A vertical section of soil that displays all of its horizons and parent material.

Solid. Matter with a definite shape and volume and some fundamental strength.

Solifluction. Mass movement in which material moves slowly downslope in areas where the soil is saturated with water. Common in permafrost areas.

Solution valley. A valley produced by solution activity either by dissolving surface materials or by removal of subsurface material such as limestone, gypsum, or salt.

Sorting. The separation of particles according to size, shape, or weight. Sorting occurs during transportation by running water or wind.

Spatter cone. A low, steep-sided cone built on a fissure or vent by the accumulation of splashes and spatter of lava (usually basaltic).

Specific gravity. The weight of a substance compared to the weight of an equal volume of water.

Spheroidal weathering. The tendency for a rock surface to become rounded as it weathers (see diagram).

Spit. A sandy bar projecting from the mainland into open water resulting from sediment moved by longshore drift (see diagram).

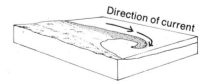

Splay. A small delta or alluvial fan deposit formed when water and sediment are diverted through a crevasse in a levee.

Spreading edges. Plate margins formed by tensional stress along the oceanic ridge. Synonymous with *diverging plate margins.*

Spring. A place where water flows or seeps naturally to the surface.

Stable platform. That part of a continent which is covered by flat-lying or gently tilted sedimentary strata and underlain by a basement complex of igneous and metamorphic rocks. Could be considered as a segment of the continent where the shield is covered with a veneer of sedimentary rocks.

Stack. A small island just offshore, formed by wave erosion of a headland.

Stalactite. An iciclelike deposit of dripstone hanging from the roof of a cave (see diagram).

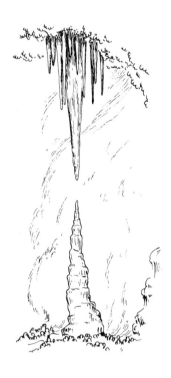

Stalagmite. A conical deposit built up from a cave floor (see diagram).

Star dune. A mound of sand with a high central point and arms radiating in various directions.

Stock. A small, roughly circular, intrusive body usually less than 100 km² in surface exposure.

Strata, sing., stratum. Layers of sedimentary rock.

Stratification. The layered structure of sedimentary rock.

Strato-volcano. A volcano built of alternating layers of ash and lava flows. Synonymous with *composite volcano.*

Streak. The color of a powdered mineral.

Stream piracy. The diversion of the headwaters of one stream into another stream. The process is accomplished by headward erosion of a stream having greater erosional activity (see diagram).

Stress. Force applied to a material that tends to change the material's dimensions or volume.

Striations. Scratches or grooves on a bedrock surface.

Strike. The bearing (compass direction) of a horizontal line on a bedding plane, fault plane, et cetera (see diagram).

Strike-slip fault. A fault in which the movement has been parallel to the strike of the fault (see diagram).

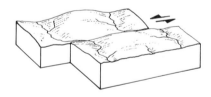

Strike valley. A valley eroded parallel to the strike of the underlying strata.

Strip mining. A method of mining in which soil and rock cover are removed to obtain the sought-after material.

Subaerial. Occurring beneath the atmosphere or in the open air; conditions or processes such as erosion that operate on the land. (Contrast with *submarine* or *subterranean.*)

Subduction. The subsidence of the leading edge of a lithospheric plate down into the mantle.

Subduction zone. An elongate zone in which one lithospheric plate descends beneath another. The zone is marked by an oceanic trench, lines of volcanoes, and crustal deformation associated with mountain-building.

Submarine canyon. A steep-sided, V-profile trench or valley cut into the continental shelf or slope.

Subsequent stream. A tributary stream eroded along a belt of weak, nonresistant rock.

Subsidence. A sinking of a part of the earth relative to its surrounding parts. Synonymous with *sinking.*

Superposed stream. A stream whose course was established, originally on a cover of rock now removed by erosion, so that the stream or drainage system was independent of the existing rock and structure. In this way, the stream pattern is superposed, or placed upon, buried ridges or other structural features (see diagram).

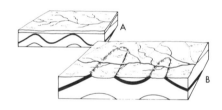

Superposition, law of. The principle that states that, in a series of sedimentary rocks that have not been overturned, the oldest rocks are at the base and the youngest are at the top.

Surface waves (seismic). A seismic wave that travels along the earth's surface. Contrasts with P and S waves, which travel through the earth.

Suspended load. That part of the total stream load that is carried for a considerable period of time in suspension free from contact with the streambed. Consists mainly of mud, silt, and sand. Contrasts with bed load and dissolved load.

Swash. Rush of water up onto the beach after a wave breaks.

S waves (secondary seismic waves). Seismic waves in which energy vibrates at right angles to the direction the waves travel. (Contrast with *P waves.*)

Symmetrical fold. A fold in which the

two limbs are essentially mirror images of each other.

Syncline. A fold in which the limbs dip toward the axis and the youngest beds are in the core after erosion.

Talus. Rock fragments that accumulate in a pile at the base of a ridge or a cliff (see diagram).

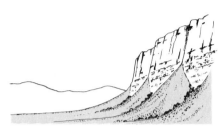

Tectonics. A branch of geology dealing with the broad architecture of the earth, i.e., the regional structural or deformational features, their origin, and histories.

Tension. Stress that tends to pull material apart.

Tephra. A general term referring to all pyroclastic material ejected from a volcano. Includes ash, dust, bombs, et cetera.

Terminal end moraine. A ridge of glacier-deposited material that accumulates at the point of maximum advance of the glacier (see diagram).

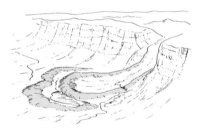

Terrace. A nearly level surface bordering a steeper slope, as in stream terrace or wave-cut terrace (see diagram).

Terrae. The rugged, light-toned highland of the moon.

Texture. The size, shape, and arrangement of the particles that make up a rock.

Thin section. A slice of rock mounted on a glass side and ground about 0.03 mm thick.

Thrust fault. A low-angle fault (45° or less) in which the hanging wall has moved up relative to the footwall (see diagram). Horizontal compression rather than vertical displacement is characteristic.

Tidal bore. A violent rush of tidal water.

Till. Unsorted and unstratified glacial deposits.

Tillite. A rock formed by lithification of glacial till (unsorted, unstratified glacial sediment).

Tombolo. A beach or a bar that connects an island to the mainland (see diagram).

Topography. The shape and form of the earth's surface.

Transform fault. A special type of strike-slip fault making up the boundary between two moving lithospheric plates usually found along an offset of the oceanic ridge (see diagram).

Transpiration. The process by which water vapor is released into the atmosphere by plants.

Transverse dune. An asymmetrical dune ridge formed perpendicular to the prevailing winds.

Travertine terrace. A terrace formed from calcium carbonate that has been deposited from water on a cave floor.

Trellis pattern. A drainage pattern in

which the tributaries are arranged in a pattern similar to that in a garden trellis (see diagram).

Trench (marine geology). A narrow, elongate depression of the deep-ocean floor oriented parallel to trends of continents or island arcs.

Tributary. A stream flowing into or joining another larger stream.

Tsunami (seismic sea wave). A long, low wave in the ocean developed by earthquakes, faulting, or landslides on the seafloor. Velocity may be up to 600 km per hour. Commonly misnamed a tidal wave.

Tuff. A fine-grained rock composed of volcanic ash.

Turbidity current. A current in air, water, or other fluid caused by differences in amounts of suspended matter such as mud, silt, or volcanic dust. Marine turbidity currents are laden with suspended sediment and move swiftly down the continental slope and spread out over the abyssal floor (see diagram).

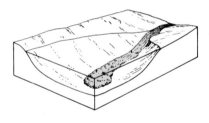

Turbulent flow. Fluid flow in which the path of motion is very irregular, with eddies and swirls.

Ultimate base level. The lowest level to which a stream can erode the earth's surface; sea level.

Ultramafic rock. An igneous rock composed entirely of ferromagnesian minerals.

Unconformity. A discontinuity in the succession of rocks in which there is a gap in the geologic record. A buried erosional surface. (See also *angular unconformity; disconformity;* and *nonconformity.*)

Uniformitarianism. The theory that the earth is a result of natural processes, many of which are operating at the present time.

Upwarp. An arched or uplifted segment of the crust.

Valley glacier. A glacier confined to a stream valley. Synonymous with *alpine glacier* or *mountain glacier.*

Varve. A pair of thin sedimentary layers, one relatively coarse and light colored and the other fine and dark colored as the result of deposition of one year in a lake. The coarse layer is formed during spring runoff and the fine layer is formed during the winter when the lake was frozen (see diagram).

Ventifact. A pebble or a cobble shaped and polished by the wind (see diagram).

Vesicles. Small holes in volcanic rock formed by gas bubbles that became trapped as the lava cooled. (see diagram).

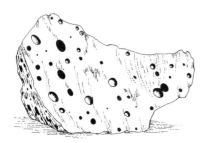

Viscosity. The tendency within a body to oppose flow. An increase in viscosity means a decrease in fluidity or ability to flow.

Volcanic ash. Dust-size particles ejected from a volcano.

Volcanic bomb. Hard fragments of lava that were liquid or plastic at the time of ejection and have forms and surface markings acquired during flight through the air. Fragments range from a few millimeters to more than a meter in diameter (see diagram).

Volcanic front. A line in the volcanic arc system (paralleling trenches) along which volcanism abruptly begins.

Volcanic neck. The solidified magma filling the vent or the neck of an ancient volcano exposed by erosion (see diagram).

Volcanism. A term referring to the processes associated with the transfer of material from the earth's interior to its surface.

Wash. A dry streambed (see diagram).

Water gap. A pass in a ridge through which a stream flows (see diagram).

Water table. The upper surface of the zone of saturation (see diagram).

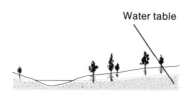

Water table

Wave base. The lower limit of wave transportation and erosion (equal to half the wave length).

Wave-built terrace. A terrace built by wave-washed sediments. Usually lies seaward of a wave-cut terrace.

Wave crest. The highest part of a wave.

Wave-cut cliff. A cliff along the coast formed by undercutting of waves and currents.

Wave-cut terrace or platform. A terrace cut across bedrock by wave erosion.

Wave height. The vertical distance between a wave crest and the preceding trough.

Wave length. The horizontal distance between similar points on two successive waves measured perpendicularly to the crest.

Wave period. The time a wave crest takes to travel a distance equal to one wave length; the time for two successive wave crests to pass a fixed point.

Wave refraction. The process by which a wave is turned from its original direction as it approaches shore. The part of the wave advancing in shallower water moves more slowly than the part moving in deeper water.

Wave trough. The lowest part of a wave between successive crests.

Weathering. The chemical and mechanical breakdown of rock with little or no transportation of the loosened or altered materials.

Welded tuff. A rock formed from volcanic ash hot enough that the particles became fused together.

Wind gap. A gap in a ridge through which a stream, which is now abandoned as a result of stream piracy, used to flow (see diagram).

Wind shadow. The area behind an obstacle where air movement is not capable of moving material.

Wrinkle ridge. A sinuous, irregular, segmented ridge on the surface of the lunar maria believed to be the result of deformation of the lava.

Yardang. An elongate ridge carried by wind erosion.

Yazoo stream. A tributary stream that flows parallel to the main stream for a considerable distance before joining it. Such a stream is forced to flow along the base of a natural levee formed by the main stream.

Zone of aeration. The zone below the earth's surface and above the water table in which pore spaces are usually filled with air (see diagram).

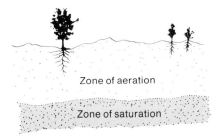

Zone of saturation. The zone in the subsurface in which all openings are filled with water (see diagram).

Credits

Plates 1-32—National Space Data Center, NASA.

Figure 2.2—Modified after Wyllie, P. J., 1976, The Way the Earth Works, New York: John Wiley and Sons.

Figure 2.3—After Wyllie, P. J., 1976, The Way the Earth Works, New York: John Wiley and Sons.

Figure 2.4—After Bullard, E., 1969, "The Origin of the Oceans," Sci. Amer. 221(3):66-75.

Figure 2.5—After Isacks, B., J. Oliver, and L. R. Sykes, 1968, "Seismology and the New Global Tectonics," J. Geophys. Res. 73(18):5855-99.

Figure 2.7—After Crittenden, D. M., 1963, New Data on the Isostatic Deformation of Lake Bonneville, U.S. Geol. Survey Prof. Paper 455-E.

Figure 5.11—U.S. Geological Survey.

Figure 5.12—Courtesy Craters of the Moon National Monument.

Figure 5.14—Courtesy Craters of the Moon National Monument.

Figure 5.17—Canadian Geological Survey.

Figure 5.20—After Williams, H., F. J. Turner, and C. M. Gilbert, 1954, Petrography, San Francisco: W. H. Freeman and Company.

Figure 6.3—U.S. Geological Survey.

Figure 6.24A—Courtesy Paul Potter.

Figure 7.1—National Air Photo Library, Department of Energy, Mines, and Resources, Canada.

Figure 7.3—Modified after Eicher, D. L., 1968, Geologic Time, Englewood Cliffs, N.J.: Prentice Hall.

Figure 9.15—After Strahkov, N. M., 1967, Principles of Lithogenesis, vol. 1, trans. J. P. Fitzemms, Edinburgh: Oliver and Boyd.

Figure 10.7—After Davis, K., and L. Leopold, 1970, Water, New York: Time-Life Books.

Figure 10.9—After Straub, L., in Meinzer, O. E. (ed.), 1942, Hydrology, New York: McGraw-Hill.

Figure 10.11—Courtesy U.S. Forest Service.

Figure 10.12—After Hjulström, F., 1935, Studies of the Morphological Activities of Rivers as Illustrated by the River Fyris, Upsala University: Geol. Inst. Bull. 25:221-527.

Figure 10.14—Courtesy Swiss Air.

Figures 10.15-10.16—After Leopold, L. B., 1968, Hydrology for Urban Land Planning, U.S. Geol. Survey Circ. 554.

Figure 11.6—Photo, U.S. Geological Survey.

Figure 11.10—U.S. Geological Survey.

Figure 11.12—Modified after Varnes, D. J., "Landslide Types and Processes," in Eckel, E. B. (ed.), 1958, Landslides and Engineering Practices, Washington, D.C.: Highway Research Board, Special Report 29, NAS-NRD 544.

Figure 11.13—Courtesy Swiss Air.

Figure 11.21—U.S. Geological Survey.

Figure 11.24—National Air Photo Library, Department of Energy, Mines, and Resources, Canada.

Figure 11.26—U.S. Corps of Engineers, Vicksburg, Miss.

Figure 11.27—U.S. Geological Survey.

Figure 11.33—Modified after Kolb, C. R., and J. R. Lopik, 1966, Depositional Environments of the Mississippi River Coastal Plain, U.S. Army Corps of Engineers Waterway Dept. Sta. Tech. Reports 3-483 and 3-484.

Figure 12.2—U.S. Geological Survey.

Figure 12.3—Canadian Geological Survey.

Figure 12.21—Modified after Shelton, J., 1966, Geology Illustrated, San Francisco: W. H. Freeman and Company.

Figure 12.22—After Strahler, A. N., 1951, Physical Geography, New York: John Wiley and Sons.

Figure 12.26—Modified after Vedeer, J. G., and R. E. Wallace, 1970, U.S. Geological Survey Map I-574.

Figure 13.12—Courtesy John Shelton.

Figure 13.15—Courtesy Onadaga Cave, Missouri.

Figure 13.16—Modified from Schneider, W. J., 1970, Hydrologic Implications of Solid Waste Disposal, U.S. Geol. Survey Circ. 601-F.

Figures 13.18-13.19—Modified after Ward, F., 1972, "The Imperiled Everglades," National Geographic 141(1):1.

Figure 14.3—National Air Photo Library, Department of Energy, Mines, and Resources, Canada.

Figure 14.5—Modified from Lake Gillian and Conn Lake topographic maps, N.W. Territories, Geological Survey of Canada.

Figure 14.6—Courtesy J. D. Ives.

Figure 14.7—Modified after Flint, R. F., 1957, Glacial and Pleistocene Geology, New York: John Wiley and Sons.

Figure 14.8—After Atwood, W. W., 1940, Physiographic Provinces of North America, Boston: Ginn and Company.

Figure 14.13—After Flint, R. F., 1971, Glacial and Quaternary Geology, New York: John Wiley and Sons.

Figure 14.14—U.S. Geological Survey.

Figure 14.15—Sketched from a photo by J. D. Ives.

Figure 14.17—National Air Photo Library, Department of Energy, Mines, and Resources, Canada.

Figure 14.18—After Strahler, A. N., 1951, Physical Geography, New York: John Wiley and Sons.

Figure 14.19—National Air Photo Library, Department of

Energy, Mines, and Resources, Canada.

Figure 14.23—Compiled from glacial map of the United States east of the Rocky Mountains, the Geological Society of America. Base map courtesy E. R. Raisz.

Figure 14.24—After Flint, R. F., 1971, Glacial and Quaternary Geology, New York: John Wiley and Sons.

Figure 14.26—Compiled from Douglas, R. J. W., 1970, Geology and Economic Minerals of Canada, Department of Energy, Mines, and Resources, Geological Survey of Canada, and Flint, R. F., Glacial and Quaternary Geology, 1971, New York: John Wiley and Sons.

Figure 14.28—National Air Photo Library, Department of Energy, Mines, and Resources, Canada.

Figure 14.29—After Hough, J. L., 1958, Geology of the Great Lakes, Urbana, Ill.: University of Illinois Press, and Douglas, R. J. W., 1970, Geology and Economic Minerals of Canada, Department of Energy, Mines, and Resources, Geological Survey of Canada.

Figure 14.30—U.S. Geological Survey.

Figure 14.32—After Flint, R. F., 1971, Glacial and Quaternary Geology, New York: John Wiley and Sons. Base map courtesy E. R. Raisz.

Figures 14.34-14.35—U.S. Geological Survey.

Figure 15.6—U.S. Geological Survey.

Figure 15.7—After Bascom, W., 1960, "Beaches," Sci. Amer. 203(2):81-94.

Figure 15.12—Courtesy Iceland Tourist Bureau.

Figure 15.15—U.S. Department of Agriculture, ASCS Western Aerial Photo Lab, Salt Lake City.

Figure 15.17—U.S. Geological Survey.

Figure 15.21—Courtesy John Shelton.

Figure 16.3—U.S. Geological Survey.

Figure 16.12—Courtesy of Aramco.

Figure 17.1—After Wegener, A., 1915, Origins of Continents and Oceans, New York: Dover. (Paperback, S1708, English translation of the fourth edition, 1929.)

Figure 17.2—After Hurley, P. M., 1969, "The Confirmation of Continental Drift," Sci. Amer. 218(4):52-64.

Figure 17.3—Modified after Takeuchi, H., S. Uyeda, and H. Kanamori, 1970, Debate about the Earth, San Francisco: Freeman, Cooper, and Company.

Figure 17.6—After American Association of Petroleum Geologists, 1928, Theory of Continental Drift: A Symposium, Tulsa, Okla.

Figure 17.9—After Cox A., G. B. Dalrymple, and R. R. Doell, 1967, "Reversals of the Earth's Magnetic Field," Sci. Amer. 216(2):44-54.

Figure 17.13—Modified after Wyllie, P.J., 1976, The Way the Earth Works, New York: John Wiley and Sons.

Figure 17.15—After Isacks, B., J. Oliver, and L. R. Sykes, 1968, "Seismology and the New Global Tectonics," J. Geophys. Res. 73(18):5855-99.

Figure 17.17—After Isacks, B., J. Oliver, and L. R. Sykes, 1968, "Seismology and the New Global Tectonics," J. Geophys. Res. 73(18)5855-99.

Figure 18.3—Modified after Zumberge, J. H., 1972, Elements of Geology, New York: John Wiley and Sons.

Figure 18.7—Modified after Sykes, L. R., 1966, "The Seismicity and Deep Structure of Island Arcs," J. Geophys. Res. 71(12): 2981-3006.

Figures 19.2-19.3—Courtesy Woods Hole Oceanographic Institute.

Figure 20.1—National Geophysical and Solar-Terrestrial Data Center, Boulder, Colo.

Figures 20.2-20.3—Courtesy John Shelton.

Figure 20.4—LANDSAT photo, U.S. Department of Agriculture, Salt Lake City.

Figure 20.18—Courtesy H. Wallace, U.S. Geological Survey.

Figures 22.1-22.2—National Air Photo Library, Department of Energy, Mines, and Resources, Canada.

Figure 22.6—LANDSAT photo, U.S. Department of Agriculture, Salt Lake City.

Figures 23.8-23.9—From Meadows, Donella H., Dennis L. Meadows, Jorgen Randers, William W. Behrens, III, 1972, The Limits to Growth: A Report for the Club of Rome's Project on the Predicament of Mankind, New York: Universe Books. Graphics by Potomac Associates.

Figure 24.1—Courtesy NASA.

Figure 24.2—After Shoemaker, E. M., 1960, Penetration Mechanics of High Velocity Meteorites, Illustrated by Meteor Crater, Arizona, International Geological Congress XXI 18:418.

Figures 24.3-24.4—Courtesy Jet Propulsion Lab.

Figure 24.5—After Weaver, K. F., 1973, "Have We Solved the Mysteries of the Moon?" National Geographic 144(3):318.

Figures 24.6-24.11—Courtesy National Space Data Center, NASA.

Figure 24.12—Modified after Weaver, K. F., 1973, "Have We Solved the Mysteries of the Moon? National Geographic 144(3):318.

Figures 24.13-24.15—Courtesy NASA and Garth H. Ladle.

Figure 24.16—Courtesy C. Klein.

Figure 24.18—After Short, N., 1975, Planetary Geology, Englewood Cliffs, N.J.: Prentice-Hall.

Figure 24.19—After Toksoz, F., et al., 1972, Lunar Science Abstracts, January 10-13, p. 670.

Figure 24.20—Courtesy NASA.

Figure 24.21—Courtesy Ames Research Center.

Figure 24.22—After Shoemaker, E. M., Geology of the Moon and Project Apollo, Contribution No. 2037 of the Division of Geological and Planetary Sciences, Calif. Inst. Tech., Pasadena, Calif.

Figures 24.24-24.26—Courtesy Jet Propulsion Lab.

Figure 24.30—After Mutch, T. A., et al., 1976, Geology of Mars, Princeton, N.J.: Princeton University Press.

Figure 24.31—Courtesy Jet Propulsion Lab.

Figure 24.32—U.S. Geological Survey.

Figures 24.34-24.35—Courtesy Jet Propulsion Lab.

Figures 24.37-24.47—Courtesy Jet Propulsion Lab.

Figure 24.45—After Mutch, T. A., et al., 1976, Geology of Mars, Princeton, N.J.: Princeton University Press.

Index

Page numbers set in boldface type indicate where a term is defined or explained. Page numbers set in italic type indicate an illustration.

Metric Conversion Table

When You Know	Multiply by	To Find	When You Know	Multiply by	To Find
Inches	2.54	Centimeters	Centimeters	0.39	Inches
Feet	0.30	Meters	Meters	3.28	Feet
Yards	0.91	Meters	Meters	1.09	Yards
Miles	1.61	Kilometers	Kilometers	0.62	Miles
Square inches	6.45	Square centimeters	Square centimeters	0.15	Square inches
Square feet	0.09	Square meters	Square meters	11.00	Square feet
Square yards	0.84	Square meters	Square meters	1.20	Square yards
Acres	0.40	Hectares	Hectares	2.47	Acres
Square miles	2.60	Square kilometers	Square kilometers	0.38	Square miles
Cubic feet	0.27	Cubic meters	Cubic meters	0.37	Cubic feet
Cubic yards	0.76	Cubic meters	Cubic meters	0.13	Cubic yards
Cubic miles	4.19	Cubic kilometers	Cubic kilometers	0.24	Cubic miles

Periodic Table of the Elements

1 H 1.0080 Hydrogen																	2 He 4.003 Helium
3 Li 6.940 Lithium	4 Be 9.013 Berilium											5 B 10.82 Boron	6 C 12.011 Carbon	7 N 14.008 Nitrogen	8 O 16.000 Oxygen	9 F 19.00 Fluorine	10 Ne 20.183 Neon
11 Na 22.991 Sodium	12 Mg 24.32 Magnesium											13 Al 26.98 Aluminum	14 Si 28.09 Silicon	15 P 30.975 Phosphorus	16 S 32.066 Sulfur	17 Cl 35.457 Chlorine	18 Ar 39.944 Argon
19 K 39.100 Potassium	20 Ca 40.08 Calcium	21 Sc 44.96 Scandium	22 Ti 47.90 Titanium	23 V 50.95 Vanadium	24 Cr 52.01 Chromium	25 Mn 54.94 Manganese	26 Fe 55.85 Iron	27 Co 58.94 Cobalt	28 Ni 58.71 Nickel	29 Cu 63.54 Copper	30 Zn 65.38 Zinc	31 Ga 69.72 Gallium	32 Ge 72.60 Germanium	33 As 74.91 Arsenic	34 Se 78.96 Selenium	35 Br 79.916 Bromine	36 Kr 83.80 Krypton
37 Rb 85.48 Rubidium	38 Sr 87.63 Strontium	39 Y 88.92 Yttrium	40 Zr 91.22 Zirconium	41 Nb 92.91 Niobium	42 Mo 95.95 Molybdenum	43 Tc (99) Technetium	44 Ru 101.1 Ruthenium	45 Rh 102.91 Rhodium	46 Pd 106.4 Palladium	47 Ag 107.880 Silver	48 Cd 112.41 Cadmium	49 In 114.82 Indium	50 Sn 118.70 Tin	51 Sb 121.76 Antimony	52 Te 127.61 Tellurium	53 I 126.91 Iodine	54 Xe 131.30 Xenon
55 Cs 132.91 Cesium	56 Ba 137.36 Barium	57 La 138.92 Lanthanum	72 Hf 178.50 Hafnium	73 Ta 180.95 Tantalum	74 W 183.86 Wolfram	75 Re 186.22 Rhenium	76 Os 190.2 Osmium	77 Ir 192.2 Iridium	78 Pt 195.09 Platinum	79 Au 197.0 Gold	80 Hg 200.61 Mercury	81 Tl 204.39 Thallium	82 Pb 207.21 Lead	83 Bi 209.00 Bismuth	84 Po (210) Polonium	85 At (210) Astatine	86 Rn (222) Radon
87 Fr (223) Francium	88 Ra (226) Radium	89 Ac (227) Actinium	104 (Russian Proposal Unofficial)														

58 Ce 140.13 Cerium	59 Pr 140.92 Praseodymium	60 Nd 144.27 Neodymium	61 Pm (147) Promethium	62 Sm 150.35 Samarium	63 Eu 152.0 Europium	64 Gd 157.26 Gadolinium	65 Tb 158.93 Terbium	66 Dy 162.51 Dysprosium	67 Ho 164.94 Holmium	68 Er 167.27 Erbium	69 Tm 168.94 Thulium	70 Yb 173.04 Ytterbium	71 Lu 174.99 Lutetium
90 Th (232) Thorium	91 Pa (231) Protactinium	92 U 238.07 Uranium	93 Np (237) Neptunium	94 Pu (242) Plutonium	95 Am (243) Americium	96 Cm (247) Curium	97 Bk (249) Berkelium	98 Cf (251) Californium	99 Es (254) Einsteinium	100 Fm (253) Fermium	101 Md 256 Mendelevium	102 No 253 Nobelium	103 Lr 257 Lawrencium

Origin of Terms Used in the Geologic Column

The Precambrian. Precambrian time is represented by a group of highly complex metamorphic and igneous rocks that form a large volume of the earth's crust. The rocks represent great thicknesses of sedimentary and volcanic rocks that were intensely folded and faulted and intruded with granitic rock. Because they contain only a very few fossils of the more primitive forms of life, arrangement of individual rock layers within this general group into their proper stratigraphic sequence is difficult, if not impossible.

The Paleozoic Era. Rocks younger than Precambrian are much less complex and contain great numbers of fossils, a fact which permits their recognition on a worldwide basis. The term *Paleozoic* means "ancient life," and these rocks contain fossils of primitive invertebrate marine organisms, primitive fish, and amphibians. The era is subdivided into periods based largely on the rock formations of Great Britain.

Cambrian comes from *Cambria,* a Latin word for "Wales," where these rocks were first studied. In most areas of the world, Cambrian rocks rest upon the highly deformed Precambrian metamorphic complex.

Ordovician is derived from *Ordovices,* an ancient tribe of Wales, and designates the strata overlying the Cambrian but differing in the types of fossils contained in the rocks.

Silurian is a term designating rocks exposed on the border of Wales, a territory originally inhabited by a British tribe, *Silures.*

Devonian is named for rocks exposed in Devonshire, England.

Carboniferous. Above the Devonian rocks lies a sequence of coal-bearing formations first studied in England and named the Carboniferous. In the United States, these rocks are subdivided into two major units: the *Pennsylvanian* (from the state of Pennsylvania) and *Mississippian* (from the upper Mississippi valley).

Permian is a term referring to rocks exposed over much of the province of Perm, Russia, just west of the Ural Mountains. Corresponding rocks in England lie above the Carboniferous.

The Mesozoic Era. *Mesozoic* means "middle life" and is used for this period of geologic time because fossil reptiles and a significant number of more modern invertebrates dominate these rocks. This era includes only three periods: the Triassic, Jurassic, and Cretaceous.

Triassic is a term which does not refer to a geographic location but to the striking three-fold division of the rocks overlying the Paleozoic in Germany.

Jurassic is the term first introduced for strata outcropping in the Jura Mountains.

Cretaceous refers to the chalk formations in France and England and is derived from the Latin *creta,* meaning "chalk."

The Cenozoic Era. *Cenozoic* refers to "recent life," and fossil forms found in these rocks include many types with close relationship to the modern forms, including mammals, modern plants, and invertebrates.

Tertiary is a term held over from the first attempts to subdivide the geologic record into three divisions referred to as *Primary, Secondary,* and *Tertiary.* The companion terms *Primary* and *Secondary* have been replaced by terms referring to types of fossilized life forms found in the rocks.

Quaternary is the name proposed for the very recent deposits that contain fossils with living representatives.

Physical Data of the Earth

Volume, Density, and Mass of the Earth

	Average Thickness or Radius in Kilometers	Volume in Millions of Cubic Kilometers	Mean Density in Grams per Cubic Centimeter	Mass in Grams Times 10^{24}
Total earth	6,371	1,083,230	5.52	5,976
Oceans	3.8	1,370	1.03	1.41
Glaciers	1.6	25	0.9	0.023
Continental crust	35	6,210	2.8	17.39
Oceanic crust	8	2,660	2.9	7.71
Mantle	2,883	899,000	4.5	4,068
Core	3,471	175,500	10.71	1,881

Areas of the Earth

	Millions of Square Kilometers	Percentage of the Earth's Surface
Land	149	29.22
Oceans	361	70.78
Glaciers	15.6	3.06
Continental shelves	28.4	5.57
Continental crust	17.7	34.7
Oceanic crust	33.3	65.3

Distribution of the Earth's Water

	Area in Thousands of Square Kilometers	Volume in Thousands of Cubic Kilometers	Percentage of Total Volume
World (total area)	510,000	—	—
Land area	149,000	—	—
Water on land			
Freshwater lakes	850	125	0.009
Saline lakes and inland seas	700	104	0.008
Rivers	—	1.25	0.0001
Soil moisture	—	67	0.005
Ground water	—	8,350	0.61
Glaciers	19,400	29,200	2.14
Atmosphere	—	13	0.001
Oceans	361,000	1,320,000	97.3
Total water volume		1,360,000	100

After Nace, R. L., 1967, U.S. Geological Survey Circular 536.

Geometry of the Earth

	Thousands of Kilometers
Equatorial radius	6,378
Polar radius	6,357
Polar circumference	40,009
Equatorial circumference	40,077

Mineral Identification

Metallic Luster

Gray Streak	**Perfect cubic cleavage;** heavy, Sp. Gr. = 7.6; H = 2.5; silver gray color	GALENA PbS
Black Streak	**Magnetic;** black to dark gray; Sp. Gr. = 5.2; H = 6; commonly occurs in granular masses; single crystals are octohedral	MAGNETITE Fe_3O_4
Black Streak	**Steel gray;** soft, marks paper, greasy feel; H = 1; Sp. Gr. = 2; luster may be dull	GRAPHITE C
Greenish Black Streak	**Golden yellow color;** may tarnish purple; H = 4; Sp. Gr. = 4.3	CHALCOPYRITE $CuFeS_2$
	Brass yellow; cubic crystals; common in granular aggregates; H = 6-6.5; Sp. Gr. = 5; lacks cleavage	PYRITE FeS_2
Reddish Brown Streak	**Steel gray;** black to dark brown; granular, fibrous, or micaceous aggregates; single crystals are thick plates; H = 5-6.5; Sp. Gr. = 5; lacks cleavage	HEMATITE Fe_2O_3
Yellow Brown Streak	**Yellow, brown, or black;** hard structureless or radial fibrous masses; H = 5-5.5; Sp Gr. = 3.5-4	LIMONITE $Fe_2O_3 \cdot H_2O$

The most diagnostic properties for each mineral are indicated in bold type.

Nonmetallic Luster—Dark Color

HARDER THAN GLASS	Cleavage Prominent	**Cleavage—2 directions nearly at 90°;** dark green to black; short prismatic 8-sided crystals; H = 6; Sp. Gr. = 3.5	PYROXENE GROUP Complex Ca, Mg, Fe, Al silicates Augite most common mineral
		Cleavage—2 directions at approx. 60° and 120°; dark green to black or brown; long prismatic 6-sided crystals; H = 6; Sp. Gr. = 3-3.5	AMPHIBOLE GROUP Complex Na, Ca, Mg, Fe, Al silicates **Hornblende** most common mineral
	Cleavage Absent	**Olive green; commonly in small glassy grains;** conchoidal fracture; transparent to translucent; glassy luster; H = 6.5-7; Sp. Gr. = 3.5-4.5	OLIVINE $(Fe, Mg)_2SiO_4$
		Red, brown, or yellow; glassy luster; conchoidal fracture; commonly occurs in well-formed 12-sided crystals; H = 7-7.5; Sp. Gr. = 3.5-4.5	GARNET GROUP Fe, Mg, Ca, Al silicates
SOFTER THAN GLASS	Cleavage Prominent	**Brown to black; 1 perfect cleavage;** thin, flexible, elastic sheets; H = 2.5-3; Sp. Gr. = 3-3.5	BIOTITE $K(Mg,Fe)_3AlSi_3O_{10}(OH)_2$
		Green to very dark green; 1 cleavage direction; commonly occurs in foliated or scaly masses; nonelastic plates; H = 2-2.5; Sp. Gr. = 2.5-3.5	CHLORITE Hydrous Mg, Fe, Al silicate
		Yellowish brown; resinous luster; cleavage—6 directions; yellowish brown or nearly white streaks; H = 3.5-4; Sp. Gr. = 4	SPHALERITE ZnS
	Cleavage Absent	**Red; earthy appearance;** red streak; H = 1.5	HEMATITE Fe_2O_3 (earth variety)
		Yellowish brown streak; yellowish brown to dark brown; commonly in compact earth masses; H = 1.5	LIMONITE $Fe_2O_3 \cdot H_2O$

The most diagnostic properties for each mineral are indicated by bold type.

Nonmetallic Luster—Light Color

HARDER THAN GLASS	Cleavage Prominent	**Good cleavage in 2 directions at approximately 90°;** pearly to vitreous luster; H = 6-6.5; Sp. Gr. = 2.5	FELDSPAR GROUP **Potassium feldspars:** $KAISi_3O_8$—Pink, white, or green **Plagioclase feldspars:** $NaAISi_3O_8$ to $CaAl_2Si_2O_8$—White, blue gray; striations on some cleavage planes
	Cleavage Absent	**Conchoidal fracture; H = 7;** Sp. Gr. = 2.65; transparent to translucent; vitreous luster; 6-sided prismatic crystals terminated by 6-sided triangular faces in well-developed crystals; vitreous to waxy	QUARTZ SiO_2 (silica) **Varieties:** Milky: white and opaque Smoky: gray to black Rose: light pink Amethyst: violet
		Conchoidal fracture; H = 6-6.5; variable color; translucent to opaque; dull or clouded luster	CRYPTOCRYSTALLINE QUARTZ SiO_2 **Varieties:** Agate: banded Flint: dark color Chert: light color Jasper: red Opal: waxy luster
SOFTER THAN GLASS	Cleavage Prominent	**Perfect cubic cleavage;** colorless to white; soluble in water; salty taste; H = 2-2.5; Sp. Gr. = 2	HALITE $NaCl$
		Perfect cleavage in 1 direction; poor in 2 others; H = 2; white; transparent; nonelastic; Sp. Gr. = 2.3; **Varieties:** Selenite: colorless, transparent; Alabaster: aggregates of small crystals; Satin spar; fibrous, silky luster	GYPSUM $CaSO_4 \cdot 2H_2O$
		Perfect cleavage in 3 directions at approximately 75°; effervesces in HCl; H = 3; colorless, white or pale yellow, rarely gray or blue; transparent to opaque; Sp. Gr. = 2.7	CALCITE $CaCO_3$ (fine-grained crystalline aggregates form limestone and marble)
		Three directions of cleavage as in calcite; effervesces in HCl only if powdered; H = 3.5-4; Sp. Gr. = 2.8; color variable but commonly white or pink rhomb-faced crystals	DOLOMITE $CaMg(CO_3)_2$
		Good cleavage in 4 directions; H = 4; Sp. Gr. = 3; colorless, yellow, blue, green, or violet; transparent to translucent; cubic crystals	FLUORITE CaF_2
		Perfect cleavage in 1 direction, producing thin, elastic sheets; H = 2-3; Sp. Gr. = 2.8; transparent and colorless in thin sheets	MUSCOVITE $KAl_2(AlSi_3O_{10}(OH)_2$
		Green to white; soapy feel; pearly luster; H = 1; Sp. Gr. = 2.8; foliated or compact masses; one direction of cleavage forms thin scales and shreds	TALC $Mg_3Si_4O_{10}(OH)_2$
	Cleavage Absent	**White to red; earthy masses;** crystals so small no cleavage visible; soft; H = 1.2; becomes plastic when moistened; earthy odor	KAOLINITE $Al_4Si_4O_{10}(OH)_8$

The most diagnostic properties for each mineral are indicated by bold type.